MW01641313

introduction to
engineering graphics

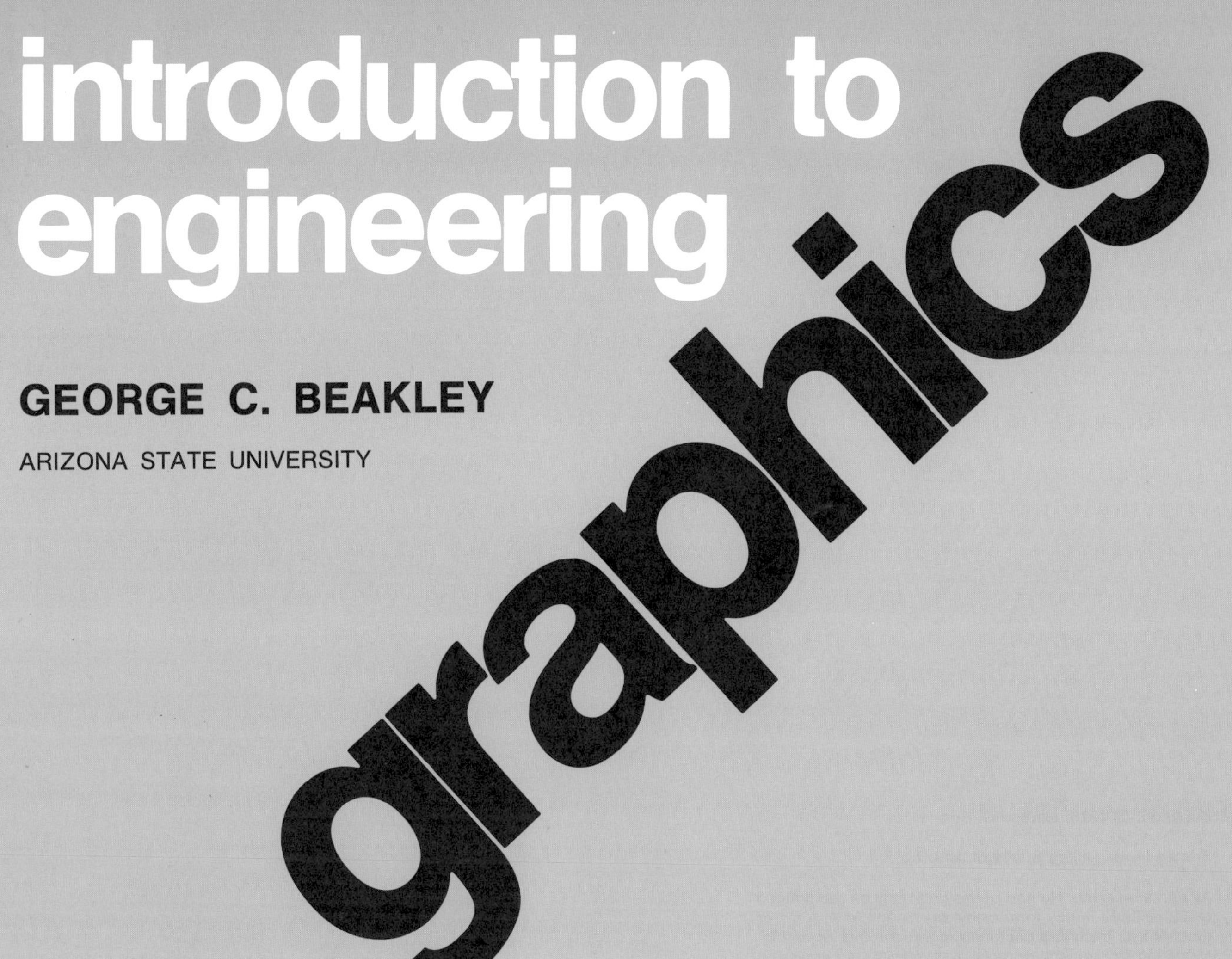

GEORGE C. BEAKLEY

ARIZONA STATE UNIVERSITY

MACMILLAN PUBLISHING CO., INC.

NEW YORK

COLLIER MACMILLAN PUBLISHERS

LONDON

Printed in the United States of America

The bulk of this book is reprinted from *Introduction to Engineering Design and Graphics,* by George C. Beakley and Ernest G. Chilton, copyright © 1973 by George C. Beakley.

Macmillan Publishing Co., Inc.
866 Third Avenue, New York, New York 10022

Collier-Macmillan Canada, Ltd.

Library of Congress Cataloging in Publication Data

Beakley, George C
Introduction to engineering graphics.

"The bulk of this book is reprinted from Introduction to engineering design and graphics by George C. Beakley and Ernest G. Chilton.
1. Engineering, graphics. I. Title.
T353.B4 604'.2 74-19410
ISBN 0-02-307210-5

Printing: 45678 Year: 7890

preface

Engineering is a creative profession because its practitioners are continuously innovative and able to communicate their thoughts to others. The international fires of commerce and industry are constantly being fed with new ideas and products, improved qualities and varieties of food, clothing, and shelter—all to achieve a more comfortable and desirable environment for life. Design is fundamental to this process. The designer's ability to communicate his ideas is enhanced greatly if he has mastered certain fundamental visual skills that are associated with technical sketching, engineering drawing, descriptive geometry, and data presentation. Traditionally instruction in these important subjects has been restricted to a study of the *theory* and practice in the *skill* involved, and too little emphasis has been placed upon the use of graphics as an indispensable tool. In fact, graphics is one of the designer's most important and useful mediums of communication.

This is an introductory text for drawing and graphics. The material has been purposely organized to allow the instructor maximum freedom in organizing his course as he chooses. Most of the chapters are free-standing and their sequence may be altered at will. They have also been written to serve the dual purpose of either formal class use or personalized self-paced instruction.

This text is a graphics abridgement of the author's larger work, *Introduction to Engineering Design and Graphics,* Macmillan, 1973. A similar abridgement of the design material was published by Macmillan in 1974 under the title, *Design: Serving the Needs of Man.*

Two workbooks have been prepared to supplement the problem assignments in the text. These are Series A and Series B of *Graphics for Design and Visualization,* Macmillan, 1973 and 1975 (Beakley-Autore-Hawley).

Several consultants and associates have assisted in the preparation of the larger work. Those most closely associated with this abridged portion are Michael J. Nielsen, Donald D. Autore, and Ernest G. Chilton. Again, their contributions are gratefully acknowledged.

G. C. B.

acknowledgments

Chapter Opening, 1 Texas Instruments Incorporated
Figure 1-1 Cessna Aircraft Company
Figure 1-5 Gramercy
Figure 1-7 Zenith Radio Corporation
Figure 1-11 Keuffel & Esser Company
Figure 1-13 Gramercy
Figure 1-16 General Dynamics, Convair Division
Figure 1-17 Gramercy
Figure 1-21 RapiDesign, Inc.
Figure 1-35 Hewlett-Packard, Medical Electronics Division
Figure 1-55 Gramercy

Chapter Opening, 2 General Motors Research Laboratories
Figure 2-1 Ford Motor Company
Figure 2-16 Travelodge International, Inc.
Figure 2-17 Robertshaw Controls Company, Milford Division
Figure 2-39 Koppers Company, Inc.
Figure 2-40 M. W. Kellogg Company
Figure 2-70 Texas Instruments Incorporated
Figure 2-72 RapiDesign, Inc.
Figure 2-80 GraphiCraft
Figure 2-81 United States Gypsum Company

Chapter Opening, 3 Electron Optics Laboratory, Engis Equipment Company

Chapter Opening, 4 General Motors Research Laboratories
Figure 4-1 Reynolds Metals Company
Figure 4-2 American Iron and Steel Institute
Figure 4-3A A. W. Faber, Castell
Figure 4-5A Mobil Oil Corporation
Figure 4-14A Reynolds Metals Company
Figure 4-23A General Motors Research Laboratories
Figure 4-32 Bell Telephone Laboratories
Figure 4-37 Inland Steel Company

Chapter Opening, 5 Carl Zeiss, Inc., New York
Figure 5-7 ANSI (Y 14.5-1966)
Figure 5-29 ANSI (Y 14.5-1966)
Figure 5-32 ANSI (Y 14.5-1966)
Figure 5-49 FMC Corporation
Figure 5-55 ANSI (Y 14.5-1966)
Figure 5-56 ANSI (Y 14.5-1966)
Figure 5-57 ANSI (Y 14.5-1966)
Figure 5-60 ANSI (Y 14.2-1957)
Figure 5-94 Los Alamos Scientific Laboratories
Figure 5-99 Texas Instruments Incorporated

Chapter Opening, 6 General Motors Research Laboratories
Figure 6-54 Bell Telephone Laboratories
Figure 6-55 Bell Telephone Laboratories
Figure 6-56 The Boeing Company
Figure 6-57 International Business Machines

A-I Photograph by William G. Hyzer. Reprinted from *Research/Development,* August 1971. © 1971 by Technical Publishing Company
A-II General Motors Corporation
A-II-Metric System U. S. Dept. of Commerce
A-III Reprinted from the January 19, 1970, issue of *Design News,* a Cahners publication.
A-III-Material Specification Systems Society of Automotive Engineers
A-V General Motors Corporation
A-VI Chicago *Tribune*
A-VII Salton
A-VIII Vermont Research Corporation. Design by Hill, Holliday, Connors, Cosmopulos Advertising, Boston
A-VIII-1 Gillette Safety Razor Company
A-VIII-2 Gillette Safety Razor Company
A-VIII-3 Gillette Safety Razor Company
A-VIII-4 Gillette Safety Razor Company
A-VIII-5 Gillette Safety Razor Company
A-VIII-6 Engineering Case Program, Stanford University
A-VIII-7 Engineering Case Program, Stanford University
A-VIII-8 Engineering Case Program, Stanford University
A-VIII-9 Engineering Case Program, Stanford University
A-VIII-10 Engineering Case Program, Stanford University
A-VIII-11 Engineering Case Program, Stanford University
A-VIII-12 Engineering Case Program, Stanford University
A-VIII-13 Engineering Case Program, Stanford University
A-VIII-14 Engineering Case Program, Stanford University

contents

graphic tools 1

Hybrid circuitry for a transistor.

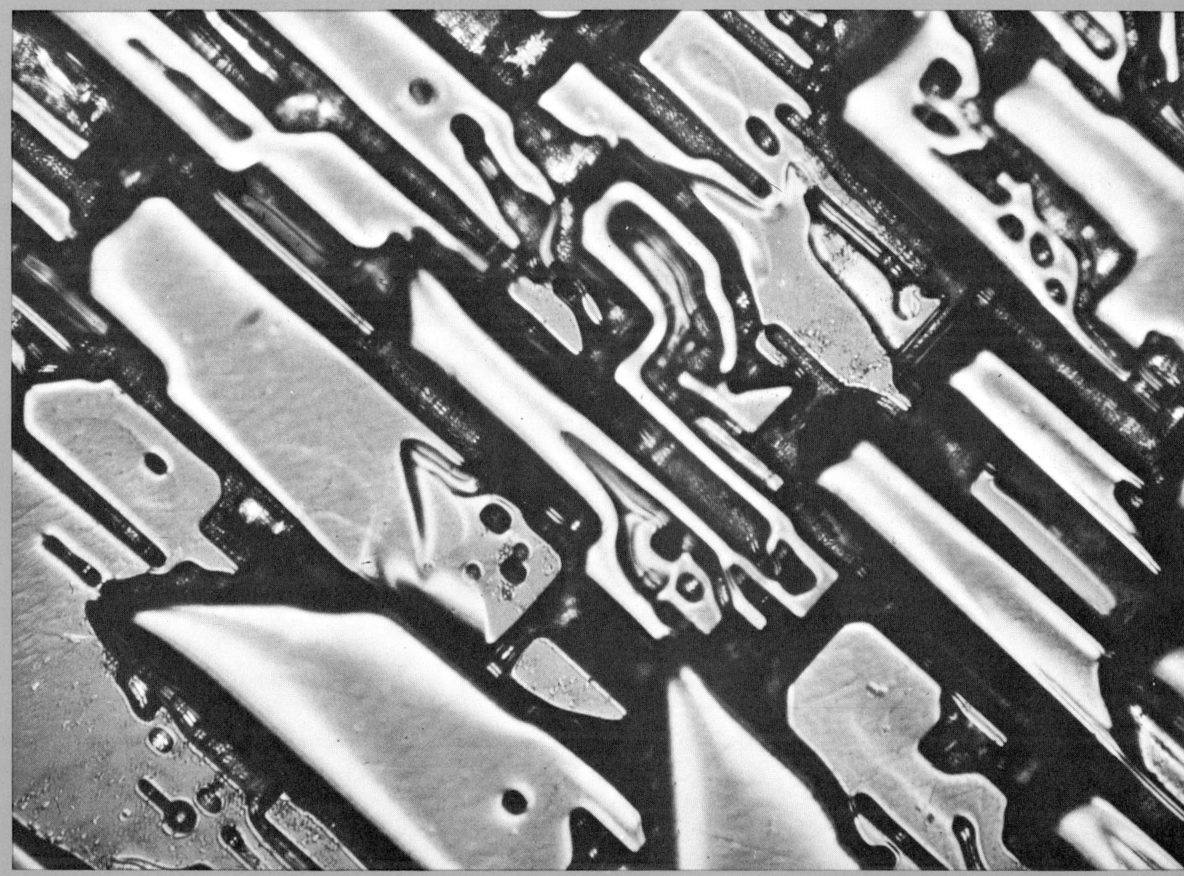

as long as a design is only in the mind of the designer, it is of little use to anyone. It must be communicated to others by word or picture. Notes or descriptions combined with sketches, drawings, and diagrams are most useful for this purpose.

Effective visual communication requires special skills so that the concepts presented are visually accurate and understandable. *Engineering graphics* is the name given to these skills. This chapter and the following five chapters will describe the techniques and tools used in engineering graphics.

There are three general types of activity in engineering graphics which require special tools. The first of these activities is *marking,* that is, making a visible trace or impression on a surface. The tools used for marking are pencils, pens, erasers, and drawing papers.

The second general type of activity is *guiding* a marking tool so that it will consistently and accurately produce the desired visible trace. At one time the only guiding tools were the straight edge and the compass. Now with the growing level of standardization and the requirement

1-1
The designer makes use of many graphic tools.

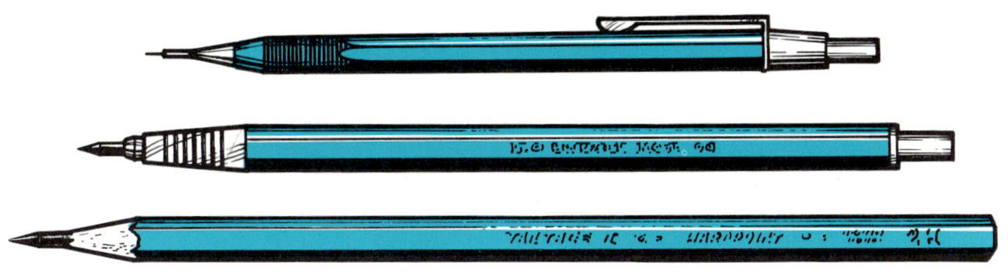

1-2
Drawing pencils.

to reduce production time, the number and types of available guides are quite large. There are circle and ellipse guides, French or irregular curves, and templates for all types of special symbols.

The third general type of activity is *measuring*. Examples of tools for measuring are scales, dividers, and protractors.

marking tools

Of the various types of marking tools used in engineering graphics, the wood-encased graphite lead *pencil* or the mechanical *lead holder* are the most common. These marking devices are used primarily for original sketches and drawings in which frequent corrections will have to be made. The "lead" is made of different mixtures of graphite and clay and ranges from very soft 7B to very hard 9H. The soft leads, 7B to about a B, contain a high percentage of graphite, produce a dark line, and are used primarily for art work. The medium grades of lead, starting at HB and ending at about 3H, have a lower percentage of graphite and are used in technical sketching, lettering, and most mechanical drawing. Finally, the hard leads, starting at 4H and continuing through 9H, have the lowest percentage of graphite and are used in very precise technical work where very fine lines are needed, such as graphical computations and charts, descriptive geometry, and some engineering drawings. The grade is written on the pencil or lead and care should be taken not to sharpen the end that is so marked. In recent years, pencil leads formed from various plastic compounds have become very popular for use on plastic drawing films.

Always conscientiously maintain a properly sharpened point on your pencil or lead. If there is any one thing that will destroy the effectiveness of a drawing, it is a dull point. Expose about $\frac{3}{8}$ to $\frac{1}{2}$ inch of the lead and

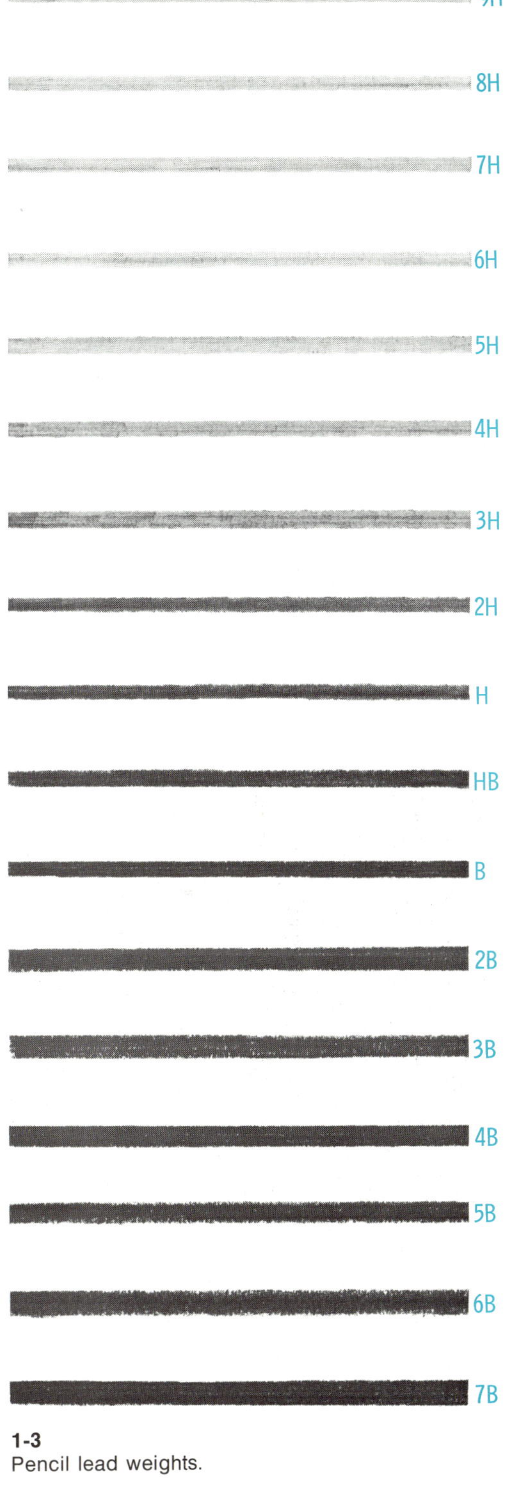

1-3
Pencil lead weights.

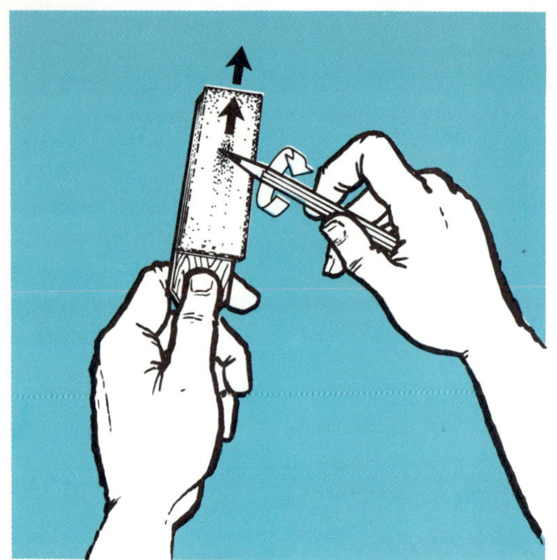

1-4
Sandpaper pad and pencil.

1-5
Pencil pointer.

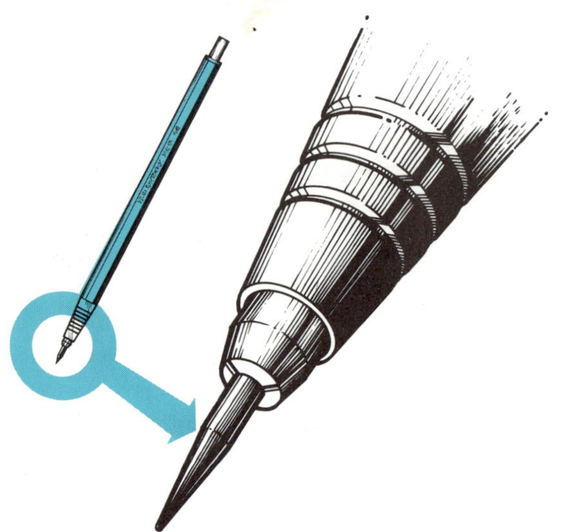

1-6
Pencil point shape.

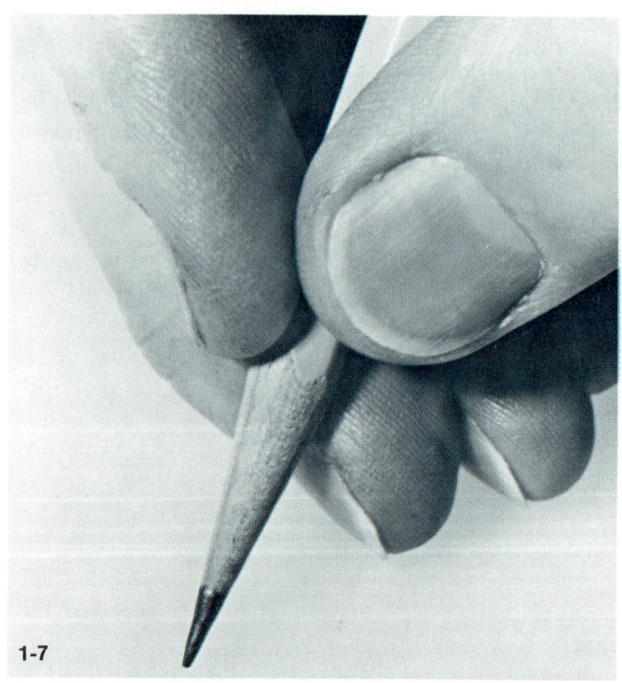

1-7
Pencil grip.

sharpen this exposed lead by rotating it as it is stroked across a pencil pointing pad [1-4], or by rotating it within a pencil lead pointer [1-5]. The point is shaped into a cone and slightly rounded at the tip [1-6]. Do not leave an extremely sharp point as this is easily broken. Also be sure to wipe off loose graphite particles created by the sharpening process.

The pencil should be held naturally and firmly but not cramped. The thumb and fingers should grasp the pencil about $1\frac{1}{2}$ inches from its point in a manner that allows the pencil to be rotated between the fingers when a line is drawn [1-7]. This produces a line of consistent weight, because the pencil point wears evenly.

Inking tools are used to make drawings of record and tracings of original pencil work. There are two types of marking pens on the market. One type is called a ruling pen and consists of two steel "nibs" that can be adjusted to produce lines of various widths [1-8]. The second type of pen uses hollow tubes of varying diameters to emit ink at fixed line widths [1-9]. Pens of this type usually feature a large ink reservoir to reduce the need to continually fill the pen. Because of the many hours of practice required for good inking skills and the limited use of inked drawings in the design phases of engineering practice, a discussion of the procedures involved has been omitted.

Another marking device useful in engineering is the *felt-tip pen* [1-10]. This is usually an inexpensive plastic pen with a small point made of Nylon or similar material.

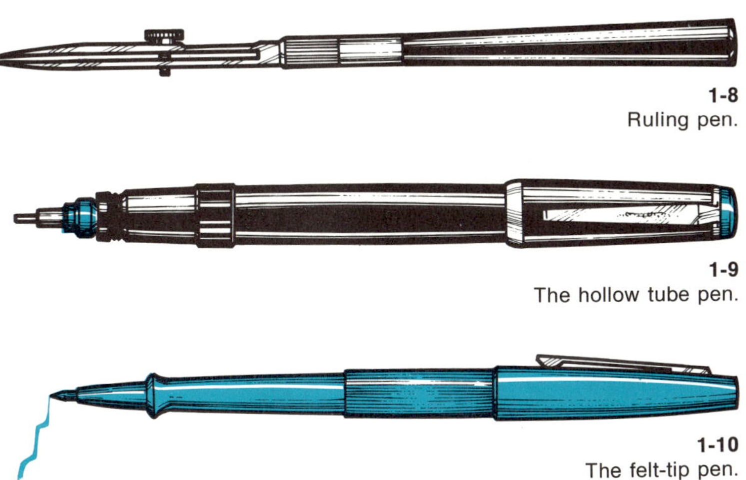

1-8
Ruling pen.

1-9
The hollow tube pen.

1-10
The felt-tip pen.

1-11
Press letters and symbols.

These pens come in a large variety of colors and are very useful to the designer for sketching, emphasizing parts of drawings, and differentiating curves and data points on charts. A word of caution: felt-pen lines are difficult to erase!

One of the most effective new graphic aids is the *transfer symbol*. Letters and symbols of all styles and types are printed on low-adherence paper sheets. By properly locating the desired symbol on the drawing paper and rubbing it in accordance with instructions, one is able to transfer the symbol to the paper [1-11]. The effects that can be achieved with these materials in graphic presentations are limited only by the imagination of the designer.

Erasers

If it is necessary to remove a mark from a sketch or drawing because of a mistake or a change in the concept, the fastest method is to erase the mark. However, erasing should be held to a minimum because it usually detracts from the final appearance of the graphic materials. Often, if large areas of a sketch or drawing need to be erased, it is better to trace the usable portion and continue from there rather than to do an extensive amount of erasing.

Erasers have different hardness and abrasiveness. For most engineering graphics one needs a fairly soft eraser that will remove a mark without causing damage to the paper surface. Erasers that fray the paper should be avoided. Soft white or pink erasers are quite satisfactory for most requirements. In addition to these standard erasers, several newer plastic-based varieties have become available that are especially useful on plastic drawing film. Every means possible should be employed during the creation of a drawing to maintain overall cleanliness. Any attempt to "clean up" an entire drawing when finished (using art gum or erasing compounds) will reduce the sharpness and reproducibility of the lines.

A very useful aid in erasing is an erasing shield [1-13]. It is used to protect the acceptable parts of a graphic presentation while erasing the errors. When used properly, this device can save many sketches and drawings that might otherwise have to be drawn again.

1-12
The eraser.

1-13
The erasing shield.

Drawing Surfaces

The surfaces used in engineering graphics range from thin paper to polyester film. The choice depends primarily on how the final graphic material is to be reproduced.

Many reproducing processes, such as diazo printing, require that the surface on which the drawing is produced be translucent. Tracing papers, tracing cloths, and plastic drawing films satisfy this requirement. Graphs, idea sketches, and preliminary layout drawings that are intended to be used without quantity reproduction can be made on opaque drawing paper.

Whether transparent or opaque, it is important that the surface have a grain that will accept readily whatever marking tool is used. It is a good practice to test the marking tool on a small piece of the surface before starting. Look to see if the surface takes a sharp, clean mark without smudges, runs, or indentations. In the case of pencils, check the point wear to see that an excessive amount of lead is not being taken away with each stroke. Test your eraser to see that the marks are removed completely without damaging the working surface.

guiding tools

One of the oldest, simplest, and still most commonly used of all the guiding tools is the *straightedge*. The straightedge can be any smooth-edged object, but the plastic triangle and the T square are the most used. As its name implies, the straightedge aids in drawing straight lines. Properly used, it helps create a precise, sharply defined line. Place the point of your pencil precisely on one of the end points of the line to be drawn. Move the straightedge into contact with the pencil point. Then align the other end of the straightedge with the other end point of the line and draw the straight line.

When using a straightedge, or any other guide, hold the pencil so that the tapered side of the point is parallel to the vertical edge of the guide. This reduces the amount of lead scraped off by the sharp corner of the guide [1-14]. Avoid using a straightedge with small indentations or printed marks along its side, such as a scale. Also avoid using any rulers with a thin metal edge. These indented and sharp-edged tools scrape off small amounts of the pencil lead with each stroke and soon produce smudges on your drawing.

The *T square* is the basic straightedge in most graphics work [1-15]. It establishes and maintains the horizontal orientation of the work and acts as a foundation for other drawing tools. T squares come with various blade lengths.

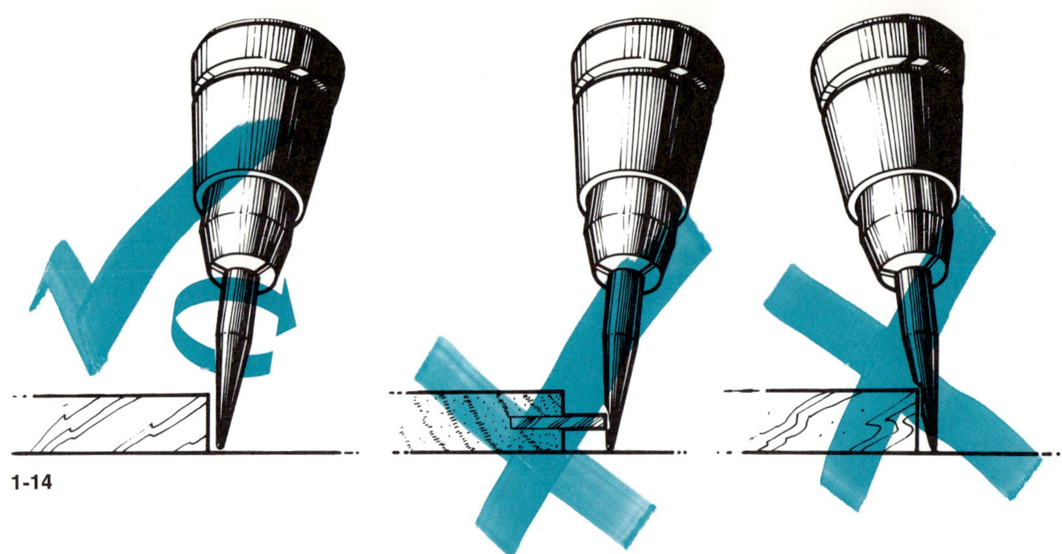

1-14

1-15
The T square.

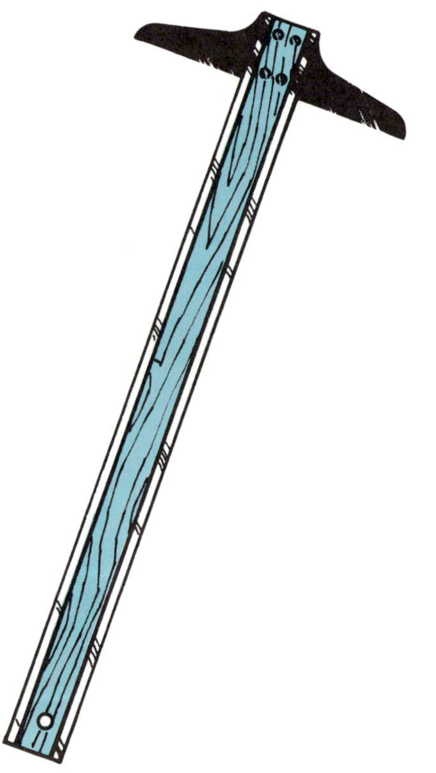

1-16
The use of triangles.

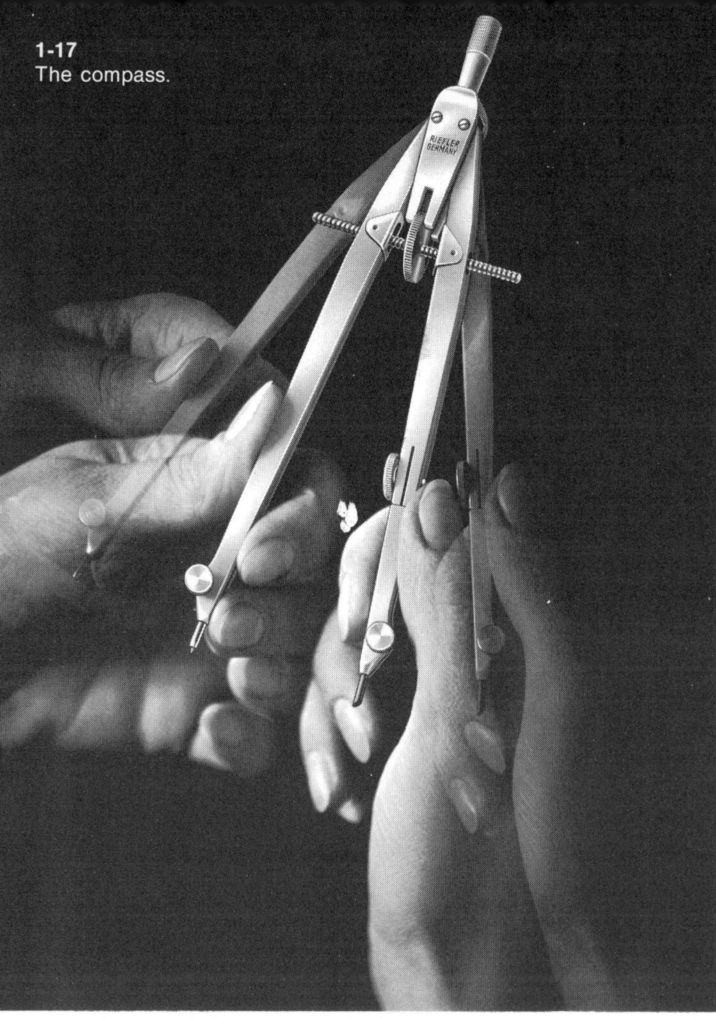

1-17
The compass.

The fixed head and hardwood blade with transparent plastic edges is the most popular.

The next most useful straightedge-type guiding tools are the *triangles*, generally made from transparent acrylic plastic [1-16]. The two most popular triangles are the 30°–60° triangle and the 45° triangle. They are right triangles with the remaining angles 30° and 60° in the first instance and 45° in the second.

The *compass* is a combination marking and guiding tool used to draw circular arcs. It holds its own lead or pen and can adjust to different radii as required. A common form of compass consists of two legs pinned at one end. They can be spread either manually or through a center wheel [1-17]. Different compasses are made to accommodate arcs of different radii. Most common engineering graphics procedures can be accomplished with a 6-inch bow-type compass. Large circles, with radii of 6 inches and more, are made with a beam compass [1–18]. Note

1-18
The beam compass.

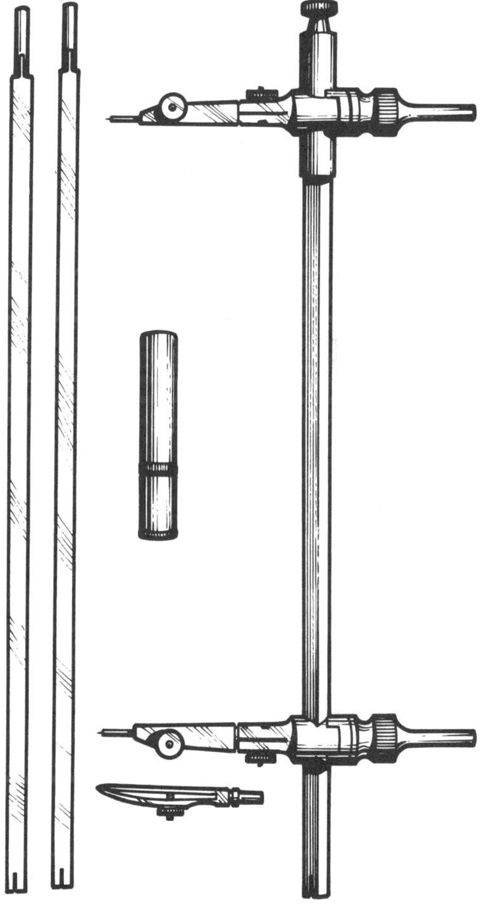

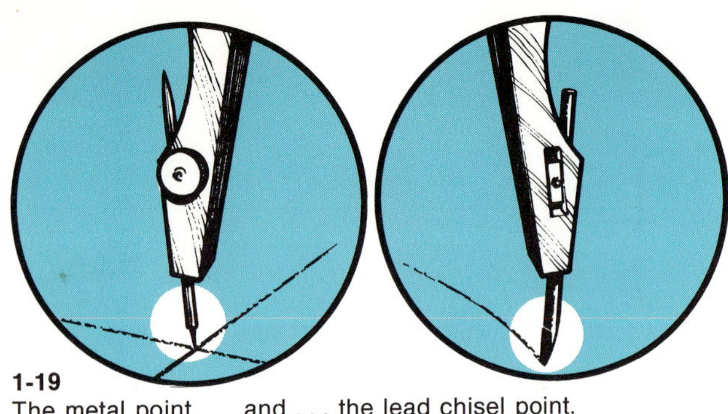

1-19
The metal point . . . and . . . the lead chisel point.

1-20
Mechanical guides.

the special metal point used on a compass and the lead point sharpened to a chisel shape [1-19].

Mechanical guides are used as an aid in drawing circles, ellipses, various irregular curves [1-20], and special symbols [1-21]. Their value comes not only in making draw-

1-21
Special symbol guides.

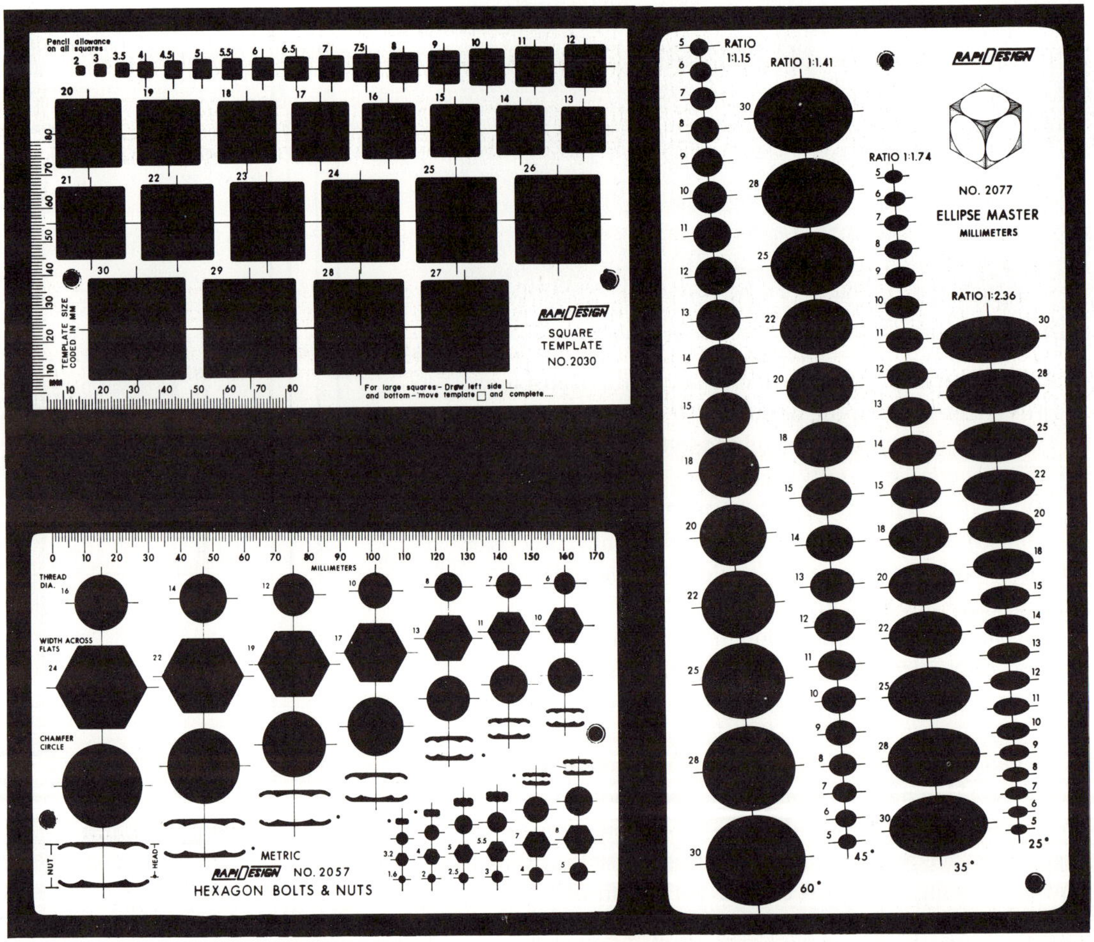

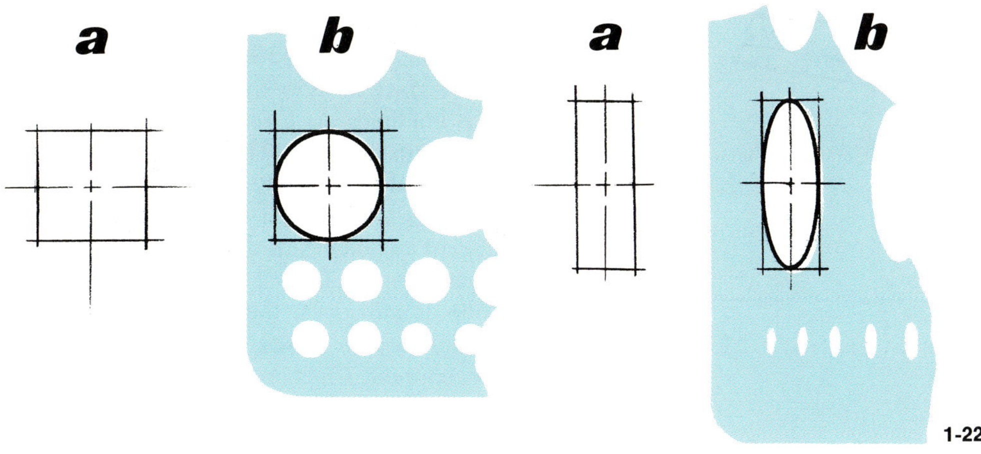

1-22

1-23

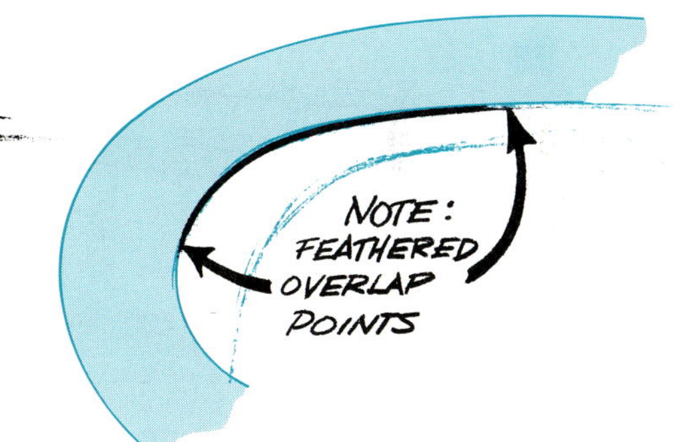

ing easier, but also in the time savings and the professional appearance added to a sketch or drawing.

With circle and ellipse guides, one needs only to indicate the center point and orientation of the figure and then trace the appropriate size [1-22].

When using mechanical guides for irregular shapes, lightly sketch the desired line or shape without using the guide. Then select the portion of the guide that fits the sketched curve closely, and formally draw in the desired line [1-23].

measuring tools

The third major class of tools consists of the measuring tools: those that measure a length or distance, and those that measure an angle.

Linear Measurement

Tools for linear measurement are called *scales*. There are several types of scales available for use in engineering graphics, of which the architect's, engineer's, and mechanical engineer's scales are the most common.

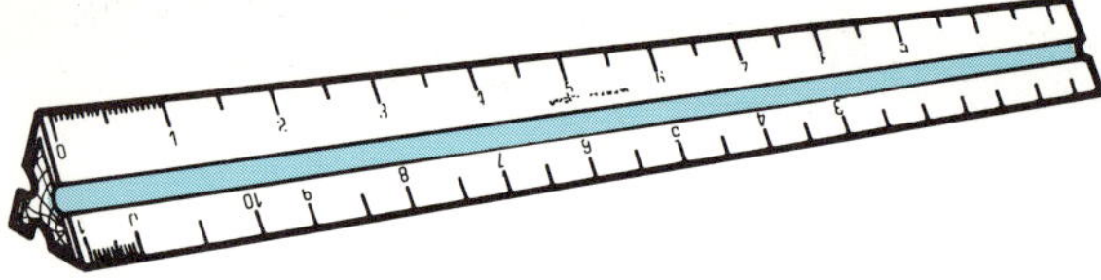

1-24
The architect's scale.

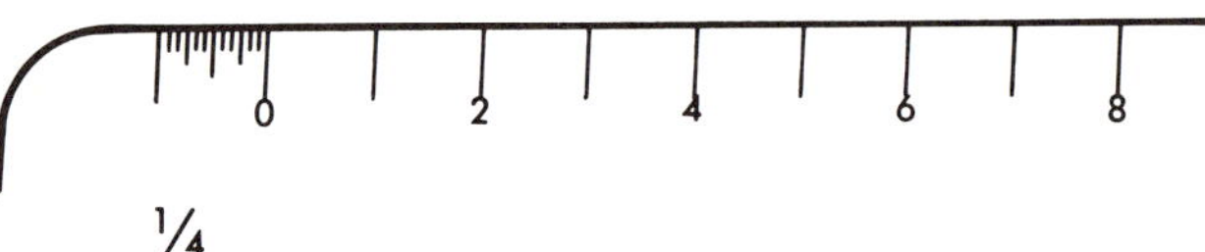

1-25
Scale graduations—architect's scale.

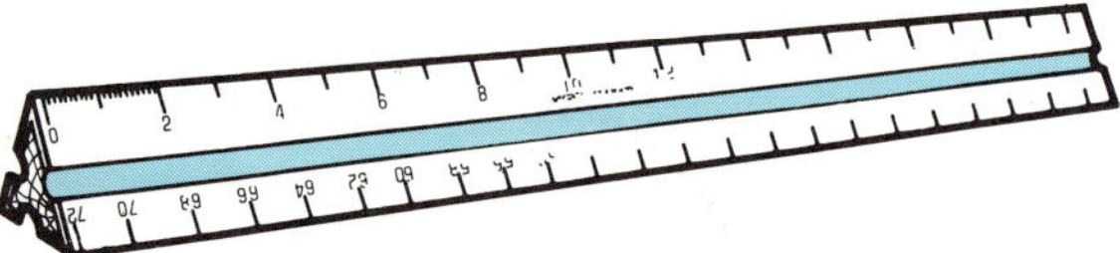

1-26
The engineer's scale.

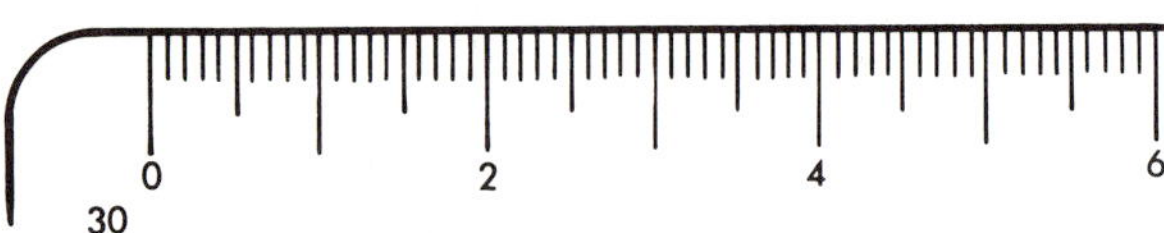

1-27
Scale graduations—engineer's scale.

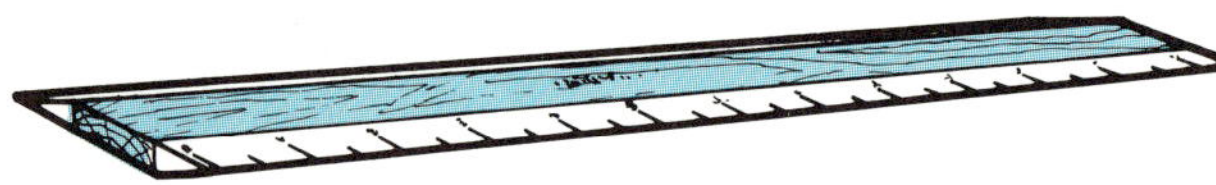

1-28
The mechanical engineering scale.

1-29
Scale graduations—mechanical engineering scale.

The *architect's scale* is a collection of scales usually inscribed on a triangular-shaped tool [1-24]. The architect's scale is designed to show the dimensions of large objects, such as buildings. As a result, all the scales are marked in increments of 1 foot. The large numbers, usually fractions, printed at the ends of each scale indicate what part of an inch is equivalent to 1 foot. As an example, $\frac{1}{4}$ at the end of a scale means that a $\frac{1}{4}$-inch increment on this scale represents 1 foot of actual structure. The very short group of marks near the 0 point represent the subdivision of 1 foot into inches. For example, the 12 small divisions represent 1-inch intervals. The other fractional numbers and the whole numbers on the architect's scale are read in a similar manner. On an engineering drawing the architectural scale used to lay out a design is expressed as $\frac{1}{4}'' = 1'\text{-}0''$, $\frac{1}{2}'' = 1'\text{-}0''$, $1'' = 1'\text{-}0''$, etc.

The *engineer's scale* is used primarily for laying out on a drawing long distances measured in feet or miles, such as might be encountered in roads or pieces of land. Like the architect's scale, it is usually constructed with several scales inscribed on a triangular-shaped tool [1-26]. The scales on the engineer's scale are divided into units of 10, 20, 30, 40, 50, and 60 graduations per inch. By appropriate location of the decimal point, 1 inch on one of these scales can be used to represent 50, 500, or 5,000 feet, etc. On an engineering drawing, the engineer's scale used is expressed as $1'' = 300'$, $1'' = 50$ miles, etc. The scale graduations of tenths make simple work of the decimal system often used in engineering drawings and of plotting computed values on graphs.

Mechanical engineer's scales are usually flat, are marked as half-size, quarter-size, etc., and are used to lay out machinery designs [1-28]. These scales are stated on engineering drawings as $\frac{1}{2}'' = 1''$ or half-size, $\frac{1}{4}'' = 1''$ or quarter-size, $2'' = 1''$ or twice size. With the growth of microfilm records and other copying machines that change the size of drawings, it is useful to draw a scale with its typical graduations on the engineering drawing. The scale commonly found on maps is a good example of this procedure. When the drawing is reduced or enlarged, the scale remains true and is more representative than any written notation that might be found in a title block. Although most mechanical engineer's scales are subdivided into fractions of an inch, decimally divided scales are preferred in many industries.

The growing interest in the metric system, especially

in scientific and some military applications, has led to scales based on the meter. The ease of conversion and the decimal base associated with the metric system are quite attractive, and serious effort is under way to accept the metric standard. If this comes about, the metric scales will become the basis for all engineering graphics [1-30].

When several measurements are to be taken along a given direction, it is best to place the scale carefully and firmly in place and measure all distances at once. Avoid moving the scale for each new measurement since an accumulation of measurement errors is likely to result. For maximum accuracy in layout or measurement of distances, dividers should be used to transfer the distances from the scale to the paper.

The *divider* is similar to a conventional compass, but it does not have a lead holder on either leg. Instead, both legs have very sharp points [1-32]. Dividers are most often used to transfer a known distance from one place to another or to mark off several measurements of the same length.

Angular Measurement

The simplest device for measuring angles is a *protractor.* A protractor is a circular or semicircular plastic or metal disc whose circumference is inscribed with marks representing the degrees of a circle. The protractor is used by placing its base line, indicated by the 0 and 180° marks, on one of the legs or sides of an angle with its center at the vertex of the angle and reading or marking the appropriate degrees from the graduations [1-33].

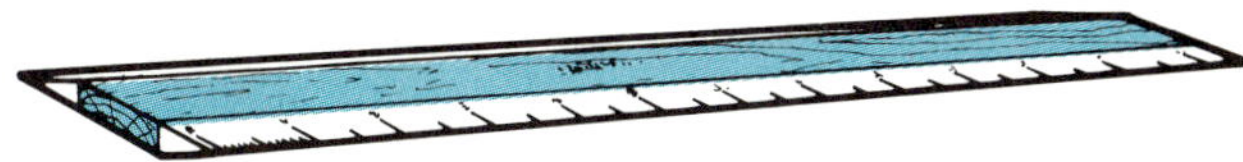

1-30
The metric scale.

1-31
Scale graduations—metric scale.

1-32
The dividers.

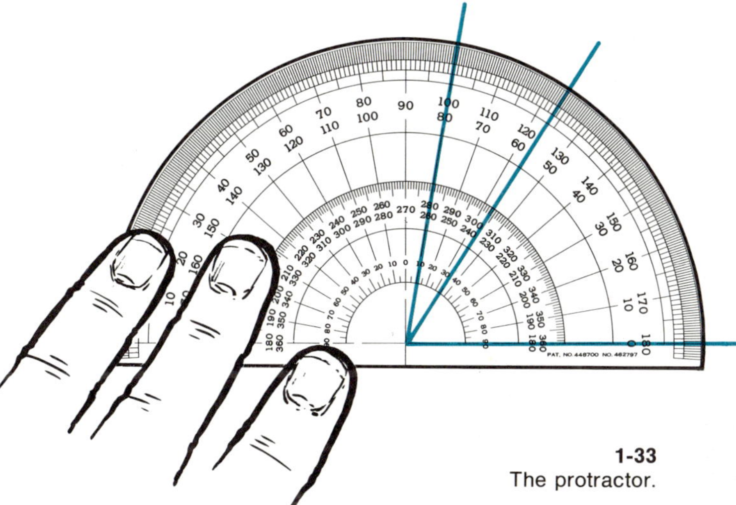

1-33
The protractor.

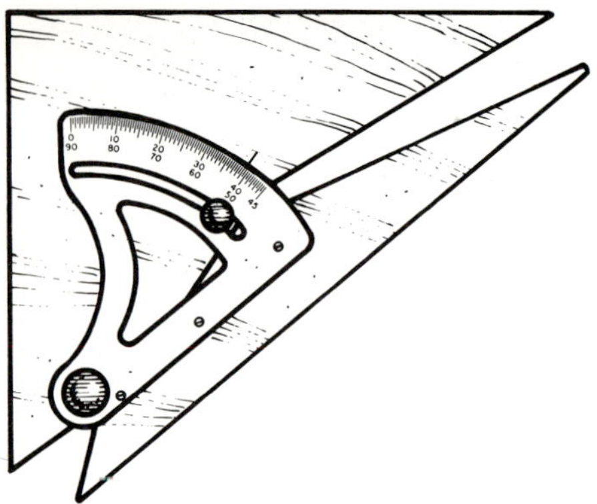

1-34
The adjustable triangle.

Angles of 30°, 45°, and 60° (or other angles that are multiples of 15°) can be laid out with standard triangles.

The *adjustable triangle* has a protractor incorporated into it, thus allowing the two angles that are not 90° to be varied [1-34].

Finally, there is the *drafting machine*. It is a combination guiding and measuring tool that incorporates a T square, triangles, and protractors. It consists of two arms at right angle to each other that can be moved all over the drafting board without change in orientation. Scales can be attached to the arms [1-35]. The orientation can be adjusted or changed, and a protractor is provided to tell the amount of angular adjustment. These machines are very effective in reducing the time required to produce good engineering drawings.

1-35
The drafting machine.

the work place

The *ideal* environment includes an adjustable desk or table of adequate size. Sketching materials, reference materials, and tools should be conveniently placed. A white, neutral light of high intensity is preferred. Fluorescent lights are best because they minimize shadows. A

1-36
The work place.

good quality flexible-arm drawing lamp is often used to supplement overhead lighting. Unfortunately, such an ideal environment is not always available when the need to make a sketch arises.

types of lines

We have discussed some of the tools useful in engineering graphics. Now we shall discuss techniques for using these tools, beginning with the types of lines used by engineers and draftsmen and how to draw them. Standard definitions for the types of lines may be thought of as a visual alphabet. You need to understand and recognize letters before you can understand words; similarly, you must understand the meaning of lines in order to understand the object being represented.

One important characteristic of lines is the weight or width of the line. Lines are classified as thin, medium, or thick. These are relative terms. It is recommended that the ratios of these widths be 1:2:4; in other words, the medium-width line is twice as wide as the thin line [1-37]. The proper width for a group of lines is governed by the style and purpose of the drawing to be produced. *There must be a distinct contrast between the different widths of lines on a drawing and these contrasts and widths must be consistent and uniform throughout the drawing.* All the lines constructed on a drawing must be clean, dense, and opaque black for legible reproduction.

When constructing lines on a drawing, orient the pencil in the direction of the stroke at an angle of approximately 60° with the paper [1-38]. Draw the line with an even, consistent pressure and, when using a conical point, rotate the pencil slowly between the thumb and forefinger as the line is being drawn. This will maintain a symmetrical point and produce a uniform line weight. If it is necessary, draw over a line several times to get the proper width and contrast.

Lines used in engineering graphics are of two general classes: lines that represent features of actual physical objects, and lines that present information but are not part of the object. Lines used to represent the physical characteristics of an object are called visible and hidden lines. They indicate the intersection of two surfaces, the edge view of a surface, or the contour of a curved surface. The visible line is usually called the *object line* and shows

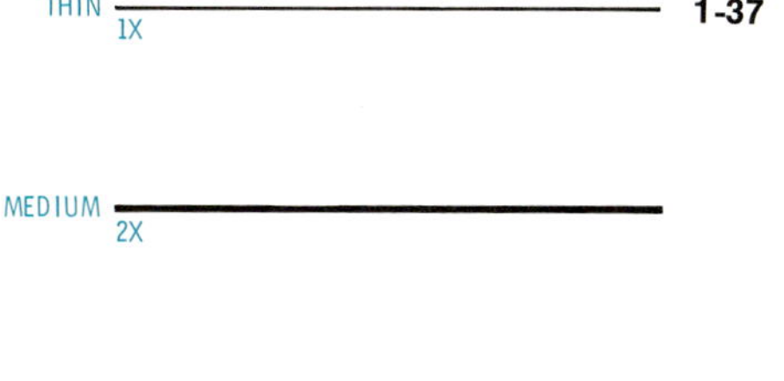

1-37

1-38

1-39
Types of lines.

TYPE	WEIGHT	
OBJECT	THICK	
HIDDEN	MEDIUM	
CENTER	THIN	
PHANTOM	THIN	
EXTENSION & DIMENSION	THIN	
LEADER	THIN	
SECTION	THIN	
CUTTING PLANE	THICK	
SHORT BREAK	THICK	
LONG BREAK	THIN	

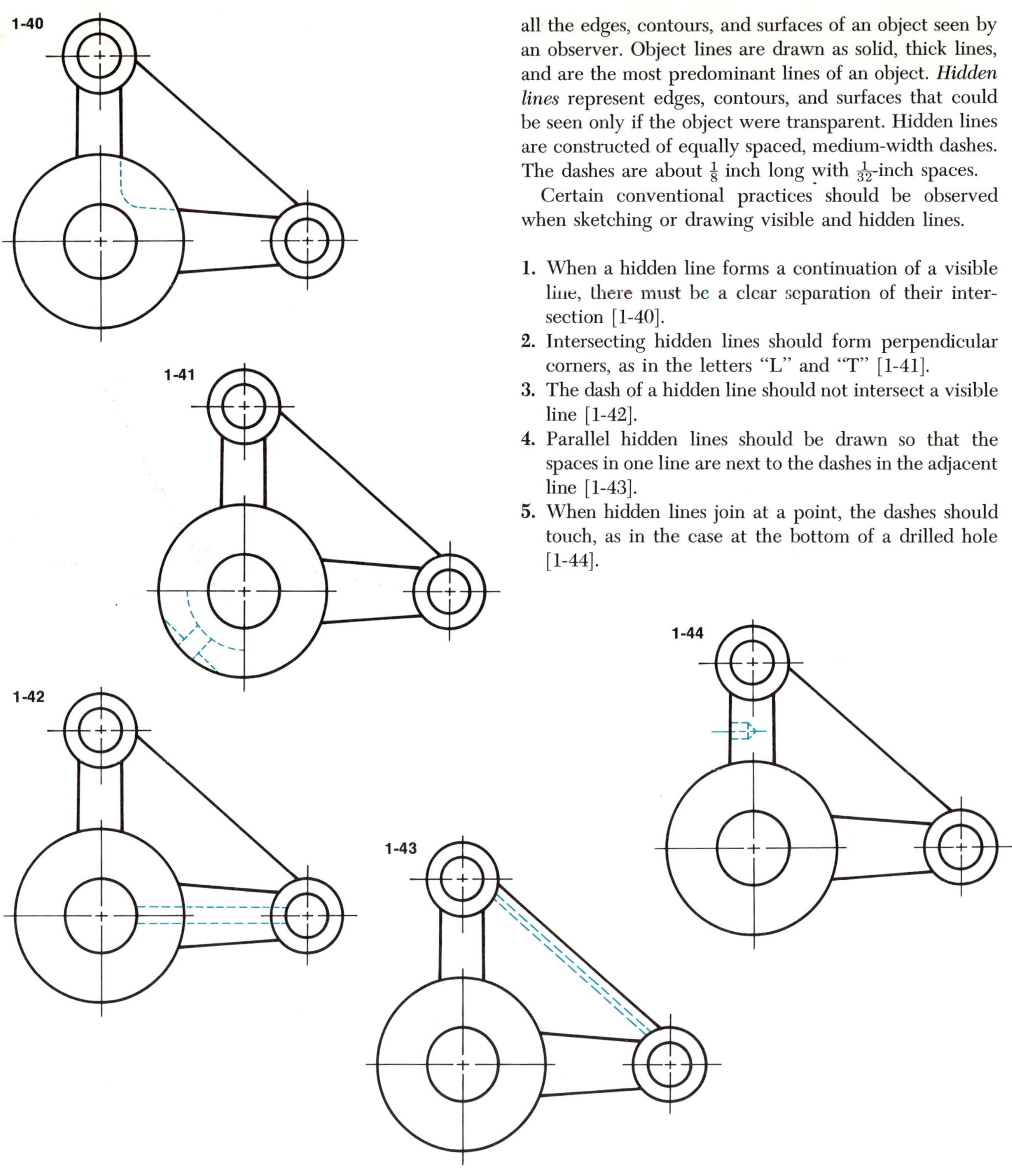

all the edges, contours, and surfaces of an object seen by an observer. Object lines are drawn as solid, thick lines, and are the most predominant lines of an object. *Hidden lines* represent edges, contours, and surfaces that could be seen only if the object were transparent. Hidden lines are constructed of equally spaced, medium-width dashes. The dashes are about $\frac{1}{8}$ inch long with $\frac{1}{32}$-inch spaces.

Certain conventional practices should be observed when sketching or drawing visible and hidden lines.

1. When a hidden line forms a continuation of a visible line, there must be a clear separation of their intersection [1-40].
2. Intersecting hidden lines should form perpendicular corners, as in the letters "L" and "T" [1-41].
3. The dash of a hidden line should not intersect a visible line [1-42].
4. Parallel hidden lines should be drawn so that the spaces in one line are next to the dashes in the adjacent line [1-43].
5. When hidden lines join at a point, the dashes should touch, as in the case at the bottom of a drilled hole [1-44].

The second class of lines includes center lines, extension lines, leaders, dimension lines, section lines, cutting-plane lines, viewing-plane lines, short-break lines, long-break lines, and phantom lines.

The *center line* is composed of thin alternate long and short dashes. As the name implies, center lines define a central axis of a hole, or indicate the origin of a radius. The center line is also used to indicate an axis of symmetry of an object.

The *extension line* is constructed as a thin solid line. It is used to extend the significant features of an object to a point where dimensions can be applied without cluttering the visual image of the object. When constructing extension lines, leave a gap of about $\frac{1}{32}$ inch between the extension line and the object line being extended. *Leaders* are closely related to extension lines. They are thin solid lines constructed from a descriptive note to a point on the object. *Dimension lines* are also drawn as thin solid lines. More detail about dimension lines, leaders, etc., will be found in Chapter 5.

Section lines are thin solid lines commonly referred to as *cross-hatching*. They represent a surface of solid material that has been cut through by a cutting plane in the drawing. In professional quality work, section lines should be evenly spaced and of uniform width. The *cutting-plane line* is a thick line composed of a series of medium-length dashes. The cutting-plane line is used to show the location where an object has been sliced with an imaginary knife to expose hidden details or internal parts. The *viewing-plane line* is drawn similar to the cutting-plane line and is used to indicate special directions for viewing an object.

The two types of *break lines*, the thick *short-break* and the thin *long-break*, are used to eliminate from the drawing parts of an object that are not needed in the particular view. They represent an intentional disruption of a view of the object and removal of some segments. Break lines are used as a convenience to eliminate unnecessary work and to clarify the drawing. The *phantom line* is used to draw the object in an alternative position, to show length of travel, or to show direction of motion. It is a thin line of two short dashes alternating with one long dash.

Construction lines are not included in the two groups mentioned above. They are composed of very thin, lightweight lines barely visible at arm's length. They are used to lay out the drawing, and are the first lines used when constructing a technical drawing. They are extremely light and need not be erased. *Guidelines* for lettering also fall in this category.

Often two or more lines coincide on a drawing. In such cases, show those lines that present the greatest information to the observer. Visible lines, hidden lines, and center lines require a system of priorities to effectively convey meaning. Visible lines always predominate. In any given drawing, if the visible line is constructed in the same position as a hidden line or center line, the visible line will have precedence. Next in importance are the hidden lines, and finally the center line.

When constructing views of an object, select the view in which the major details of the object are visible and the fewest number of hidden lines are required. It is desirable to keep the number of lines to describe an object to a minimum, and to use consistent methods of depicting visible, hidden, and center-line relationships. This aids in eliminating error on the part of the designer and assists the reader in his understanding of the drawing.

When you have learned to identify these lines and to understand their meaning, you will have an easy time comprehending the shape of the object presented in an engineering drawing.

constructions

A few important techniques are used frequently in basic drawing. The simplest and most effective way to transfer a line parallel to an original line is by using two triangles. Place one of the triangles on the work and carefully align one of its sides with the line to be transferred. When the alignment is true, hold the triangle firmly and place the second triangle in contact with one of the remaining sides [1-45]. (Be sure that no eraser crumbs or other material are caught between the two triangles.) While holding the second triangle firmly in position, slide the first triangle to a new location where a parallel line is desired [1-46]. Any line drawn along the side that was aligned with the original line will be parallel with that line. This technique is helpful in constructing lines parallel with planes, the sides of rectilinear figures, or guidelines for lettering.

To construct one line perpendicular to another, orient the first triangle so that a side other than the hypotenuse is aligned with the line to which a perpendicular line

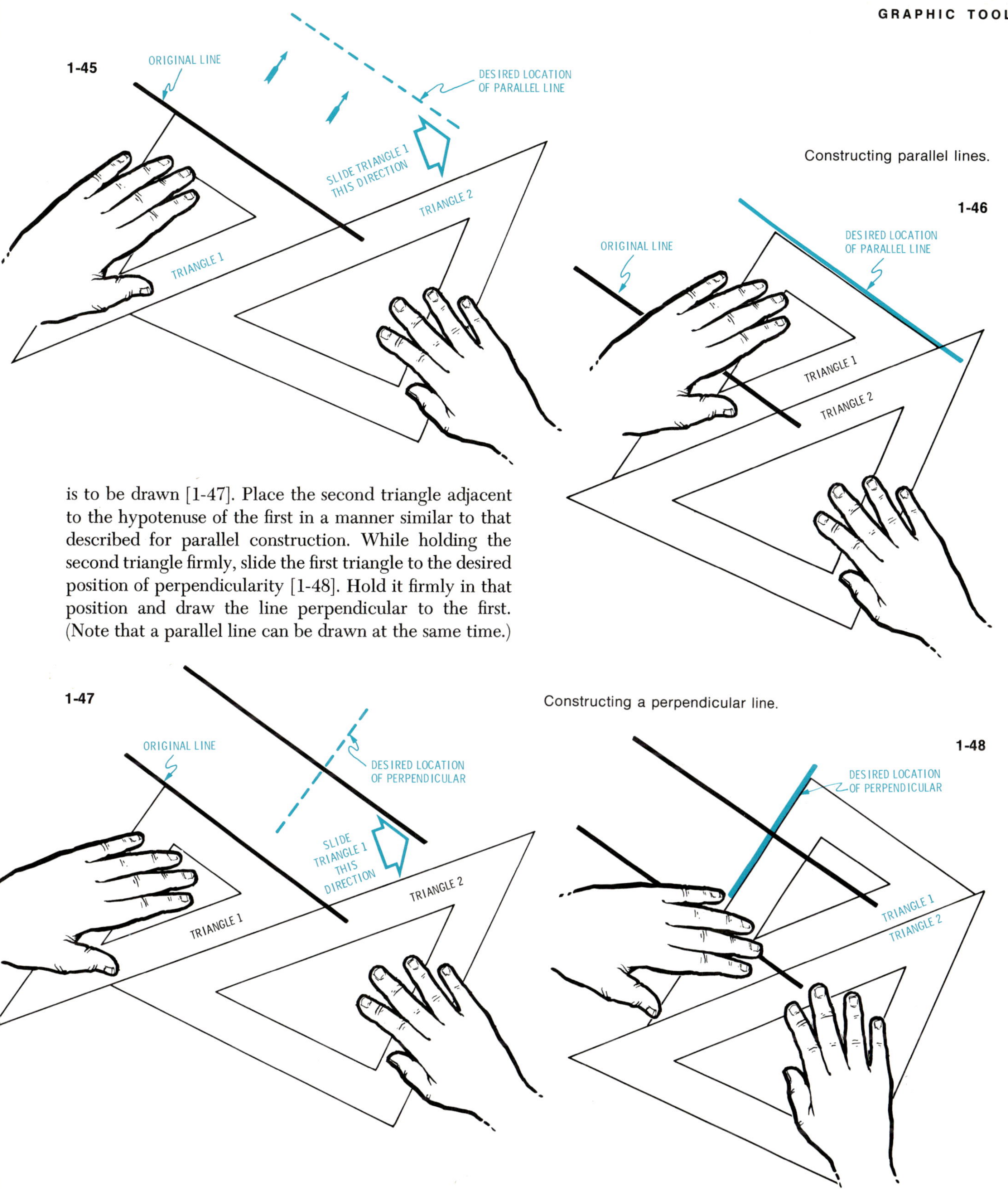

Constructing parallel lines.

is to be drawn [1-47]. Place the second triangle adjacent to the hypotenuse of the first in a manner similar to that described for parallel construction. While holding the second triangle firmly, slide the first triangle to the desired position of perpendicularity [1-48]. Hold it firmly in that position and draw the line perpendicular to the first. (Note that a parallel line can be drawn at the same time.)

Constructing a perpendicular line.

This is an effective and accurate method for constructing right-angle corners of rectilinear objects, perpendicular planes, and vertical guidelines for lettering.

1-49

Basic Vertical Strokes

Basic Inclined Strokes

lettering

A new design can be described by drawings of its components, but many details of the design are best communicated through words and symbols. The lettering used in engineering graphics is more uniform and more easily read than script. Lettering can be considered as an application of freehand drawing. Great care must be exercised in lettering, because poorly made letters and numbers can be misread and may cause expensive errors in production or fabrication. The preferred style of lettering used in engineering graphics is called single-stroke Gothic. It was

1-50
Vertical lettering.

A B C D E F G H I J K L M N
O P Q R S T U V W X Y Z
0 1 2 3 4 5 6 7 8 9

Upper-Case Vertical Gothic Lettering

a b c d e f g h i j k l m n
o p q r s t u v w x y z

Lower-Case Vertical Gothic Lettering

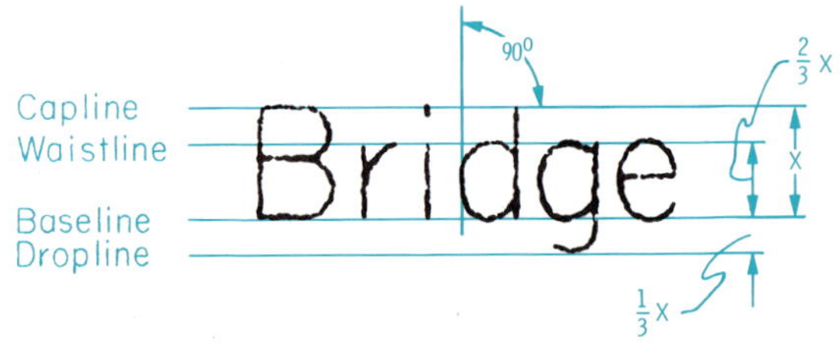

ABCDEFGHIJKLMN
OPQRSTUVWXYZ
0123456789

Upper-Case Inclined Gothic Lettering

abcdefghijklmn
opqrstuvwxyz

Lower-Case Inclined Gothic Lettering

1-51
Inclined lettering.

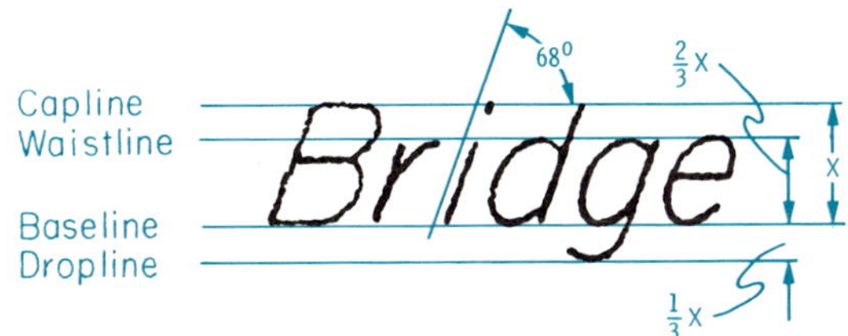

designed and standardized for maximum legibility and easy, rapid execution.

To letter well in the Gothic style, the engineer should be thoroughly familiar with the characteristics of every letter and numeral in the Gothic alphabet [1-50 and 1-51]. He should learn

1. The shape and general proportions of each letter.
2. The recommended order of strokes to draw each letter.
3. The way in which letters are spaced to make words.

Some simple guidelines will assist in the development of lettering skill. Beginners tend to make letters too narrow. As a rule, it is the width of the letters that make them legible. All capital or uppercase letters are almost as wide as they are high, except the letters "I" and "W." The bodies of the small or lowercase letters are approximately two thirds the height of the capitals. Fractional

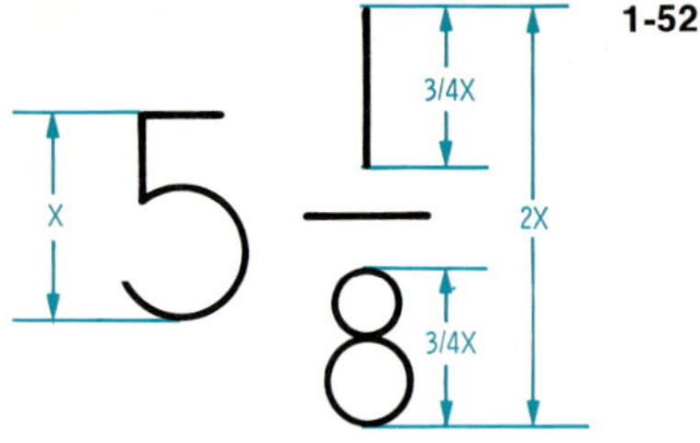

1-52

numbers are about twice the height of uppercase letters, with the numerator and denominator each about three fourths the whole number size. Normally, engineering lettering is $\frac{1}{8}$ inch high. If a reduction of size is anticipated, such as in microfilming, it is desirable to make the lettering somewhat larger, say $\frac{5}{32}$ inch or more. The space between lines of lettering should be at least one half the letter height—more if space is available.

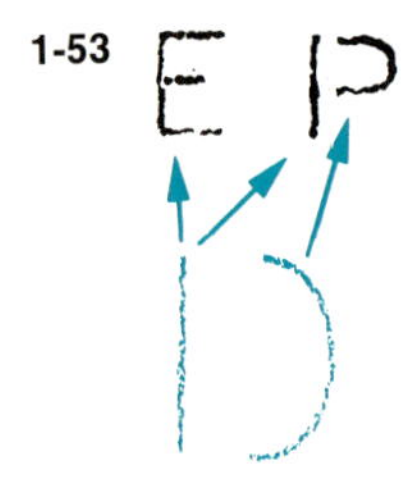

1-53

Lettering is composed of two basic graphical elements: the straight line and the curve. Vertical strokes should be made downward with finger movement, and horizontal strokes from left to right with a wrist movement. The forearm should be at approximately a right angle to the line of lettering and resting on the writing surface. Uniformity is important and is best achieved by constructing and using horizontal and vertical lettering guidelines. All guidelines should be drawn *very lightly* with a sharp, hard 4H or 6H pencil. There are four horizontal guidelines: the capline, waistline, baseline, and dropline. The first three of these guidelines are commonly used in lettering practice. The dropline is used only with lowercase letters, and there are but five lowercase letters that require it. Consequently, it is often omitted. The drawing of accurately spaced guidelines is greatly facilitated by the use of an Ames or Braddock lettering guide [1-55].

Vertical guidelines are either truly vertical or inclined about 68° depending on the choice of vertical Gothic or inclined Gothic lettering. Vertical guidelines are spaced randomly throughout a line of lettering and are constructed parallel to each other. The spacing of these guidelines has nothing to do with the spacing of the letters—they are simply a convenience to the letterer.

If lettering is not Uniform

It is unprofessional

1-54

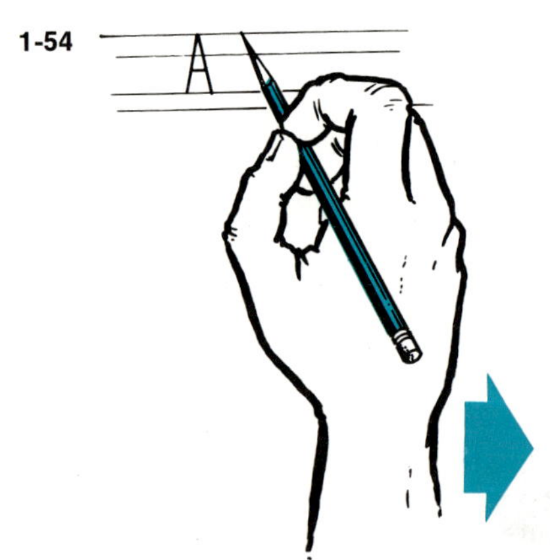

It usually makes little difference whether vertical or inclined lettering is used, but once established on a drawing, the lettering must be consistent, uniform, and parallel. Generally, it is easier to produce acceptable vertical lettering, because one's eye is not often used to judge accurately adherence to a specific angle of inclination.

Uniformity in lettering is further emphasized by consistent line thickness or weight. When using a pencil, this is accomplished by evenly applied pressure and frequent sharpening. Use a relatively soft pencil (H or softer) to produce a good black line. A conical point, slightly rounded, is best, and the pencil should be rotated a little before each letter is started. This practice will promote even wearing of the point and maintain its symmetry. The strokes of each letter should be one continuous motion.

Spacing of letters and words on a drawing is a visual and subjective art that is developed only through continued practice. Letter spacing is a function of the blank areas between adjacent letters. Good spacing cannot be accomplished by measuring. It is the visual areas between letters that must appear to be equal, and this changes with each combination of letters. The space between words obviously must be greater than between letters. A good rule is to space words at least the height of a capital letter or just slightly more; in other words, leave at least enough room for a capital "O" between words. Remember that good lettering involves the artistic designing of the black lines and white areas to create a balanced and pleasing appearance.

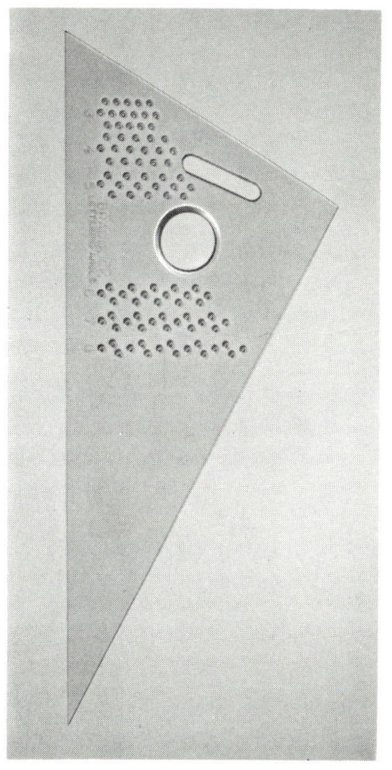

1-55
The lettering guide.

1-56

A slightly rounded, conical point is best.

Good spacing is related to visual area... not measured distances.

ALLOW EQUAL AREAS BETWEEN LETTERS IN WORDS.

problems

1-1. Using words only, describe the shape or form of one of the following mechanisms:
(a) desk-top stapler
(b) wall-mounted pencil sharpener
(c) fractional horsepower electric motor
(d) door latch
(e) hand-operated can opener

1-2. Make a freehand sketch of the object's shape that was verbally described in Problem 1-1. Sketch only the most descriptive view of the object. Use unruled $8\frac{1}{2}$ by 11 inch white paper.

1-3. There is an activity of engineering graphics called *marking* that uses a marking tool, an eraser, and a surface on which to mark. A common example is a 2H pencil, an art gum eraser, and tracing paper. List five other reasonable combinations of marking tools that might be used by the engineer.

1-4. On an $8\frac{1}{2}$-by 11-inch sheet of unruled paper, randomly place six points at least 3 inches apart. Using a straightedge guiding tool and a 2H pencil, draw lines such that each point is connected to the other five. Be particularly mindful of the proper procedure for using the marking tool with the guiding tool and the method of positioning the guiding tool.

1-5. Using a curved guiding tool, such as a French curve, follow the instructions of Problem 1-4.

1-6. Using the compass as a marking and guiding tool, construct seven concentric circles, starting with a 1-inch radius and increasing each successive radius by $\frac{1}{2}$ inch.

1-7. Using all capital letters $\frac{1}{8}$ inch high, print a list of all the engineering graphics tools that you have collected for use in this class. Indicate on your list the primary use for each tool (marking, measuring, or guiding).

1-8. Using all capital letters, print a list of all the engineering graphics tools that you own that are multiple-activity tools. Include in this list any tools that are not used for marking, measuring, or guiding.

1-9. Using all capital letters, print the names of 6 of the types of *marking* tools that are available for sale in the engineering graphics section of the school supply store.

1-10. List by name 6 of the *guiding tools* as instructed in Problem 1-9.

1-11. List by name 6 of the *measuring tools* as instructed in Problem 1-9.

1-12. Using an architect's scale, accurately lay off a full-size $4\frac{1}{2}$-inch line. In a column under this line, list all the scales available on the architect's scale, and beside each listed scale place the *measured* equivalent of the $4\frac{1}{2}$-inch line.

1-13. Using an engineer's scale and a full-size $3\frac{3}{4}$-inch line, measure and list the measurements as instructed in Problem 1-12.

1-14. Using a protractor or an adjustable triangle, divide a 6-inch-diameter circle into five equal segments.

1-15. Using a T square, a 30°–60° triangle, and a 45° triangle (singularly and in combination), construct 15° segments of a 6-inch-diameter circle from 0° to 360°. Do not use any other measuring tools.

1-16. Draw three series of 12 parallel lines at any angle other than horizontal or vertical. Make one of the series *thin* weight lines, another *medium* weight, and the third *thick* weight. Draw each line in the series 4 inches long and parallel with $\frac{1}{8}$-inch space between the lines. (Use the parallel-line construction technique involving two triangles. See page 19.)

1-17. Using three or more figures from [1-57] as assigned, place a dot about $\frac{1}{16}$ inch in diameter on every object line shown.

1-57

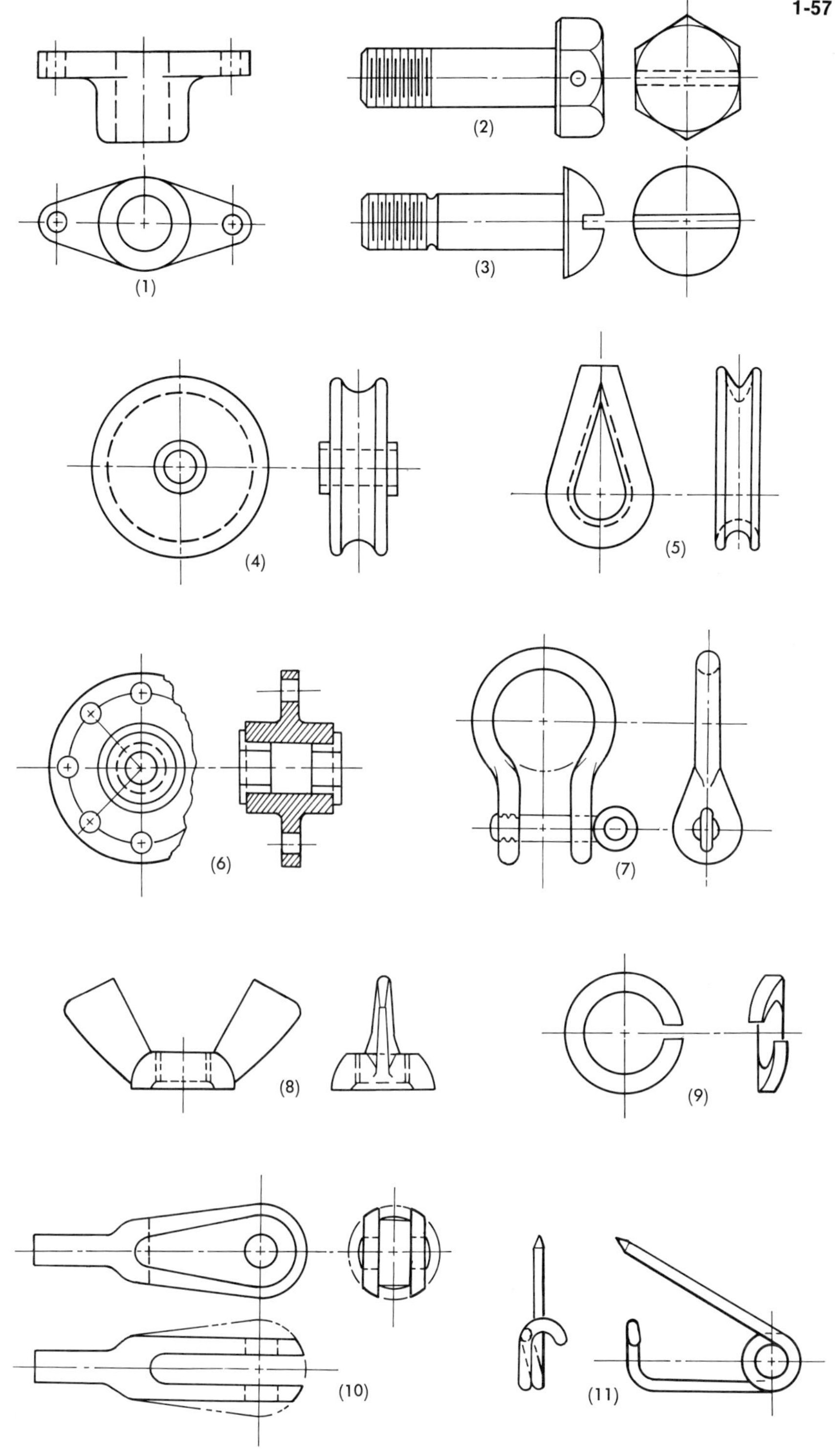

1-18. Identify all the hidden lines as instructed in Problem 1-17.

1-19. Identify all the center lines as instructed in Problem 1-17.

1-20. Using vertical Gothic lettering, uppercase and lowercase, letter the first paragraph under the subtitle "Lettering," which begins with "A new design . . ." Use guidelines and make the letters $\frac{1}{8}$ inch high. (See page 20.)

1-21. Using slant lettering, uppercase and lowercase, letter the same paragraph as instructed in Problem 1-20.

sketching and conceptualization 2

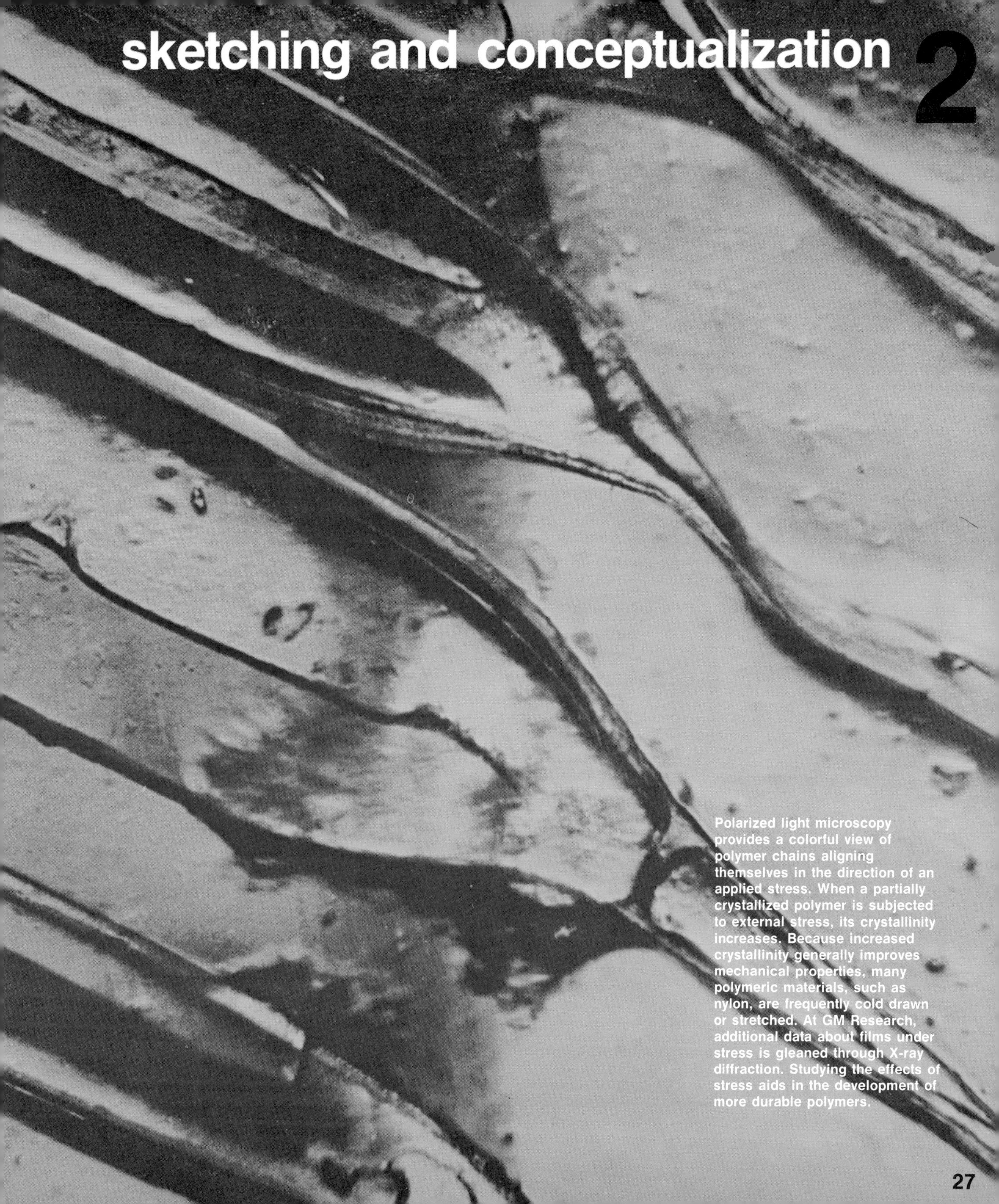

Polarized light microscopy provides a colorful view of polymer chains aligning themselves in the direction of an applied stress. When a partially crystallized polymer is subjected to external stress, its crystallinity increases. Because increased crystallinity generally improves mechanical properties, many polymeric materials, such as nylon, are frequently cold drawn or stretched. At GM Research, additional data about films under stress is gleaned through X-ray diffraction. Studying the effects of stress aids in the development of more durable polymers.

The ability to make clear, concise sketches is important to the engineer, but this does not imply that engineers must have artistic talents. Rather, proficiency in sketching is acquired through practice. Engineering sketches conform to uniform standards and conventions so that people of all disciplines and languages can understand the meaning intended by the designer. This chapter considers both the conventions and techniques of sketching.

Sketching helps to stimulate one's imagination as well as to convey information. For this reason it is ideally suited for the engineer's communication with others *and with himself*. A sketch frequently serves to clarify an original idea and at the same time suggest additional versions, adaptations, and improvements. One often finds that he is able to organize his thoughts more effectively

2-1
The ability to sketch is an essential skill for those in design work.

if he can present his mental impressions in graphic form, whether it be a flow chart summarizing the main points of a presentation or a detailed sketch explaining a complicated mechanism.

In general, technical sketching is a mechanical process that conforms to specified guidelines. However, one's physiological makeup does play an important role—particularly the interplay between brain, eye, hand, and arm. A simple experiment will illustrate the remarkable precision of the eye when used as a proportioning instrument. On the line below place a mark at the point you estimate to be the midpoint. Do not use a scale or any other measuring device except your eye.

Now measure the line and locate the midpoint exactly. How closely did you estimate the midpoint? This is just one of many functions of your eye–brain–hand system, *and it becomes more proficient with practice.*

Emphasis in this chapter will be directed toward the freehand sketch which is ideal for conceptualization and ideation and which also can be used effectively in design development and presentation.

One cannot learn to sketch and express himself graphically solely by reading about it, any more than one can learn to swim while standing by the pool. Several examples are shown here [2-2 to 2-12] to illustrate how sketches of all types are used to communicate ideas.

2-2

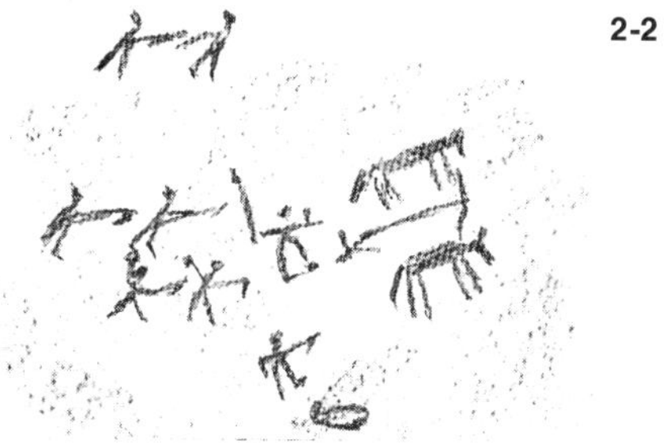

2-3

2-4

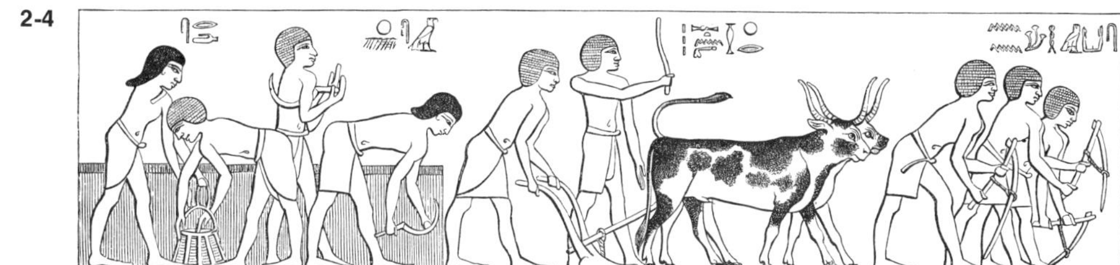

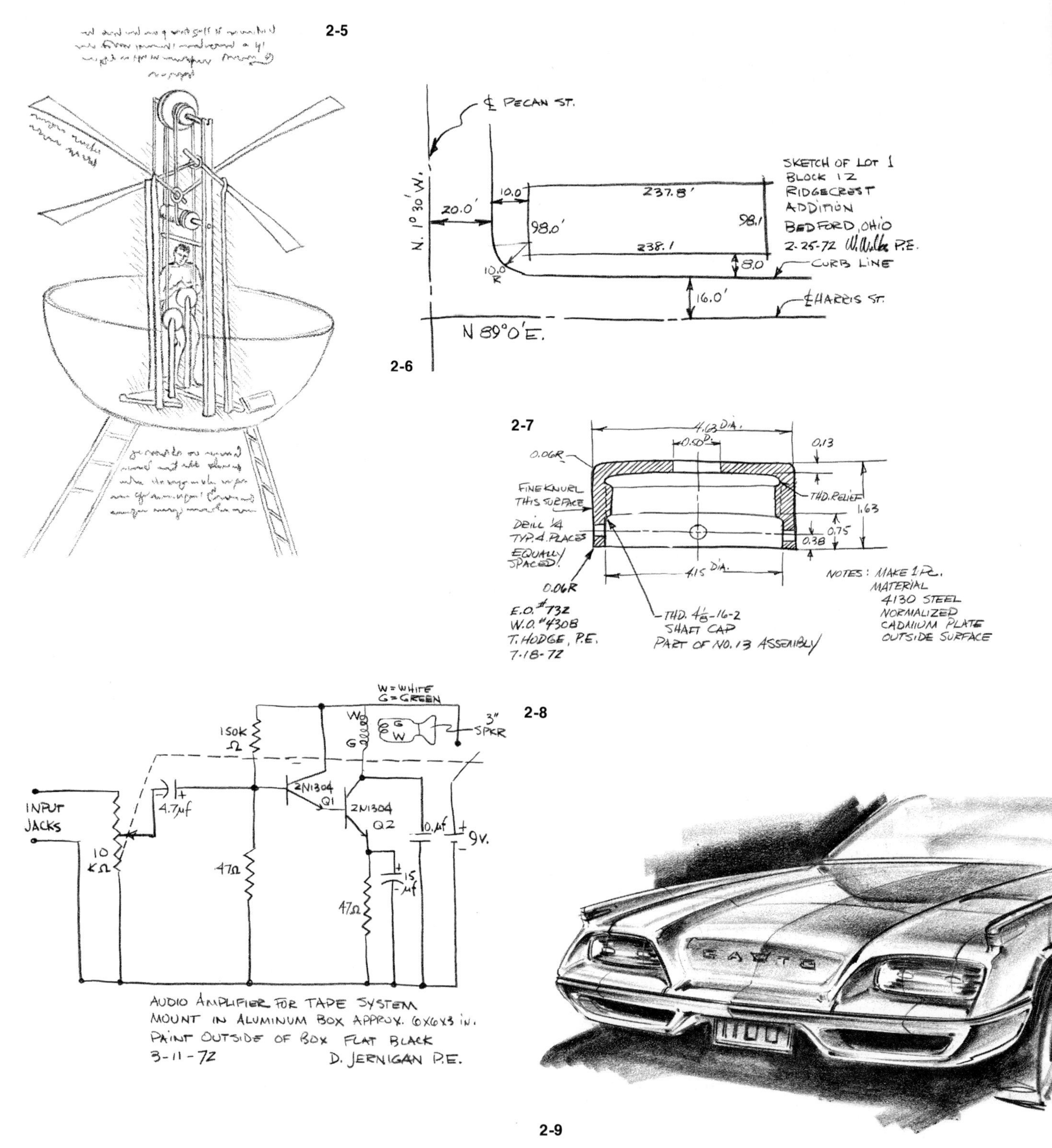

2-5
2-6
℄ PECAN ST.
N. 1° 30' W.
20.0'
10.0
237.8'
98.0'
98.1
238.1
10.0 R
8.0
CURB LINE
16.0'
℄ HARRIS ST.
N 89°0'E.
SKETCH OF LOT 1
BLOCK 12
RIDGECREST
ADDITION
BEDFORD, OHIO
2-25-72
P.E.
2-7
4.63 DIA.
0.50 D.
0.13
0.06R
FINE KNURL
THIS SURFACE
THD. RELIEF
1.63
DRILL 1/4
TYP. 4. PLACES
EQUALLY
SPACED.
0.75
0.38
4.15 DIA.
0.06R
NOTES: MAKE 1 PC.
MATERIAL
4130 STEEL
NORMALIZED
CADMIUM PLATE
OUTSIDE SURFACE
E.O. #732
W.O. #430B
T. HODGE, P.E.
7-18-72
THD. 4 1/8-16-2
SHAFT CAP
PART OF NO. 13 ASSEMBLY
W = WHITE
G = GREEN
2-8
150K Ω
3" SPKR
INPUT
JACKS
10 KΩ
4.7μf
2N1304
Q1
2N1304
Q2
10.μf
9V.
47Ω
15 μf
47Ω
AUDIO AMPLIFIER FOR TAPE SYSTEM
MOUNT IN ALUMINUM BOX APPROX. 6x6x3 in.
PAINT OUTSIDE OF BOX FLAT BLACK
3-11-72
D. JERNIGAN P.E.
1100
2-9

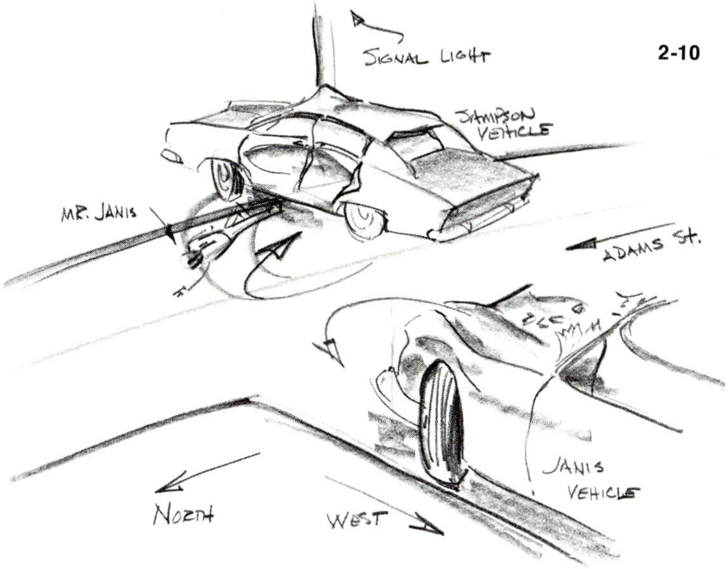

2-10

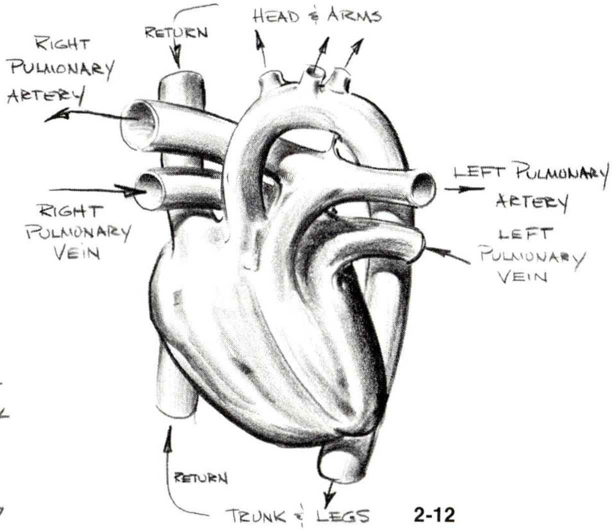

2-12

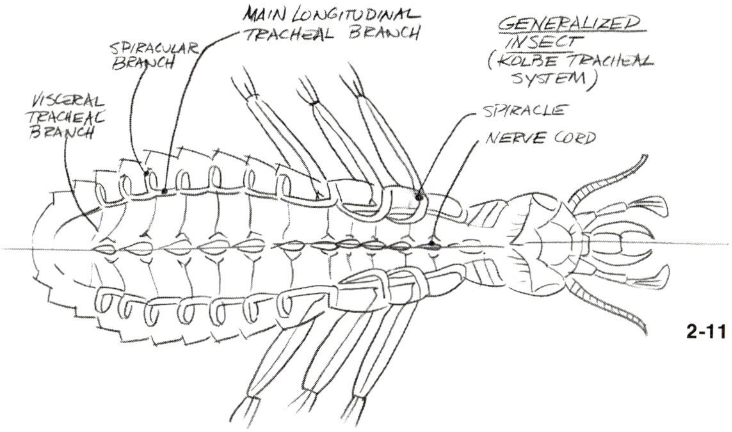

2-11

tools

Sketching requires only a marking tool and a surface upon which to mark. Almost any combination will suffice—from a ball-point pen on a tablecloth to a piece of chalk on a structural I beam.

The ideal tools are a *pencil* and good quality *paper*. The pencil point must be sharpened and shaped according to the thickness of line desired. A conical shape, slightly rounded at the tip, is best. A sharp point will produce a thin line, a dull point a thick line. If the lead is too sharp or pointed, it tends to break when used. Sharpen and shape the pencil point as often as is necessary, and remove its sharp edges on a pad of rough paper or sandpaper.

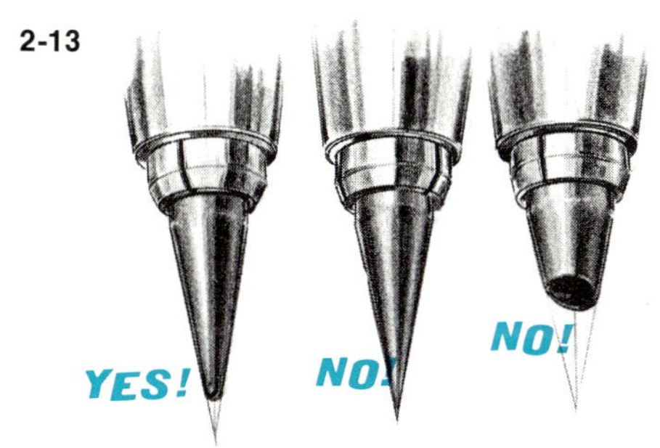

2-13

The most desirable sketch paper is a good-quality medium-weight transparent vellum. This eliminates the need for most erasing because alterations can be quickly retraced on an overlay. Thin paper is difficult to use for sketching because it tears easily. Smooth or slick paper will not be abrasive enough to give the pencil a "bite" and enable it to form a solid dark line. When vellum is unavailable, a good-quality bond paper is acceptable, as long as its surface has a slightly rougher texture than the paper used in this book. For rough field sketching, many engineers prefer pads of paper having a light cross-hatched background.

Ideally, one should *not* use an *eraser* while sketching. An important objective is confidence—the confidence to act surely and boldly when putting ideas in graphic form. If one knows that he can erase and has an eraser handy, there is a temptation to sketch without conviction. In addition, erasing destroys the paper's surface so that the corrected line on the sketch is poorly defined and has an imprecise appearance. Such techniques produce weak sketches that are unconvincing and devoid of character. An eraser should be used only after the sketch has been completed to remove smudges and extraneous construction lines that may detract from the appearance or clarity of the drawing. For this purpose art gum or other soft erasers are quite effective.

The guiding tools described in Chapter 1 may be especially useful for "finished" sketches, but a good freehand sketch is most often preferred.

basic sketching techniques

Initial sketches are most often discarded, and no effort should be wasted in finalizing the drawing. Rather, the first sketch should be drawn as a trial sketch to present only the essence of the idea. This practice can actually save time, as well as build confidence and proficiency. When one is learning to sketch, he must be willing to use a lot of paper and to discard unsuccessful efforts without remorse.

Lines in sketching are not intended to be comparable with instrument-drawn work, and their smoothness and uniformity are not as necessary as in instrument drawings. Rather the quality of a line drawn freehand is judged by its accuracy of direction and its density or weight. It is good practice to look where you are going when sketch-

2-14

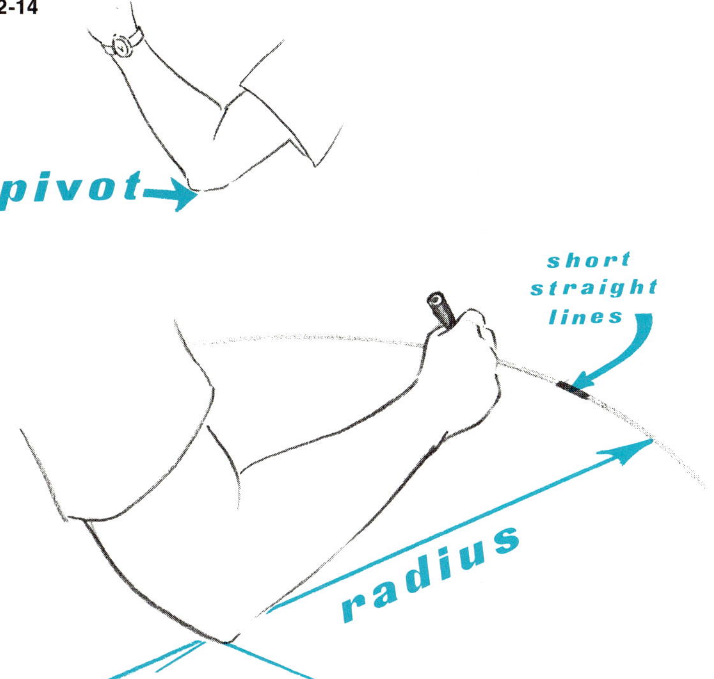

ing. Place the pencil point at the start of the intended line, then look to the point where the line will be finished. Without shifting your eyes, draw directly to the finish point in a *single* smooth, even stroke. This technique will eliminate one of the most obvious characteristics of an amateur—short, scratchy lines. When drawing lines, always move boldly to the objective. If an error appears, stop and return to the beginning. Do not erase and do not try to repair the wayward line. Final adjustment can be made after the overall sketch has been drawn.

The arm can serve as a large compass with the point of the elbow (the "funny bone") as the pivot [2-14]. With a smooth motion, small portions of a large arc can be sketched as short straight lines. The arm is rather heavy and provides the necessary stability for control. For this reason arm motions should be used when sketching, just as "follow-through" is used in tennis or golf—where smoothness of motion is essential to good performance.

Variations in width and density of lines produce sketches with vitality and interest. Variations lead the eye from one area to another and emphasize important points. Rough construction lines should be very thin, light lines. Dimension lines, center lines, and extension lines should also be lightweight [2-15]. The other lines, how-

2-15
Types of lines.

TYPE	WEIGHT	
OBJECT	THICK	
HIDDEN	MEDIUM	
CENTER	THIN	
PHANTOM	THIN	
EXTENSION & DIMENSION	THIN	
LEADER	THIN	
SECTION	THIN	
CUTTING PLANE	THICK	
SHORT BREAK	THICK	
LONG BREAK	THIN	

ever, should be sharp, distinct, and of a heavier weight. The ends of all lines, including dashes, are slightly accented. The object lines on the sketch should have the heaviest weight. Contrast among the different line weights should be maintained.

Layout

Plan the sketch *before* you begin to draw, and arrange the subject of the sketch in a pleasing and logical composition on the paper. One of the most useful procedures for planning a sketch is to first make a series of *thumbnail sketches*. These are small sketches drawn in a minimum amount of time with little attention to detail. The thumbnail sketch that best satisfies the overall requirements intended by the engineer will become the basis for the more detailed sketches to follow. Remember, vellum or paper is less expensive than time, so use as many sheets as necessary to achieve the most effective layout. Avoid crowding notes and details, and allow ample paper to surround or frame the sketch. This emphasizes the importance of the work and adds to its appearance.

2-16
A crowded condition . . .

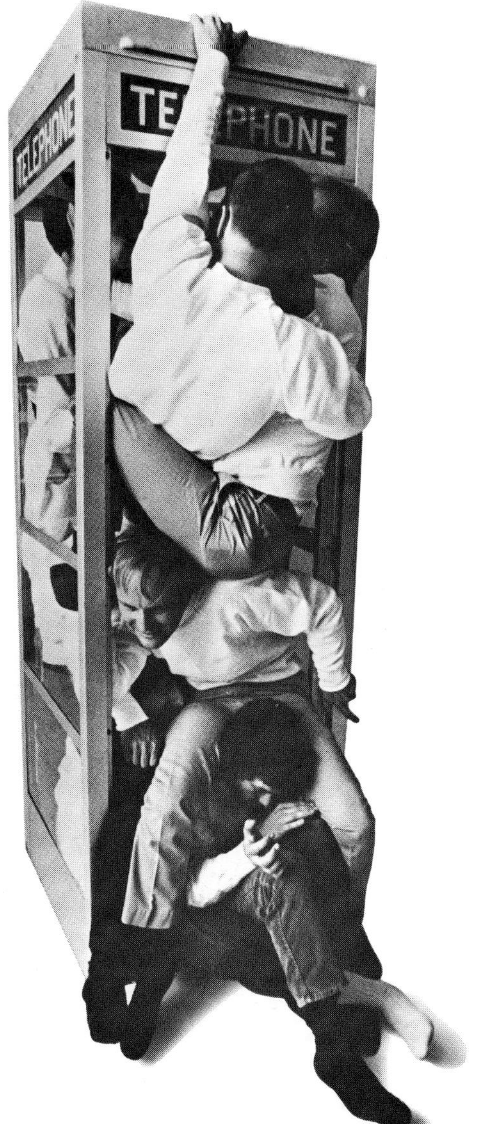

2-17
. . . creates confusion and frustration.

Proportioning

Technical sketches are not drawn to an exact scale, but must maintain the *proportions* of the object being sketched. The overall size of the sketch is optional and depends upon its intended use, its complexity, and the size of the paper available. Usually, small objects are sketched oversize, whereas large objects are sketched to some reduced scale.

There are several ways to develop a properly proportioned sketch. One is called *scaffolding*. Scaffolding establishes the foundation for a sketch.

Scaffolding. Using lightweight lines, construct a rectangle that will enclose the overall width and height of the object [2-18]. This rectangle serves in much the same way as the scaffolding used for constructing a building. It can be eliminated when the sketch is finished, just as construction scaffolding is removed from the site once the job is completed.

Subdivide the scaffold into smaller sections representing the major proportions of the object [2-19 and 2-20]. When objects are symmetrically formed about some axis or plane, use center lines [2-21].

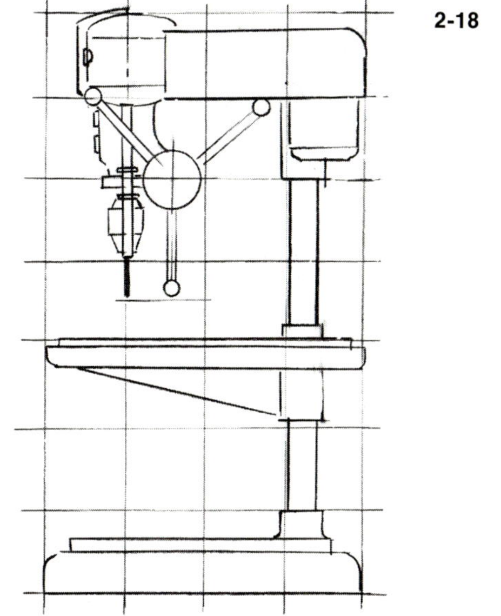

2-18

2-19

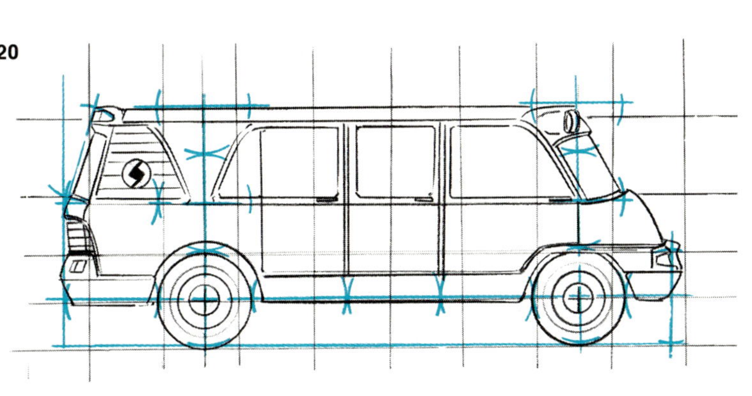

2-20

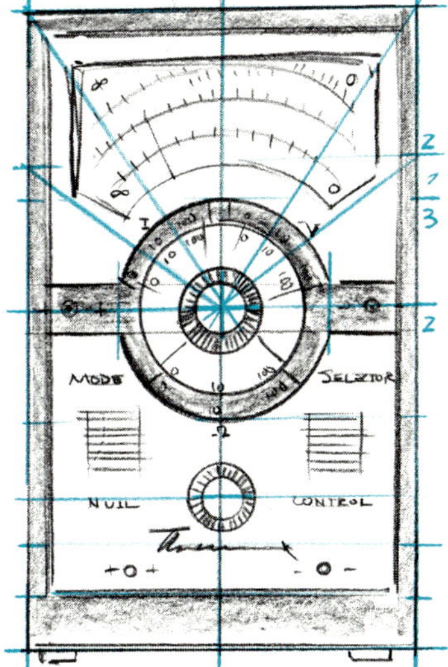

2-21

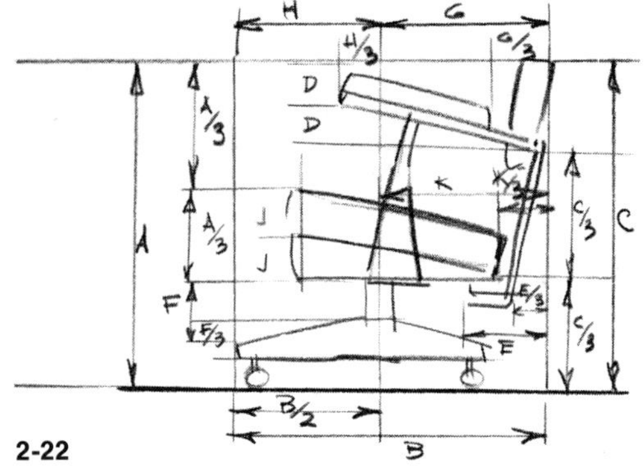

2-22

Sketch in the remaining details to define the final form of the object. After the rough sketch is completed [2-22], it is traced onto another vellum sheet. This procedure eliminates the need for time-consuming and messy erasing [2-23].

In summary, when sketching an idea, the overall width and height should first be estimated and "blocked," as in [2-24]. Then establish the major proportions of the object and the shapes of the curved parts. Finally, develop and darken the object lines of the completed sketch, and trace the sketch onto another piece of vellum.

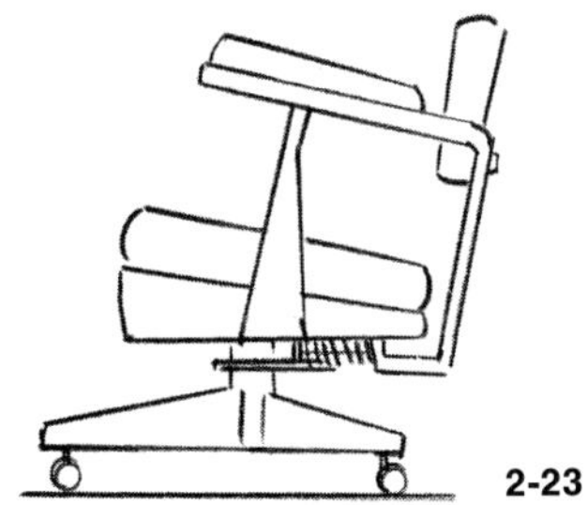
2-23

Although we have related scaffolding specifically to two-dimensional multiview sketches, the technique can be used also in the more complicated pictorial systems. The main difference is the need for three-dimensional scaffolding boxes in place of the two-dimensional rectangles used for multiview sketches.

Curvilinear Elements. In most instances all curvilinear forms can be simplified to a circle (or portions of circles). Small circles or arcs can be sketched in progressive steps. A scaffold square, whose sides equal the diameter of the circle, is sketched first with very lightweight lines. The center lines are sketched through the midpoints of the sides, and diagonal lines are drawn across the square. The circle or arc is then drawn tangent to the sides of the square [2-25].

Another acceptable technique uses a series of radial lines from the center of the circle. The radius of the circle

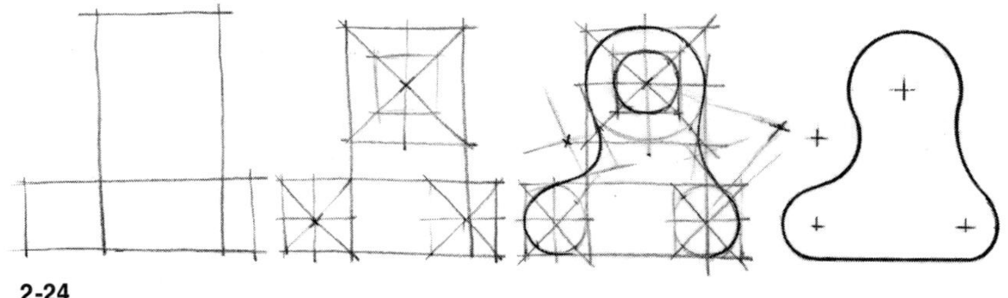
2-24

2-25
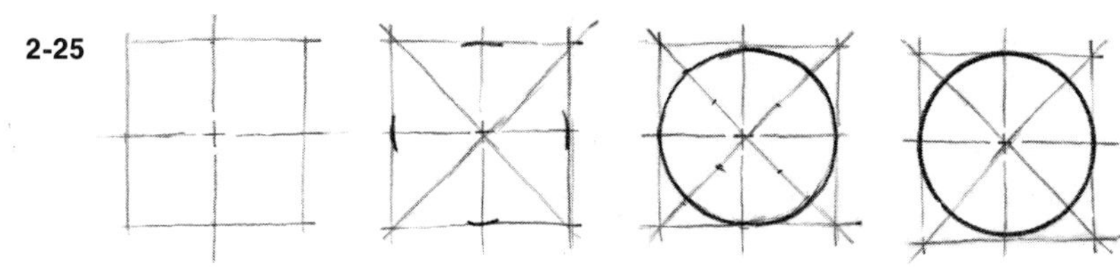

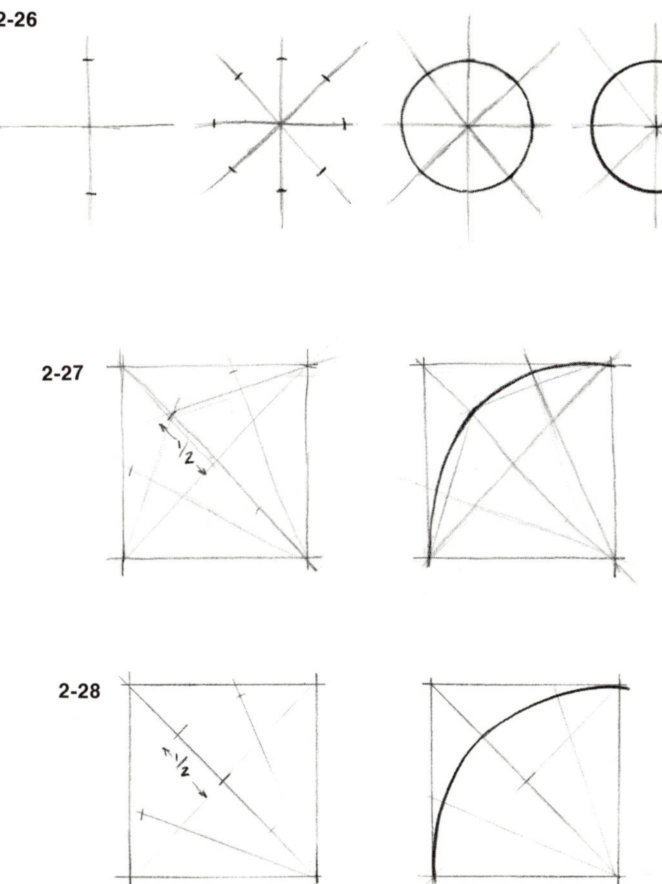

2-26

2-27

2-28

is then estimated or measured along each radial line and lightly marked. The circle is drawn through each of the radius points [2-26]. As many radial lines as desired may be drawn, but eight lines are generally enough.

Segments of circles can be drawn freehand using the same techniques [2-27 and 2-28]. Scaffolding lines or radial lines may be used, but it is important that the points of tangency be carefully observed in either case.

In several types of pictorial systems, circles are viewed at an angle. When this happens the circles *appear as ellipses* to the observer. There are several techniques for constructing accurate ellipses, but the following approximate method will suffice for sketching. This technique, known as the four-center ellipse, will work for a large range of varying elliptical axes. Proceed as shown in [2-29 to 2-33].

Step 1 Sketch a rhombus whose sides are equal to the diameter of the true circle that is to be represented; sketch the diagonal between the two apexes farthest apart [2-29].

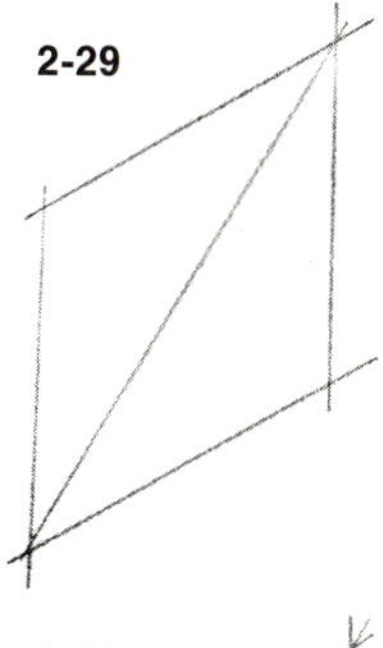

2-29

Step 2 Estimate the midpoints of each side of the rhombus, and connect the midpoints of the sides with the rhombus apexes on the opposite side [2-30].

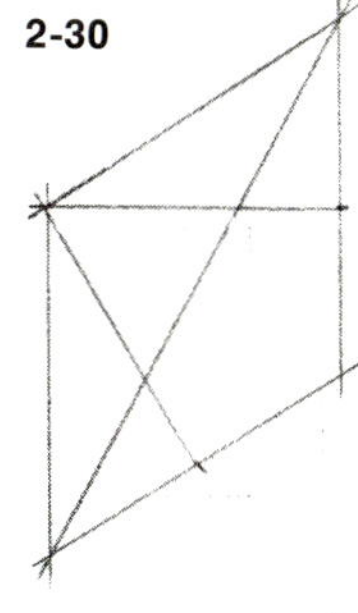

2-30

Step 3 The two intersections of the bisectors will serve as centers from which circular arcs can be sketched at the two apexes farthest apart [2-31].

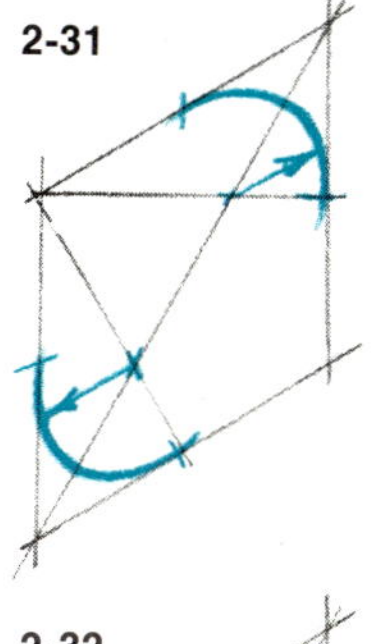

2-31

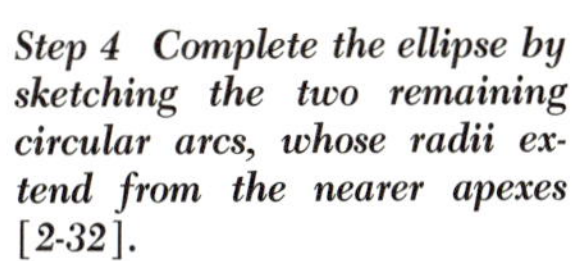

Step 4 Complete the ellipse by sketching the two remaining circular arcs, whose radii extend from the nearer apexes [2-32].

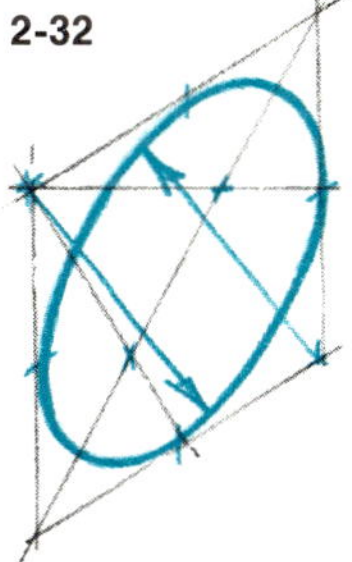

2-32

Step 5 Erase the construction lines and darken the outline of the ellipse or retrace the ellipse on clean paper [2-33].

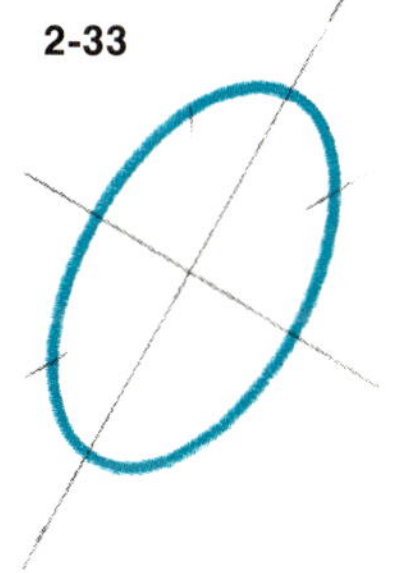

2-33

2-34

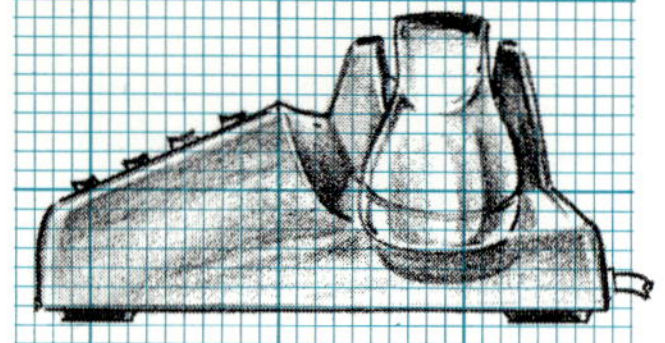

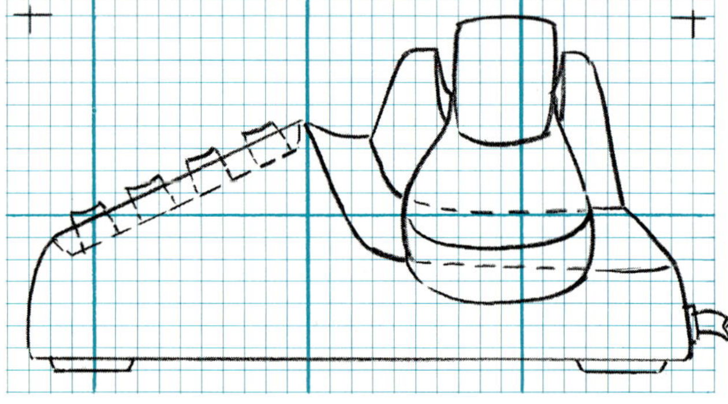

Sources of Visual Material

Frequently, you will want to sketch an object at reduced or enlarged scale from a photograph or drawing. In such instances, construct a grid across the original picture [2-34]. Then make another grid, larger or smaller than the original, as needed. Draw the final sketch into the new grid, using the grid squares to proportion lengths of lines and radii of arcs.

Pictures may be selected from magazines or journals whenever they illustrate a point or an idea that one wants to show. Pictures of people, cars, trees, trucks, and equipment may be used—anything which illustrates or amplifies the idea that one wishes to sketch. There is no reason to make a sketch if a picture is readily available. Photocopy the picture and use it if it is what you have in mind. Such techniques save time and add realism to the sketch. When translucent vellum is used, they may be traced directly. When it is only a close approximation, such as the handle on an appliance [2-35], trace the parts that are relevant to your design and then complete the remainder of the sketch from your own imagination [2-36]. This fundamental technique can be mastered if one learns to *think* graphically. *Think in terms of pictures rather than words.*

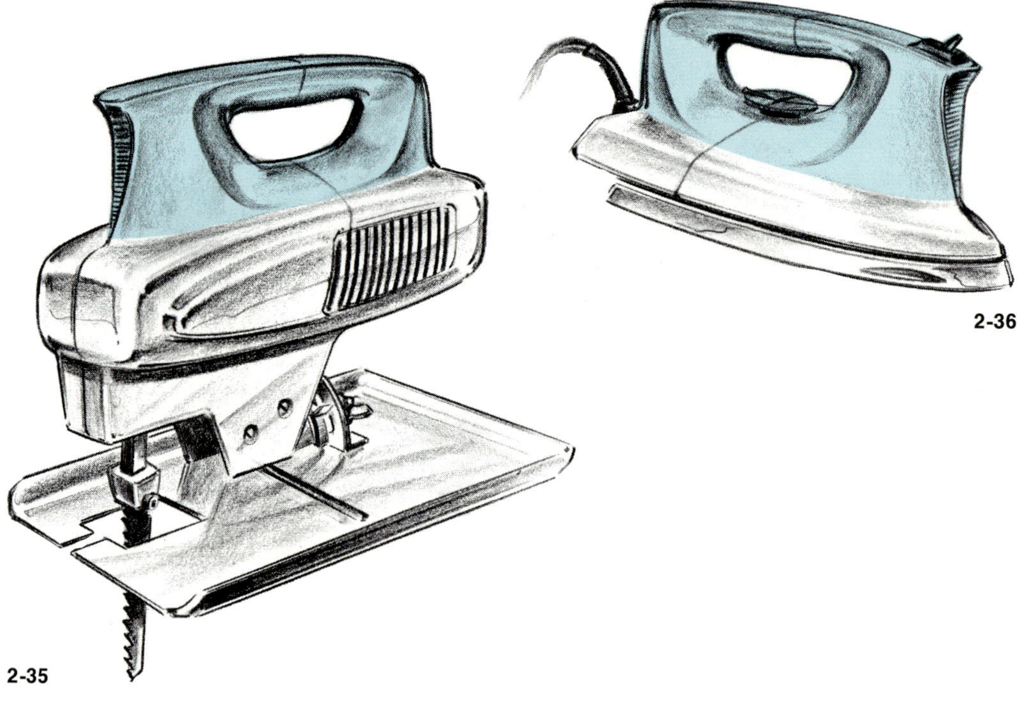

2-36

2-35

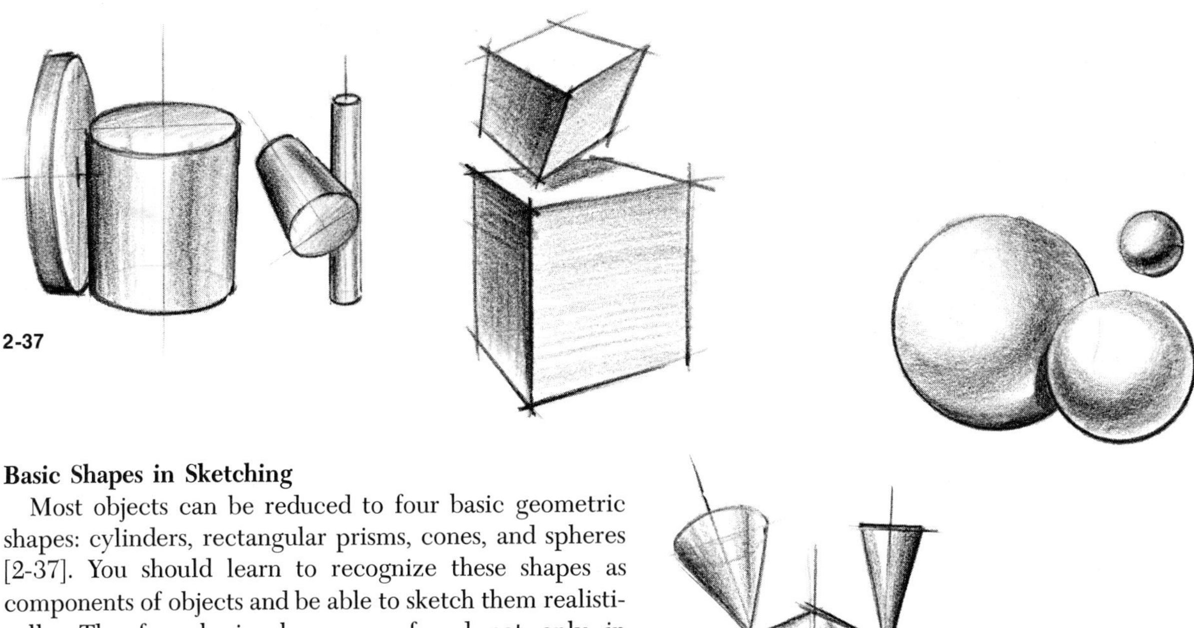

2-37

Basic Shapes in Sketching

Most objects can be reduced to four basic geometric shapes: cylinders, rectangular prisms, cones, and spheres [2-37]. You should learn to recognize these shapes as components of objects and be able to sketch them realistically. The four basic shapes are found not only in man-made designs [2-39 and 2-40], but also in the structure of living things. If you can draw these four shapes as viewed from various angles and depict their lines of intersection, you can draw almost any object.

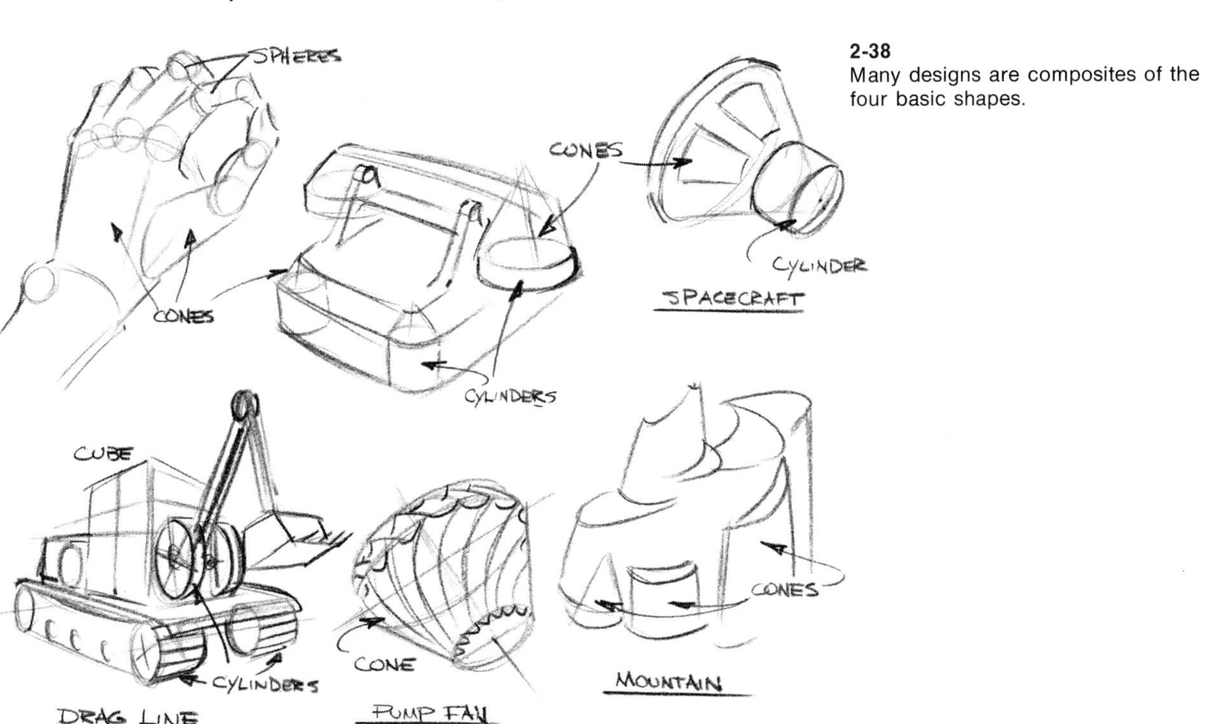

2-38
Many designs are composites of the four basic shapes.

2-39

2-40

graphic portrayal systems

Several techniques can be used to sketch a three-dimensional object on two-dimensional paper: orthographic (multiview), oblique, axonometric, and perspective. Each has certain features that make it suitable to specific needs and each has its limitations.

Orthographic (Multiview) System

The *orthographic* or *multiview* is the most frequently used portrayal system in engineering graphics. In this system, two, three, or more views of an object are shown. The views are sketched as if they were projected perpendicularly on the faces of an imaginary transparent glass box that surrounds the object [2-41]. If this glass box is unfolded and laid flat, six views of the object will be seen [2-42]. In this way one can "see" simultaneously

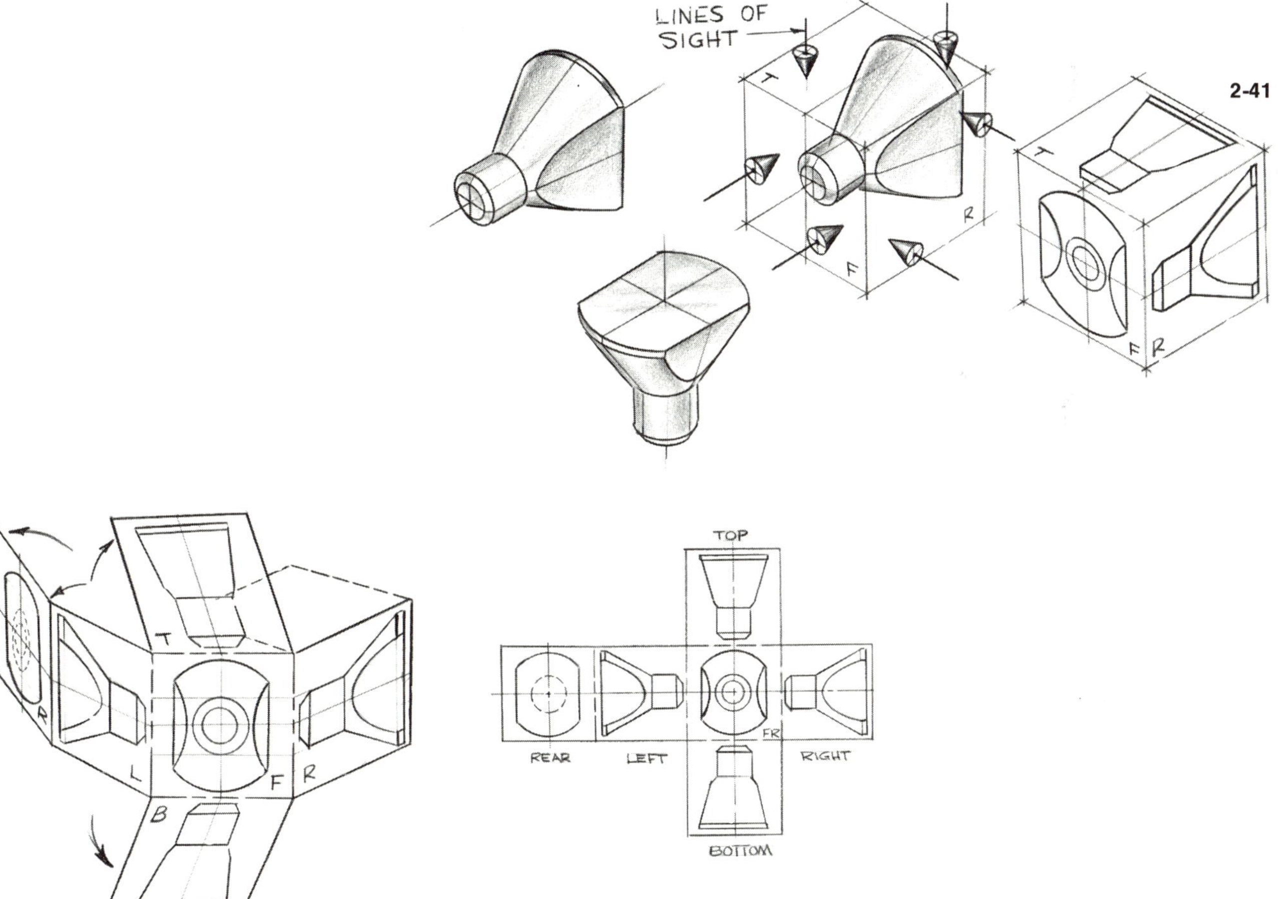

2-41

2-42

the object from several directions. For a more complete discussion of orthographic projection, refer to Chapter 3.

An orthographic sketch is relatively easy to make, has accurate dimensions, and is particularly useful as a basis for model or prototype construction. Its major limitation is the difficulty of visualization. Since the viewer can concentrate on only one view at a time, he must mentally reassemble the object from the various views in order to understand it. This can be particularly difficult for very complex objects, such as an automobile carburetor. To overcome this limitation, a pictorial sketch, such as the oblique sketch, can be used (see page 46).

Multiview sketches are most frequently used

1. When true shape views are required.
2. When dimensions must be specified.
3. When a model or finished part is to be fabricated from the sketch.
4. When angular or oblique surfaces must be described accurately.
5. To precede the preparation of accurate sectional views.
6. When an accurate description of a detail assembly is required.

Let us consider the block in [2-43]. Ordinarily, the three views used to describe this object would be the front, top, and right or left side views [2-44]. They are

2-43

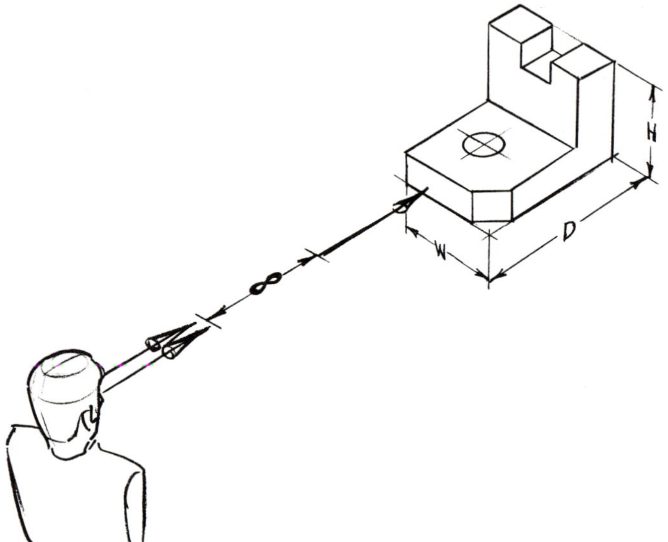

2-44

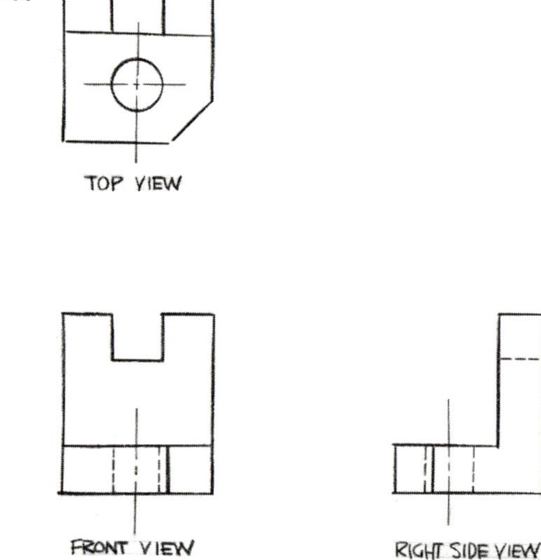

called *principal views*. After the front view is drawn, the top view is obtained by rotating the object 90° up and forward toward the observer. The right side view is obtained by rotating the right side toward the observer [2-45].

Selection of Views. Select appropriate views and minimize their number. Sketch only those views that are necessary for a clear description. The most important view is the one that shows the prominent or characteristic contour of the object. Usually, this view is designated as the front view.

The object shown here [2-46] has three distinctive features that should be described completely by the sketch:

1. The square silhouette top and the rectangular slot in the vertical surface—both of which can be seen from the front view.
2. The hole in the horizontal surface and the beveled front corner—which are described by the top view.
3. The right angle at the rear—which can be seen from the side.

Upon examination we can see that the left side, back, and bottom views are unnecessary because they repeat what is contained in the front, top, and right side views.

In many instances only two principal views are re-

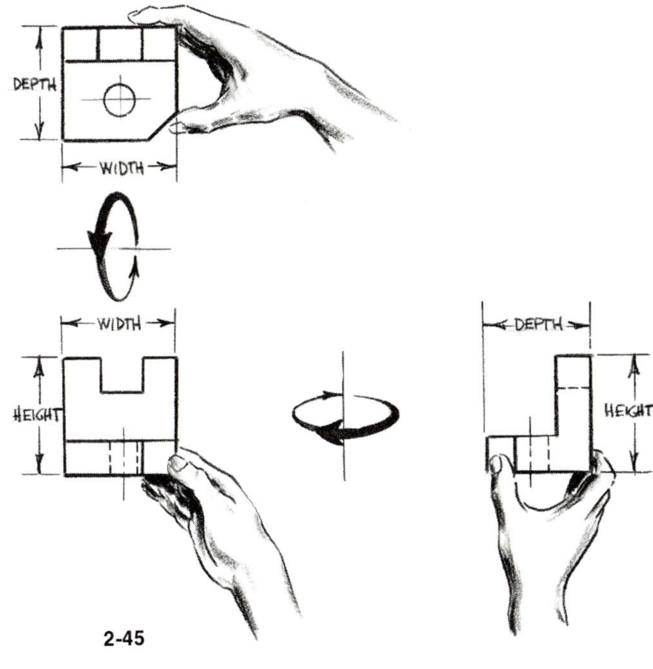

2-45

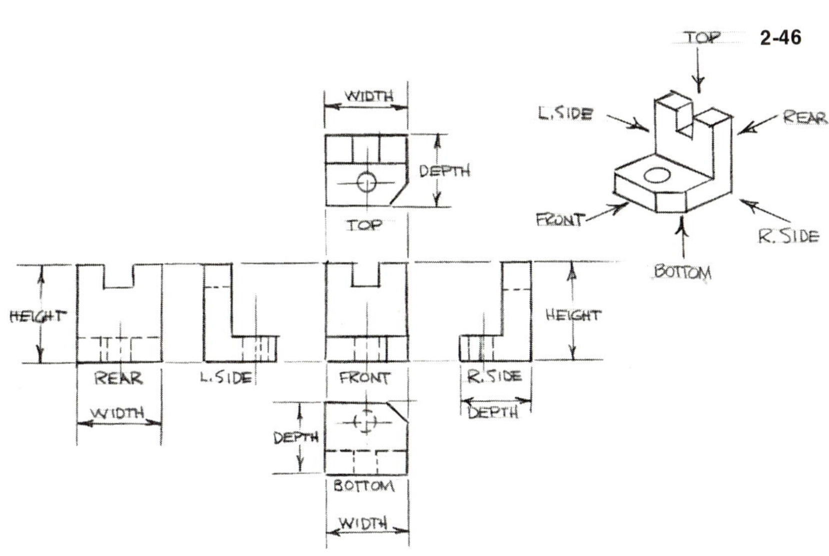

2-46

2-47

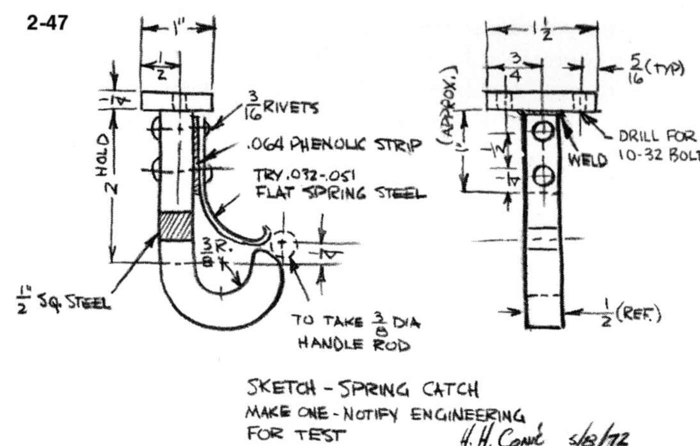

2-48

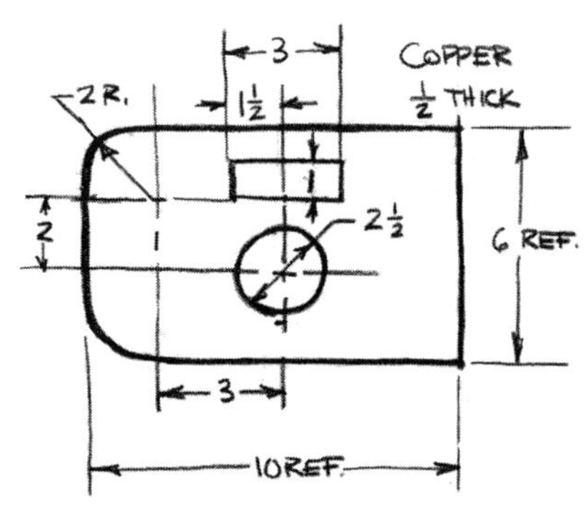

2-49

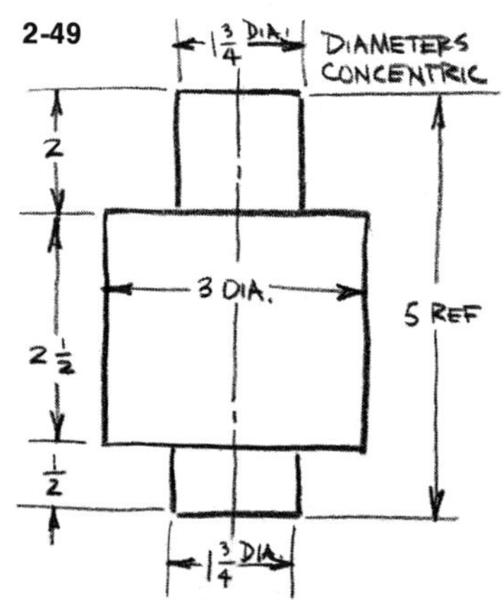

quired [2-47]. Cylindrical, conical, and pyramidal shapes and, in general, any shape that is symmetrical about an axis usually can be described adequately with only two views.

Occasionally, some objects can be described adequately with a single view. When only one view is used, supplementary descriptive notes must be added. In [2-48] the view would be incomplete without the note specifying the thickness. In [2-49] notes are needed to identify the cylindrical shapes of the various portions of the roller and their concentricity.

There are cases when three views will not be adequate and sectional or auxiliary views must be used as supplements. The use of these special types of views will be discussed in Chapters 3, 4, and 5.

Orthographic Sketching. In orthographic sketching the overall dimensions of each view should be blocked in first using the scaffolding technique mentioned earlier. Figure [2-50] shows the successive steps in preparing the multiview sketch of the support bracket. Note the vertical lines drawn between the top and front views to establish the width of the bracket. The horizontal lines drawn between the front and right side views establish the height of the bracket. Diagonals are drawn lightly to serve as locators for the center of the holes.

Grid paper aids the designer in multiview sketching. It will do much to assure proper alignment of the views and realistic proportions [2-51].

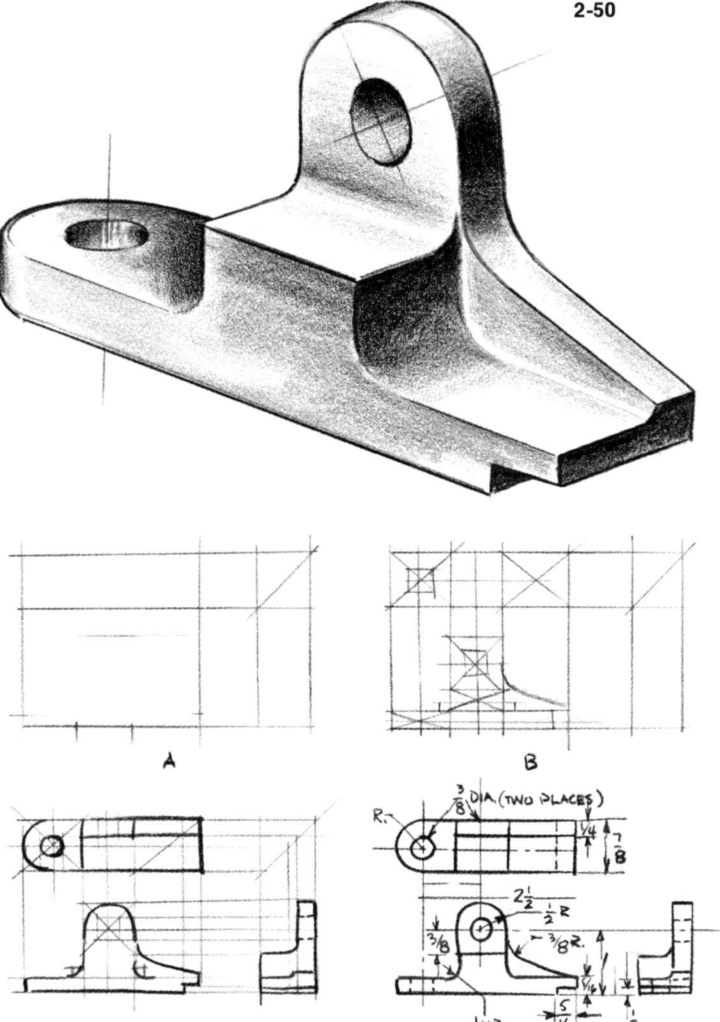

2-50

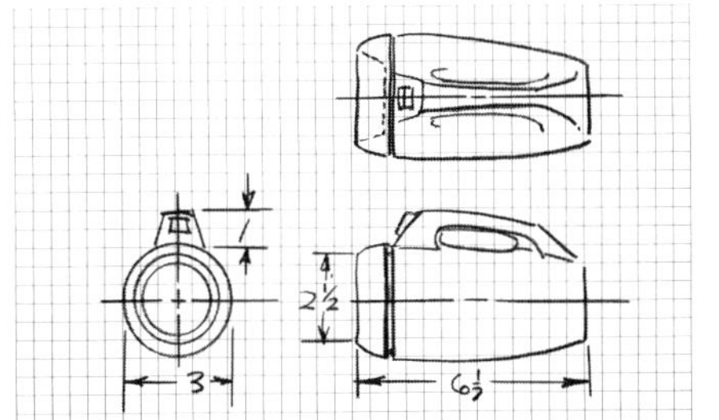

2-51

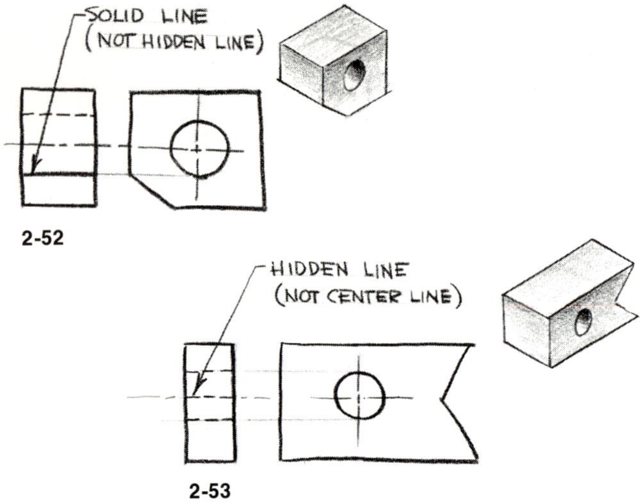

2-52

2-53

Precedence and Representation of Lines. Priority of linework is similar to that described in Chapter 1. Visible lines are necessary to establish the physical features of the object and take precedence over all other lines. A solid line can cover a hidden line, but not vice versa. The following order of preference is recommended:

1. Object (visible) lines.
2. Hidden lines.
3. Center lines.
4. Cutting-plane lines.
5. Dimension and extension lines.
6. Cross-hatch lines.

The procedures for dimensioning an orthographic sketch are the same as those discussed in Chapter 5.

Oblique System

Oblique, axonometric, and perspective systems usually are known as *pictorial* because, in contrast with the orthographic system, they provide the viewer with an illusion of depth while he looks at only one view. The *oblique* system is the simplest transition from the orthographic system because the orthographic front view can be used directly as a major portion of the oblique view.

Orthographic views are *perpendicular* projections onto the sides of an imaginary glass box enclosing the object. In oblique projection the object is viewed through a single picture plane, but the observer's line of sight is now at an angle (oblique) to the picture plane [2-54]. This enables him to see depth dimensions as well as an undistorted view of the face of the object parallel to the picture plane [2-55].

An oblique view of an object is sketched in much the same way as an orthographic view, except that the scaffold used is a three-dimensional box following the oblique axes selected [2-56]. The face of the object that best displays the most distinctive or characteristic contours of the object should be oriented parallel to the picture plane to take advantage of the ease of sketching the undistorted image that will appear [2-57]. The angle at which the depth dimension recedes from the horizontal is usually chosen as 45°, but other angles varying from 15° to 60° often provide a better view of secondary surfaces [2-58]. If the shape of an object is such that the conventional or *cavalier* oblique sketch [2-59] appears to distort the

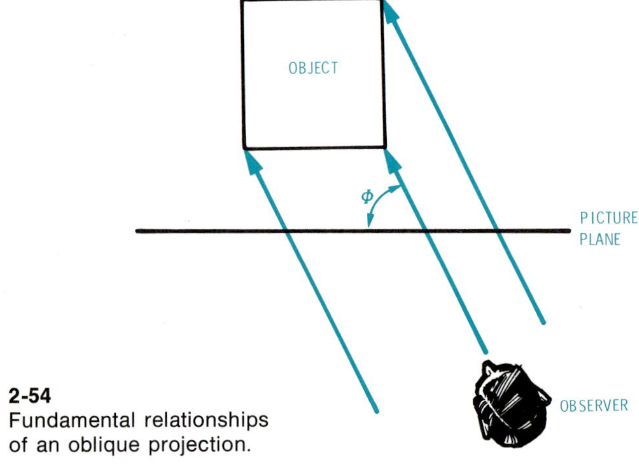

2-54
Fundamental relationships of an oblique projection.

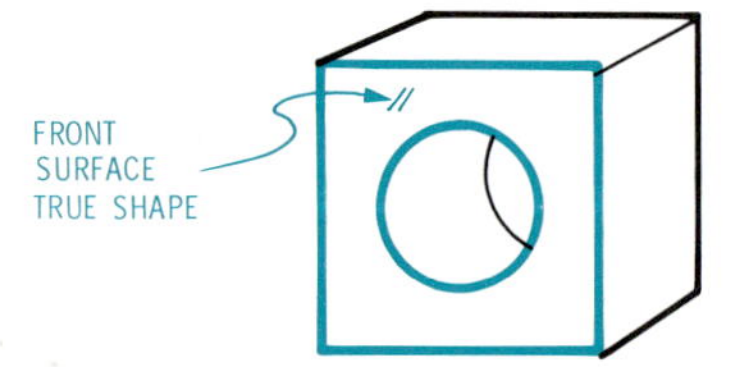

2-55
All the features of the front view of an oblique sketch are drawn as a true shape.

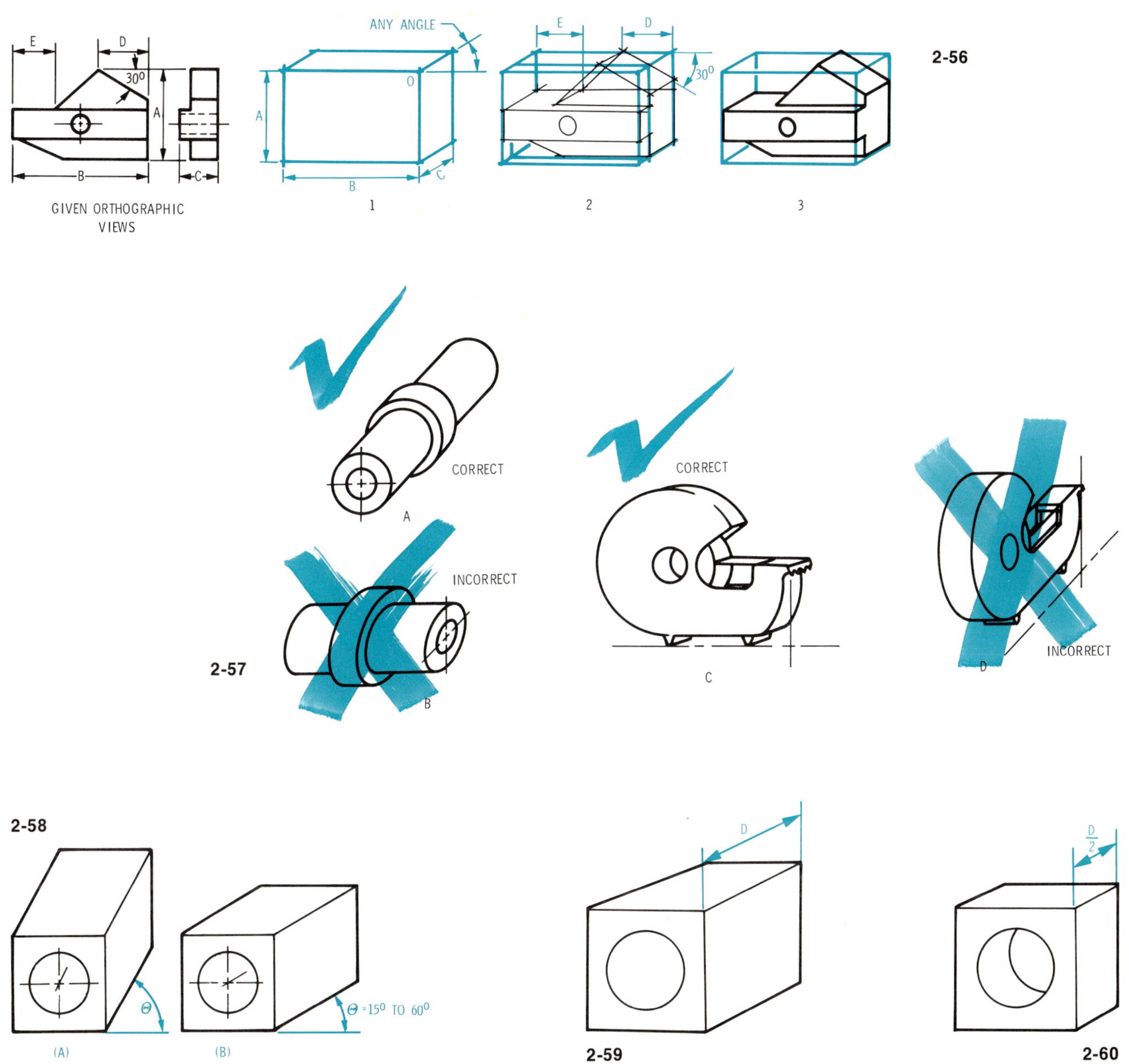

2-59
Oblique sketch: cavalier.

2-60
Oblique sketch: cabinet.

object too much, a modification known as *cabinet* oblique may be used. In this technique all depth dimensions are reduced by one half [2-60].

Axonometric System

In the *axonometric* portrayal system the object is oriented at an angle with a single picture plane, and the

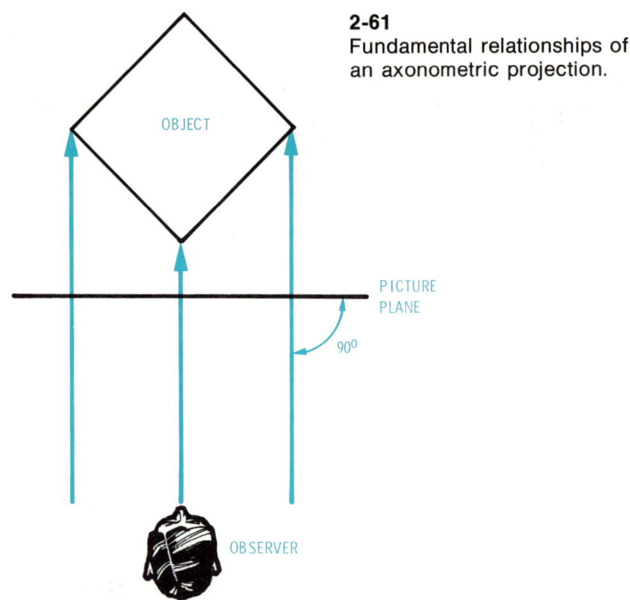

2-61
Fundamental relationships of an axonometric projection.

observer's line of sight is perpendicular to that picture plane [2-61]. As in the oblique system, the image projected on the picture plane gives the observer an illusion of depth in a single view and provides a more realistic and understandable picture than an orthographic sketch. The viewer is able to see the height, width, and depth of an object in one view and does not need to mentally assemble a composite view, as in the case of orthographic projection. The necessity of having to draw several views of an object is also eliminated [2-62].

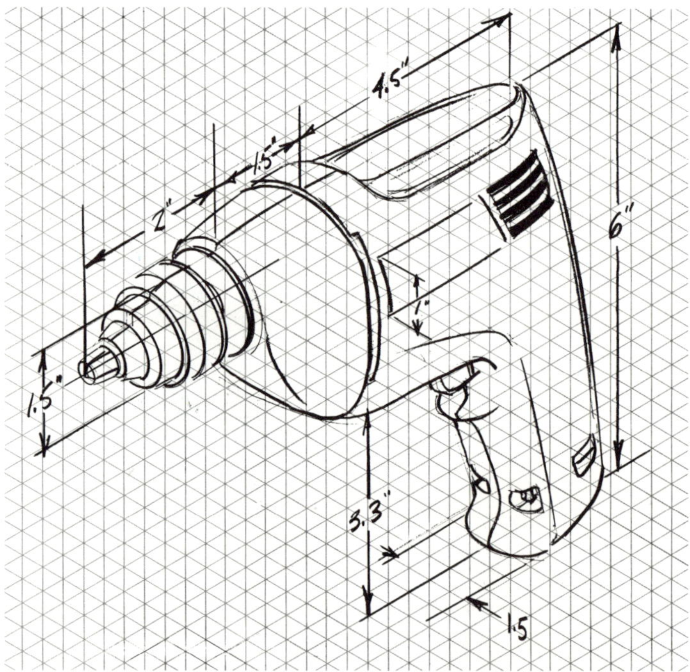

2-62

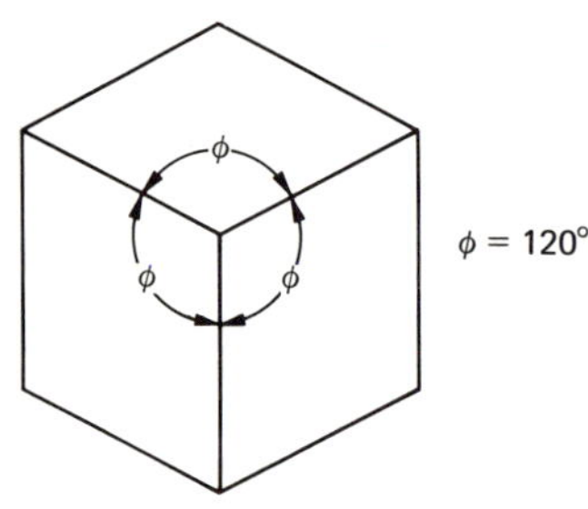

2-63
In an isometric sketch, all the angles are 120°

The three types of axonometric projection—*isometric, dimetric,* and *trimetric*—differ only in the angular orientation of the object with the picture plane. For example, a cube in the isometric system would be tilted so that all three of its principal edges would form equal angles with the picture plane. This orientation greatly simplifies the construction of isometric sketches, because the angles between the principal edges will appear at 120° with each other [2-63]. Since isometric is the most commonly used form of axonometric projection, it is the only one that will be discussed here.

The steps for making an isometric sketch are

2-64

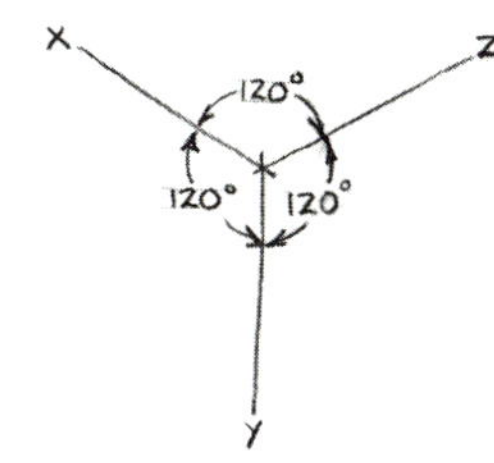

Step 1 Establish the isometric axes x, y, *and* z *at angles of 120° to each other with the* y *axis being vertical [2-64].*

2-65

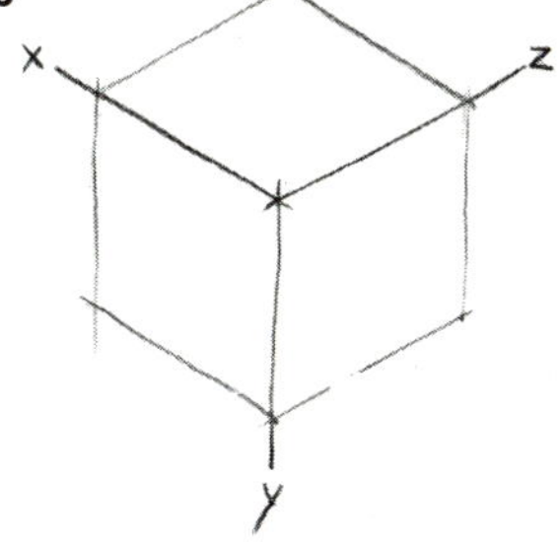

Step 2 Construct a scaffolding box large enough to enclose the object with the y *isometric axis being the front, vertical edge of the box [2-65].*

2-66

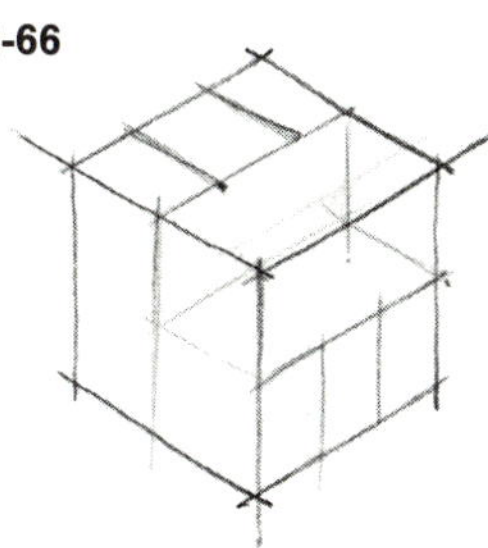

Step 3 Place the object to be drawn in the scaffolding. Surfaces and edges parallel to the xy *and* yz *planes should be sketched in their respective places within the scaffold [2-66].*

2-67

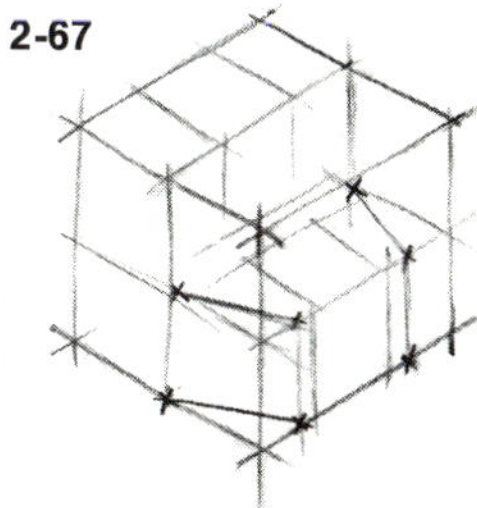

Step 4 Angular surfaces or edges should be sketched by first locating their end points on the scaffolding and then connecting them [2-67].

2-68

Step 5 Sketch in the holes and arcs as ellipses or partial ellipses; noncircular arcs by the plotting of points [2-68].

2-69

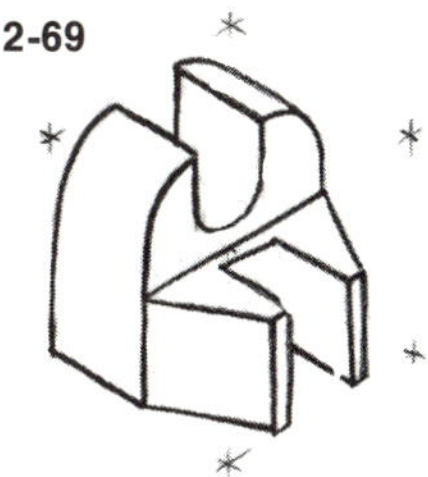

Step 6 Darken the object lines of the finished sketch [2-69] or trace on another piece of vellum. Hidden lines are normally omitted.

2-70
Circles viewed obliquely appear as ellipses.

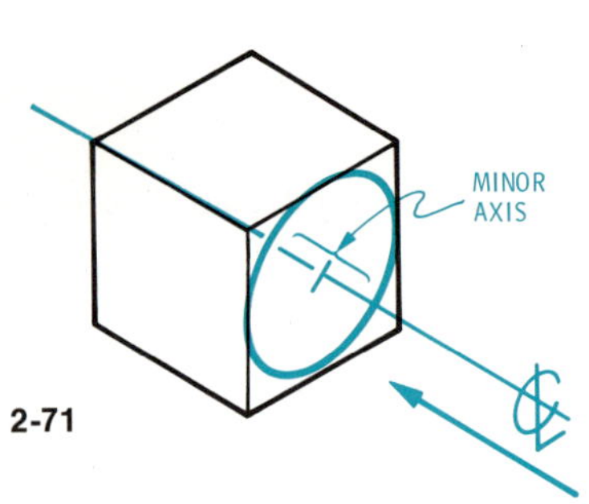

2-71

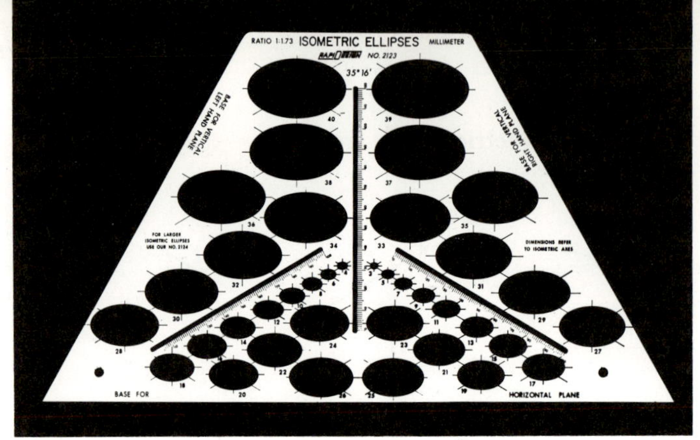

2-72
The ellipse guide can speed up your sketching.

Paper with isometric background lines is available and is an effective aid when used either directly or as a reference background under the sketching paper.

Circles viewed obliquely appear as ellipses [2-70]. To sketch such a circle, construct a rhombus with sides equal to the diameter of the circle. The ellipse is formed by sketching arcs that are tangent to the midpoints of the sides of the rhombus. The major axis of the ellipse will be in line with the longer diagonal of the enclosing rhombus—the minor axis aligns with the shorter diagonal [2-71]. An isometric ellipse guide is very useful for quick construction of isometric circles [2-72]. If one is not available, the four-center-ellipse method [2-29 to 2-33] produces a good approximation. Examples of circles in isometric orientation can be seen in [2-73] to [2-75].

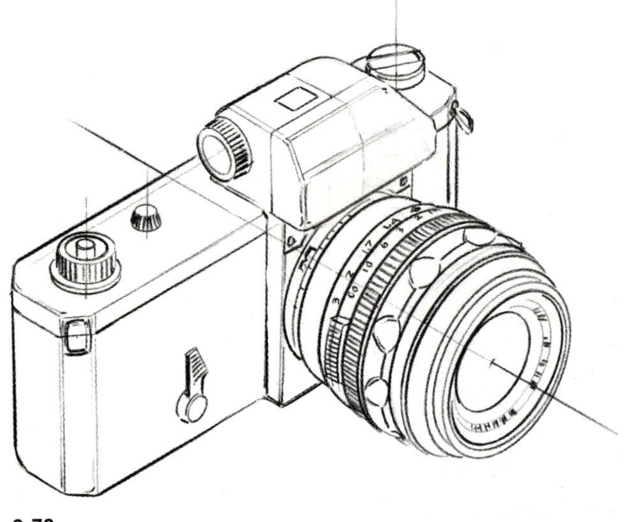

2-73

2-74

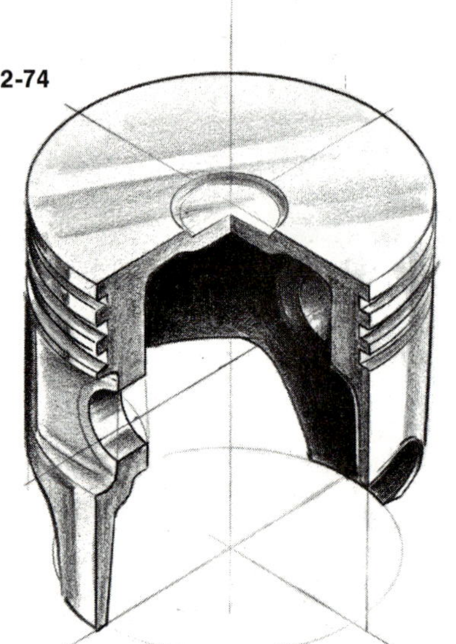

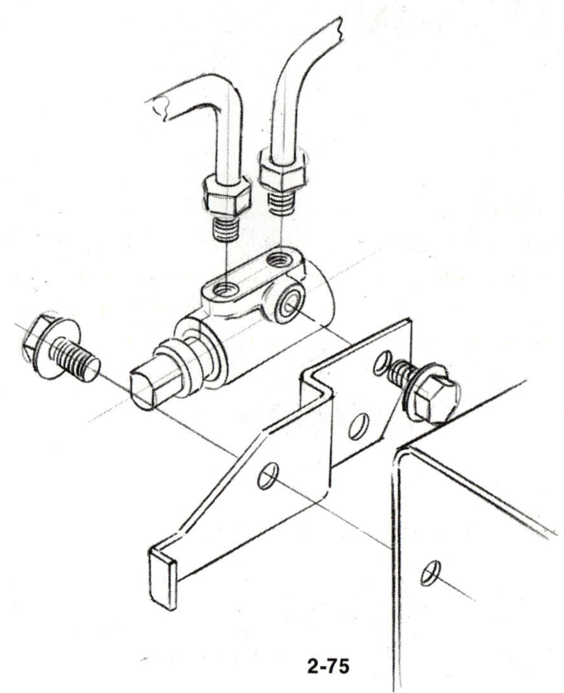

2-75

2-76

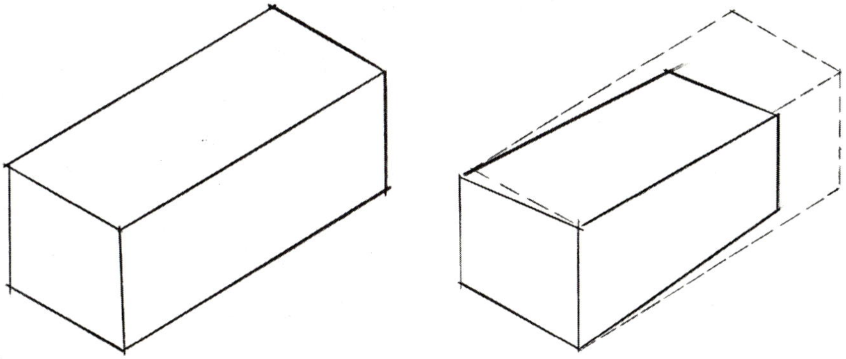

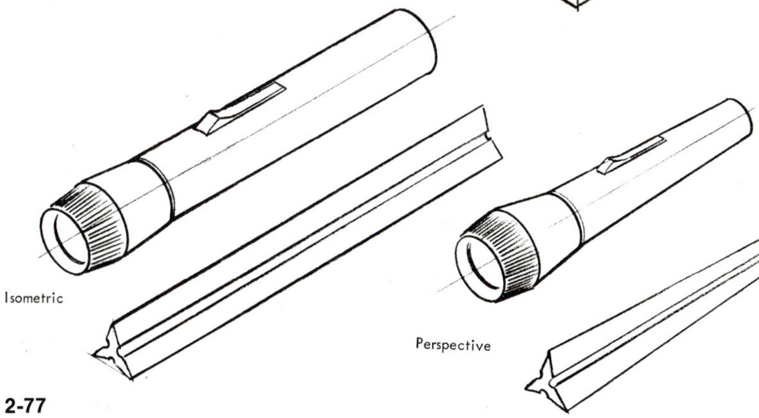

Isometric

Perspective

2-77

The most annoying limitation of isometric sketching is the visual distortion that exists because all receding lines are parallel. This distortion is especially pronounced in objects of considerable depth [2-76]. The distortion of relatively long objects can be alleviated by using perspective sketching techniques [2-77].

Perspective System

Perspective sketching is the most difficult to learn, but is preferred to all other methods when visual realism is important. Three types of perspective sketching are known as one-point, two-point, and three-point, depending upon the number of vanishing points used. For the most part, one- and two-point perspective sketches will suffice.

Perspective sketching differs from isometric sketching in that the observer is a finite distance from the object—rather than infinite; that is, the viewing lines from the corners of the object are not parallel, as in the techniques discussed before, but converge on the eye of the beholder. The view is a close approximation to that obtained by a camera [2-78]. In perspective, objects appear smaller

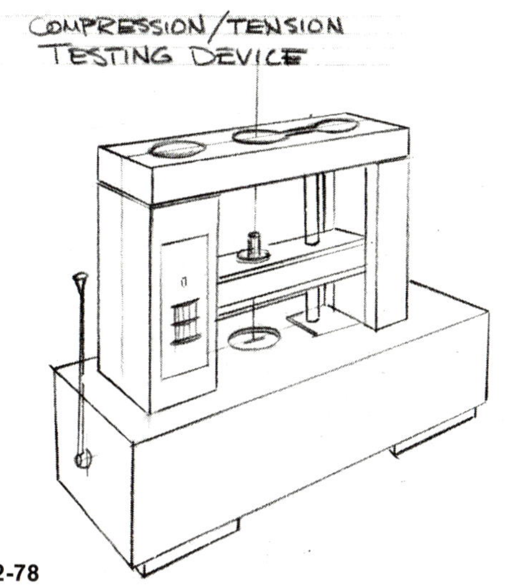

2-78

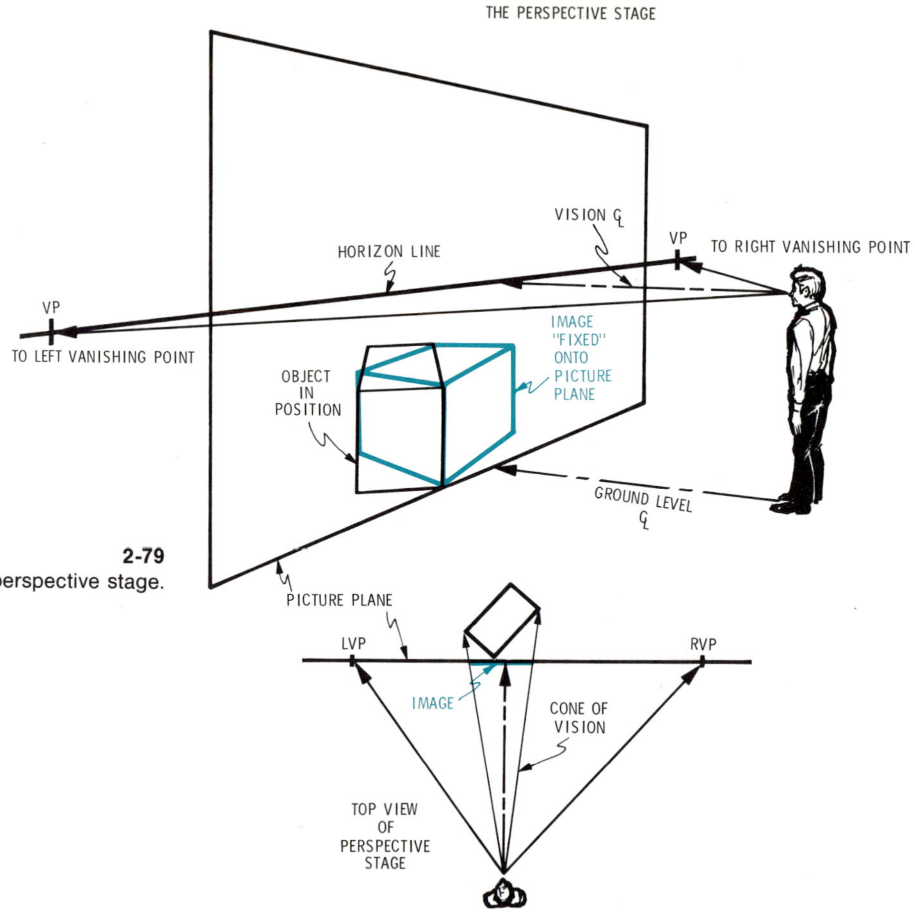

2-79
The perspective stage.

2-80
Perspective charts.

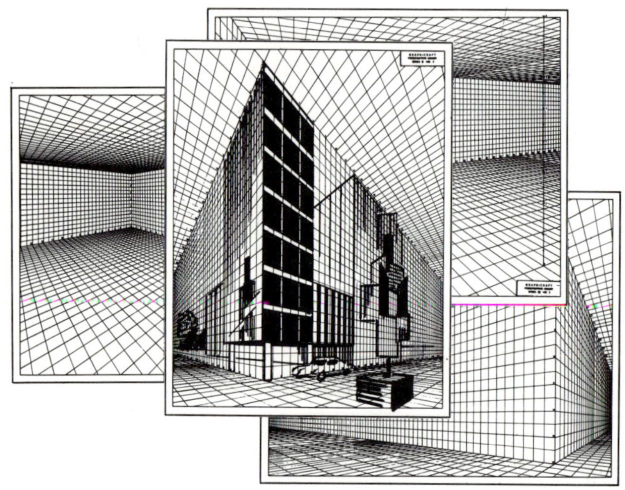

when their distance from the eye increases, parallel lines appear to converge as they recede, and horizontal lines and planes appear to vanish on the horizon. These effects are achieved by selective positioning of the object relative to the picture plane, by proper location of the horizon, and by designation of an appropriate vanishing point or points on the horizon [2-79].

One major advantage of a perspective sketch, when compared with an isometric sketch, is that the engineer can select the most appropriate view of the object. In the isometric sketch, the object must always be placed at the same angle with respect to the picture plane, but in a perspective drawing the object can be placed at *any* angle, thus providing a means to illustrate particular details. Several types of commercially prepared perspective charts are available to the engineer [2-80]. These charts minimize the overall construction requirements of

2-81
One point perspective.

a perspective drawing, but their use will not assure a realistic view if the engineer fails to adhere to the fundamentals of perspective drawing.

A *one-point perspective* sketch uses but one vanishing point. This means that all receding parallel lines in the sketch terminate at a single point on the horizon of the sketch [2-81]. The one-point perspective is most effective for sketches showing the inside of a space or the internal structure of a mechanism.

Since all the principles used in making a one-point perspective are similar to those used in making a *two-point perspective*, we shall explain the perspective technique by analyzing the more commonly used two-point perspective sketch.

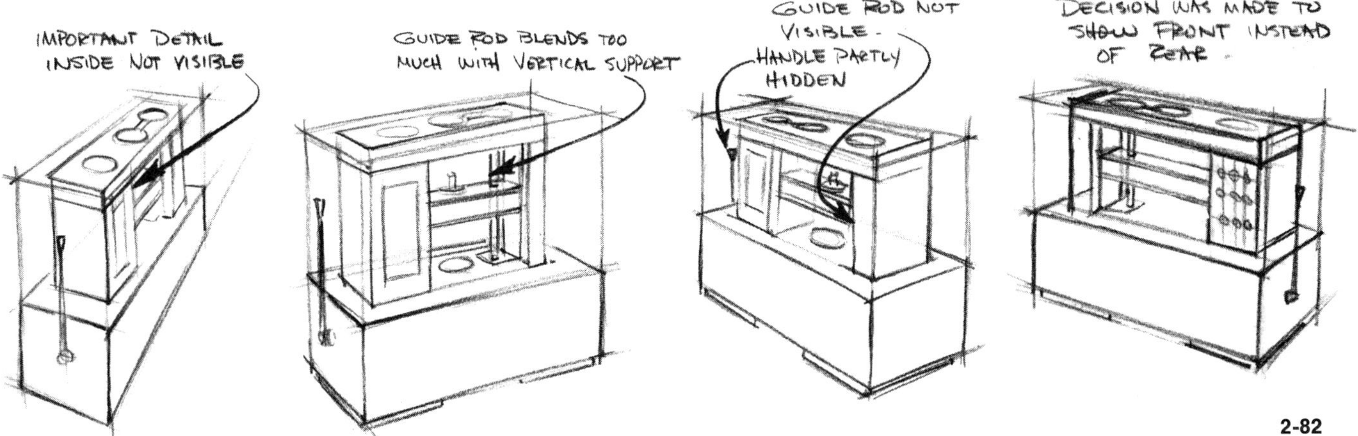

2-82

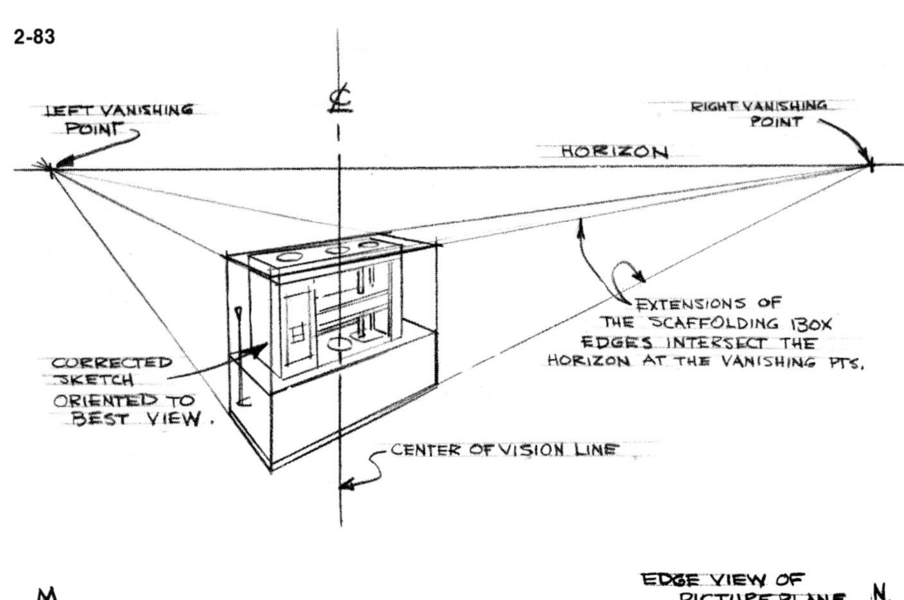

2-83

LEFT VANISHING POINT
RIGHT VANISHING POINT
HORIZON
EXTENSIONS OF THE SCAFFOLDING BOX EDGES INTERSECT THE HORIZON AT THE VANISHING PTS.
CORRECTED SKETCH ORIENTED TO BEST VIEW.
CENTER OF VISION LINE

2-84

M
N
EDGE VIEW OF PICTURE PLANE
B
90°
θ
O
OBSERVER IS HERE
LVP
HORIZON
RVP
CENTER LINE OF VISION

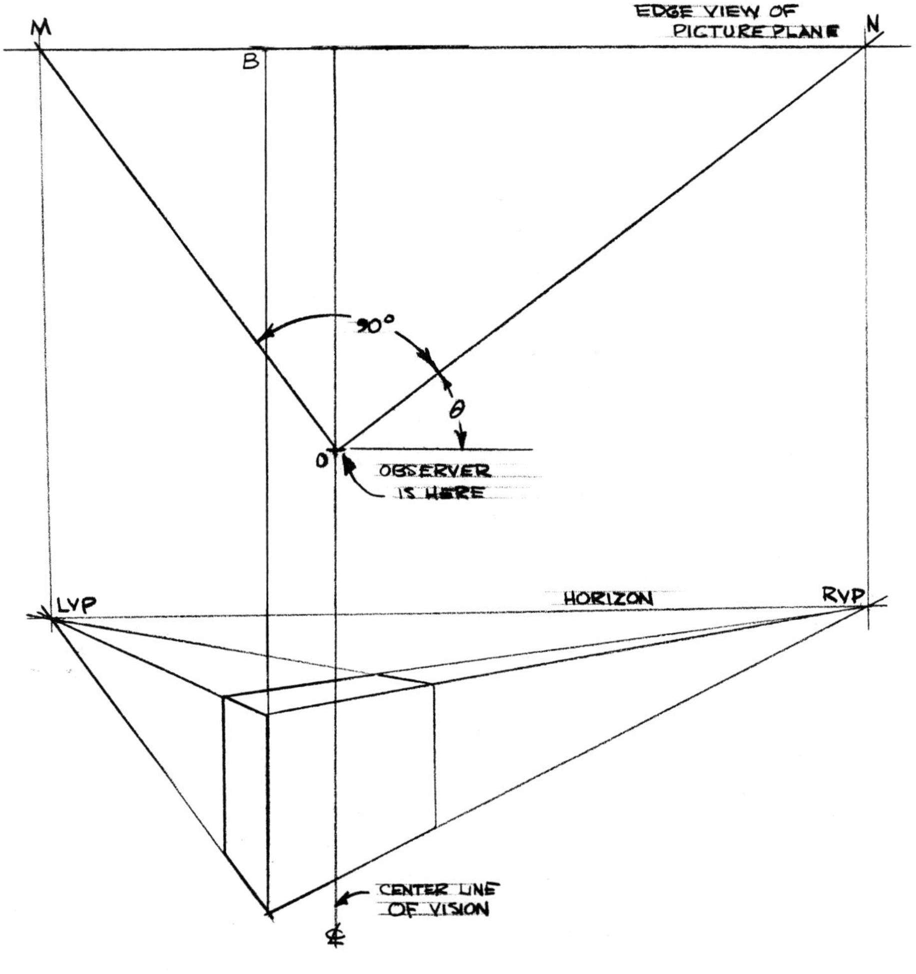

First orient the object so that all of its important features are clearly visible. Several preliminary sketches of the scaffolding box in varying orientations should be tried to achieve the most descriptive visual presentation [2-82]. After the best orientation has been selected, left and right vanishing points are obtained by projecting lines along the edges of the box until they intersect the horizon line [2-83]. The vanishing point is that point on the horizon where parallel lines appear to converge—to vanish. The next step is to locate the *center-of-vision* line. The location is purely arbitrary, but in general it should be near the visual center of the object being sketched.

Once the center-of-vision line has been selected, it is extended vertically until it intersects the edge view of the picture plane, which has been drawn immediately above and parallel to the horizon [2-84]. Similar vertical projections are made to the edge of the picture plane from the left and right vanishing points, resulting in intersections *M* and *N*. The leading edge of the scaffolding box is projected vertically resulting in the location of point *B*.

Point *O*, the position of the observer with respect to the picture plane, is determined by the observer's sight lines to the vanishing points *M* and *N*. Since the observer's lines of sight subtend an internal angle of 90°, the location of the sight lines and point *O* can be found by sliding the apex of a 90° triangle down the center-of-vision line until the two sides of the 90° angle intersect the edge view of the picture plane at points *M* and *N*.

Now construct an orthographic top view of the object with the front corner touching the picture plane at point *B* and with sides *BA* and *BD* parallel to observer sight lines *OM* and *ON*, respectively [2-85]. Another orthographic view is drawn so that it is aligned horizontally with the leading edge of the perspective view. Intermediate points on the object within the scaffolding box now can be located by extending appropriate projections from the two orthographic views.

The horizon line *always* passes through the eye level of the observer. Therefore, if one were to sketch a view of a refrigerator that is approximately 5 feet high, the horizon line of the sketch should be drawn such that it passes through the refrigerator top [2-86]. The observer's sight line to the horizon should pass approximately through the center of a large object, such as a bus. Objects that are normally viewed from above, such as typewriters, should be located below the horizon. In the case of small

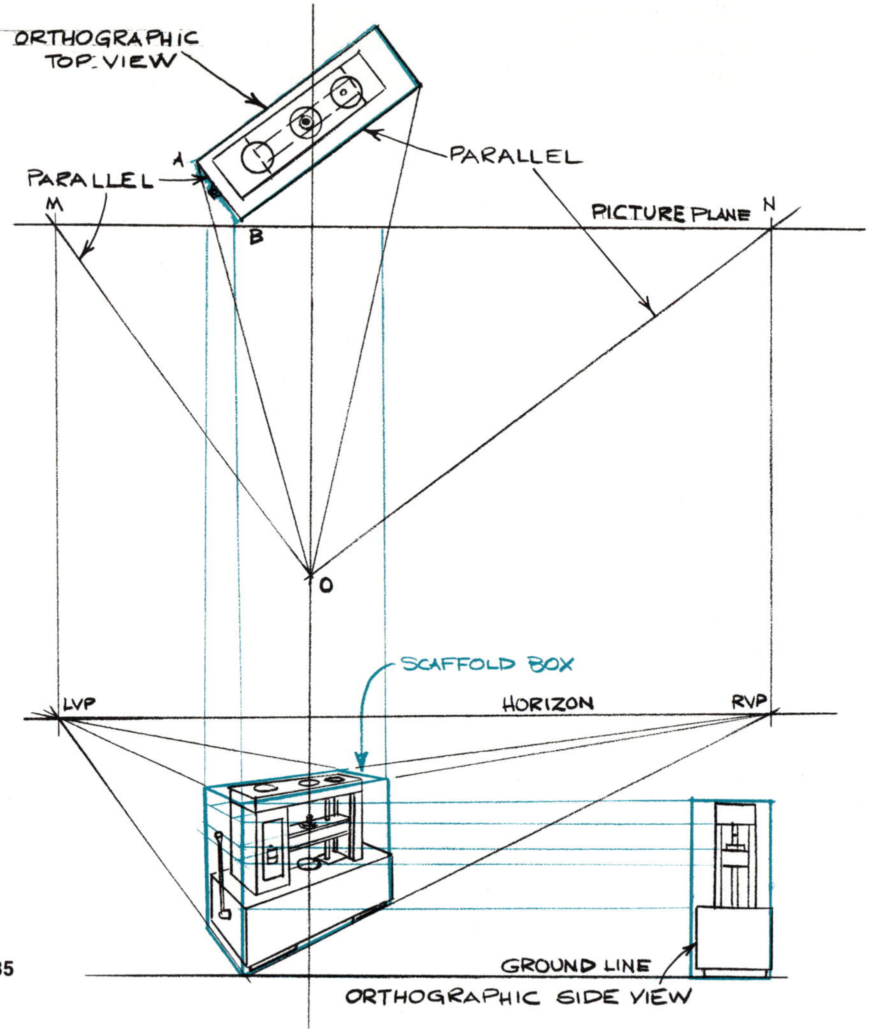

ORTHOGRAPHIC TOP VIEW

PARALLEL

A

PARALLEL

M

B

PICTURE PLANE

N

O

SCAFFOLD BOX

LVP

HORIZON

RVP

GROUND LINE

ORTHOGRAPHIC SIDE VIEW

2-85

2-86

2-87

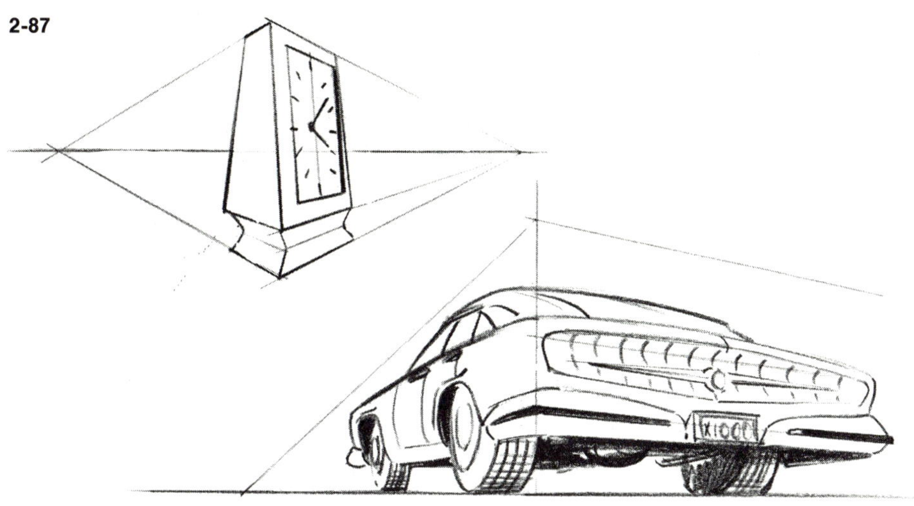

objects that rest on the ground (like a sidewalk skateboard), the horizon should be located approximately 5 feet above the object. Dramatic effects can be obtained by adjusting the eye level of the observer up or down to emphasize or enhance certain details of a design [2-87]. Additional realism and proportion in sketching can be achieved by adding some familiar object, such as the silhouette of a person [2-88].

2-88

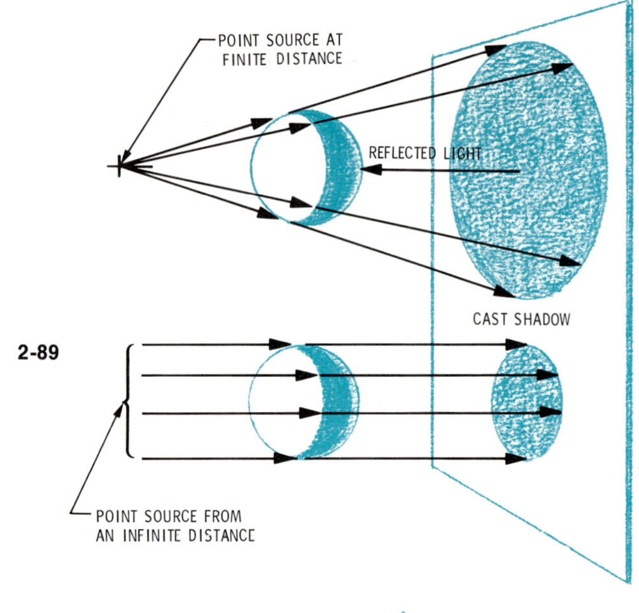

2-89

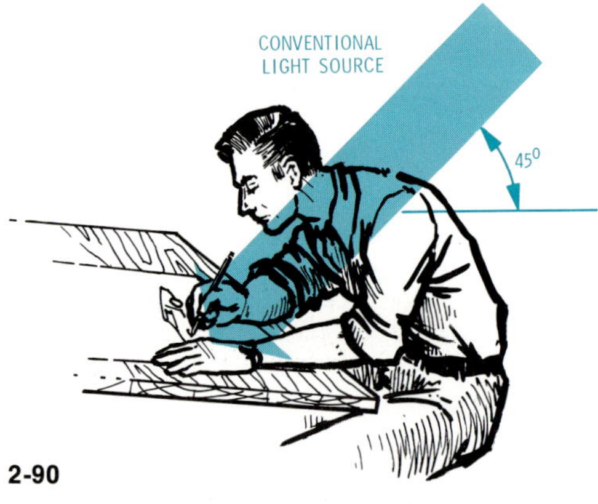

2-90

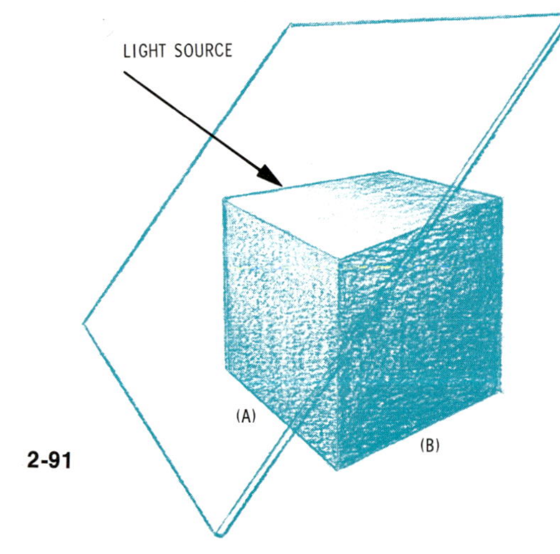

2-91

shading

Shading refers to the light and dark areas of an object created by the light pattern (or lack of it) falling directly on that object or by reflection. Shading adds clarity and realism.

Light originates either from a point source at a finite distance (such as a lamp) or from a point source at a theoretically infinite distance (such as the sun) [2-89]. In either case, light rays travel in straight lines until they are intercepted by an object. When the source of light is undetermined, we should assume that it comes from a single remote source. As a general rule the light is assumed to shine over the left shoulder of the observer at an angle of approximately 45° to the ground [2-90].

Any spot or surface on an object that intercepts the light perpendicularly will give maximum reflection, and this area will appear white, or nearly so, on the sketch. Other surfaces will be shaded in varying degrees, depending upon the amount of light that falls upon them. The gradation of shading on the surfaces will give additional emphasis to the depth of the object.

The form of an object determines the way in which it intercepts light rays [2-91]. The top surface is most directly exposed to the light source and will be the lightest, especially at its near edge. The surface becomes darker as it recedes into the distance. Face *B* (which is hidden from the light source) should receive the darkest shading. The shading becomes darker on this face as it recedes in depth, unless some light is arbitrarily assumed to be reflected from an imaginary surface behind the cube. Face *A* is shaded similarly to Face *B*, but is somewhat lighter, because it is reflecting more light to the observer.

The shading of the curved surface of a cylinder or a sphere will be darkest where the rays of light fall tangent to the surface [2-92 and 2-93]. The lightest area occurs where the light rays strike perpendicularly. If a flat surface is placed on the side of the object opposite the light source, not only will there be a dark shadow cast upon it by the object, but also the surface will return some reflected light to dilute the shading on the back side of the object.

Pencil shading can be applied in two ways. For best results use a pencil with a flattened point and lightly cover the entire area that is to receive shading. The darker shades for particular areas are built up to the desired darkness by repeated strokes. Make sure that each shade blends smoothly into the next. For shading, begin and end lines with a feather edge—a gentle landing and takeoff of the pencil point on the paper surface [2-94]. An eraser may be used to remove small "areas of light" within the dark shaded areas of a sketch.

A second method of shading involves the use of lines of varied weight and spacing [2-95]. No lines are needed

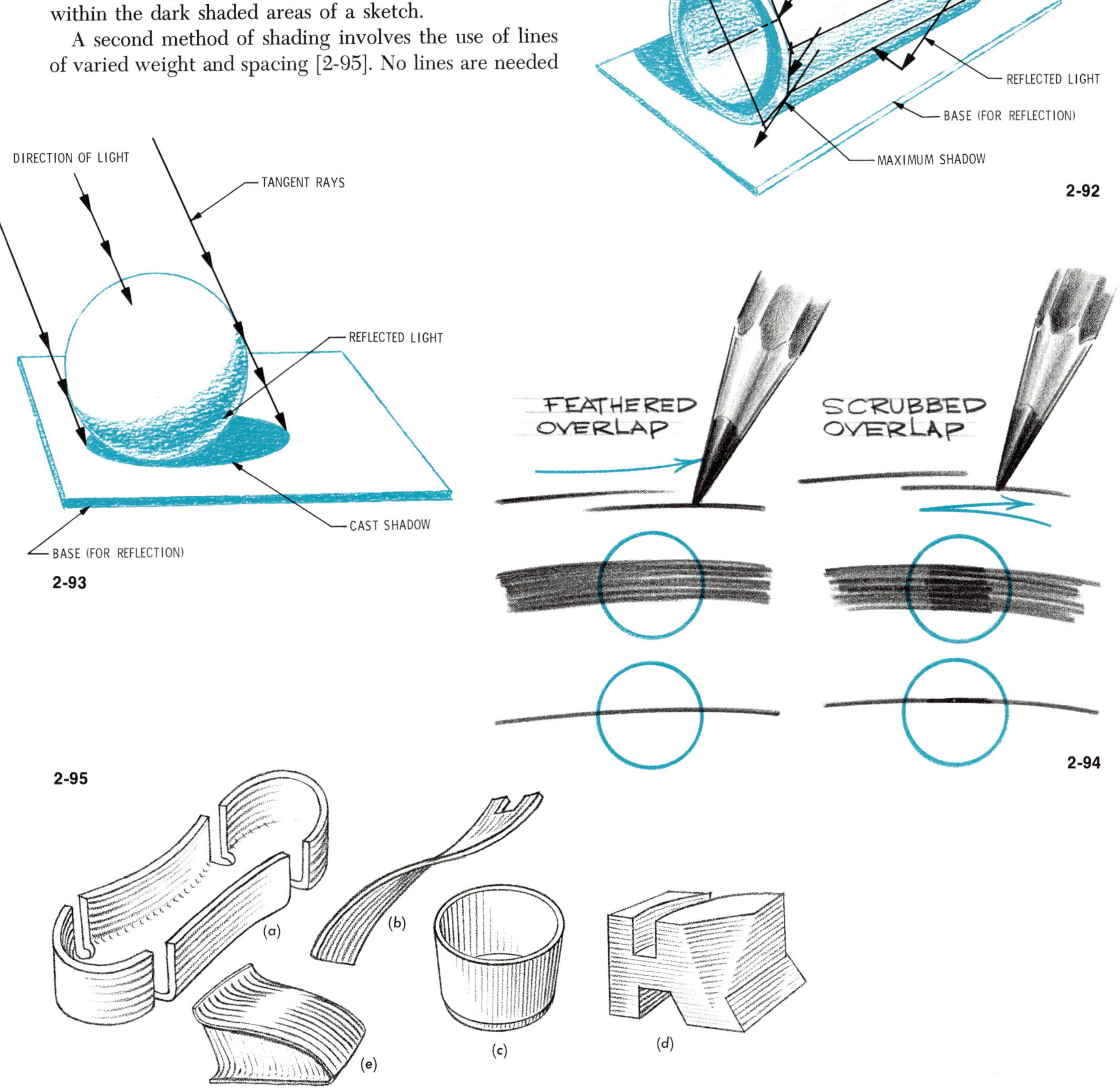

2-92

2-93

2-94

2-95

for the very light areas. This simple and effective technique is particularly useful when speed is a factor or when the sketch is drawn from the imagination.

problems

2-1. Sketch eight straight freehand lines on a sheet of unruled paper. Make the lines of various lengths and angles to each other, but do not allow them to intersect. Segment two of the lines into two equal parts, two of the other lines into three equal parts, two others into four equal parts, and the last two into five equal parts. Accomplish the segmentation by eye; then verify the accuracy of the segments with a scale.

2-2. Using only a pencil and an unruled sheet of paper, sketch 25 *medium* weight lines about 5 inches long. Sketch the lines so that they are all parallel with the top edge of the paper and about $\frac{1}{4}$ inch apart. Sketch each line as a *single stroke* and do not erase.

2-3. Using only a pencil and an unruled sheet of paper, sketch 25 *thin* weight lines about 6 inches long. Sketch the lines parallel with the left-hand edge of the paper and about $\frac{1}{8}$ inch apart. Sketch each line as a *single stroke* and do not erase.

2-4. Sketch 25 *thick* lines about 4 inches long and at a 45° angle using the applicable instructions of Problem 2-2.

2-5. Using a pencil and unruled paper, make four different one-point perspective layout (thumbnail) sketches of one of the figures in Figure [2-96]. After completing the sketches, draw a circle around the layout sketch that best illustrates the main features of the object in a one-point perspective.

2-6. Quickly sketch one of the following objects from memory. Show the proportionate elements of the figure by using scaffolding techniques (1) in the front view of the orthographic system, (2) in the isometric pictorial system, and (3) in the one point perspective system:

(**a**) coffee cup showing the handle
(**b**) three legged stool
(**c**) wall mounted pencil sharpener
(**d**) side view of a door knob
(**e**) front view of a TV set
(**f**) bicycle

2-7. Sketch six circles, starting with a $\frac{1}{2}$-inch diameter. Increase each successive circle diameter by $\frac{1}{2}$ inch. Use the scaffold-square technique for each circle and do not use any measuring tools. Verify the diameters after all six circles have been sketched.

2-8. Sketch six ellipses all having a 2-inch major axis. Make the first ellipse with a minor axis of $\frac{3}{4}$ inch. Increase the minor axis of each successive ellipse by $\frac{1}{4}$ inch. Use the four-center-ellipse technique to construct each ellipse.

2-96

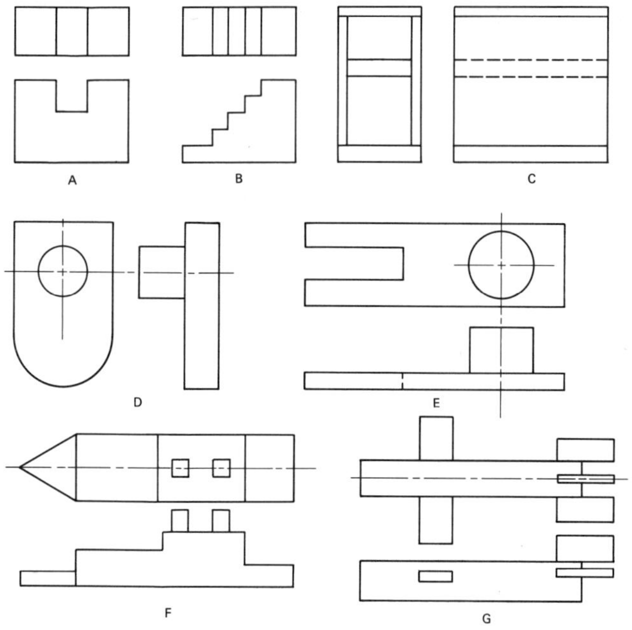

A B C D E F G

2-9. Trace Figure [2-34] on a piece of tracing paper. Do not trace the grid. Construct a $\frac{1}{4}$-inch grid over the tracing. On a second piece of paper construct a $\frac{1}{2}$-inch grid; then sketch the telephone on the $\frac{1}{2}$-inch grid.

2-10. Find a picture in a popular magazine that is basically a single orthographic view, such as front or side view. Construct a grid on the picture; then make a new sketch of the picture that is three times as large as the original, using the grid technique.

2-11. Find pictures of three different products in a magazine. Using a felt tip or a ball point pen, analyze the three products and sketch the relationships of the four basic shapes from which each product is formed. (Sketch the location of each of the basic shapes directly on the picture.)

2-12. In [2-97] the circles indicate the location of missing views. For each problem select a correct view from the 30 choices shown at the bottom of the page and place its number in the appropriate circle.

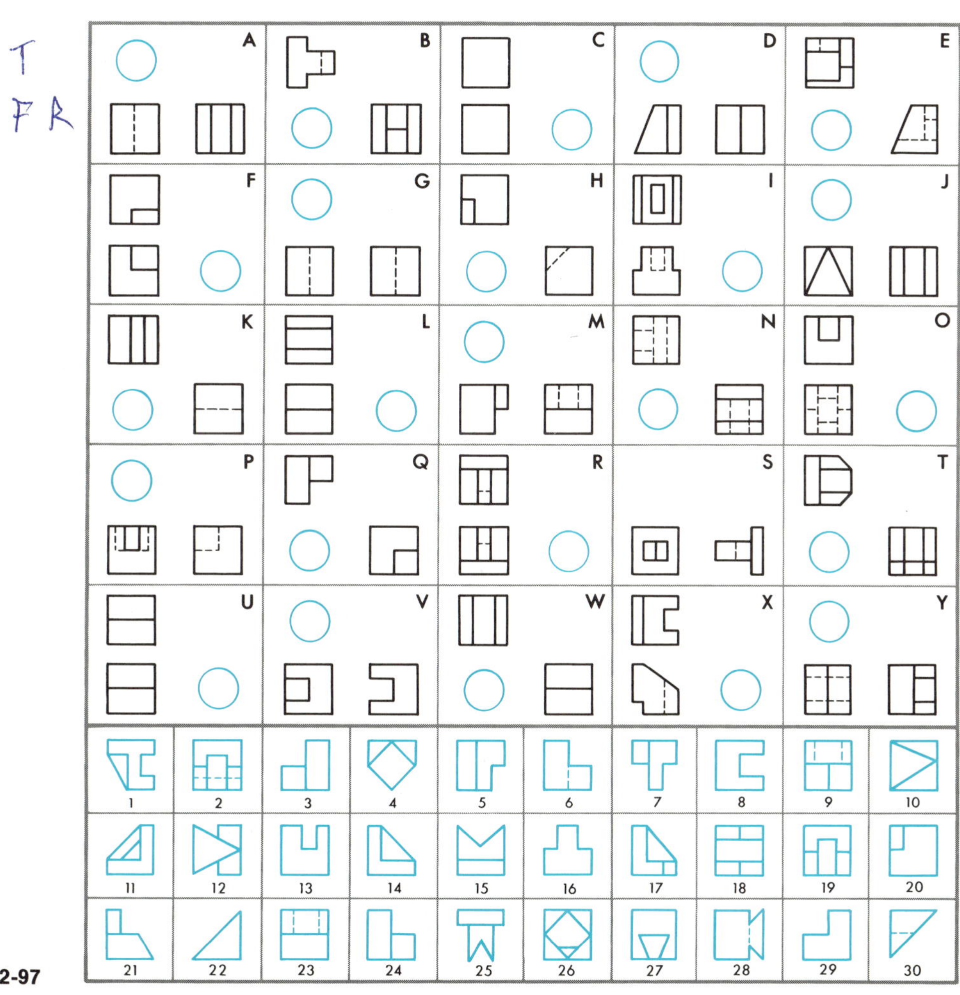

2-97

2-13. In [2-98] the circles indicate the location of missing principal views. For each problem select a correct view from the 30 choices shown at the bottom of the page and place its number in the appropriate circle.

2-98

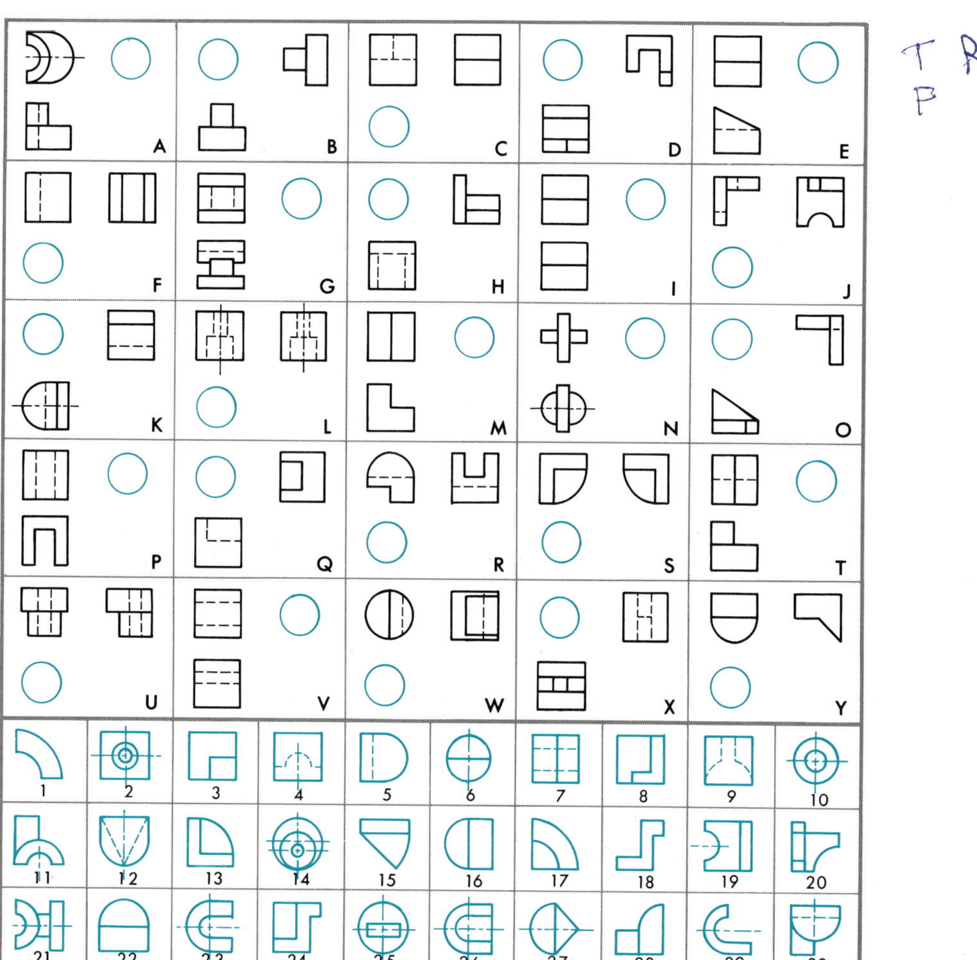

2-14. In [2-99] for each object select one of the 20 views that correctly shows an appropriate principal view having one or more hidden lines. Place its number in the circle provided.

2-99

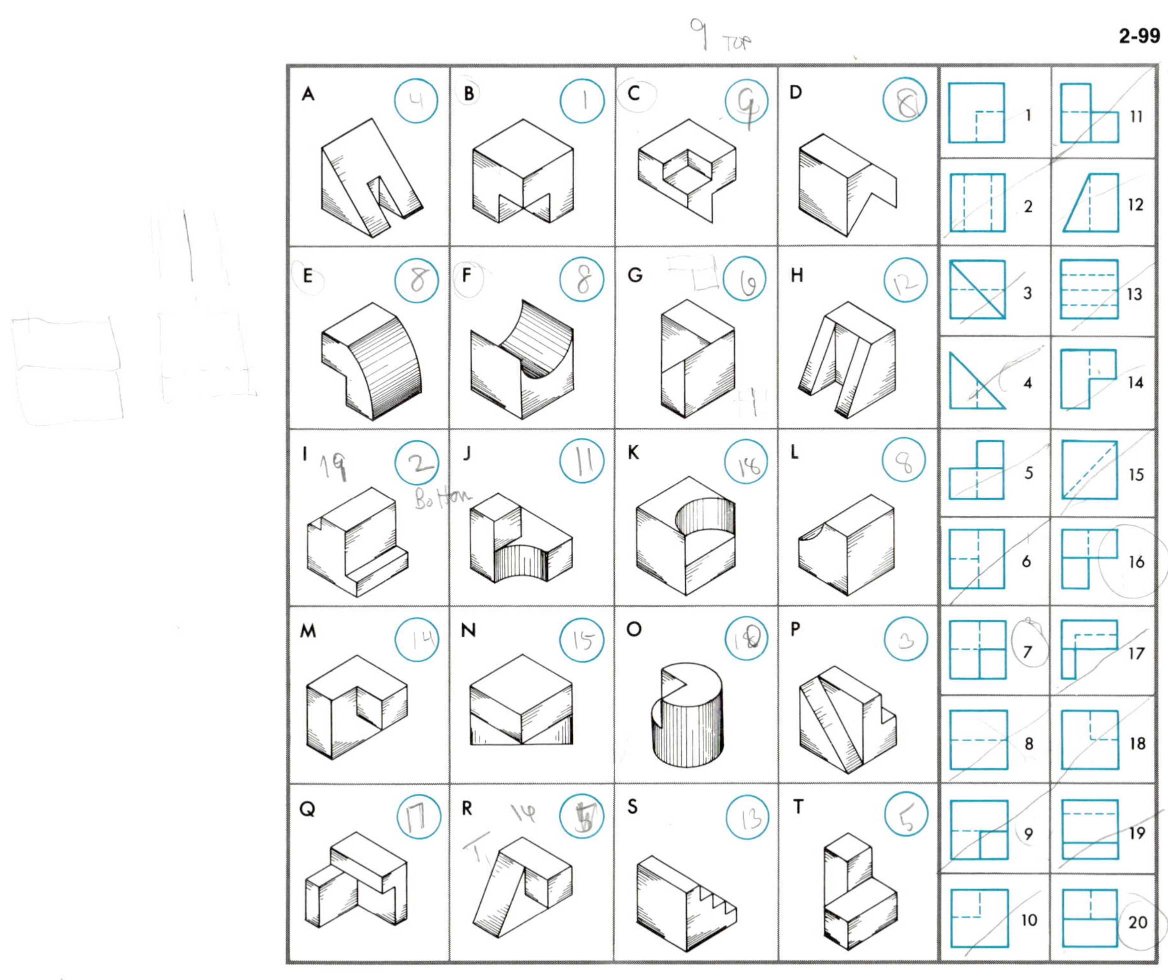

2-15. Sketch the three principal orthographic views of the objects shown in [2-100]. The arrows represent the line of sight of the viewer.

2-100

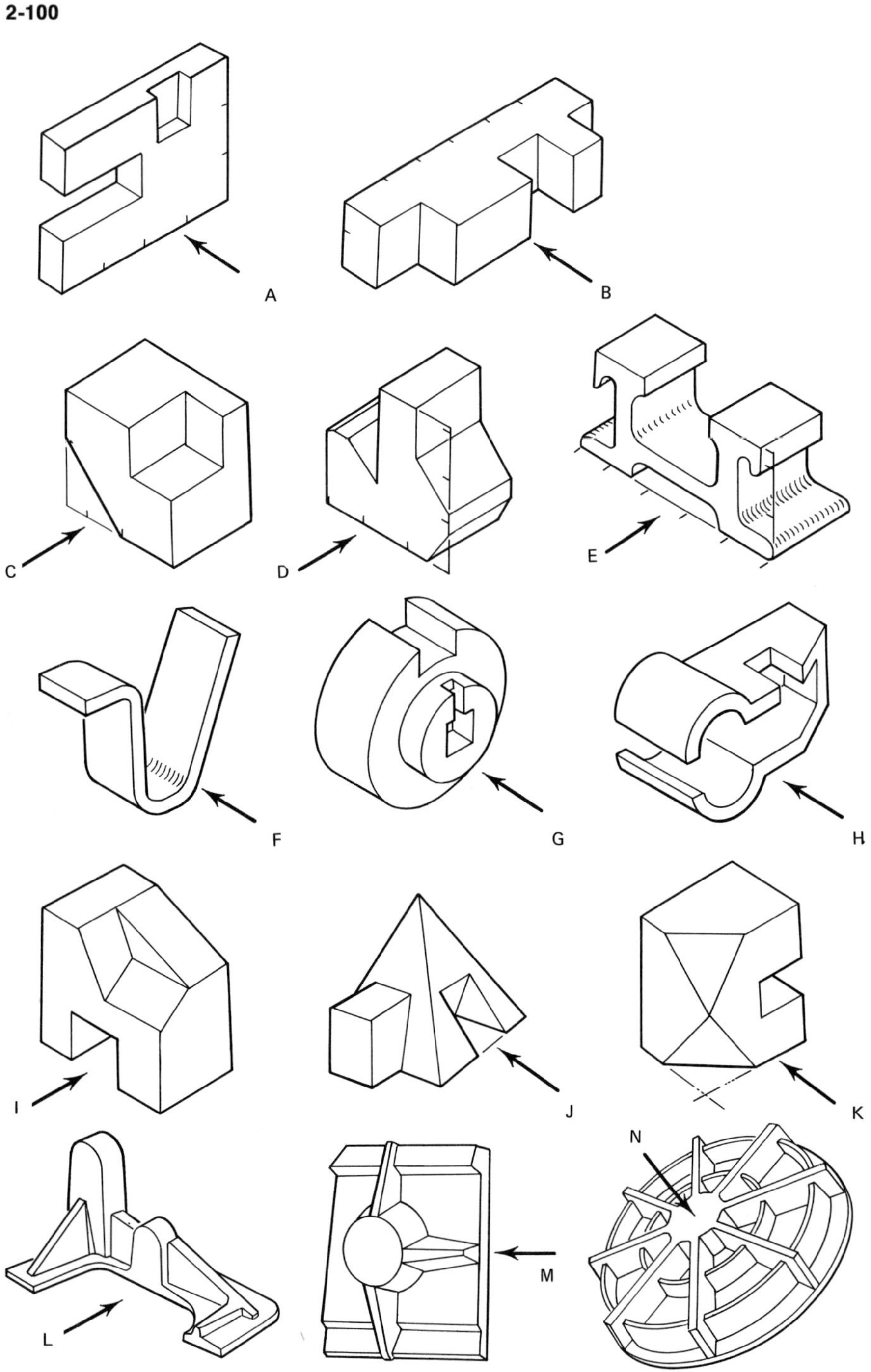

2-16. Sketch three principal orthographic views of the objects shown in [2-101].

2-101

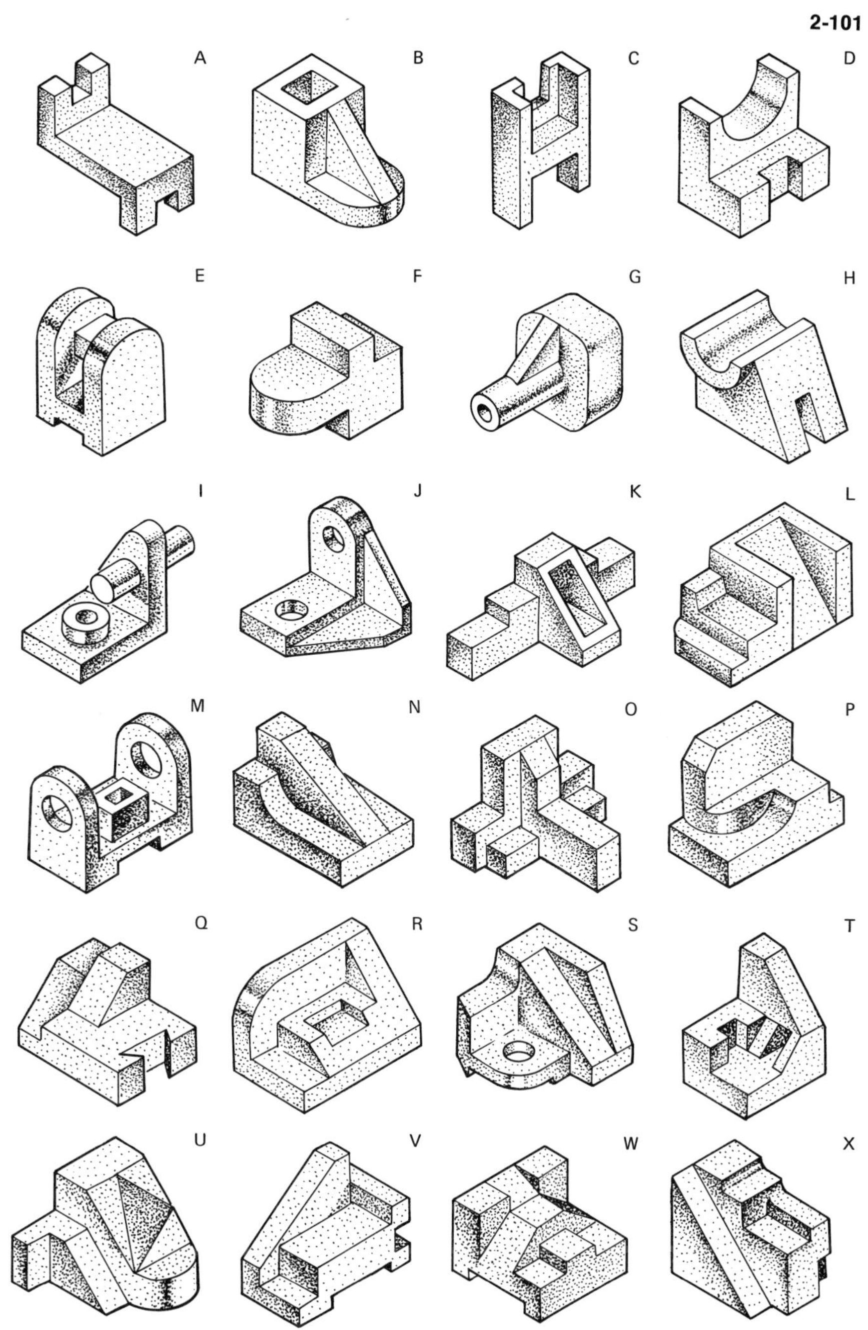

2-17. Make a *cavalier oblique* sketch of two of the objects shown in [2-102]. Using the same two objects make a *cabinet* sketch of each. Label each sketch according to the type of oblique used.

2-18. Make an isometric sketch of each object shown in [2-103].

2-102

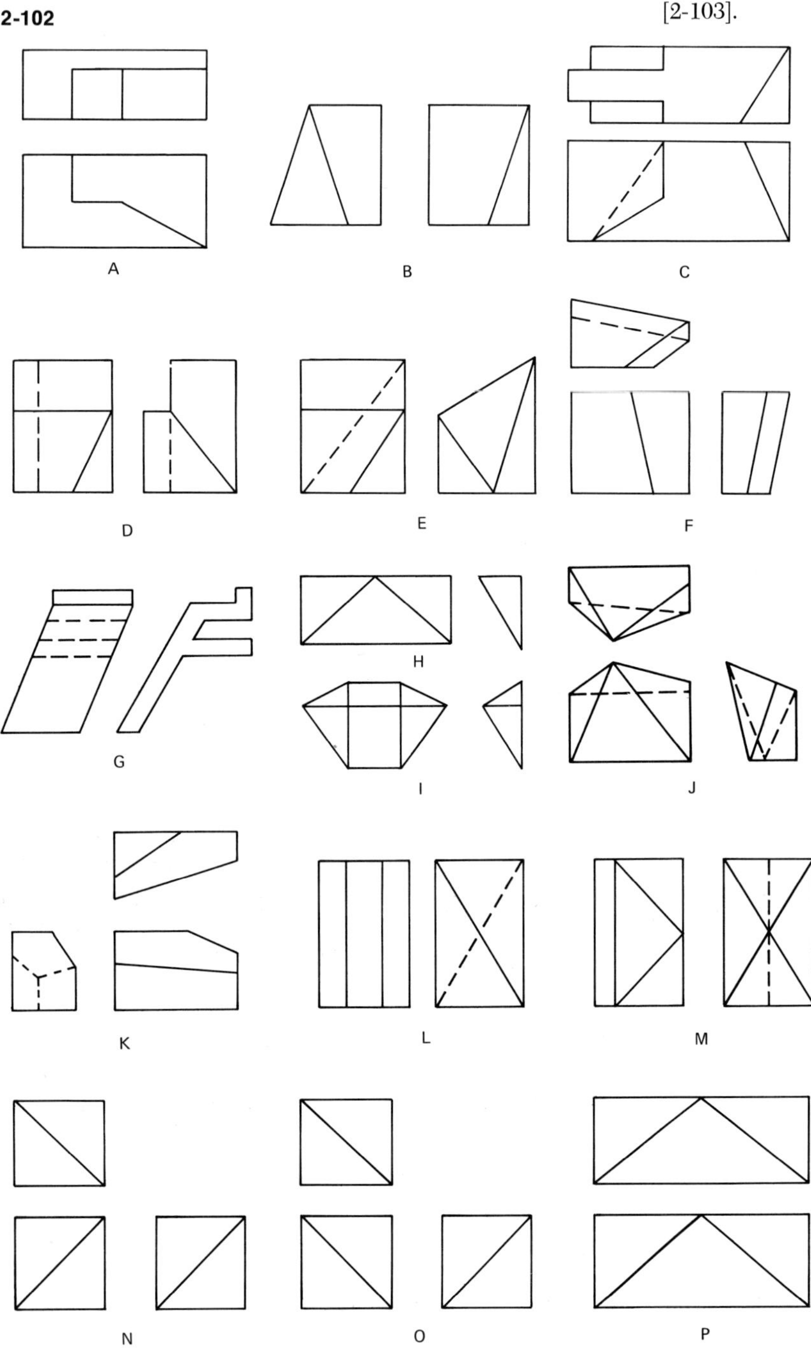

2-19. Using the conventional 45° light source, apply shading to isometric sketches of the objects shown in [2-104]. Use the technique of lines of varied weight and spacing to achieve shading.

2-20. Make a two-point perspective sketch of one of the objects in Figure [2-96].

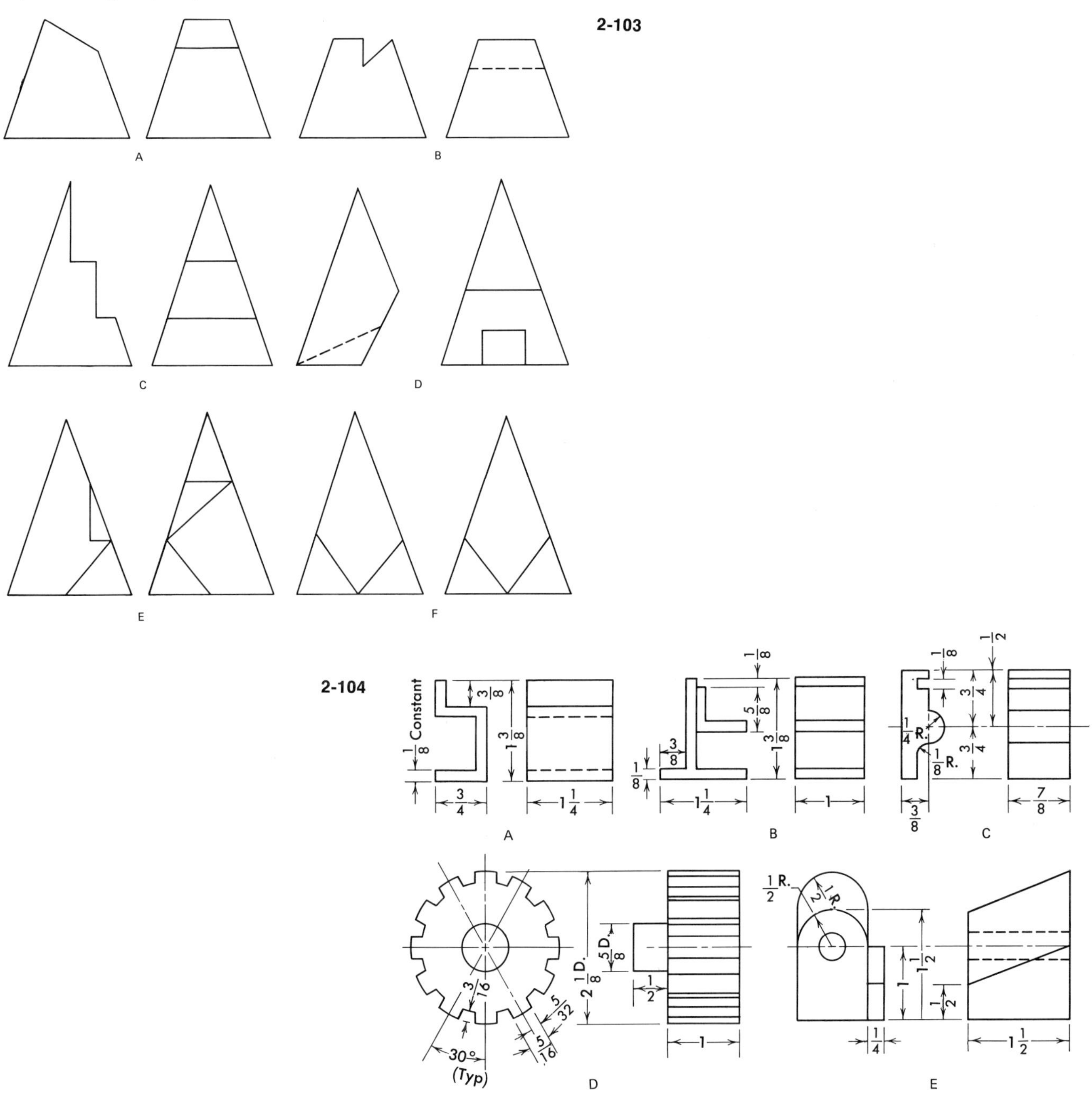

bibliography

French, T. E., and C. J. Vierck, *Engineering Drawing and Graphic Technology,* 11th ed., New York: McGraw-Hill, 1972.

French, T. E., and C. J. Vierck, *Graphic Science,* 3rd ed., New York: McGraw-Hill, 1970.

Giesecke, F. E., Alva Mitchell, H. C. Spencer, I. L. Hill, and R. O. Loving, *Engineering Graphics,* 2nd ed., New York: Macmillan, 1975.

Giesecke, F. E., Alva Mitchell, H. C. Spencer, and I. L. Hill, *Technical Drawing,* 6th ed., New York: Macmillan, 1974.

Hammond, R. H., C. P. Buck, W. B. Rogers, G. W. Walsh, Jr., and H. P. Ackert, *Engineering Graphics,* 2nd ed., New York: Ronald Press, 1971.

Hoelscher, R. P., C. H. Springer, and J. S. Dobrovolny, *Graphics for Engineers,* New York: Wiley, 1971.

Katz, H. H., *Technical Sketching and Visualization for Engineers,* New York: Macmillan, 1949.

Levens, A. S., *Graphics,* 2nd ed., New York: Wiley, 1968.

Luzadder, W. J., *Basic Graphics,* 2nd ed., Englewood Cliffs, N.J.: Prentice-Hall, 1968.

Luzadder, W. J., *Fundamentals of Engineering Drawing,* 6th ed., Englewood Cliffs, N.J.: Prentice-Hall, 1971.

McKim, R. H., *Experiences in Visual Thinking,* Belmont, Calif.: Brooks/Cole, 1972.

Mochel, M. G., *Fundamentals of Engineering Graphics,* Englewood Cliffs, N.J.: Prentice-Hall, 1960.

Rule, J. T., and S. A. Coons, *Graphics,* New York: McGraw-Hill, 1961.

Schneerer, W. F., *Programmed Graphics,* New York: McGraw-Hill, 1967.

Svensen, C. L., and W. E. Street, *Engineering Graphics,* New York: Van Nostrand Reinhold, 1962.

Wellman, B. L., *Introduction to Graphical Analysis and Design,* New York: McGraw-Hill, 1966.

the orthographic system 3

Oxide growth formed during corrosion of metal surface—magnification 3200×.

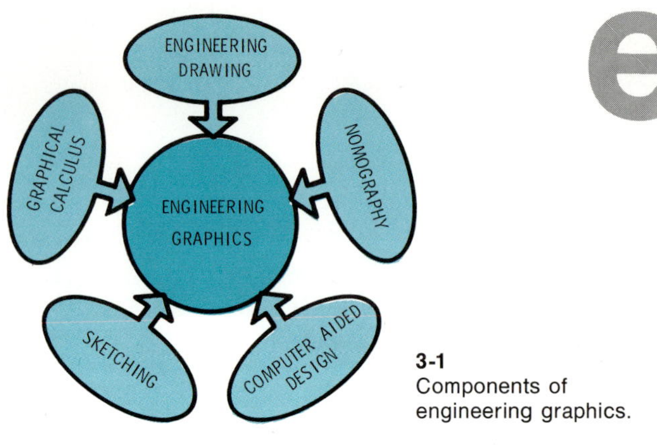

3-1
Components of engineering graphics.

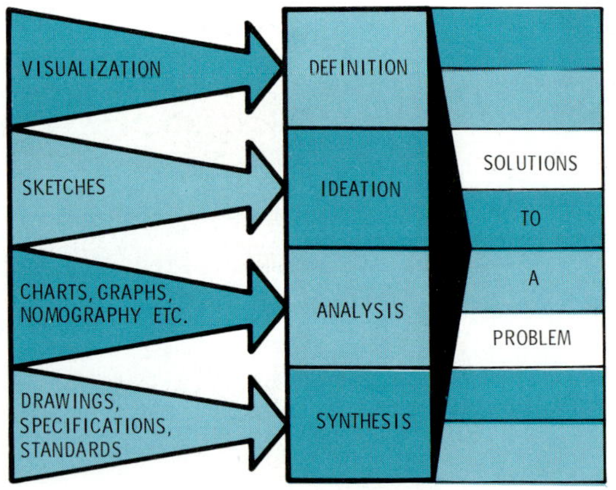

3-2
Graphical inputs to an engineering problem.

Πάντες ἄνθρωποι τοῦ εἰδέναι ὀρέγονται φύσει.

ARISTOTLE

3-3

e*ngineering graphics,* a relatively new term, combines areas of specialized knowledge and visual skill, such as *engineering drawing, graphical calculus, nomography, computer aided design* and *technical sketching* [3-1]. Most engineering problems demand the application of some, if not all, of these skills: to define the problem, to clarify ideas for solutions, to present analytical and test results, and to draw the details of the final design [3-2].

Engineering graphics uses certain accepted symbols and conventions just as our written communication system. In writing, the symbols are letters and punctuation marks that can be combined to make words, sentences, paragraphs, and books. The Declaration of Independence, for example, can be understood by all who know the meaning of the symbols. The symbols used in engineering graphics are the point, the line, and their numerous combinations. Onc must learn and understand these symbols to comprehend the ideas they express, just as one would have to learn to read the Greek written language to appreciate fully the ideas of Aristotle.

engineering drawing

Engineering drawing as used in this text includes the fundamentals of the traditional areas of *mechanical drawing* and *descriptive geometry.* The techniques and skills of mechanical drawing are necessary to draw accurate, understandable pictures of physical objects. Descriptive geometry shows how two or more objects are related to one another in space; it is discussed in Chapter 4. These two areas complement each other in engineering drawing.

Reference Planes

Engineering drawing is a visual communication system based entirely on the existence of the three *dimensional axes—height, width,* and *depth.* Since every physical object occupies space and consequently has volume, it is quite logical that the basis for describing physical objects should be related to volume and the dimensions associated with volume [3-4]. All the axes are mutually perpendicular. This means that they are located at right angles to each other; they are *orthogonal* [3-5]. It is this simple orthogonal relationship of the three major axes of volume that constitutes the basis for most technical drawing systems.

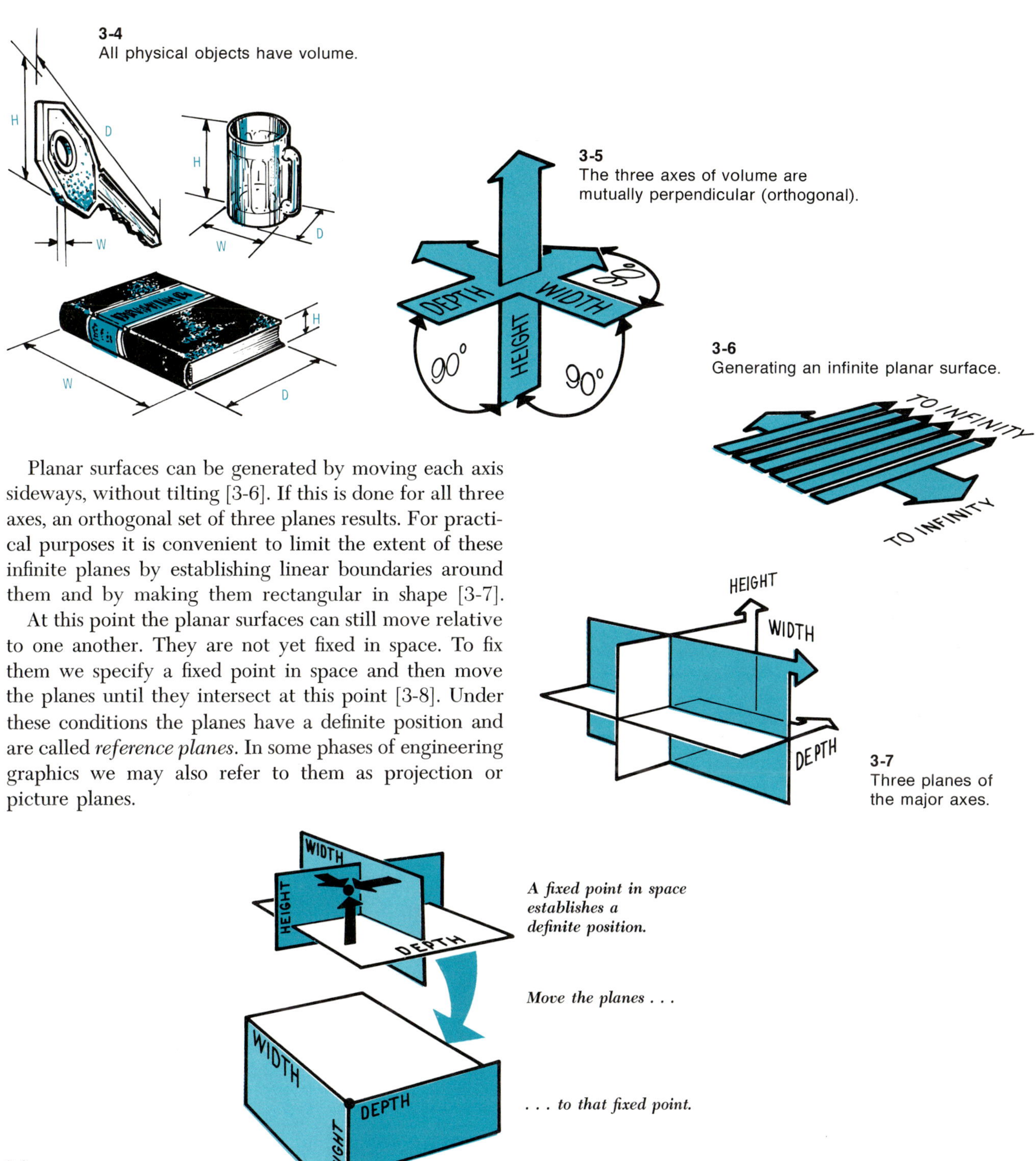

3-4
All physical objects have volume.

3-5
The three axes of volume are mutually perpendicular (orthogonal).

3-6
Generating an infinite planar surface.

Planar surfaces can be generated by moving each axis sideways, without tilting [3-6]. If this is done for all three axes, an orthogonal set of three planes results. For practical purposes it is convenient to limit the extent of these infinite planes by establishing linear boundaries around them and by making them rectangular in shape [3-7].

At this point the planar surfaces can still move relative to one another. They are not yet fixed in space. To fix them we specify a fixed point in space and then move the planes until they intersect at this point [3-8]. Under these conditions the planes have a definite position and are called *reference planes*. In some phases of engineering graphics we may also refer to them as projection or picture planes.

3-7
Three planes of the major axes.

A fixed point in space establishes a definite position.

Move the planes . . .

. . . to that fixed point.

3-8
Establishing a fixed planar relationship.

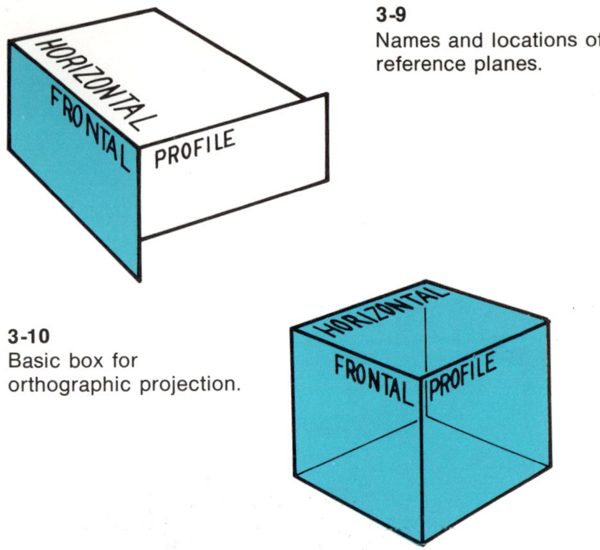

3-9
Names and locations of reference planes.

3-10
Basic box for orthographic projection.

The plane parallel to the surface of the earth is called the *horizontal* reference plane; the plane that is perpendicular to that surface and faces forward is the *frontal* reference plane; the remaining vertical reference plane is perpendicular to both the frontal and horizontal reference planes and is known as the *profile* reference plane [3-9].

The three reference planes occupy the same positions as do the sides of a cube or a rectilinear box. Such a box would have as its top the horizontal reference plane, as its front the frontal reference plane, and as its right side the profile reference plane [3-10].

3-11
Comparative directions in point location systems.

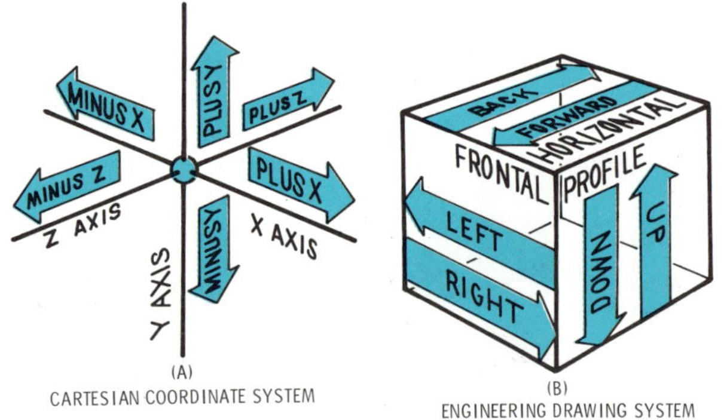

Locations in Space

A strong similarity exists between the use of reference planes and the Cartesian coordinate system found in mathematics. The fixed point that was used to establish the reference plane positions is analogous to the point of origin, or zero, in Cartesian coordinates. The intersections of the orthogonal planes correspond to the axes of the Cartesian system. Width corresponds to the x axis, height to the y axis, and depth corresponds to the z axis [3-11A].

In engineering drawing, directions are described somewhat differently than in mathematics. Instead of describing a point as being plus or minus a certain number of units along a given axis, the point is described as being so many units up or down, right or left, and forward or back of the appropriate reference planes [3-11B]. For example, it may be necessary to locate a point A in space. By the Cartesian coordinate system, its location might

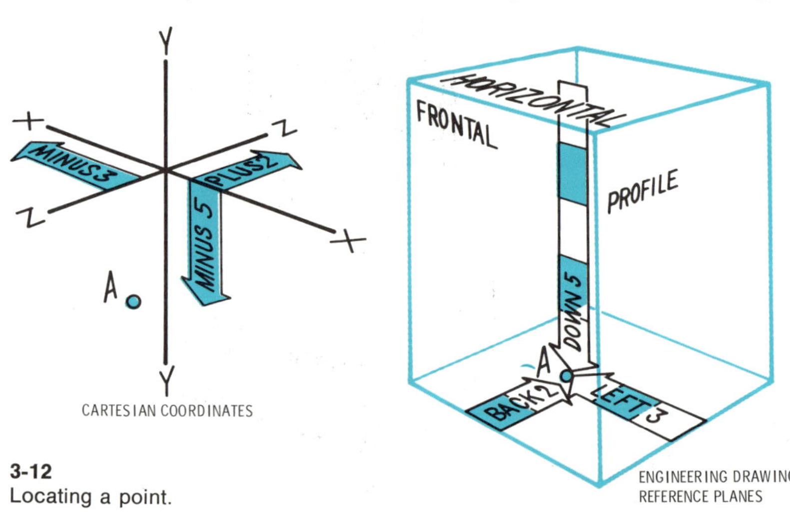

3-12
Locating a point.

3-13

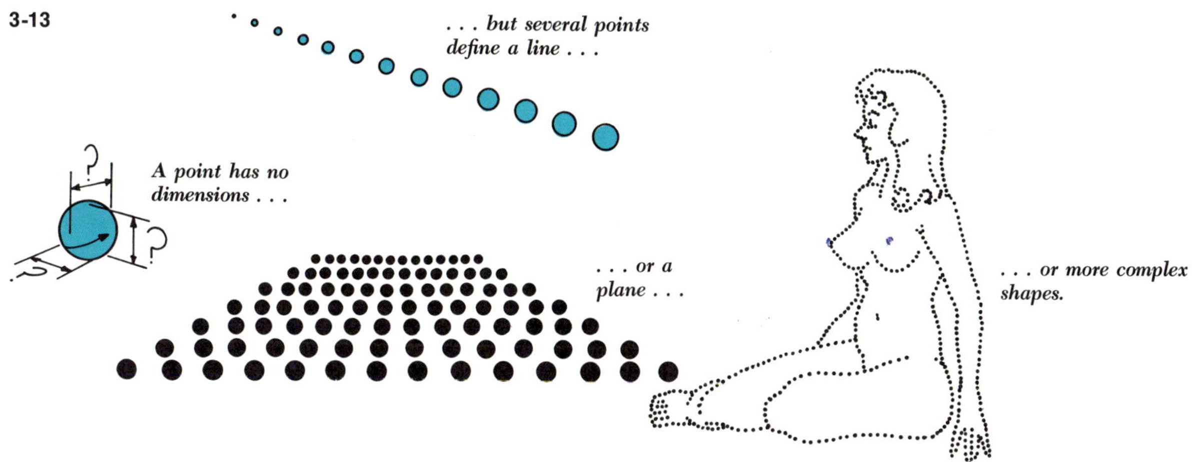

be described as −3 units on the x axis, −5 units on the y axis, and +2 units on the z axis. In engineering drawing this same point would be described as 5 units *below* the horizontal reference plane, 3 units *left* of the profile reference plane, and 2 units *in back of* the frontal reference plane [3-12].

The path of a moving point can generate very complex lines, surfaces, and volumes [3-13]. Learning to draw an object—to communicate visually—might be considered as learning to locate or follow the points that describe the object.

orthographic projection

Orthographic projection is based on the premise that the image of an object should always be viewed at right angles (perpendicularly) to the major planes of that object. As an example, an orthographic view of a tabletop would assume that the observer is looking down from directly above the table [3-14]. In another example, the observer stands directly in front of a painting so that his line of sight is perpendicular to the front plane of the painting. The term *major planes* is used here to suggest that more than one orthographic view of an object is possible. In fact, the orthographic projection system has six *principal views,* which relate directly to the horizontal, frontal, and profile principal reference planes [3-15].

A second and equally important premise of the orthographic-projection system is that the observer must be

3-14
In orthographic projection the line of sight is perpendicular to a plane.

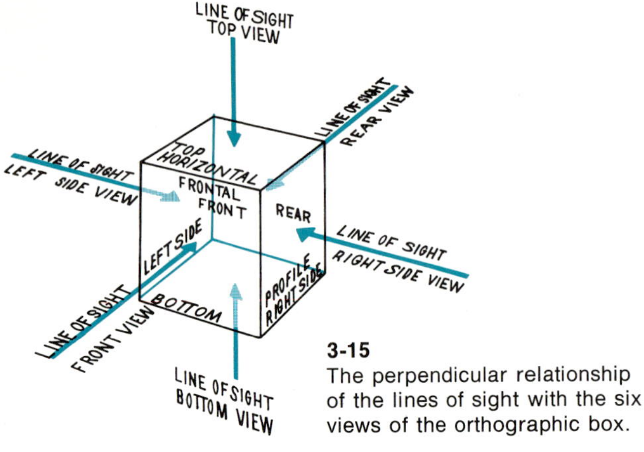

3-15
The perpendicular relationship of the lines of sight with the six views of the orthographic box.

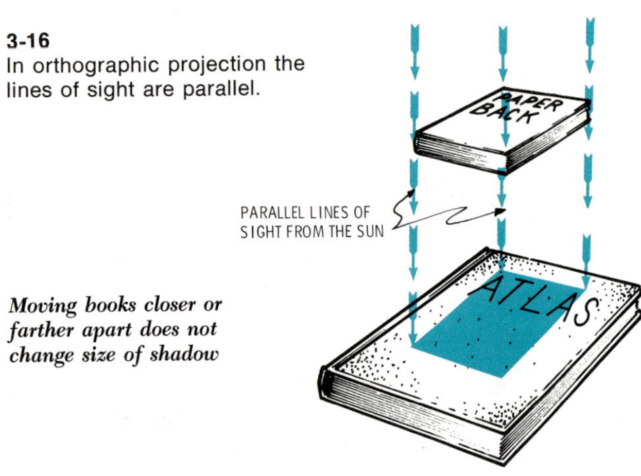

3-16
In orthographic projection the lines of sight are parallel.

Moving books closer or farther apart does not change size of shadow

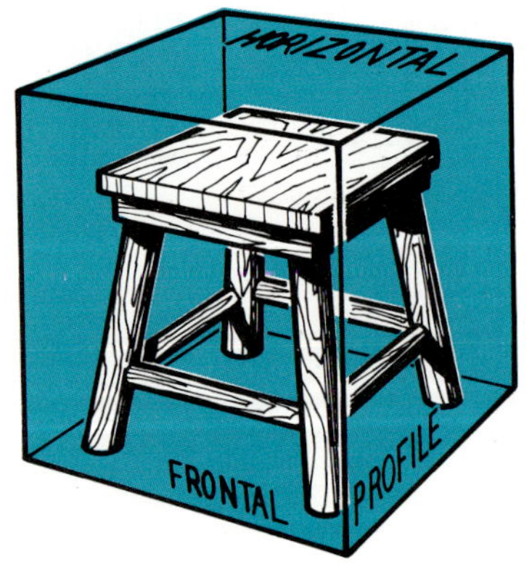

3-17
Orthographic box concept.

located an infinite distance from the object. This premise forms the basis for projecting objects in their true size and shape to an intervening plane, because the lines of sight, when emanating from an infinite point in space, are virtually parallel. A good example of this principle can be demonstrated on a bright, sunny day by holding a small book so that it is perpendicular to the sun's rays. Now hold a larger book at some distance behind and parallel with the smaller book. Notice the size of the shadow that is cast on the larger book. While maintaining the perpendicular relationship with the sun's rays and the parallel relationship between the books, move the books farther apart or closer together. The size and the shape of the shadow does not change, because the light rays (lines of sight) from the sun are essentially parallel [3-16]. These two principles, the perpendicular orientation of the line of sight to a plane and the parallel lines of sight, are basic to the theory of orthographic projection.

The Orthographic-box Concept

Recall the six-sided box that was developed from the three major orthogonal axes. Imagine this box as being transparent, as if it were made of clear plastic or glass, and that some object, for example a stool, is located within the box [3-17].

The orientation of the object should relate to its natural or most useful position. If the object has an obvious base, as most objects do, it should be facing toward the bottom plane of the orthographic box. The object should be oriented within the box so that its longest horizontal dimension is parallel with the frontal plane. Frequently, this will make the dominant or most characteristic contour of the object visible through the frontal plane. The importance of this orientation will be amplified later in the chapter.

This requirement may, in some cases, present the engineer with a puzzling contradiction. For example, it is obvious that an automobile has a front end. However, because the side of the automobile has a longer horizontal length and more important contours, we would normally orient the side view parallel with the frontal plane [3-18].

So it is with other objects, such as a telephone. Since the longer horizontal dimension is found on the "side" of the telephone, it may be best to orient the side parallel to the frontal plane. This results in another dominant

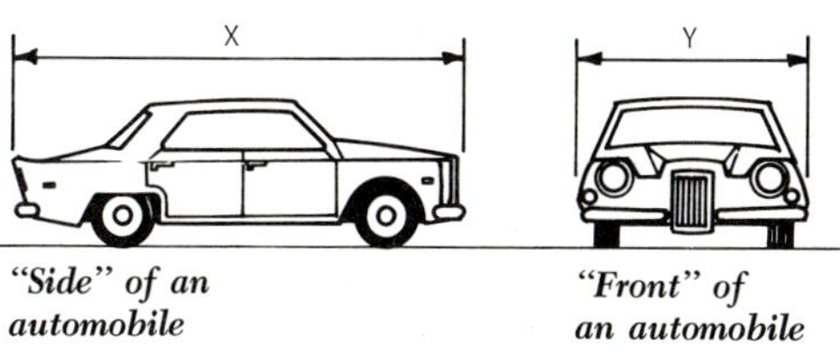

3-18
Preferred orientation of an automobile within an orthographic box.

"X" is greater than "Y" therefore . . .

. . . the "side" should be oriented parallel to the frontal plane

"Side" of a telephone *"Front" of a telephone*

X Y

"X" is greater than "Y" therefore the "side" is oriented parallel to the frontal plane.

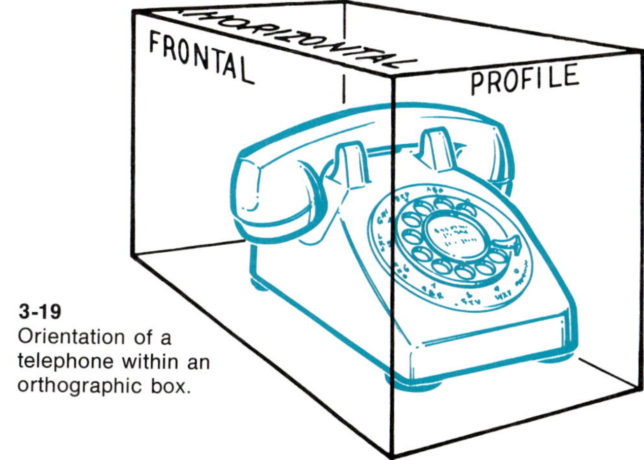

3-19
Orientation of a telephone within an orthographic box.

contour of the telephone (the conventional "front") being oriented parallel with the profile plane [3-19].

In orthographic projection the observer must be located so that his line of sight is *perpendicular* to one of the six principal planes of the box [3-20]. He may be in front or back, above or below, or to the right or left of the box. The second principle states that the lines of sight to the object are *parallel* to one another when they intercept the principal planes [3-21].

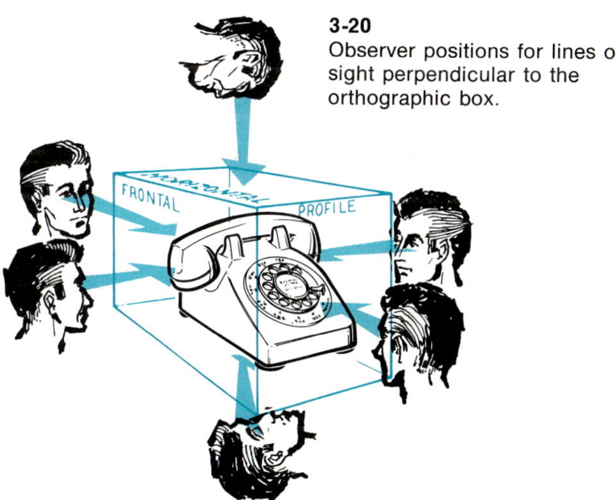

3-20
Observer positions for lines of sight perpendicular to the orthographic box.

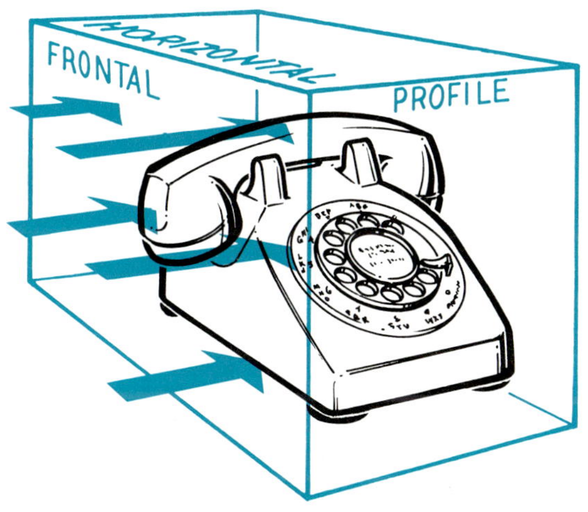

3-21
All lines of sight are parallel to one another and perpendicular to a principal plane.

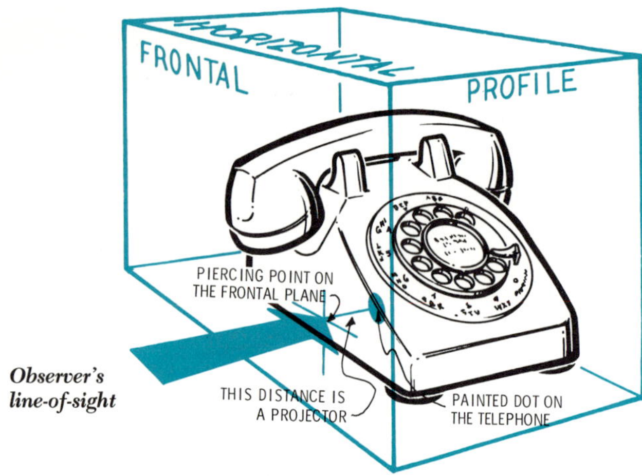

Observer's line-of-sight

3-22
A point is projected to a principal plane along the line of sight.

3-23
Projecting points on an object to the principal plane.

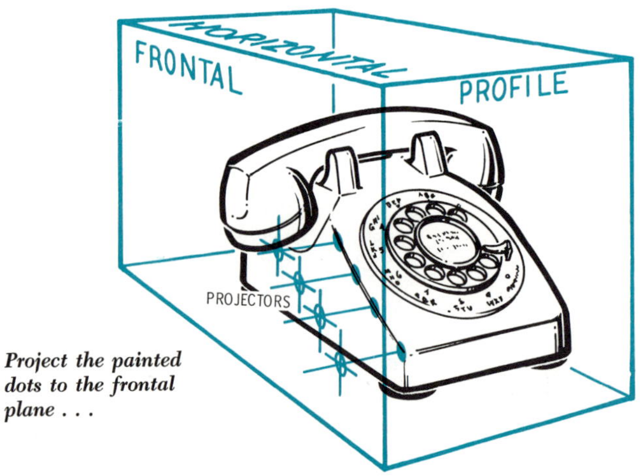

Project the painted dots to the frontal plane . . .

. . . and continue the process with all the points on the object.

Projecting the Image

Let us see how a three-dimensional object can be transformed into a two-dimensional picture. Imagine that a very small dot has been painted on a corner of the telephone. The observer, who is located at the ideal *observation position* (where his lines of sight are perpendicular to the frontal plane) will see the designated point on the telephone through the frontal plane. The line of sight passes through the frontal plane of the orthographic box. If we isolate the segment of the line of sight beginning at the point where it pierces the frontal plane and ending at the dot on the telephone, we will have a *projector* that extends from the object to the frontal plane. Imagine sliding the painted dot along this projector back toward the observer until it contacts and adheres to the frontal plane [3-22]. This process is known as *projecting a point to a plane.*

Continuing the process, place a second dot at a different position on the telephone, generate a line of sight to the second point, and project the point back to the frontal plane. Now place another dot, and another, and yet another until all the lines that define the telephone as seen through the frontal plane have been covered with dots, and these dots have been projected to the frontal plane of the orthographic box—thus creating a two-dimensional image—an *orthographic view* [3-23]. Of course, it should be understood that in actual practice one need not laboriously transfer the infinite number of points necessary to generate a complete image of the object. Instead, one makes use of the principle that *two points define a straight line* to simplify and accelerate the projection process. For example, one needs only to project the end points of the line that defines the bottom edge of the telephone to the frontal plane and then draw a line on the frontal plane to connect these two end points.

principal views

Remember that the points could be projected to any of the six principal planes of the orthographic box. The picture obtained by projecting all the points of the object to the frontal plane of the orthographic box is called the *front view* [3-24]. The same process used with other principal planes of the box produces other *principal views* of the object.

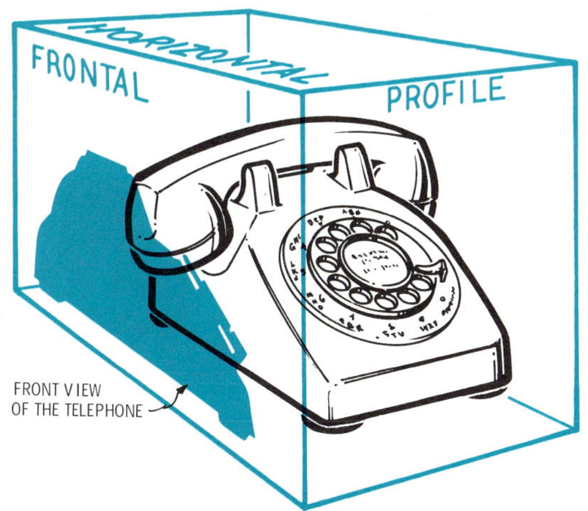

3-24
A projected view of an object on the frontal plane.

Where is depth?

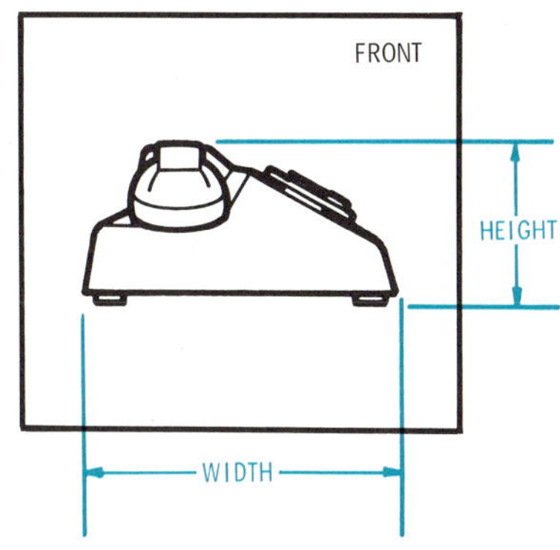

3-25
A single orthographic view shows only two dimensions of space.

Unfortunately, two-dimensional orthographic views, such as the front view just described, have one inherent weakness: a two-dimensional view can show only two dimensions of a three-dimensional object [3-25]. With this view alone, it is impossible for the observer to ascertain the third major dimension of the object. To obtain that, he needs another view at right angles to the front view. Any of four possible positions in the orthographic projection system—top, bottom, left, or right side—will provide the necessary information about the third dimension.

Recall the earlier discussion in this chapter concerning the location of a point in space. The point's location within the orthographic box was described relative to the reference planes of the box. In the example used, a point was located two units behind the frontal plane. It is important to remember that if this same point is properly projected to *any* plane perpendicular to the frontal plane, it *still* will be two units behind the frontal plane [3-26].

Most dimensions of objects can be depicted on the principal views, but more than one principal view is needed [3-27]. Since it isn't practical to build a transparent box when we want to make a drawing, a way must be found to show all principal views on one sheet of paper. This is not only practical, but it allows an observer to view all the principal planes of the orthographic box from a single position.

Imagine that all the edges of the transparent orthographic box are hinged [3-28]. Some of the hinges may be removed from certain edges. By this means it would

3-26

A point always projects the same distance from a reference plane . . .

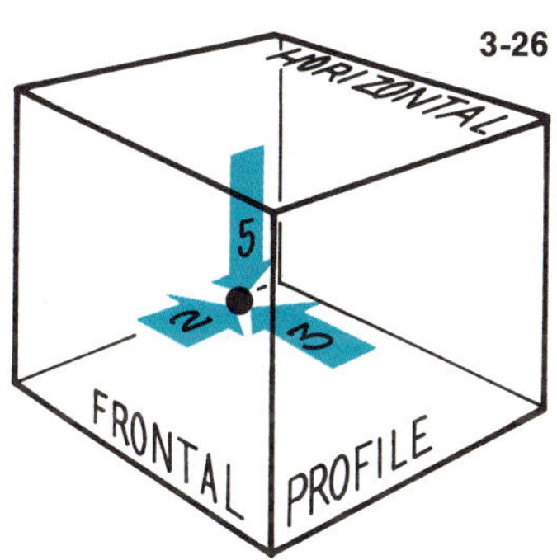

. . . on all the planes that are perpendicular to that reference plane.

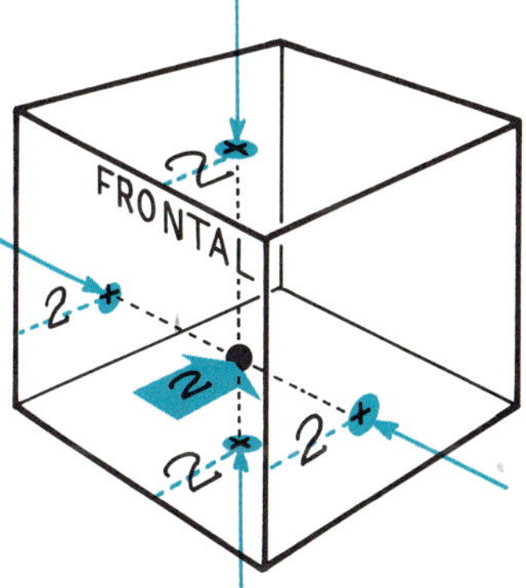

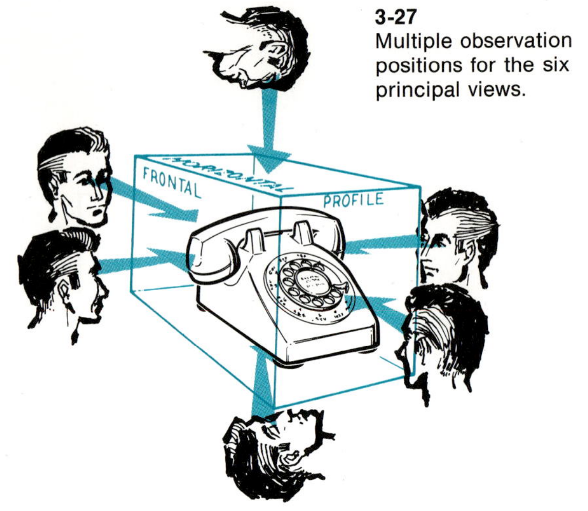

3-27
Multiple observation positions for the six principal views.

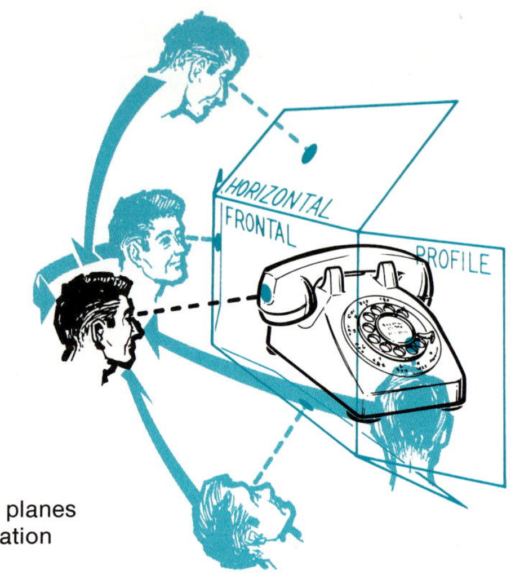

3-28
Rotating principal planes to a single observation position.

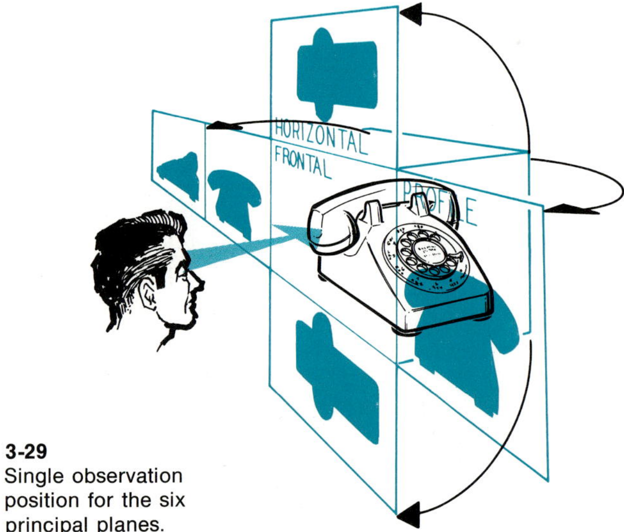

3-29
Single observation position for the six principal planes.

be possible to unfold the box. In this way the images that had been projected onto the principal planes of the box would lie flat on a two-dimensional surface [3-29].

Standard Arrangement of Views

There are many possible ways that the box might be unfolded, but convention has established a preferred arrangement. This arrangement, known as the *six principal views,* starts with the *front view* of the object. Located immediately above the front view is the *top view* and below it is the *bottom view.* To the immediate right of the front view is the *right side view.* To the immediate left of it is the *left side view.* The *rear view* normally is positioned to the left of the left side view [3-30]. To avoid error in interpretation always follow this convention faithfully, and align the views correctly.

Choice of Views

Although six principal views are possible, it is rarely necessary to draw all six to describe an object adequately. Usually, all essential information can be shown in three of the views. The views most commonly used are the top, front, and right or left side views [3-31]. This particular choice is easily understood when one realizes that the remaining three views are basically reversed images of the first views, and do not usually provide new informa-

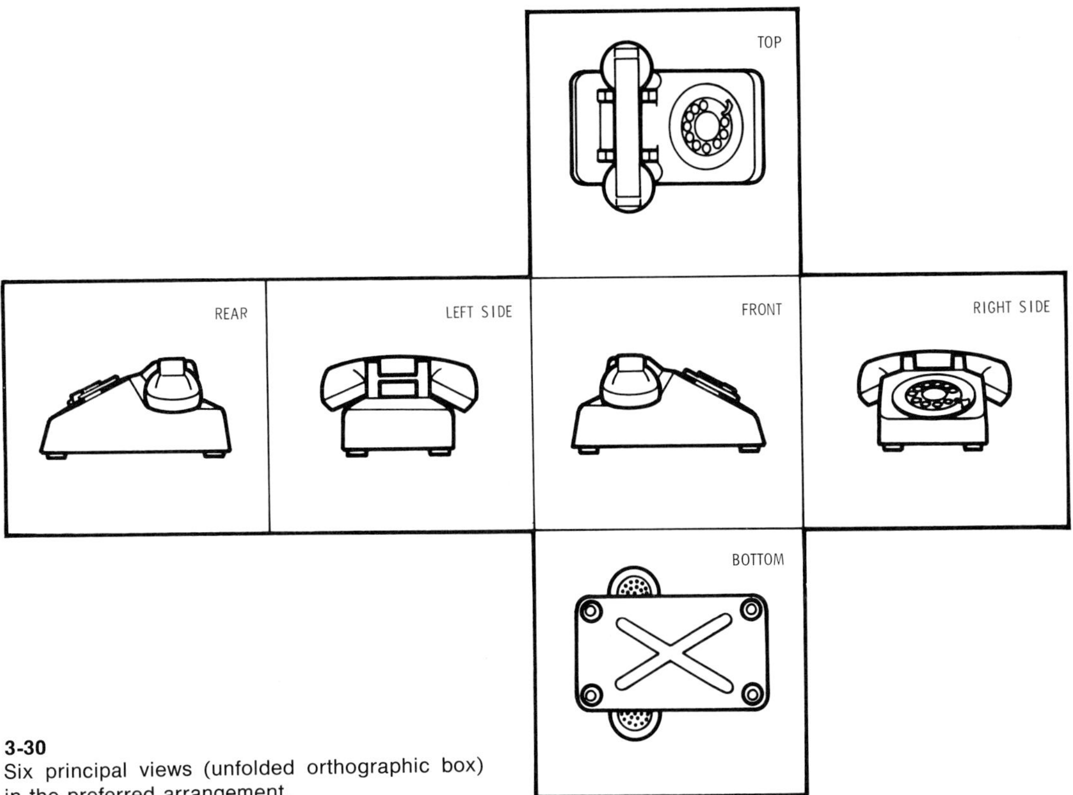

3-30
Six principal views (unfolded orthographic box) in the preferred arrangement.

tion about the object (especially when all hidden lines are shown). Sometimes it is possible to completely describe an object with only two views, and in special cases with only one view and appropriate notes (see p. 44).

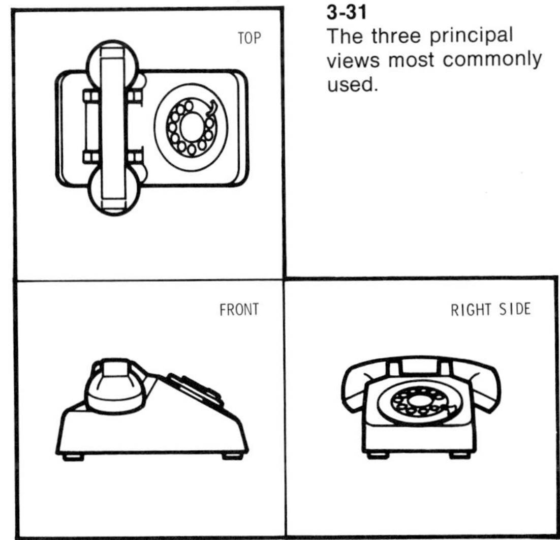

3-31
The three principal views most commonly used.

Reference Lines

In the example describing the unfolding of the orthographic box [3-28], attention was directed to the hinged edges of the box. The box was flattened by turning its sides about these hinges. A closer examination of the flattened box reveals that the hinged edges have become the lines that separate the views from one another. However, it is not that the lines separate the views which is important, but rather that the lines *represent the edge views* of the reference planes. For this reason they are called *reference lines* [3-32]. These reference lines (sometimes called *fold lines* or *hinge lines*), are important for locating a point in a particular view, because they perform the same function on a two-dimensional sheet

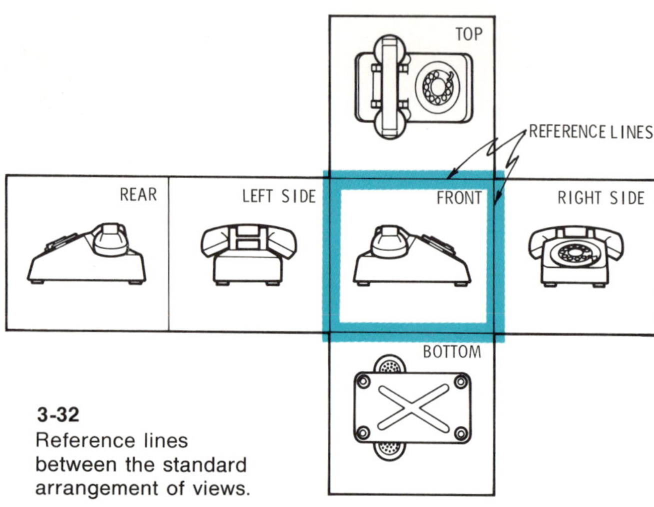

3-32
Reference lines between the standard arrangement of views.

of paper as the reference planes did in three-dimensional space.

Adjacent and Related Views

Any two views that share a common reference line and are therefore perpendicular to each other are called *adjacent views*. Using the standard arrangement of views as an example, let us consider some of the possible adjacent views. If our attention is centered upon the front view, then the left side, top, right side, and bottom views are *all* adjacent views to that view [3-33]. If, on the other hand, our attention is centered upon the right side view, *only* the front view is adjacent to it.

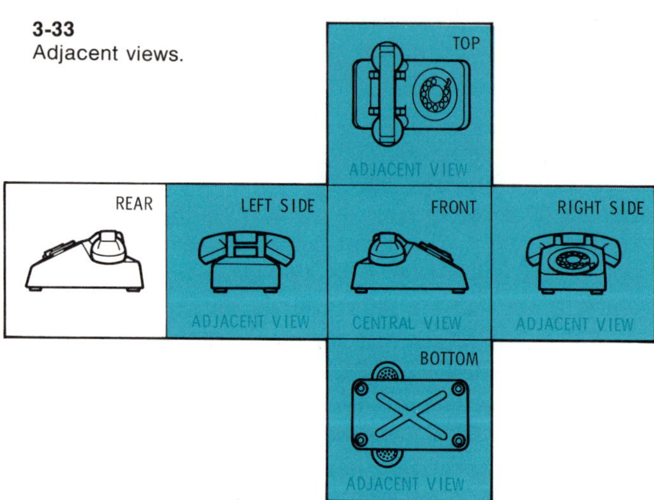

3-33
Adjacent views.

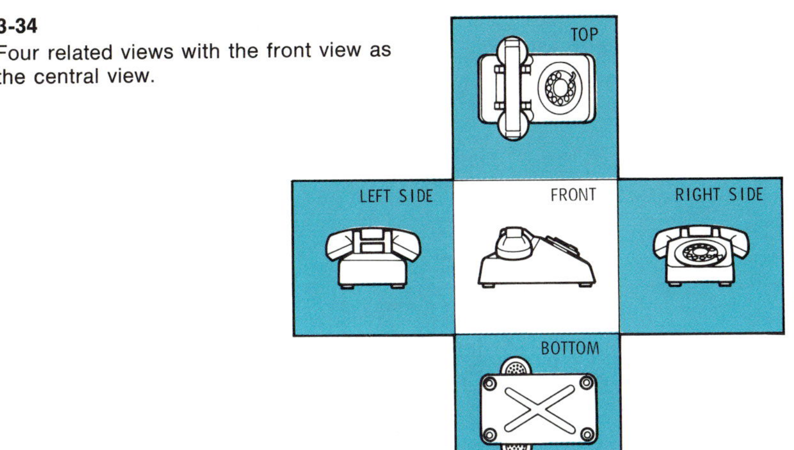

3-34
Four related views with the front view as the central view.

Views that are adjacent and perpendicular to the same central plane are called *related views*. Thus the top, bottom, left, and right side views are related views to one another, since they are all adjacent views to the front view.

Projecting Points Between Views

Imagine again the example of the telephone with a dot painted on its side [3-22]. We know that the point can be projected to any or all of the viewing planes of the orthographic box. If we look at the point in each of the planes adjacent to the front view, we find that its distance from the reference line is the same in all these views [3-35]. This is true because the frontal plane is an equal perpendicular distance, *d*, from the point in every view adjacent to the frontal plane. Notice, however, that the distance *d* is not found in the frontal view. This demonstrates again that at least two adjacent views are needed to fully describe an object.

Because points are equidistant from the reference line in related views, any point in one view can be transferred directly to any other related view. The transferral process usually is best accomplished with the aid of dividers [3-36].

Technique for Projecting Principal Views

Assume you have two adjacent views of an object (or sufficient descriptive data from which such views can be constructed), how can you find a third principal view?

3-35
A given point is equidistant from the same reference plane.

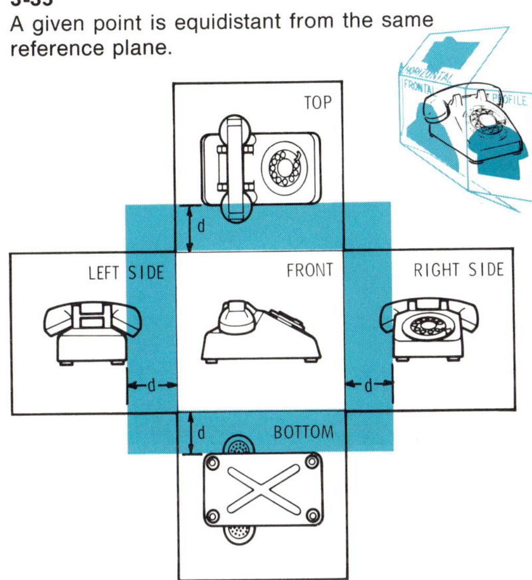

3-36

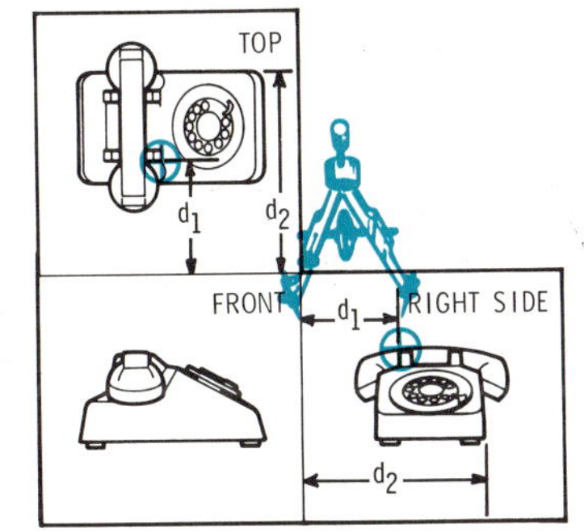

We shall illustrate the technique using a simplified sketch of the front and top views of the telephone to construct a right side view.

Construct a vertical reference line between the front view and the right side view.

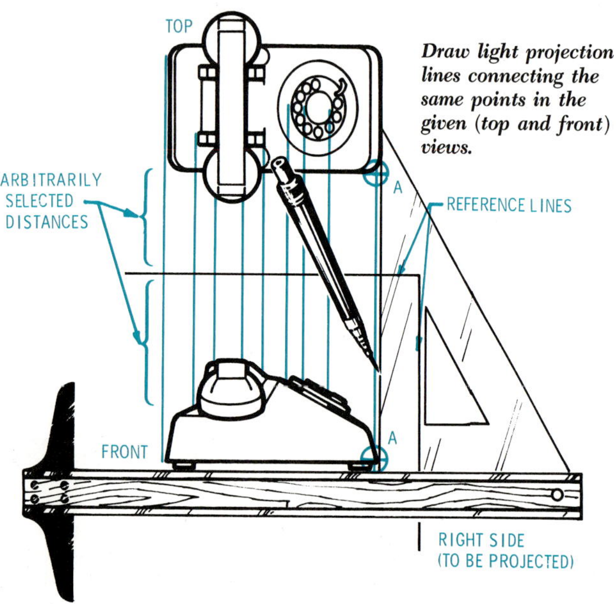

3-37

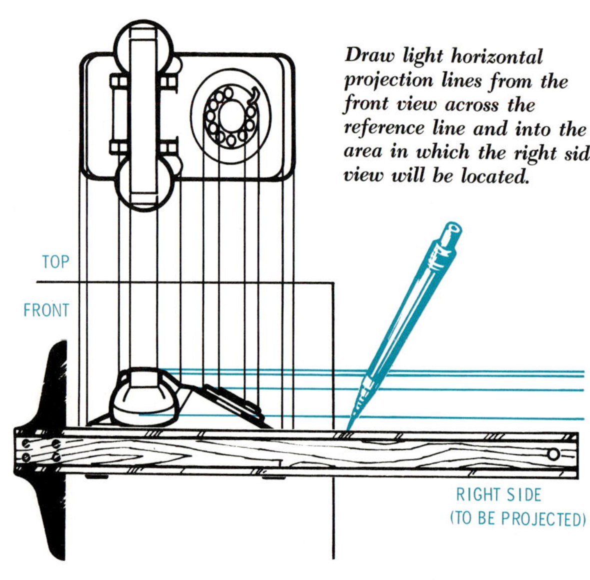

3-38

In the top view use dividers to measure how far each point is located away from the reference line separating the top and front views.

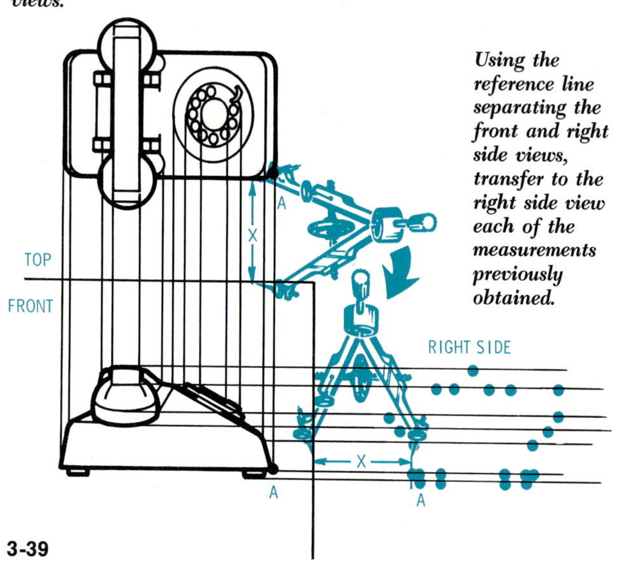

3-39

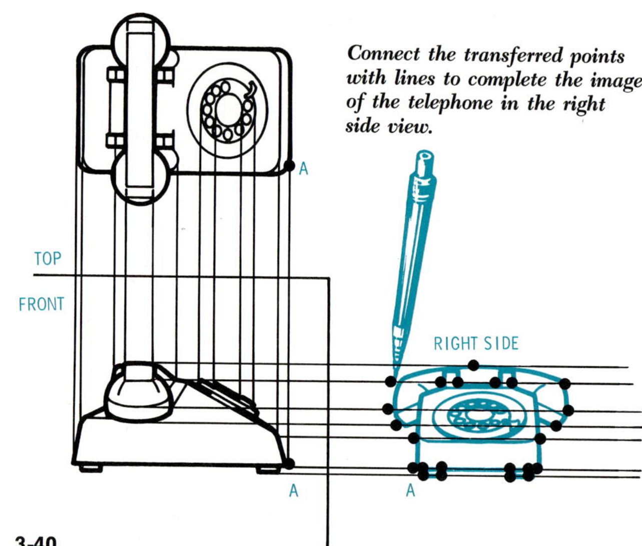

3-40

auxiliary views

With six principal views one would expect to have enough two-dimensional images for all the essential information about a physical object. In most simple, rectilinear objects this is true, but look again at the telephone example. Close inspection of the right side view, for example, shows some ambiguities that are not resolved by the other principal views. The dial face is seen as an ellipse in this view and in all the other views excepting the front and rear where it shows as an edge [3-41]. However, dials on telephones are not elliptical. In fact, all the circular shapes on the telephone appear as ellipses in the principal views. Also, notice the dialing stop. What is its true shape? Obviously, there are many things about the physical characteristics of the telephone that are not explained in the principal views. This lack of visual information exists because the plane of the telephone that contains the dial is not parallel to a principal plane. Since the lines of sight to the dial plane are not perpendicular to any existing plane, an accurate image of the dial cannot appear [3-42].

All circles appear to be elliptical.

What is the true shape of the dialing stop?

3-41

Line-of-sight perpendicular to the horizontal plane . . .

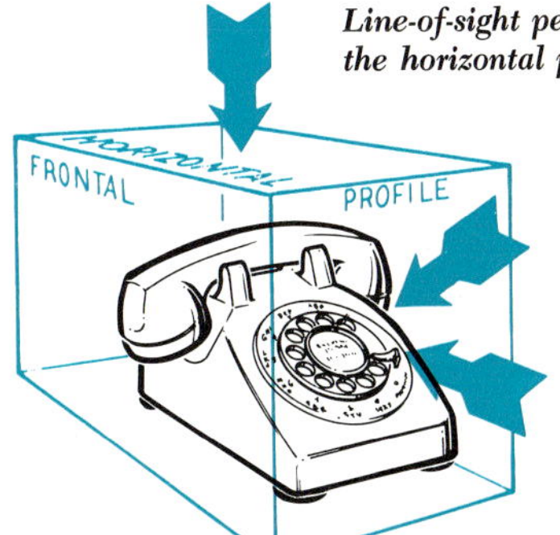

. . . and a line-of-sight perpendicular to the profile plane . . .

. . . but there is no plane parallel to the dial face for a perpendicular line-of-sight.

3-42

Location of Auxiliary Planes

To project the true physical characteristics of the telephone dial, we must construct a new plane within the transparent box that permits the line of sight to be perpendicular to the telephone dial. We simply use a viewing plane parallel to the dial of the telephone [3-43]. Such a parallel plane within the orthographic projection box is called an *auxiliary plane.* Lines of sight pierce the

The lines-of-sight are perpendicular to the auxiliary plane.

Auxiliary plane is parallel to the dial face plane.

3-43
Location of an auxiliary view.

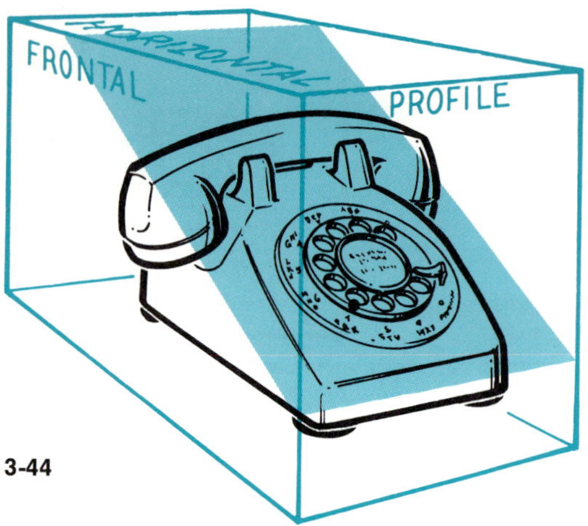

3-44

The points of the object may be projected to an auxiliary plane.

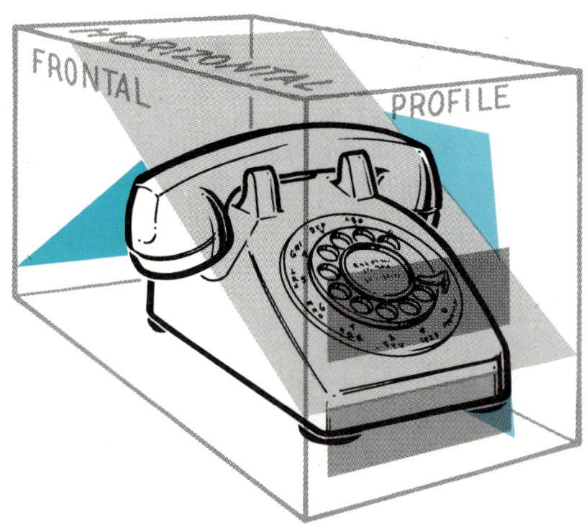

3-45
Many auxiliary planes are possible.

auxiliary plane perpendicularly and terminate on the object in the same manner as they would with a principal plane. The points on the object are then projected back to the auxiliary plane, and the resulting image transferred to it is called an *auxiliary view* [3-44].

In this example an auxiliary plane was constructed to show the view of the dial on the telephone in its true shape. To show the details of the sloping back of the telephone in its true shape, another auxiliary view would have to be constructed using another auxiliary plane parallel to the desired surface. In other words, unlike the six principal views, there is an unlimited number of possible auxiliary views of an object [3-45]. One need construct only those views that will aid in clarifying those parts of an object that need further explanation.

Construction of Auxiliary Views

To construct a typical auxiliary view find the edge view of the plane for which an auxiliary view is needed and then draw an edge view of the auxiliary plane parallel to it.

The *edge view* of a plane, just as the name implies, shows the entire plane as an edge—a straight line. To an observer of the frontal plane, the horizontal and the

3-46

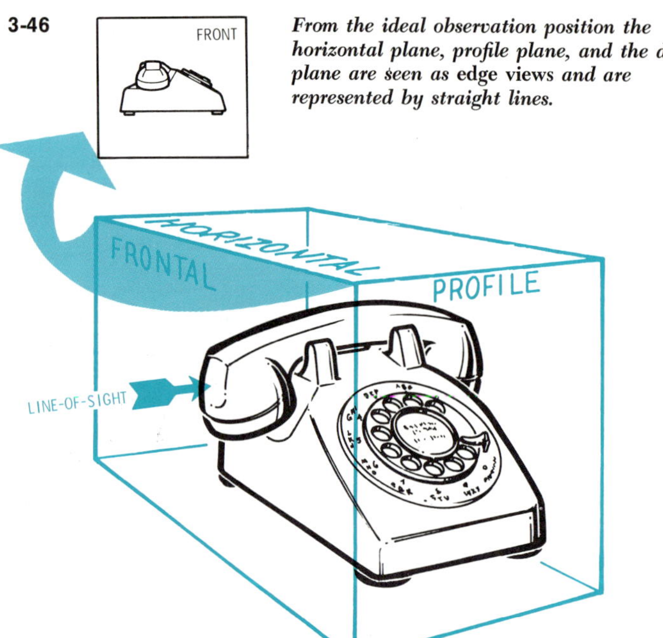

From the ideal observation position the horizontal plane, profile plane, and the dial plane are seen as edge views and are represented by straight lines.

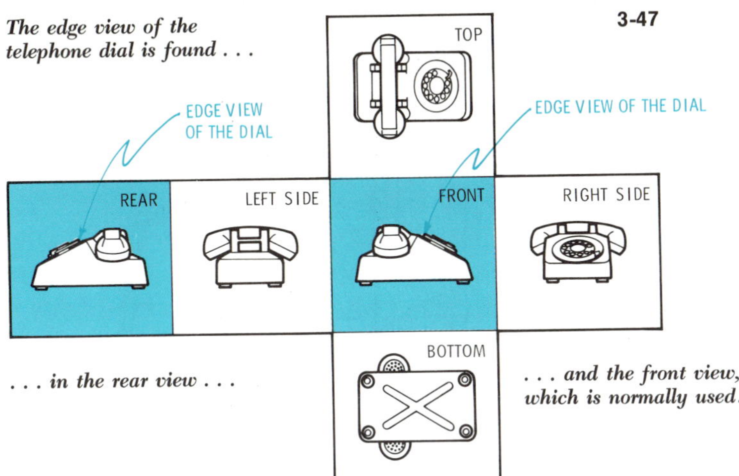

3-47

The edge view of the telephone dial is found . . .

. . . in the rear view . . .

. . . and the front view, which is normally used.

3-48
Unfolding an orthographic box having an auxiliary plane.

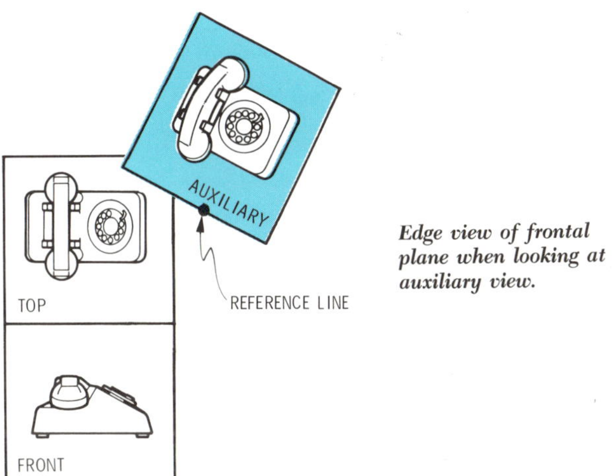

Edge view of frontal plane when looking at auxiliary view.

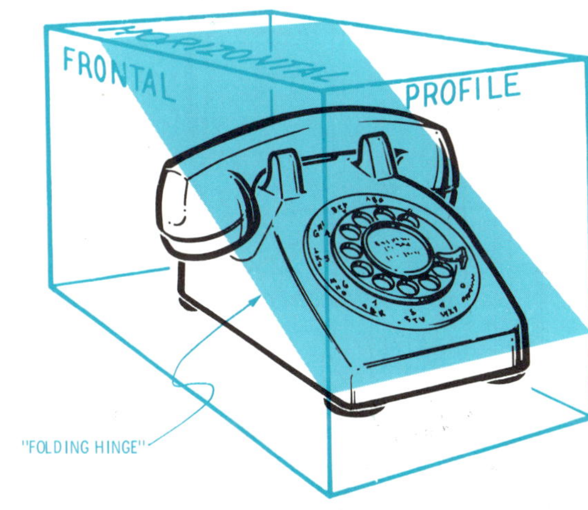

profile planes are seen as edge views; and when the telephone is viewed from this same position, the dial face appears as a straight line—an edge view [3-46]. A line is drawn parallel to this straight line to form the edge view of the auxiliary plane.

With the telephone, the edge views of the dialing plane appear in the front and rear principal views [3-47]. Since both views provide an edge view of the dial, either can be used to establish the auxiliary plane. Normally, the front view would be used to construct the auxiliary view.

The straight-line edge view of the auxiliary plane is the *folding hinge* by which the auxiliary view may be unfolded. In the flattened configuration, the newly constructed parallel line which was the edge view of the auxiliary plane, is now the *reference line* between the front view and the auxiliary view of the telephone [3-48].

We have constructed here a *primary auxiliary view*—primary because the auxiliary plane is perpendicular to one of the principal planes of the orthographic box. In the example of the telephone, the primary auxiliary plane is perpendicular to the frontal plane and parallel to the dial plane.

Technique for Projecting a Primary Auxiliary View

As with the principal views, a primary auxiliary view can be projected if two adjacent views are given (or if sufficient data to describe the object are available.) The procedure is as follows:

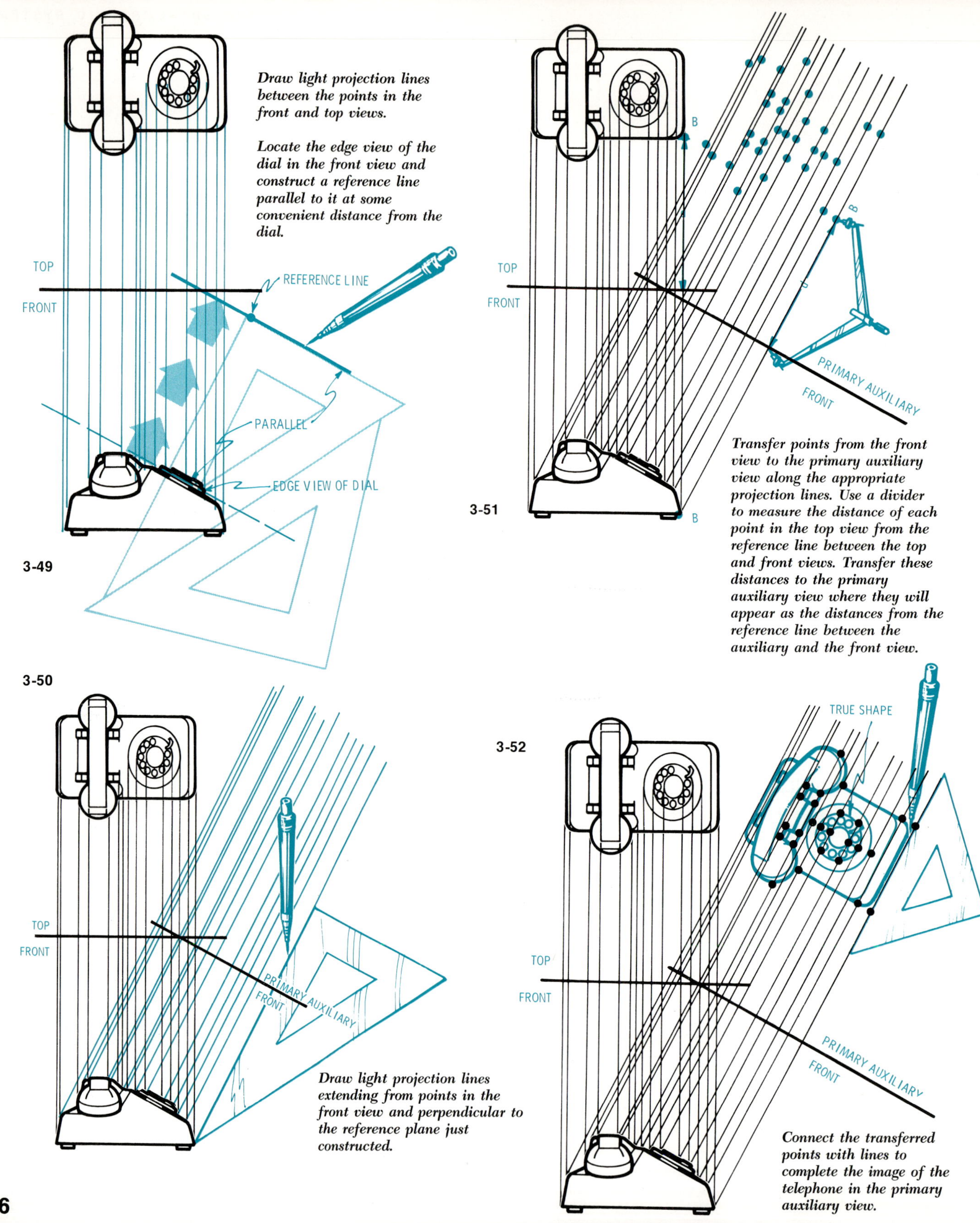

3-49

3-51

3-50

3-52

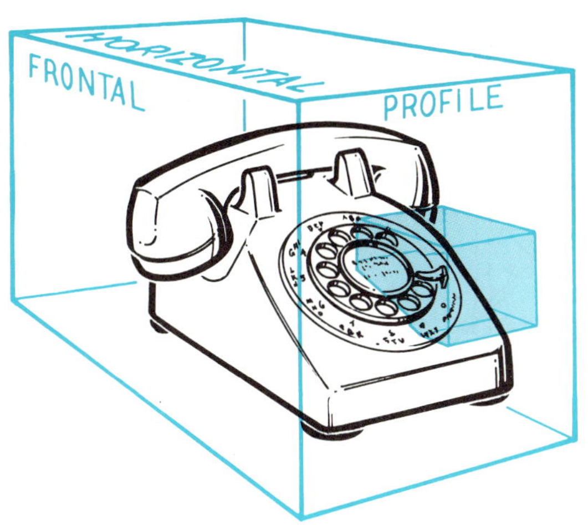

3-53
Reduce the size of the orthographic box to concentrate on the details of the dialing stop.

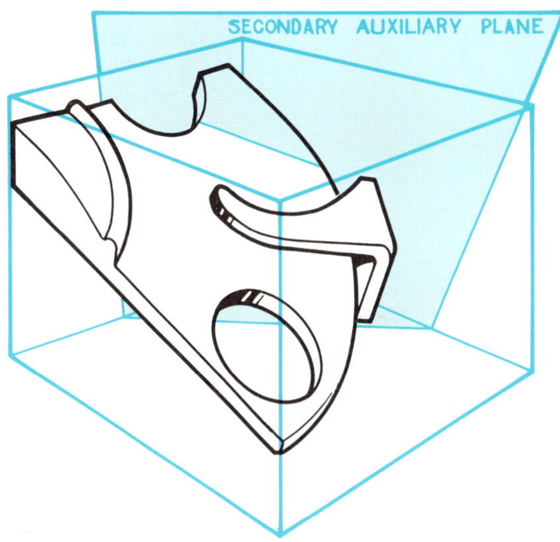

3-54
An enlarged view of the dialing stop showing placement of a secondary auxiliary plane perpendicular to the two planes of the dial stop.

Successive Auxiliary Views

Although it is seldom necessary to use more than principal views and primary auxiliary views, occasionally there is a need for successive auxiliary views to analyze the geometric relationships between specific features of an object. The *secondary auxiliary view* is the image that is projected to the *secondary auxiliary plane* from a primary auxiliary view. The secondary auxiliary plane is adjacent and perpendicular to a primary auxiliary plane; that is, it has the same relation to a primary auxiliary view that the primary had to the principal view.

Like primary auxiliary views, there is an infinitely large number of possible secondary auxiliary views. Furthermore, there are additional auxiliary views that may be constructed in sequence from the secondary auxiliary views. The principles and techniques for the development of these additional auxiliary views are similar to those for construction of the secondary auxiliary views.

As an example of the successive-view concept, let us again refer to the telephone; however, this time let us examine only one of the smaller parts—the dialing stop [3-53]. To see this part effectively, it will be necessary to enlarge our view of the part by considering a smaller orthographic box surrounding the immediate area of the dialing stop [3-54]. Now unfold the box and observe the three principal views of the dialing stop [3-55]. None of the principal views shows any of the meaningful dimensions of the part because the dialing stop is oriented at an angle to all the principal planes. Auxiliary views must therefore be constructed to obtain accurate information.

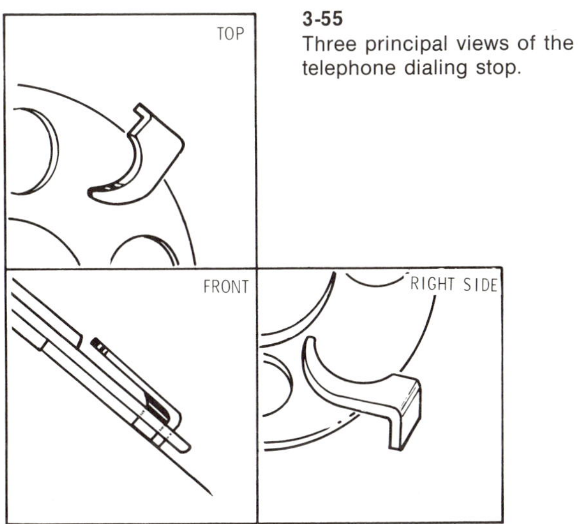

3-55
Three principal views of the telephone dialing stop.

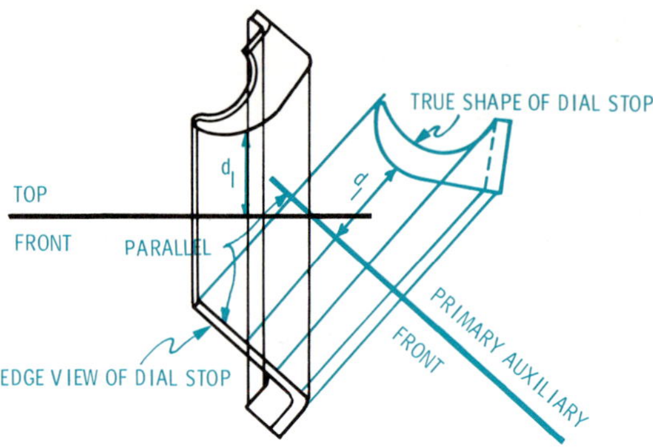

3-56
Construct a primary auxiliary from the front view to establish true shape of the dialing stop.

To establish the true shape of the finger recess in the dialing stop, a primary auxiliary view must be constructed parallel to the edge view of the dialing stop that is seen in the front view [3-56]. This primary view shows the true shape of the finger recess, but it does not show the true angle between the planes of the dialing stop. This information is obtained from a secondary auxiliary view perpendicular to and projected from the primary auxiliary view just constructed. This secondary auxiliary must be perpendicular to the line of intersection of the two planes of the stop. Our line of sight in the secondary auxiliary will then produce an image showing the edge views of the two planes and the true angle between them [3-57]. Note that in transferring the points from the primary auxiliary view to the secondary, the required distances are transferred from the front view, because the secondary auxiliary view and the front view are related views with respect to the primary auxiliary view to which they both are adjacent. The principles of projecting true angles are discussed more fully in Chapter 4.

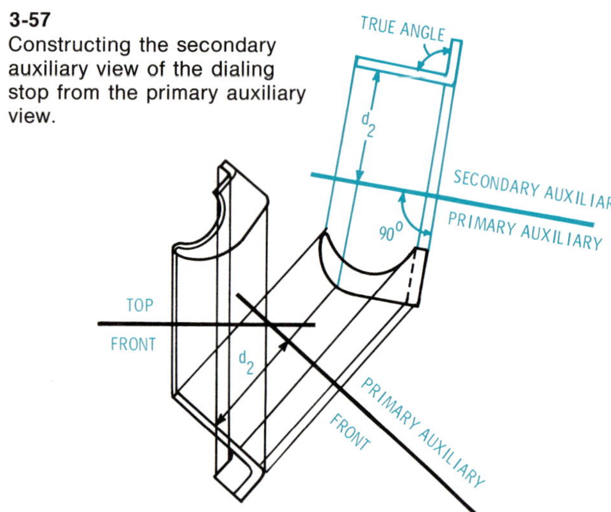

3-57
Constructing the secondary auxiliary view of the dialing stop from the primary auxiliary view.

problems

3-1. Using the objects found on page 65, sketch the outline front view of five of the objects as assigned. Select the front view in accordance with the principles of orientation of an object within an orthographic box. Estimate the units of each dimension of the view.

3-2. Using a scale of $\frac{1}{8}$ inch = 1 inch, draw the six principal views in the standard arrangement of two of the following objects. Take the measurements necessary to produce the drawing directly from the object, and use the necessary tools to produce an accurate drawing. *Draw only the object lines in each view.*

(**a**) door knob (one side of the door)
(**b**) wall-mounted pencil sharpener

(**c**) giant bow compass
(**d**) *Introduction to Engineering Design and Graphics* textbook
(**e**) an object whose maximum dimension does not exceed 9 inches

3-3. A series of points can describe the form of an object. Select the front view of three of the objects listed below. Describe them by placing a point at all the important intersections and other important features that are *visible.* Connect the points together using the appropriate guiding and marking tools to complete the form.
(**a**) straight-back chair
(**b**) eyeglasses
(**c**) portable television set
(**d**) pocket knife
(**e**) shoe for right foot
(**f**) ink bottle

3-4. Find a window that is easily accessible and has a view of a house, water tower, car, or other stationary object. Recalling the principles of orthographic projection, draw the object lines of the selected object on the window glass with a marking tool such as a grease pencil, whose mark is easily removable.

3-5. List five objects that you think have no obvious base, such as a door knob. Briefly explain the rationale that would be used to orient each object within an orthographic box. Quickly make a proportional drawing of the front view. It is not necessary to draw to exact scale.

3-6. From the following list of objects, name the view that best illustrates the dimension of the object that is in parentheses. Draw that view only in the scale indicated. *Show only the object lines.*
(**a**) scissors (length; full scale)
(**b**) paper clip (height; 3 times full scale)
(**c**) desk chair (width; $\frac{1}{10}$ scale)
(**d**) wall-mounted doorstop (length; full scale)
(**e**) ball-point pen (height; full scale)

3-7. Using the six principal views of the objects in Problem 3-2, identify the views that are adjacent to the right side view of the first object by placing a small "a" in the lower left hand corner of the appropriate views. In the same way identify the related views by using a small "r" in the lower right corner. Using the same procedure with the second object, identify the adjacent and related views to the *top* view.

3-8. Visualize a transparent cubical orthographic box, 2 inches on a side. Unfold the box into the 6 principal views in the standard arrangement. Locate and mark in all 6 principal views the 3 points whose positions within the box are described below. Use the symbol indicated in the parentheses.
(**a**) $1\frac{1}{2}$ inches to the right of the left side view, $\frac{3}{4}$ inch forward of the rear view, and $\frac{7}{8}$ inch below the top view (⊙).
(**b**) $\frac{1}{4}$ inch to the left of the right side view, $\frac{5}{8}$ inch above the bottom view, and $1\frac{3}{8}$ inches behind the front view (*).
(**c**) 1 inch to the left of the right side view, $\frac{3}{8}$ inch behind the front view, and $1\frac{1}{4}$ inches below the top view (⊡).

Using the techniques for projecting principal views, complete the appropriate third principal view in the following problems. Measure the various dimensions from the objects in [3-58 to 3-61] and transfer them to your paper at double size.

3-9. Object in [3-58].
3-10. Object in [3-59].
3-11. Object in [3-60].
3-12. Object in [3-61].

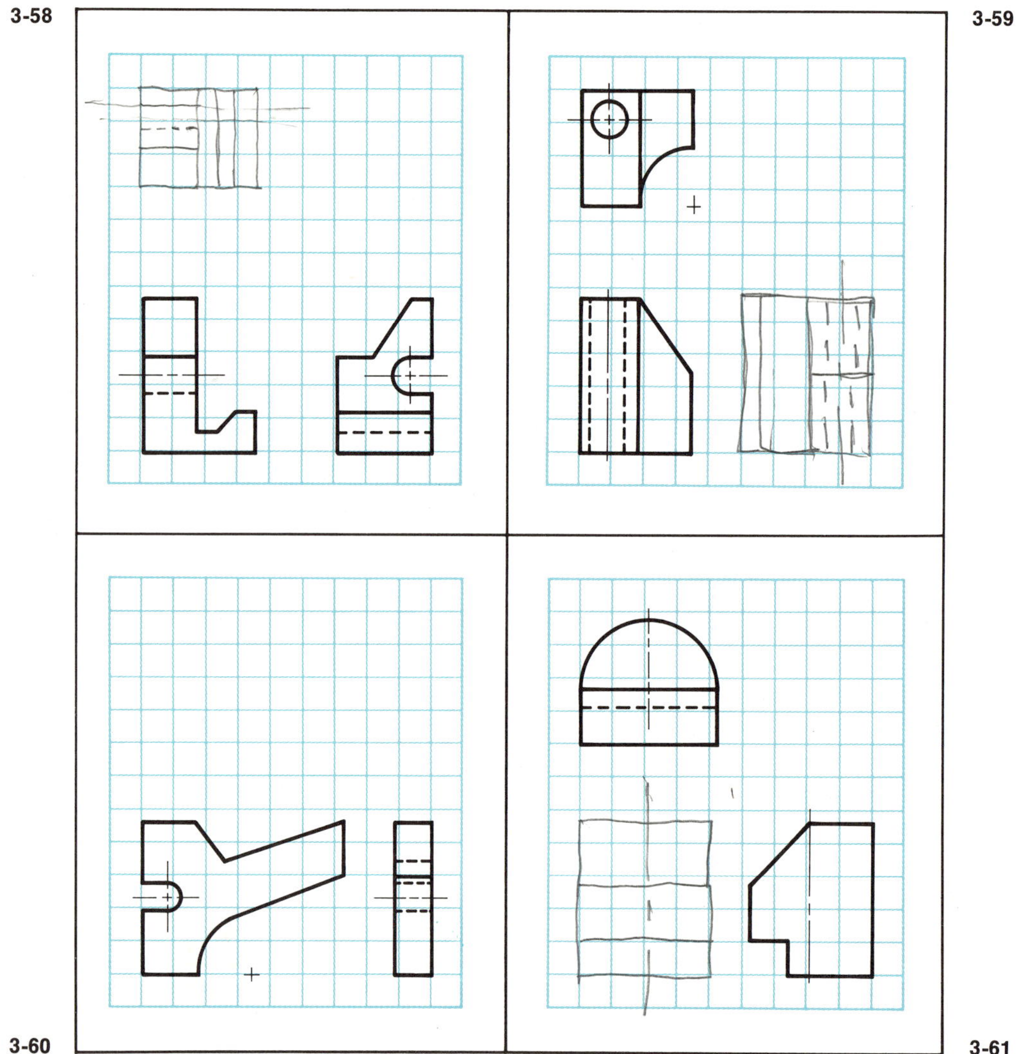

3-13. In [3-62] the circles indicate the location of missing views. For each problem select a correct view from the 30 choices shown at the bottom of the page and place its number in the appropriate circle.

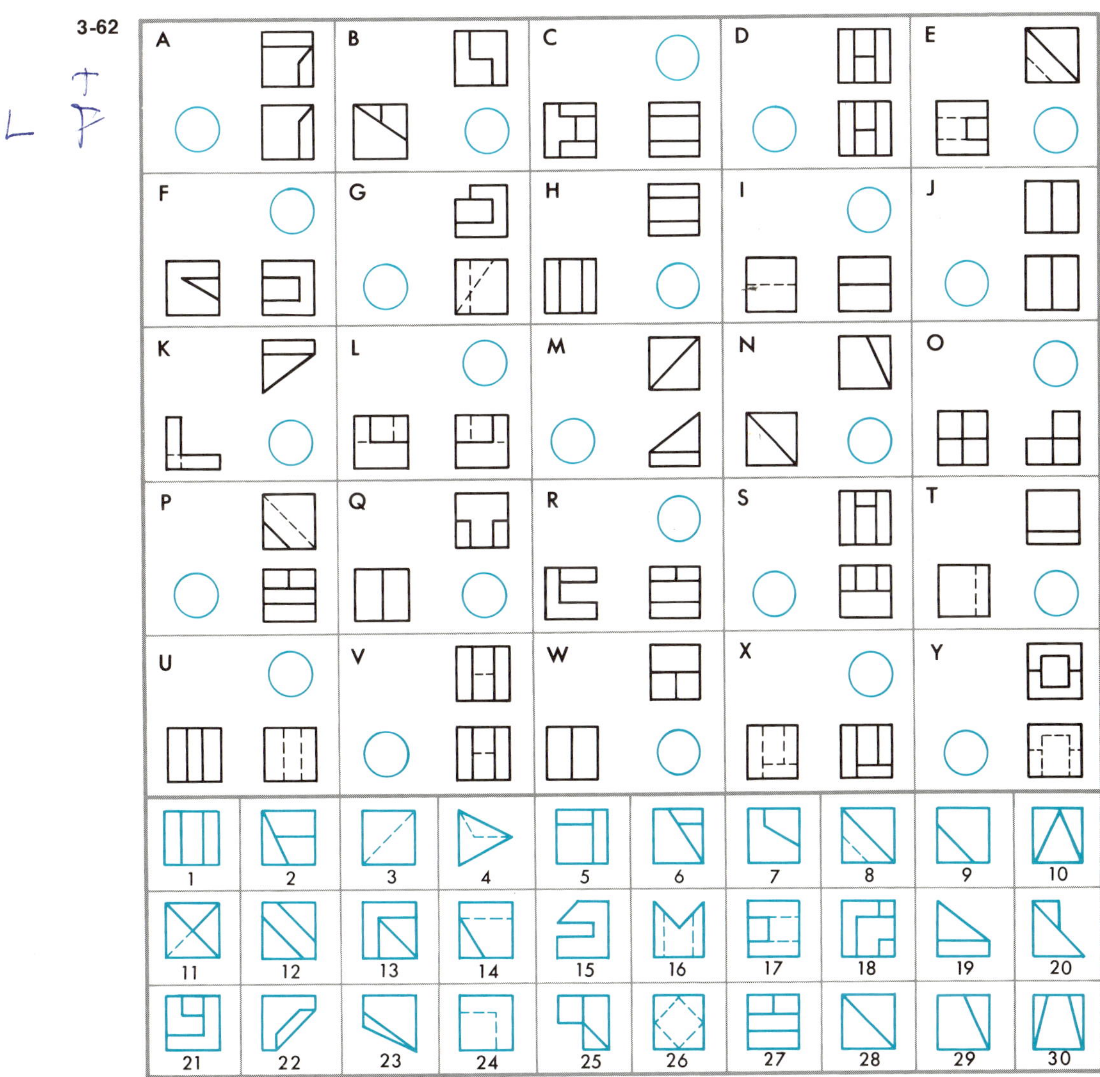

3-14. In [3-63] the circles indicate the location of missing views. For each problem select a correct view from the 30 choices shown at the bottom of the page and place its number in the appropriate circle.

3-63

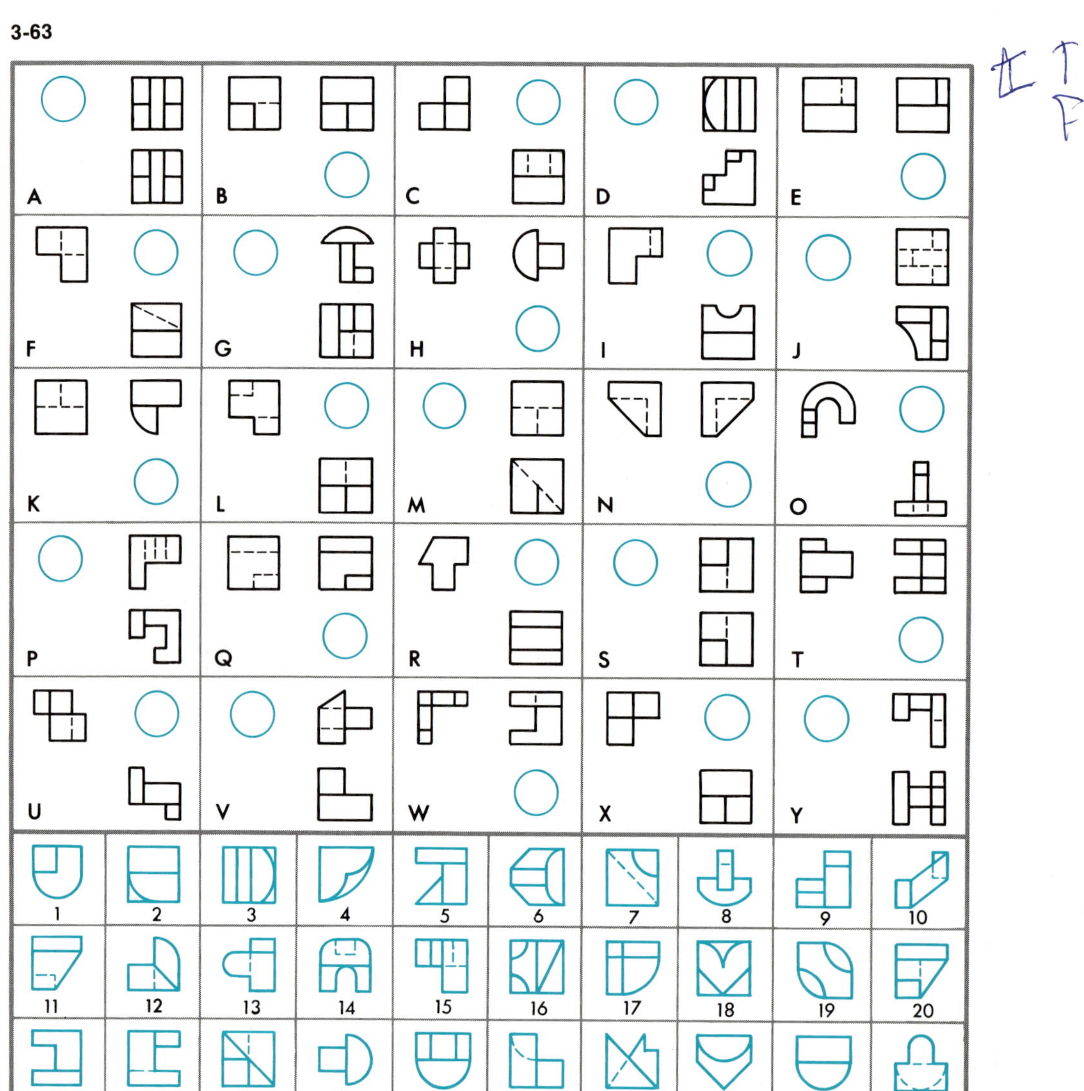

In the following problems, referring to [3-64 to 3-67] and using the designated isometric view, draw the three principal views. Measure the dimensions directly from the isometric view and construct the principal views in true scale.

3-15. Object in [3-64].
3-16. Object in [3-65].
3-17. Object in [3-66].
3-18. Object in [3-67].

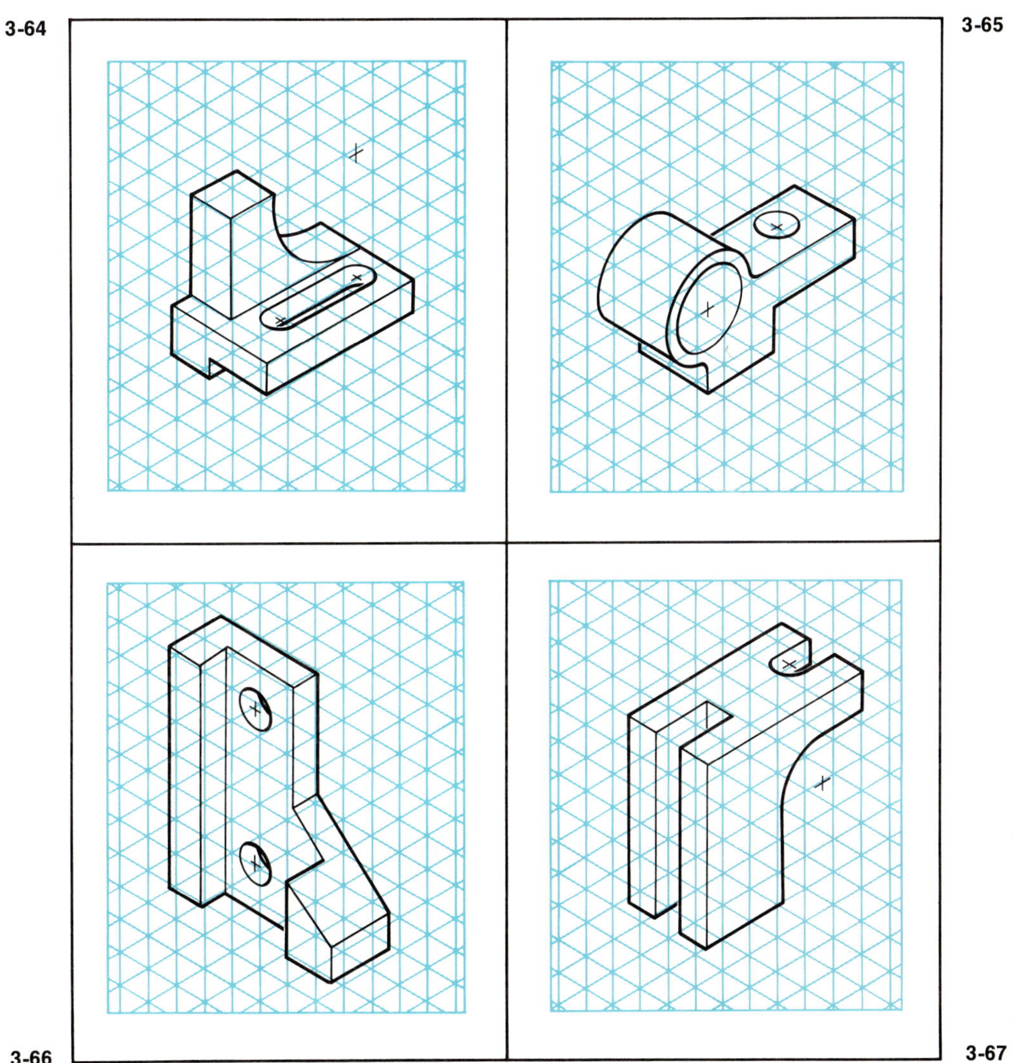

3-19. List five familiar objects that require an auxiliary view to show the true shape of the planes of the object. Make an orthographic drawing showing the necessary auxiliary view for one of the objects on this list.

In the following problems, using the designated isometric view, draw the three major orthographic views, and include any auxiliary views, to show the true shape of the surfaces labeled "A" in [3-68 to 3-71]. Project only the partial auxiliary view of the plane.

3-20. Object in [3-68].
3-21. Object in [3-69].
3-22. Object in [3-70].
3-23. Object in [3-71].

Construct the auxiliary view showing the details of the "A" surfaces.

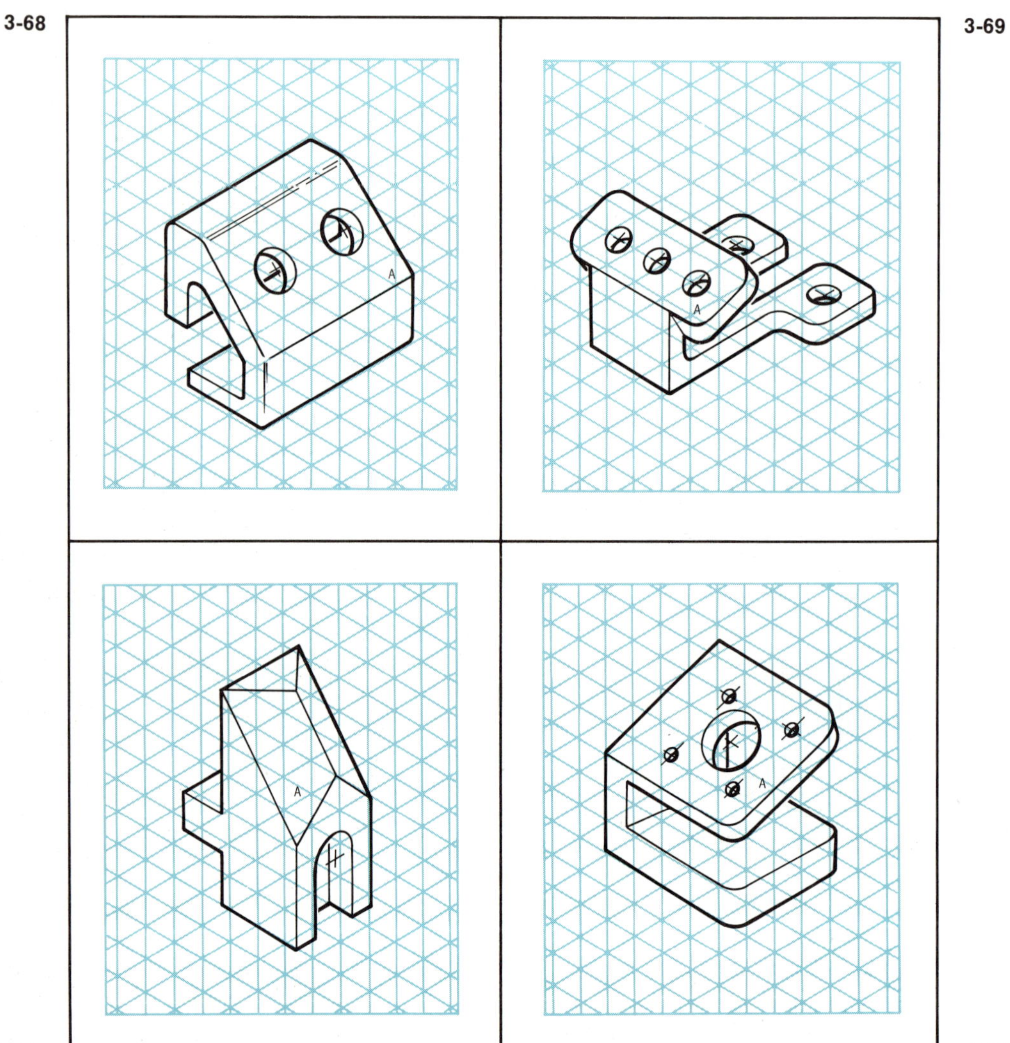

spatial relationships of geometric elements 4

Rolling friction on copper crystals.

engineering drawings often are thought of simply as pictures to be used in the cost estimating, production planning, and fabricating of an object. Although these are important uses, drawings are also used for engineering analysis.

The engineer often must determine if the distances between parts, the angles between planes, or the shapes of surfaces are really the way he wants them. It is frequently impossible, or at least economically undesirable, to build a scale model. Graphic means may be his only practical alternative. Imagine the difficulties an engineer would have determining the locations for the rivets of a bridge truss if he could not graphically show the true angles of the truss [4-1]. The problems of specifying the lengths of pipes and the location of their intersections in a chemical processing plant would be considerable if the engineer could not quickly and easily find the distances, lengths, and intersections by using graphical techniques [4-2].

4-1
A bridge truss requires a graphical analysis to determine the true angles.

To assist the engineer in the analysis of physical forms, there are methods in engineering drawing that will provide information about complex forms. These methods are known as *descriptive geometry*. Often they are the only way to solve very complex physical relationships in a direct and efficient manner.

the four fundamental concepts (views)

Four concepts are essential to the graphic analysis of physical objects. These are known as the *true length* of a line, the *point view* of a line, the *edge view* of a plane, and the *true size* of a plane.

Let us examine the two views associated with a line—the *true length* of a line and the *point view* of a line. By definition a line is the shortest distance between two points in space. This definition describes the length of a given line, but it tells the observer nothing about how the line looks to him. The visual image received is directly related to the angle between the observer's line of sight and the position of the line in space [4-3].

A *true-length* view is the most important view of a line. To see true length of a line, the observer must be located so that his line of sight intercepts the line at right angles—perpendicularly. As a practical example, hold your pencil at arm's length in front of you [4-4]. Orient the pencil so that your line of sight is perpendicular to

4-2
A chemical processing plant.

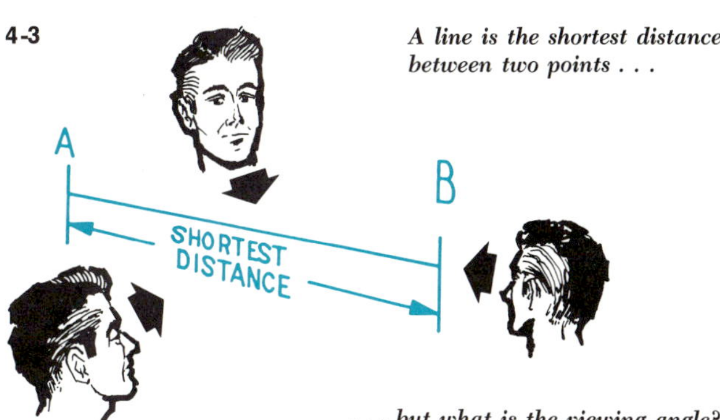

4-3
A line is the shortest distance between two points . . .

. . . but what is the viewing angle?

4-4
A true-length view of a line.

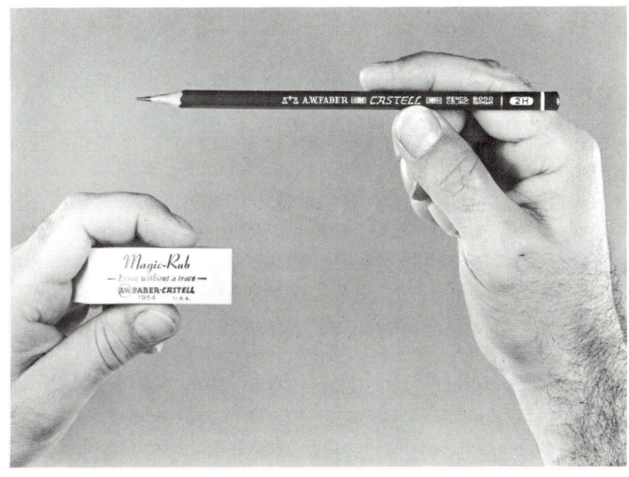

it [4-5]. The view that is presented to your eye is a true-length view of the pencil and tells you how long the pencil is. There are many true-length views possible. With a pencil, some true-length view shows the brand name and its grade. Other true-length views will not. This is so because there are alternative perpendicular positions about a line (or pencil).

Recall that a reference plane is perpendicular with the line of sight just as is the true-length view of the line. Therefore the two must be parallel [4-6].

4-5
Hold a pencil so that your line of sight is perpendicular to the long axis.

4-6
The true-length view is parallel to the reference plane because both the plane and the view are perpendicular to the line of sight.

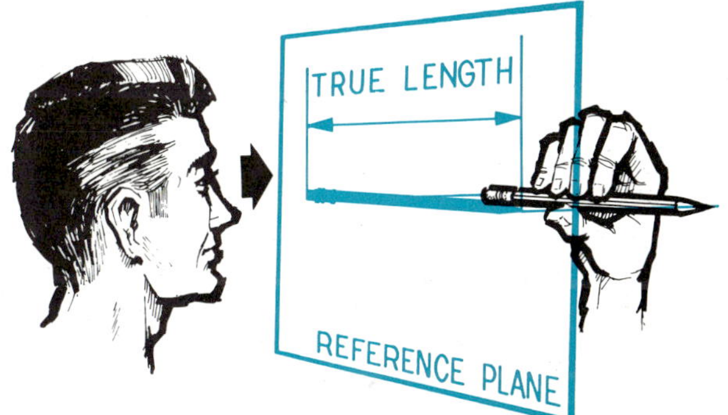

4-7

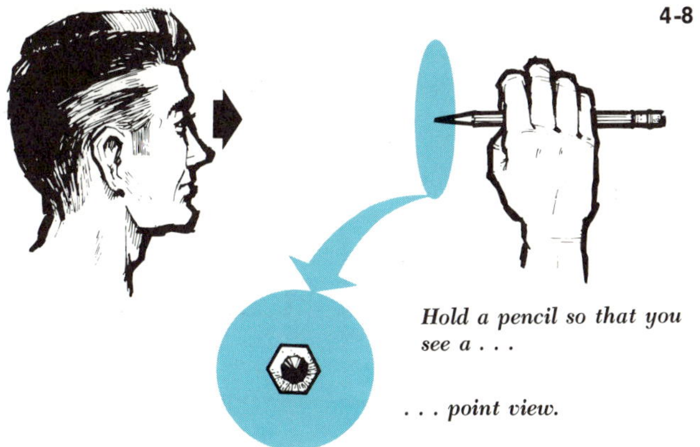

4-8

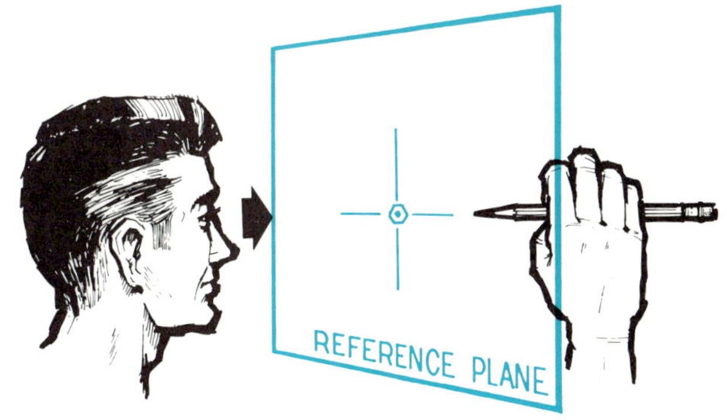

4-9
The reference plane is perpendicular to the line of sight.

To obtain the *point view* of a line, the observer should be positioned so that his line of sight is parallel to the line [4-7]. The line will then appear to be only a point. To illustrate the point view of a line in a practical way, hold your pencil at arm's length from your eye. Close one of your eyes and, using the other to sight, orient the pencil so that all that is visible to your eye is the graphite point surrounded by the wood case [4-8].

Since the reference plane is always perpendicular to the line of sight, a line viewed as a point is always perpendicular to the reference plane [4-9].

The remaining two relationships deal with views of a plane—the *edge view* of a plane and the *true-size* view of a plane. The edge view of a plane is seen when the line of sight is parallel to the surface of the plane. Using a drawing triangle as an example, hold it so that the line of sight from your eye is parallel with the surface of the triangle. All that will be visible is a line as wide as the thickness of the triangle [4-10]. This is the edge view of the plane of the triangle.

Projected onto the conventional reference plane, the edge view of the triangle would appear as a line [4-11]. Since the line of sight and the plane surface of the triangle are parallel, the plane (of the triangle) is also perpendicular to the reference plane.

The edge view of a plane is similar to the point view of a line in that the edge view can be considered as an infinite number of point views [4-12].

The final concept is the *true-size* view of a plane. As with the true-length view of a line, the true-size view of a plane is oriented parallel with the reference plane [4-13]. The observer's line of sight is perpendicular to the plane being viewed and is perpendicular, by defini-

4-10
Hold a triangle so that the line of sight is parallel to the triangle. This will produce an edge view of a plane.

4-11
A plane seen as an edge view is perpendicular to the reference plane.

4-12
The edge view of a plane and the point view of a line in the plane are both parallel with the line of sight; therefore, both are perpendicular to the reference plane.

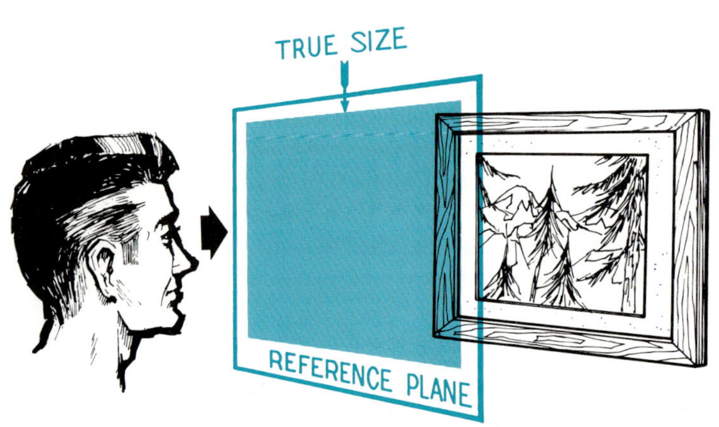

4-13
The true size view of a plane is oriented parallel with the reference plane; both are perpendicular to the line of sight.

tion, to its reference plane. The similarity between a true-size view of a plane and a true-length line is further emphasized by the fact that all the lines that compose the plane are seen in true length in the true-size view of the plane.

The four fundamental views just discussed form the basis for the analytical capabilities of engineering graphics.

notation systems

The four fundamental views are all related in either a parallel or perpendicular manner with a reference plane. The reference plane used may be one of the three principal reference planes—horizontal, frontal, or profile—that were described in Chapter 3. Or, more often, the plane may be a primary or secondary auxiliary plane, such as those that were developed in the latter part of the same chapter.

Graphic Notation of Reference Planes

A system of planar notation has been developed to tell the user quickly which position or plane he is observing. This system uses letters and numerals related to the reference line common to the adjacent reference planes. The three basic reference planes are denoted by the first letter of their name—capital *H* for the horizontal reference plane, *F* for the frontal, and *P* for the profile [4-14]. Reference planes associated with successive auxiliary views are numbered in the order in which they are constructed. For example, the first auxiliary plane constructed is number 1, the second auxiliary plane is 2, and so on [4-15].

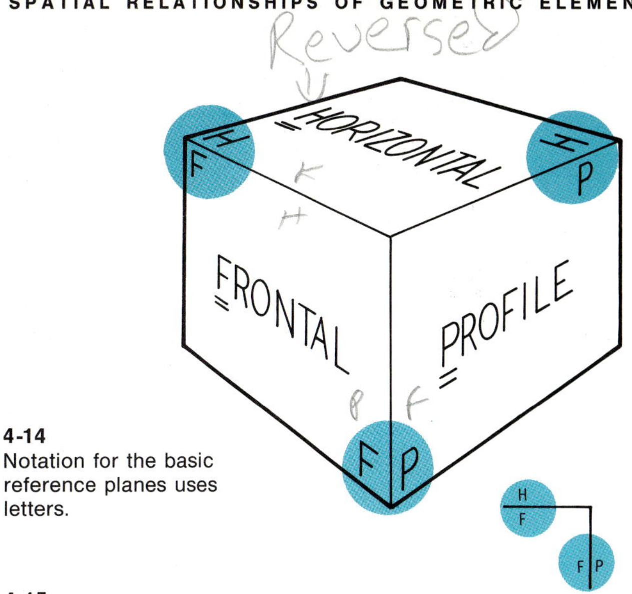

4-14
Notation for the basic reference planes uses letters.

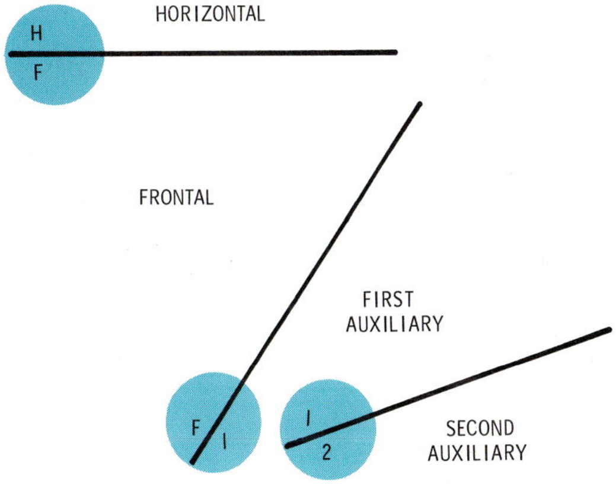

4-15
Notation for the auxiliary planes uses numbers.

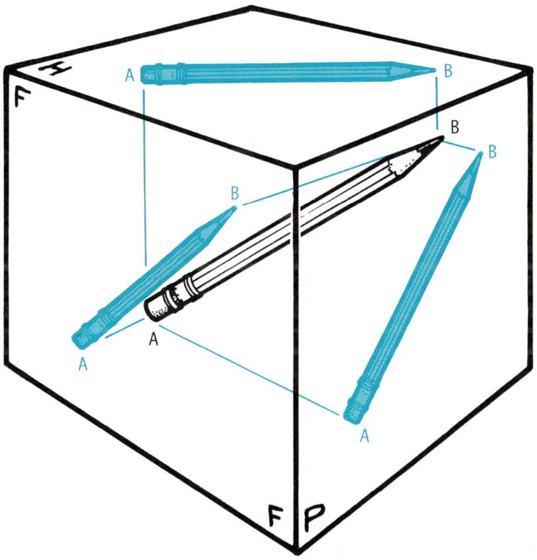

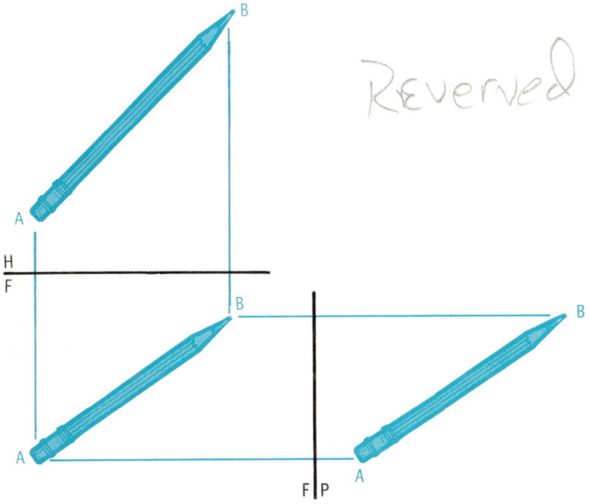

Graphic Notation of Elements

It is also desirable to label the lines and planes of the object being viewed. For the purposes of this text, single points in space, end points of a line, and boundary points of planes will be identified with different capital letters [4-16].

4-16
Points that define lines and planes are noted with letters.

4-17
A line appears as true length in a view that is parallel to the line.

techniques for constructing fundamental views

True-Length View of a Line

A line appears as true length (TL) in a view that is parallel to the line.

In the first case, if a line is seen in a principal view as being parallel to a principal reference plane, then the adjacent view will show the true length of the line [4-18 and 4-19].

In the second, more general case, the line is seen as an oblique or foreshortened line in all the principal reference planes. In other words, the line is not parallel to any of the principal reference planes [4-20]. Consequently, an auxiliary view must be constructed to show

When a line is parallel to the reference line, it is seen in true length in its adjacent view.

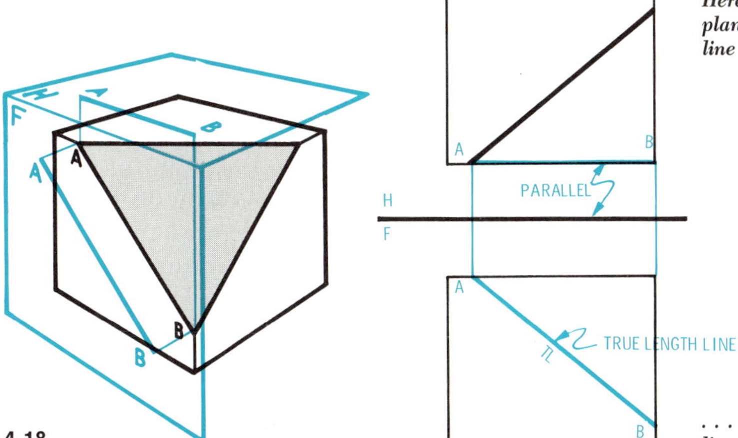

Here line AB is seen through the horizontal plane as being parallel to the H/F reference line

. . . therefore, AB appears as a true length line when viewed through the frontal plane.

4-18

Here line BC is seen through the frontal plane as being parallel to the F/P reference line; so . . .

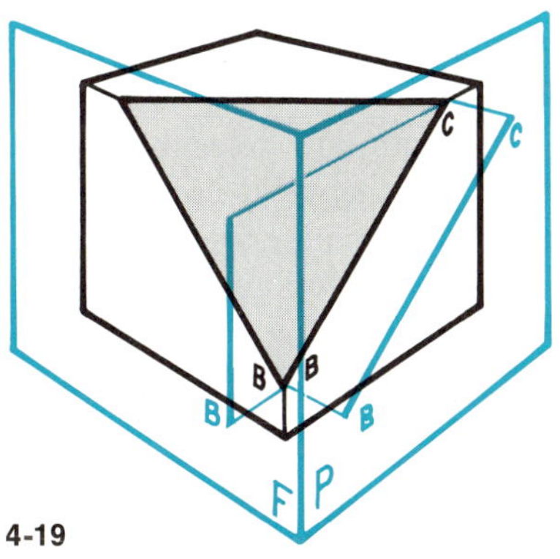

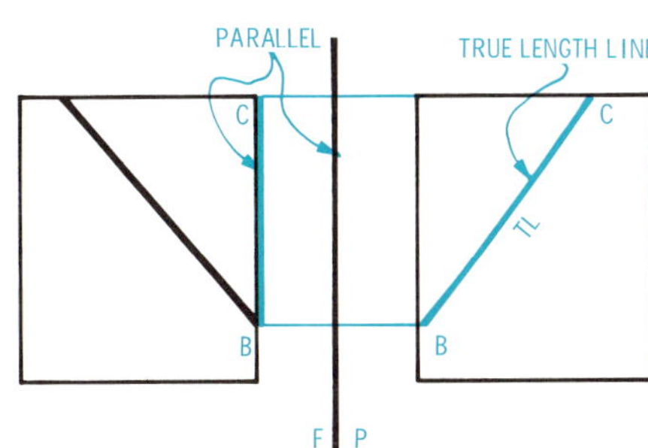

. . . it will be seen on the adjacent profile plane as a true length line.

4-19

An oblique line is not parallel to any of the principal planes.

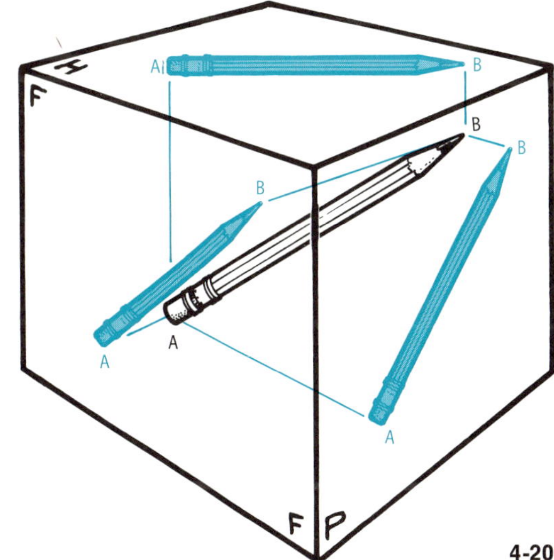

4-20

4-21

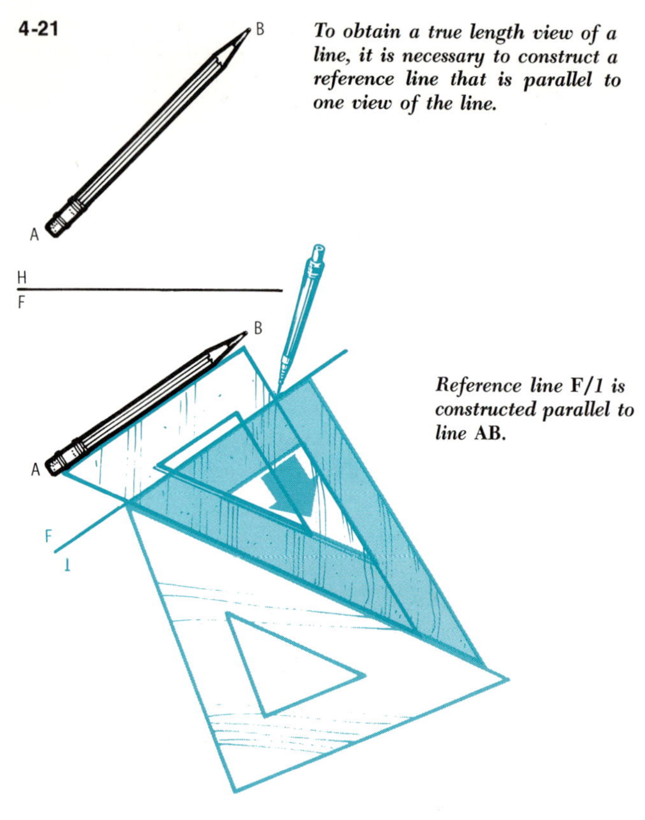

To obtain a true length view of a line, it is necessary to construct a reference line that is parallel to one view of the line.

Reference line **F/1** *is constructed parallel to line* **AB.**

the true length of the line. This auxiliary view must be parallel with the line (in this case the pencil). To achieve this relation, the auxiliary plane is constructed so that the mutual reference line is parallel to a view of the line on one of the reference planes [4-21]. The end points *A* and *B* of the line are projected across the reference line into the auxiliary view. In the related view, the distances d_A and d_B are measured from the reference line and transferred to the auxiliary view along the appropriate projection lines [4-22]. After the points have been transferred, they are connected and a true-length view of the line is drawn [4-23].

Even though this example used the horizontal and frontal planes to construct the true-length view, it is possible to use this technique with any pair of adjacent views that are available. Except in the case, in which one view of the line is already parallel to a reference plane, two adjacent views and one auxiliary view are necessary to show the true length of a line.

The true-length-line concept is basic to further developments in the analytical uses of engineering graphics. It is used in determining such relationships as the angles between lines, the slope of a line, and the construction of a point view of a line.

4-22

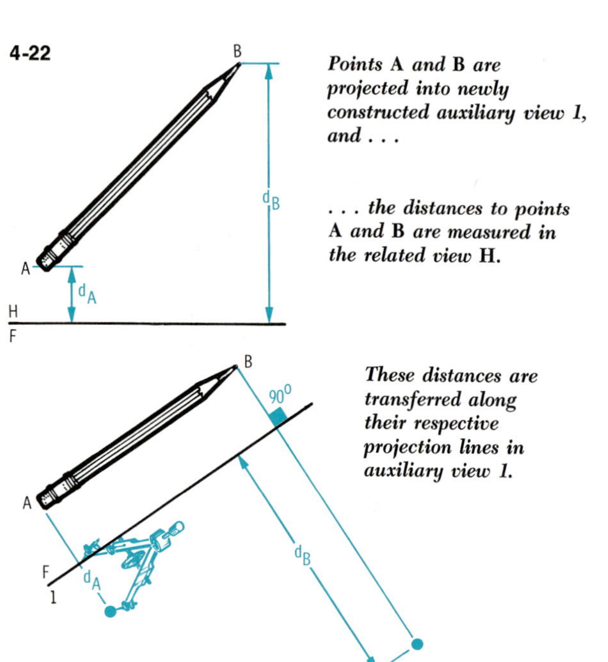

Points **A** *and* **B** *are projected into newly constructed auxiliary view 1, and . . .*

. . . the distances to points **A** *and* **B** *are measured in the related view* **H.**

These distances are transferred along their respective projection lines in auxiliary view 1.

4-23

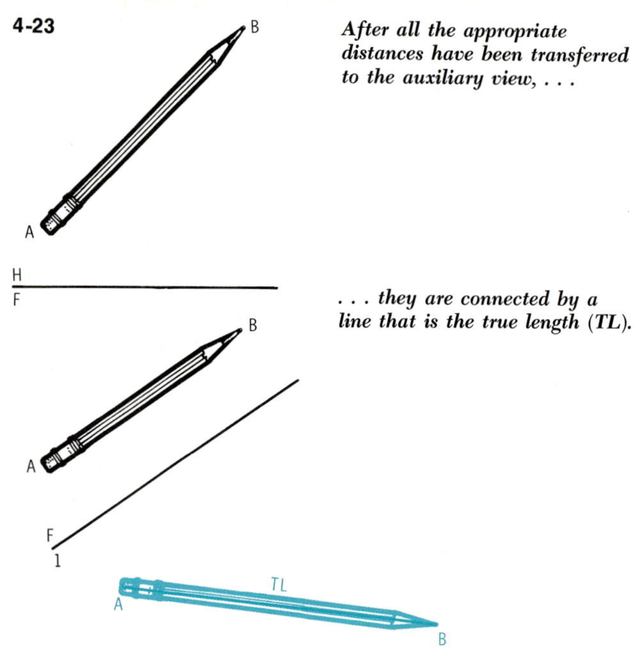

After all the appropriate distances have been transferred to the auxiliary view, . . .

. . . they are connected by a line that is the true length (TL).

4-24

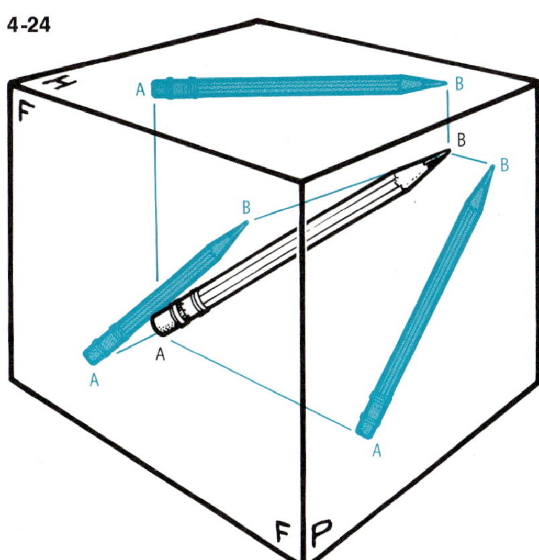

To develop the point view of the line **AB**, *it is necessary to construct an auxiliary view perpendicular to a true-length view of the line.*

The true length of line **AB** *is developed. A reference line 1/2 is constructed perpendicular to the axis of the true length line.*

A projection line is drawn from auxiliary view 1 into auxiliary view 2.

4-25

Point View of a Line

A *line appears as a point in a view that is perpendicular to a true-length view of the line*. Constructing a point view of a line is begun by constructing a true-length view of the line. Using the example of the pencil oriented obliquely with the principal reference planes [4-24], let us now construct a point view from the true-length line in auxiliary view 1 [4-23]. Since the point view is perpendicular to the true-length view, a reference line should be constructed perpendicular to the axis of the true-length line. A projection line is then constructed across this reference line into the newly constructed auxiliary view 2 [4-25]. The point view of the line will be located on this projected line. Its location is determined by measuring the distance *d* to the line (pencil) from the reference line in the related view—in this case the front view. Notice that this related view is the one which had the *F*/1 reference line constructed parallel to it thus making the end points of the line *AB* equidistant from the reference line. The distance is transferred to auxiliary view 2 and the desired point view plotted on the projection line [4-26].

The distance from the reference line **F/1** *to parallel line* **AB** *is measured, and . . .*

. . . is transferred into auxiliary view 2 on the projection line. The newly located point is the point view of line **AB**.

4-26

4-27
An edge view of a plane.

Edge View of a Plane

A plane appears as an edge view (EV) in a view that contains a point view of a line within that plane.

If it is desired to obtain an edge view of a plane *ABC* [4-28], which is obliquely oriented with respect to the three principal reference planes, it is first necessary to select an appropriate line in the plane. Since a true-length line is necessary to find the point view of the line, the line selected in the plane should be one that will appear true length in a principal view. In plane *ABC* establish a true-length line by drawing a line parallel with the reference line *H/F* in one of the views. In this example, construct such a line *CD* in the horizontal view [4-29].

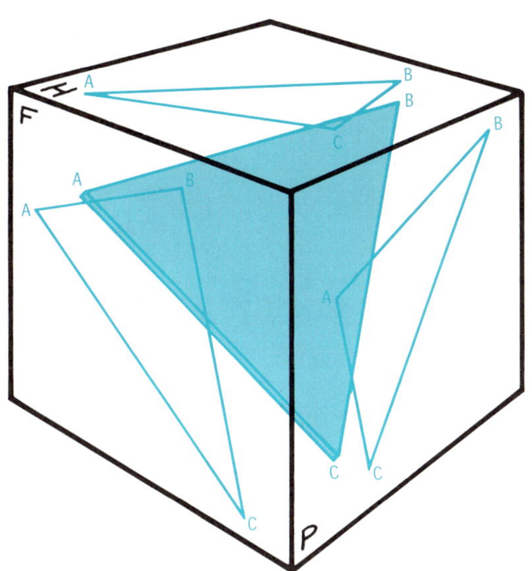

4-28
Construct the edge view of the plane *ABC*.

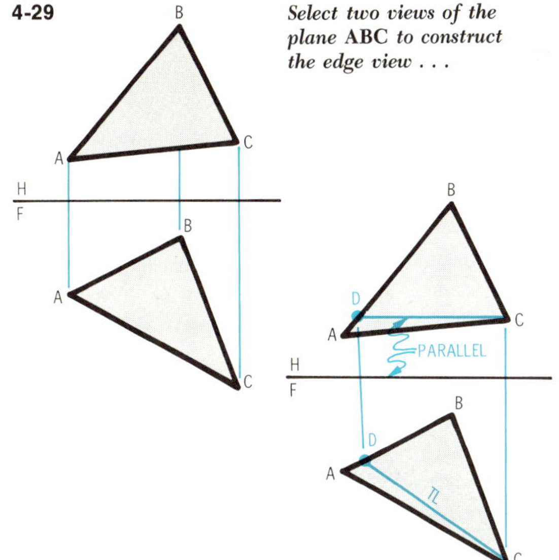

4-29

Select two views of the plane **ABC** *to construct the edge view . . .*

. . . then construct a line in the plane parallel with the reference line **H/F.** *Project this line to the frontal view to obtain a true length line.*

After the parallel line *CD* is constructed in the horizontal view, project point *D* to locate the true-length frontal view of the line. Because this line connects two points of the plane *ABC*, it is within the plane.

Now to get a point view of line *CD*, construct a reference line *F*/1 perpendicular to the true length line and project the points *C* and *D* across the reference line into auxiliary view 1, just as in [4-26]. Return to the horizontal view and measure the related distance *d* between the reference line and the parallel line. Transfer this distance along the projection line in auxiliary view 1 and establish the location of the point view of the line *CD* [4-30].

After the point view of the line *CD* within the plane has been located, project the remaining points *A* and *B* that define the plane into the auxiliary view. Measure the distances d_A and d_B to these points from the reference line of the related view and transfer them to the auxiliary view. Connect these transferred points in the auxiliary view to get the edge view (EV) of the plane *ABC* [4-31]. The points *ABCD* in the auxiliary view must fall in a straight line or else there is an error in the construction. Using the method just described, one auxiliary view is sufficient to obtain an edge view of a plane.

The edge view of a plane is required for the construction of the fourth fundamental view—the true size of a plane.

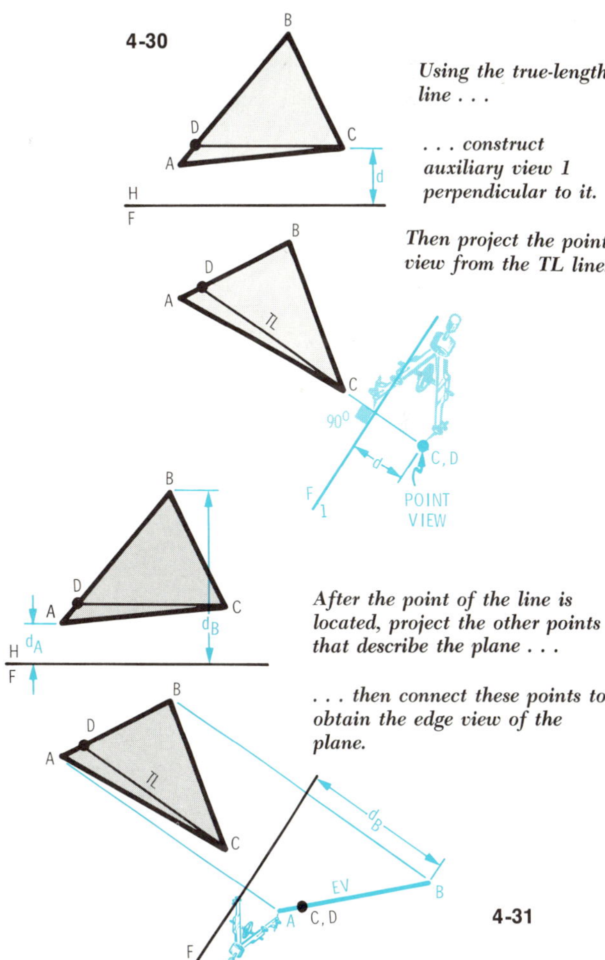

4-30

Using the true-length line . . .

. . . construct auxiliary view 1 perpendicular to it.

Then project the point view from the TL line.

After the point of the line is located, project the other points that describe the plane . . .

. . . then connect these points to obtain the edge view of the plane.

4-31

4-32

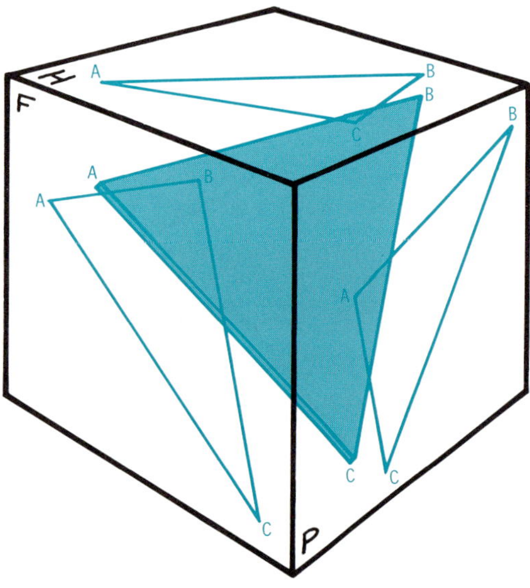

4-33
Construct a true-size view of the plane.

4-34

Construct an edge view (EV) of plane ABC . . .

True Size of a Plane

A plane appears as true size (TS) in a view that is parallel to the edge view of that plane. In our earlier discussion of the true size of a plane, it was stated that the line of sight must be perpendicular to the plane being viewed. Therefore, if an auxiliary view is constructed parallel with an edge view of a plane and the plane is projected into this auxiliary view, the plane must appear as true size.

Since the true size of the plane is usually developed from an edge view that has been found in a first auxiliary view, a second auxiliary view must be established. If it is desired to obtain a true size of plane *ABC* [4-33], an edge view must be constructed as before [4-34]. A new reference line 1/2 for a second auxiliary view is drawn parallel to the edge view of the plane [4-35]. The projection lines for the points *A*, *B*, and *C* are drawn across the reference line 1/2 and some distance into the second auxiliary view.

The distances d_A, d_B, and d_C in the frontal view from reference line *F*/1 to the points *A*, *B*, and *C* of the plane are then measured. These distances are plotted from the reference line 1/2 along their respective projection lines in the second auxiliary view. When lines are drawn connecting the transferred points, the resulting figure is the true size of the plane *ABC* [4-36].

The dependence of each view on the preceding one

4-35

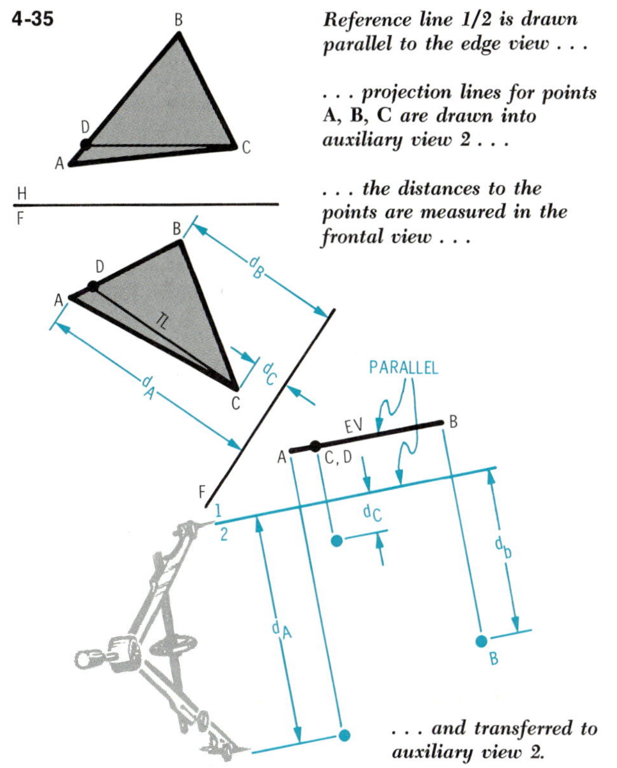

Reference line 1/2 is drawn parallel to the edge view . . .

. . . projection lines for points **A, B, C** *are drawn into auxiliary view 2 . . .*

. . . the distances to the points are measured in the frontal view . . .

. . . and transferred to auxiliary view 2.

4-36

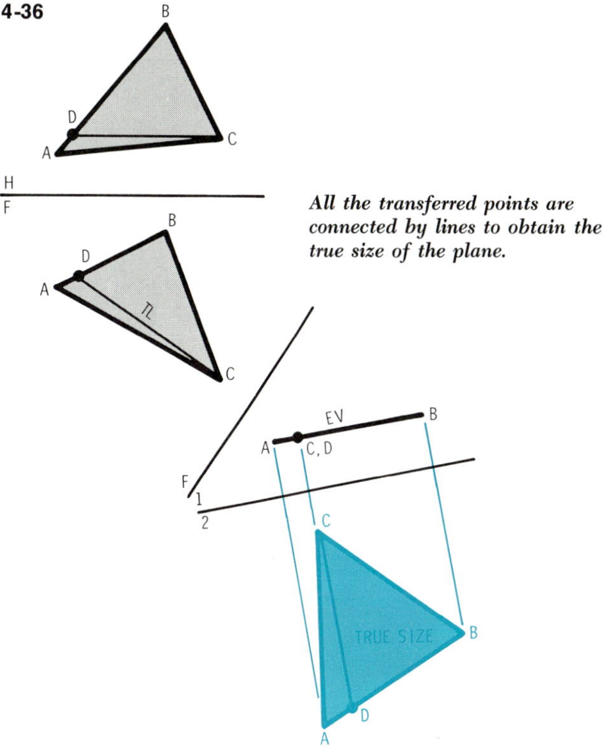

All the transferred points are connected by lines to obtain the true size of the plane.

is apparent. The true length of a line is basic and is used to develop the point view of the line. The point view of a line in a plane leads to an edge view of that plane. And finally, the edge view is used to construct the true size of a plane. The development of these four fundamentals is like the parts of a Chinese puzzle: one will not work alone. All the concepts depend on each other. They are also required to develop some of the concepts of graphical analysis that follow.

relationship of geometric elements

The four fundamental views previously discussed often may be applied directly to the solution of engineering problems. For example, the techniques for finding the true length of a line may be useful in determining the length of the control cables, tubing, and wiring in an airplane. The techniques for finding the true size of a plane may be useful in designing the various body and wing panels of the airplane.

4-37
The many intersections of this hanger construction may be analyzed using engineering graphics.

The fundamental views become especially valuable when used as tools to solve problems involving the relationships between geometric elements. The remainder of this chapter uses our knowledge of the four fundamental views to develop, first, the techniques required to analyze the angular relationships between elements, such as the structural members seen in the hanger roof [4-37], and, second, the distance relationships between the elements of objects, such as those of the airplane and the surrounding hangar structure.

Most analytical problems associated with engineering graphics can be simplified into relationships that exist between lines and lines, lines and planes, or planes and planes. Perhaps the most basic relationship between these geometric elements is that they are intersecting or nonintersecting.

Line-to-Line Relationships

Intersecting lines meet at a point. That point is mutually shared by each line [4-38]. If lines intersect, it is usually important that we be able to determine their angle of intersection. The special case in which the two lines are perpendicular to each other [4-39] is used to determine the distance between a point and a line, or between two lines. Unless stated otherwise, the *distance* between elements of objects is always considered to be the *shortest* distance—the perpendicular distance.

Although two lines may appear to intersect in one orthographic view, it is important to check an adjacent view to determine if the point of apparent intersection in one view is really the mutual point of intersection shared by both lines. If projection to an adjacent view shows the apparent intersection point to be really two points, then the lines are *nonintersecting* [4-40]. A spe-

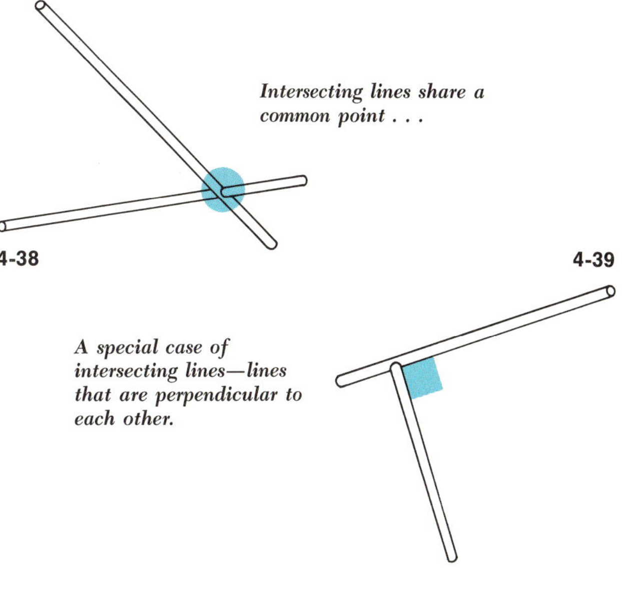

4-38

Intersecting lines share a common point . . .

4-39

A special case of intersecting lines—lines that are perpendicular to each other.

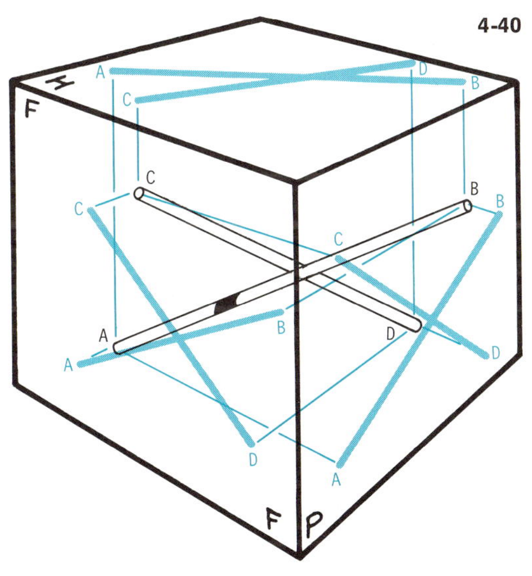

4-40

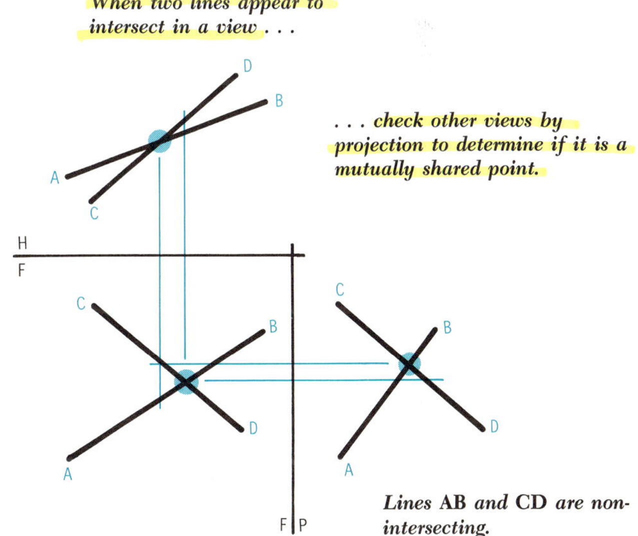

When two lines appear to intersect in a view . . .

. . . check other views by projection to determine if it is a mutually shared point.

Lines AB and CD are nonintersecting.

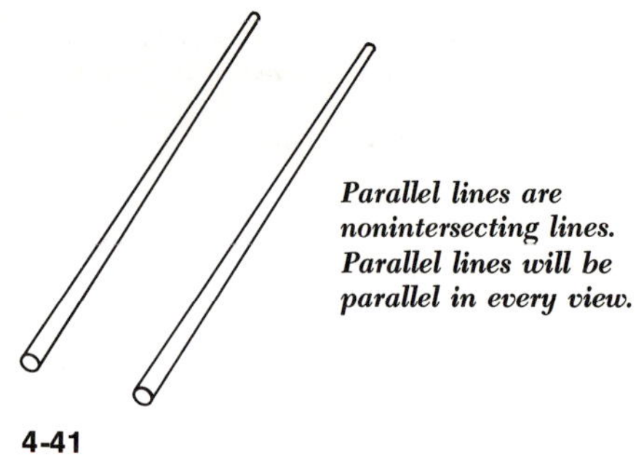

Parallel lines are nonintersecting lines. Parallel lines will be parallel in every view.

4-41

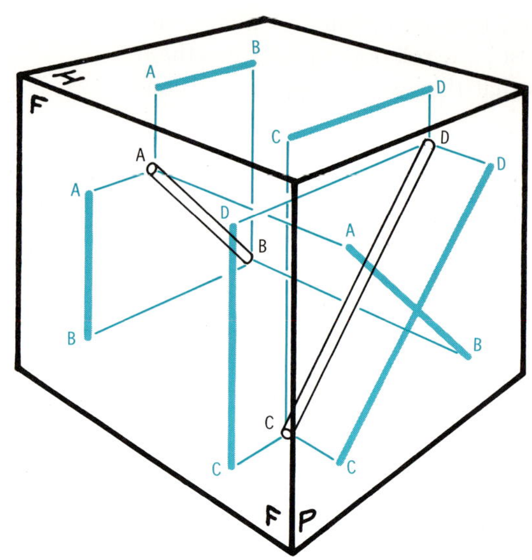

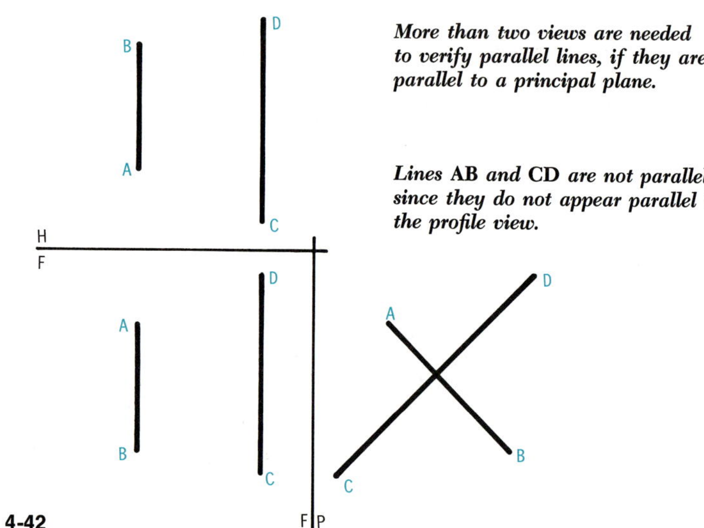

More than two views are needed to verify parallel lines, if they are parallel to a principal plane.

*Lines **AB** and **CD** are not parallel since they do not appear parallel in the profile view.*

4-42

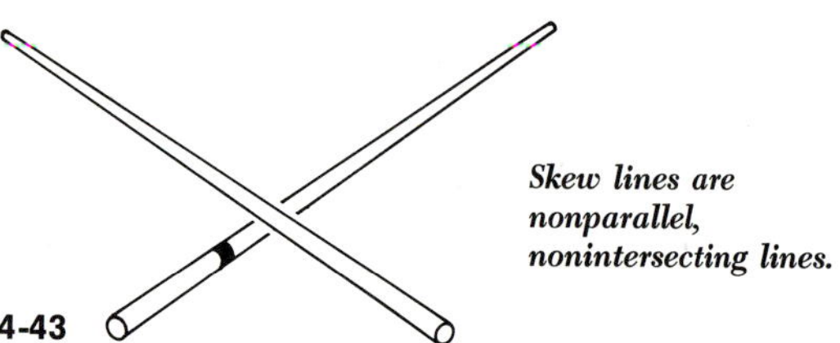

Skew lines are nonparallel, nonintersecting lines.

4-43

cial case of nonintersecting lines occurs when lines are parallel to one another [4-41]. *Parallel lines will appear to be parallel regardless of the viewing position of the observer.* Generally, only two orthographic views are required to verify parallelism. However, if the lines are parallel to a principal reference plane, a third view is usually required to verify parallelism [4-42]. *Skew* lines are lines that do not intersect and are not parallel to one another [4-43].

Line-to-Plane Relationships

Just as a line may intersect a line, a line may also intersect a plane. The angle between a line and a plane is often of interest to the engineer. The special case in which the line is perpendicular to the plane is useful for establishing and measuring the true distances between a point and a plane, between a line and a plane, or between two planes.

One way to verify if a line does intersect a plane is to obtain a view showing both the edge view of the plane and the line [4-44]. This view will show whether the line intersects the plane or is parallel to it. The techniques for locating such an intersection will be described in Chapter 5.

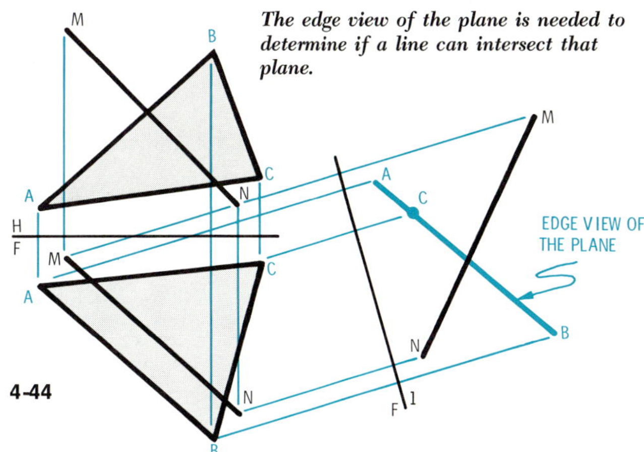

4-44 *The edge view of the plane is needed to determine if a line can intersect that plane.*

Plane-to-Plane Relationships

The final analytical relationship of interest is that of one plane to another plane. Just as lines may intersect each other, so may planes. However, instead of having a point of intersection, intersecting planes have a line of intersection [4-45]. The most common angle of intersection formed is 90°, but other angles are of equal importance to the engineer. Planes may also be parallel to each other. Often these conditions of intersection or parallelism may be verified by finding the edge views of the planes.

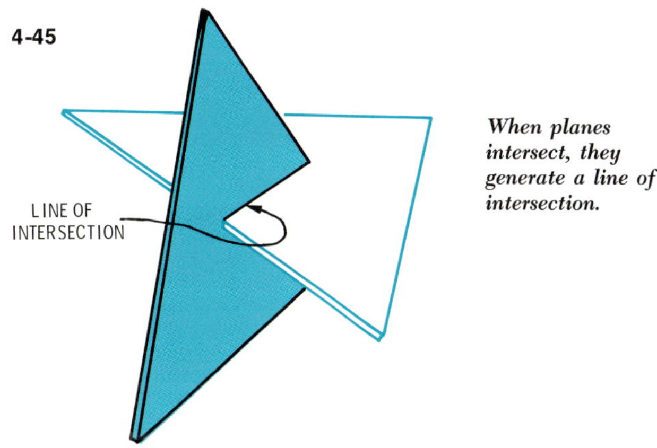

4-45 *When planes intersect, they generate a line of intersection.*

true angles

One general area of geometric analysis of great interest to the engineer is the determination of *true angles* at the intersection of one geometric element (such as a line or a plane) with another [4-46].

True Angle Between a Line and a Line

The true angle between two lines appears when both lines are true length in the same view. This applies to the true angles between both intersecting and nonintersecting lines.

In the case of intersecting lines, both can be seen as true length in the view that shows the true size of the plane formed by these lines. Therefore, the method necessary to establish the true angle between two intersecting lines is simply that of finding the true size of a plane.

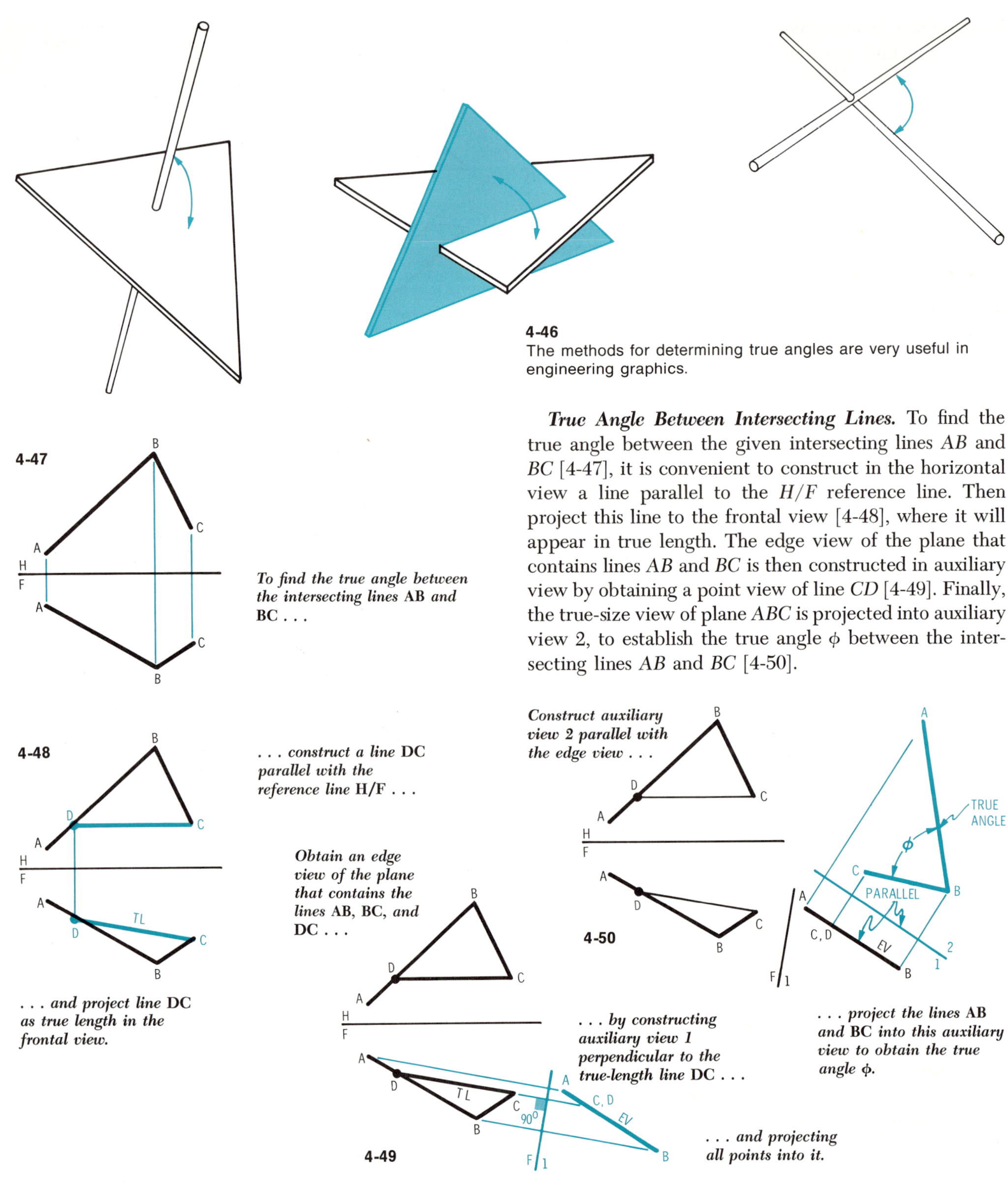

4-46
The methods for determining true angles are very useful in engineering graphics.

True Angle Between Intersecting Lines. To find the true angle between the given intersecting lines *AB* and *BC* [4-47], it is convenient to construct in the horizontal view a line parallel to the *H/F* reference line. Then project this line to the frontal view [4-48], where it will appear in true length. The edge view of the plane that contains lines *AB* and *BC* is then constructed in auxiliary view by obtaining a point view of line *CD* [4-49]. Finally, the true-size view of plane *ABC* is projected into auxiliary view 2, to establish the true angle ϕ between the intersecting lines *AB* and *BC* [4-50].

4-47
To find the true angle between the intersecting lines **AB** *and* **BC** *. . .*

4-48
. . . construct a line **DC** *parallel with the reference line* **H/F** *. . .*

. . . and project line **DC** *as true length in the frontal view.*

4-49
Obtain an edge view of the plane that contains the lines **AB, BC,** *and* **DC** *. . .*

. . . by constructing auxiliary view 1 perpendicular to the true-length line **DC** *. . .*

. . . and projecting all points into it.

4-50
Construct auxiliary view 2 parallel with the edge view . . .

. . . project the lines **AB** *and* **BC** *into this auxiliary view to obtain the true angle* ϕ*.*

4-51

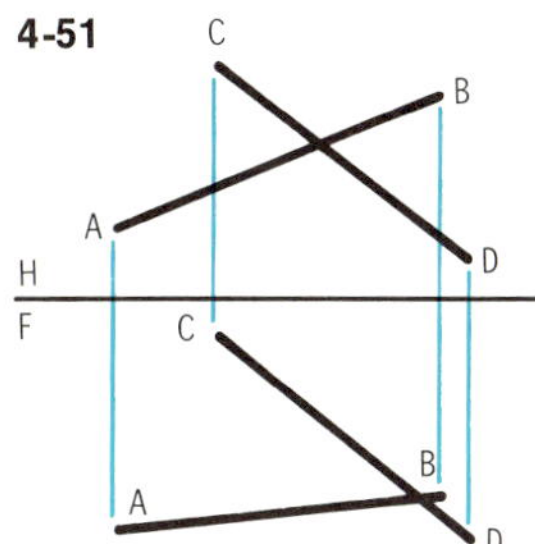

To find the true angle between the skew lines **AB** *and* **CD**, *it is necessary to see both lines as true length in the same view.*

4-52

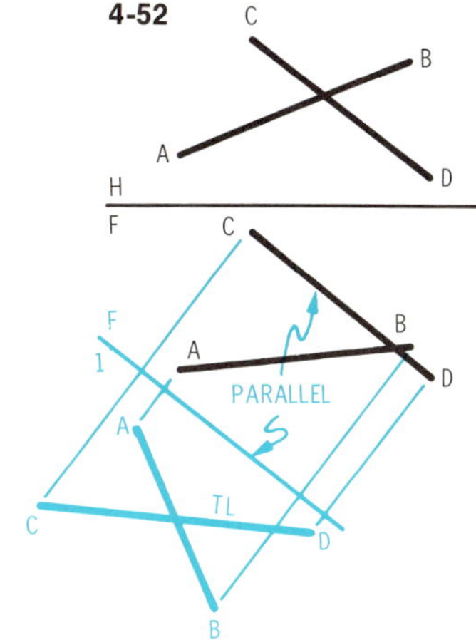

Establish the true length of line **CD** *by constructing auxiliary view 1 parallel to line* **CD** . . .

. . . *project the line* **CD** *into the view as true-length* . . .

. . . *project line* **AB** *into auxiliary view 1.*

4-53

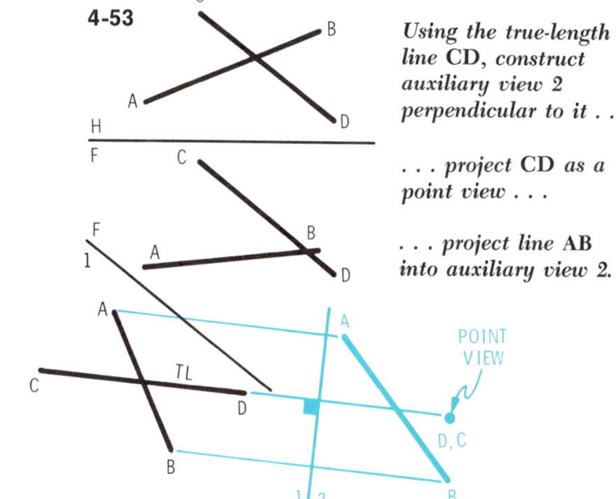

Using the true-length line **CD**, *construct auxiliary view 2 perpendicular to it* . . .

. . . *project* **CD** *as a point view* . . .

. . . *project line* **AB** *into auxiliary view 2.*

True Angle Between Nonintersecting (Skew) Lines. The procedure to establish the true angle between skew lines is a bit more complex than determining the true angle between intersecting lines, even though the same principle is used—that of finding the true length of both lines in the same view. The method used requires that we first establish the point view of one of the lines. Since any view adjacent to a point view will be true length, a view created parallel to the second line will then show the true length of both lines in the same view.

To find the true angle between the skew lines *AB* and *CD* [4-51], first project *CD* in true length into auxiliary view 1. Then project *AB* into the same auxiliary view [4-52]. Now find the point view of line *CD* by constructing auxiliary view 2 and projecting *CD* into it. Line *AB* is also projected into the third view [4-53]. Now that the point view of line *CD* is established, we must develop a true-length view of line *AB*. This is accomplished by the use of a third auxiliary view constructed parallel with line *AB*. This line is then projected into auxiliary view 3 in its true length. Because line *CD* is in a point view in auxiliary view 2, it will be projected into the third auxiliary view as a true-length line. Now both lines show in true length in the same view so we can measure the true angle ϕ that exists between them [4-54].

True Angle Between a Line and a Plane

The true angle between a line and a plane appears when the line is true length and the plane is an edge view in the same view. The method used to accomplish this involves finding the true size of the plane. Since any view

4-54

To obtain the true lengths of both **AB** *and* **CD**, *construct auxiliary view 3 parallel with line* **AB** . . .

. . . *project lines* **AB** *and* **CD** *into auxiliary view 3 as true-length* . . .

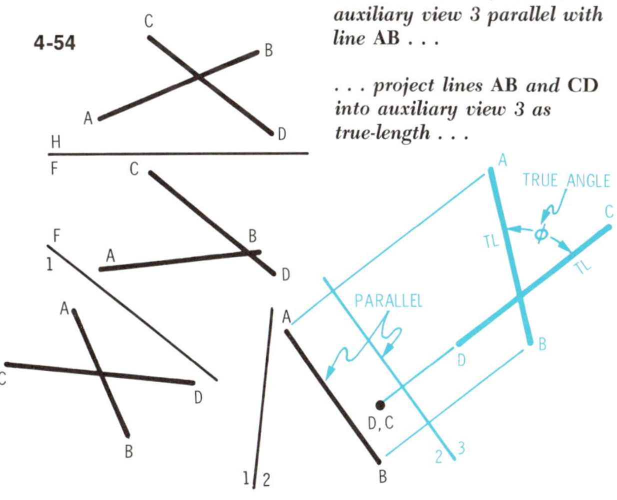

. . . *measure the true angle* ϕ *between lines* **AB** *and* **CD**.

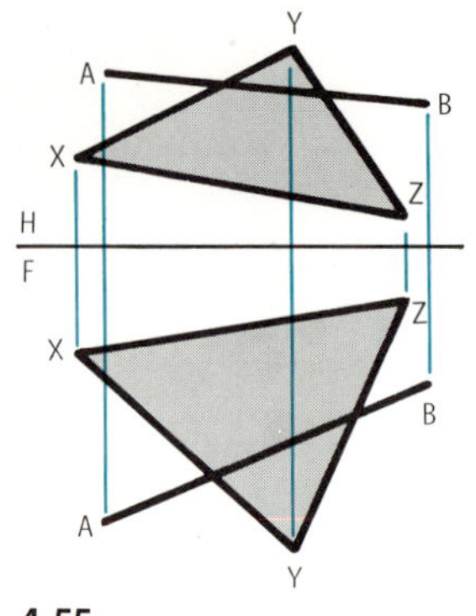

4-55

To find the true angle between the line **AB** *and the plane* **XYZ** . . .

. . . *the line must be true length and the plane must be an edge view in the same view.*

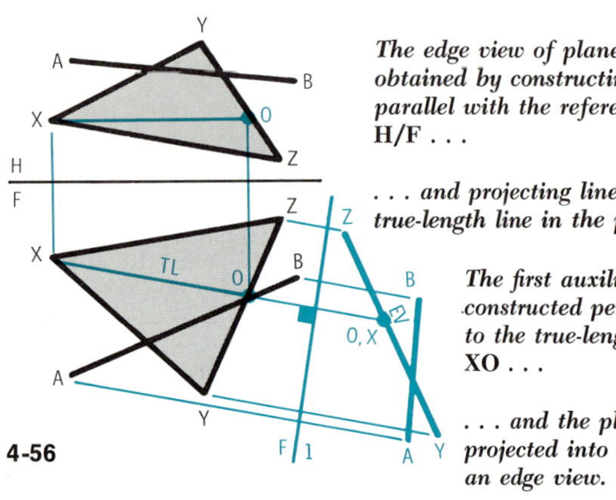

4-56

The edge view of plane **XYZ** *is obtained by constructing line* **XO** *parallel with the reference line* **H/F** . . .

. . . *and projecting line* **XO** *as a true-length line in the plane* **XYZ.**

The first auxiliary view is constructed perpendicular to the true-length line **XO** . . .

. . . *and the plane* **XYZ** *is projected into view 1 as an edge view.*

The line **AB** *is projected into auxiliary view 1.*

adjacent to a true-size view will be an edge view, a view constructed parallel to the line will then show the true length of the line and the edge view of the plane in the same view.

To find the true angle between the line *AB* and the plane *XYZ* [4-55], it is first necessary to develop the true size of the plane. This is accomplished by line *XO* in the horizontal view parallel with the reference line *H/F*. This line in plane *XYZ* then appears as true length in the frontal view. A first auxiliary view is constructed perpendicular to the true length to develop the edge view of the plane. Although an edge view has been established, the line *AB* in this view is not true length [4-56]. Therefore, additional auxiliary views are necessary. Auxiliary view 2 is constructed parallel with the edge view of the plane *XYZ*. The plane is transferred as true size into this view along with the line [4-57]. Auxiliary view 3 is constructed parallel with line *AB* in auxiliary view 2. This construction leads to a true-length view of the line *AB* in auxiliary view 3. Since the plane *XYZ* is true size in auxiliary view 2, it will appear as an edge view again in auxiliary view 3 [4-58]. This condition satisfies the requirements of the true angle between the line and the plane, so the angle ϕ may now be measured accurately.

Rotation of a Line

Another technique convenient for establishing the true angle between a line and a plane is the *rotation* method. This method utilizes the principle that if a line is rotated about an axis, its length is not altered [4-59]. If the new

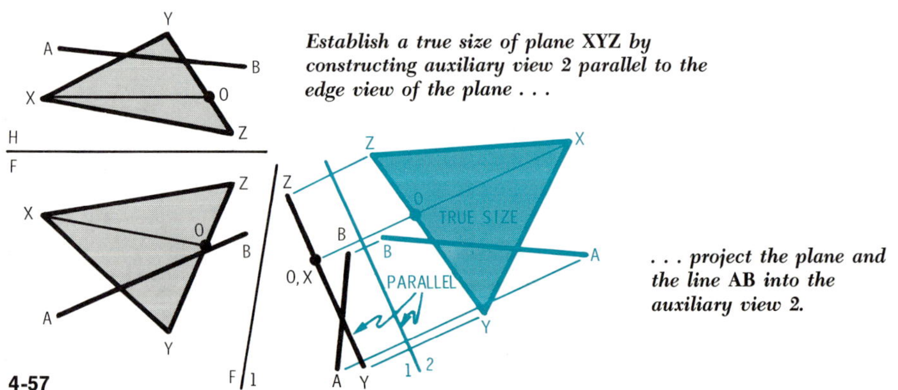

4-57

Establish a true size of plane **XYZ** *by constructing auxiliary view 2 parallel to the edge view of the plane* . . .

. . . *project the plane and the line* **AB** *into the auxiliary view 2.*

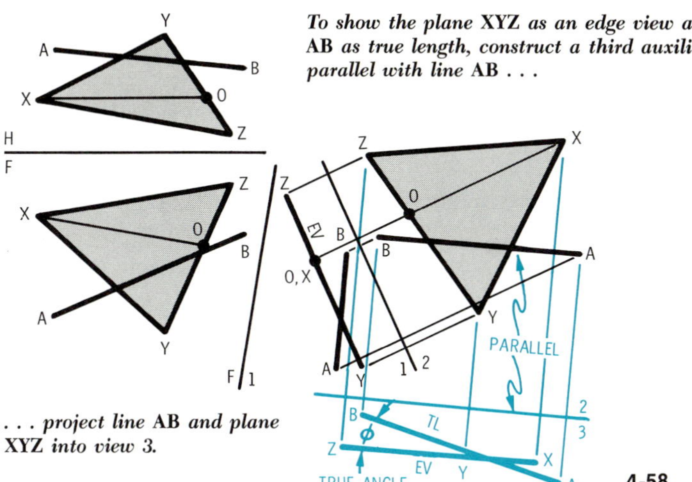

To show the plane XYZ as an edge view and the line AB as true length, construct a third auxiliary view parallel with line AB . . .

. . . project line AB and plane XYZ into view 3.

Then the true angle ϕ can be measured.

4-58

position of the line happens to be parallel with a reference plane, the line will appear as true length in this adjacent view. Also, this method of rotation about a central axis generates a cone whose base angle is the angle between the line and the plane upon whose surface one end of the line has rotated [4-60].

If it is desired to find a true-length view of the line AB in the frontal plane, an axis perpendicular to the horizontal plane is constructed through point A. This is

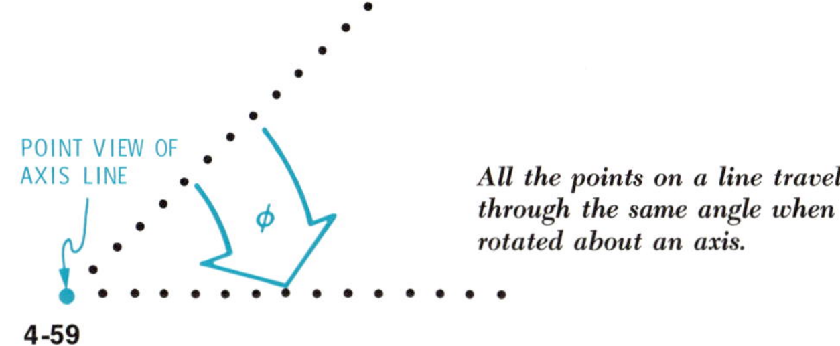

All the points on a line travel through the same angle when rotated about an axis.

4-59

4-60

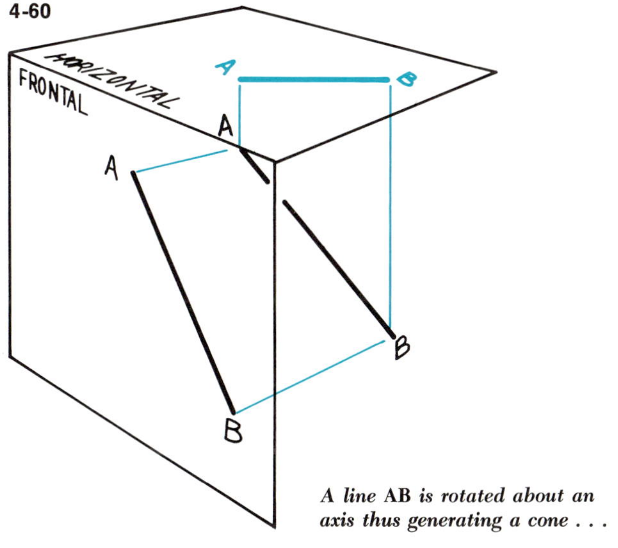

A line AB is rotated about an axis thus generating a cone . . .

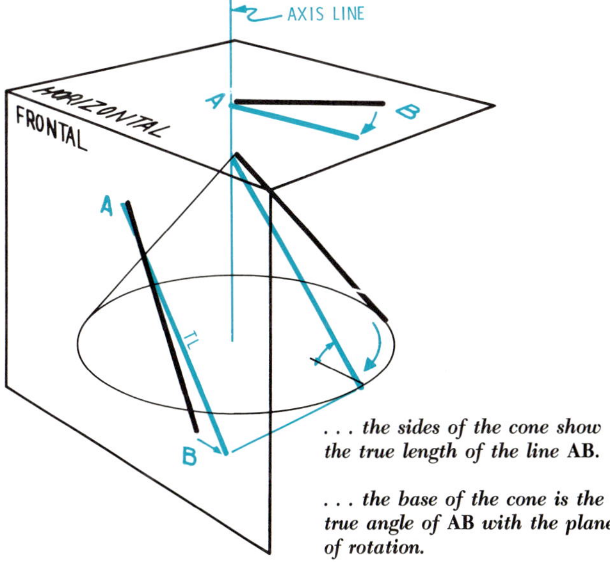

. . . the sides of the cone show the true length of the line AB.

. . . the base of the cone is the true angle of AB with the plane of rotation.

4-61

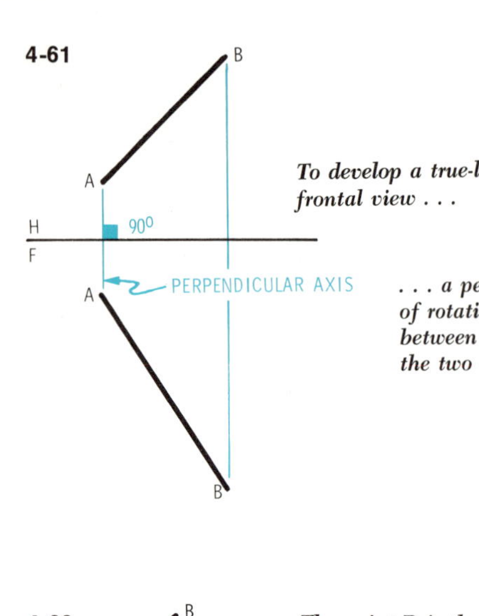

To develop a true-length line in the frontal view . . .

. . . a perpendicular axis of rotation is constructed between the points **A** *in the two views.*

4-62

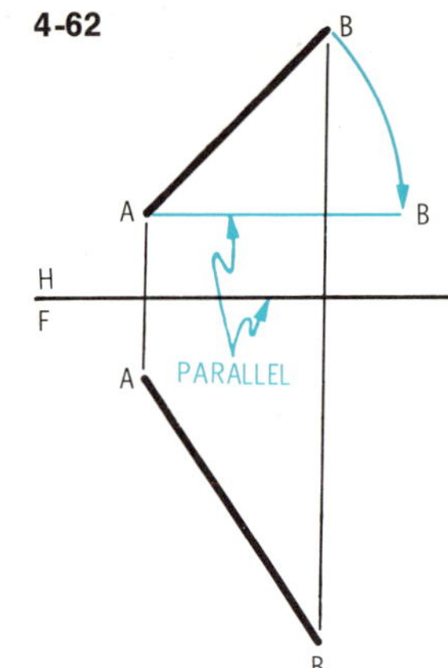

The line **AB** *is rotated about the axis at point* **A** *until it is parallel with the reference line* **H/F.**

4-63

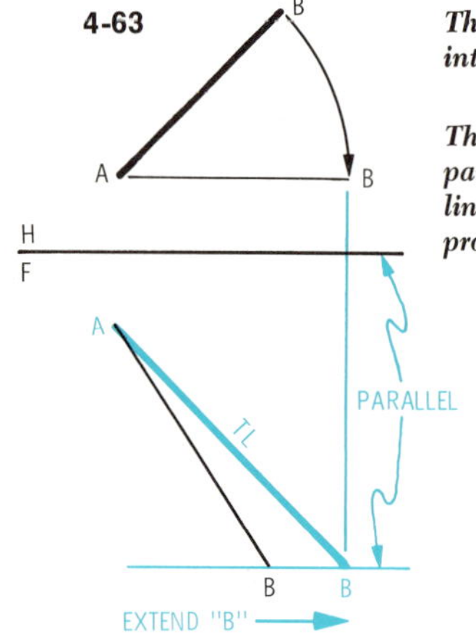

The point **B** *is then projected into the frontal view.*

The point **B** *is extended parallel with the reference line* **H/F** *to intersect the projection line.*

By connecting points A and B, a true length view of line AB is obtained.

4-64

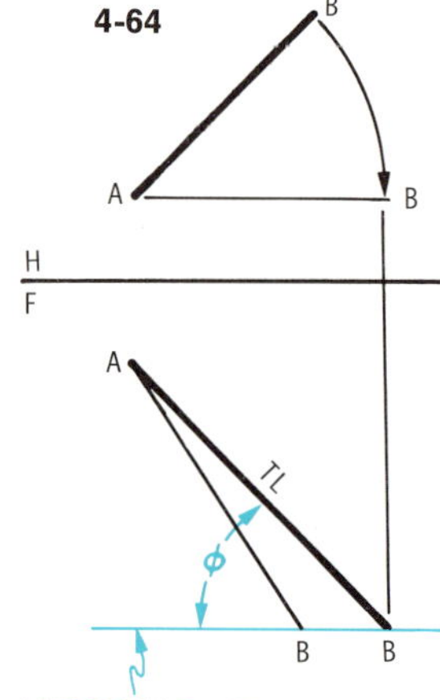

The rotated true length line **AB** *defines the true angle ϕ with the horizontal plane.*

the *axis of rotation* [4-61]. The line *AB* in the horizontal view is then rotated about point *A* until it is parallel with the reference line *H/F*, and the rotated point *B* is projected into the frontal view [4-62]. Point *B* in the frontal view is extended along a path parallel with the reference line *H/F* until it intersects the projection line. This new position of point *B* is then connected to point *A* to produce a true-length line *AB* in the frontal view [4-63].

At the same time that the true length of the line *AB* is established, the true angle between the line *AB* and the horizontal plane is available, since that is the plane upon which point *B* has been rotated. The true angle can be measured between the true-length line *AB* and the line through point *B* parallel to the horizontal plane [4-64].

4-65

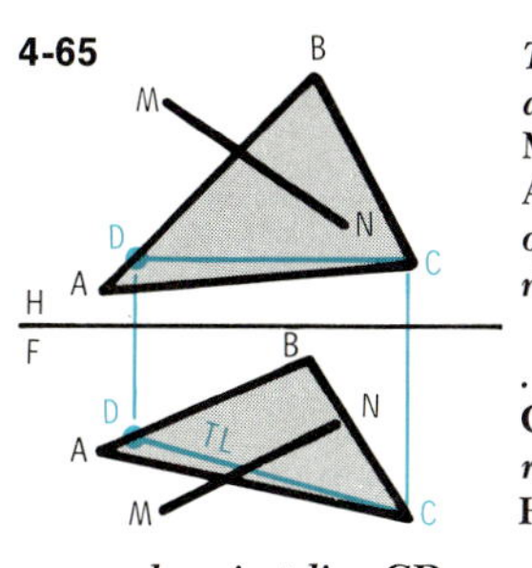

To show the true angle between line **MN** *and plane* **ABC**, *an edge view of the plane is needed.*

. . . construct line **CD** *parallel with reference line* **H/F** *. . .*

. . . and project line **CD** *into the frontal view as true length.*

4-66

Construct auxiliary view 1 perpendicular to the true-length line **CD** *. . .*

. . . and project plane **ABC** *into auxiliary view 1 as an edge view . . .*

. . . project line **MN** *into the view.*

True Angle Between a Line and a Plane (Rotation Method)

The general situation for finding the true angle between a line and plane using the rotation method is somewhat simpler than the auxiliary-view method because a smaller number of auxiliary views is necessary.

In order to obtain the angle that a line makes with a plane, it is necessary for the axis to be perpendicular to that plane. To find the true angle between line *AB* and plane *XYZ* by the rotation method, it is necessary to establish an edge view of the plane and then rotate the line into true length in the same view. In the horizontal view, construct a line *CD* in the plane *ABC* parallel with the reference plane *H/F*. Project this line *CD* into the frontal view as a true-length line [4-65].

Construct auxiliary view 1 perpendicular to line *CD* and project the plane *ABC* as an edge view. Line *MN* is also projected into this auxiliary view [4-66].

Now with an edge view of the plane established in the auxiliary view 1, it is necessary to get the line *MN* into true length in the same view using the rotation method. A second auxiliary view is constructed parallel with the edge view of the plane *ABC*. Only the line *MN* is then projected into this view. Point *N* is designated as the axis of rotation and the line *MN* is revolved until it is parallel with the reference line 1/2 [4-67].

The point *M* is then projected back to auxiliary view 1 until it intersects the parallel extension of point *M* parallel to reference line 1/2. This intersection point is then connected to point *N* to generate the true-length line *MN* and the true angle ϕ between line *MN* and the plane *ABC* [4-68].

4-67

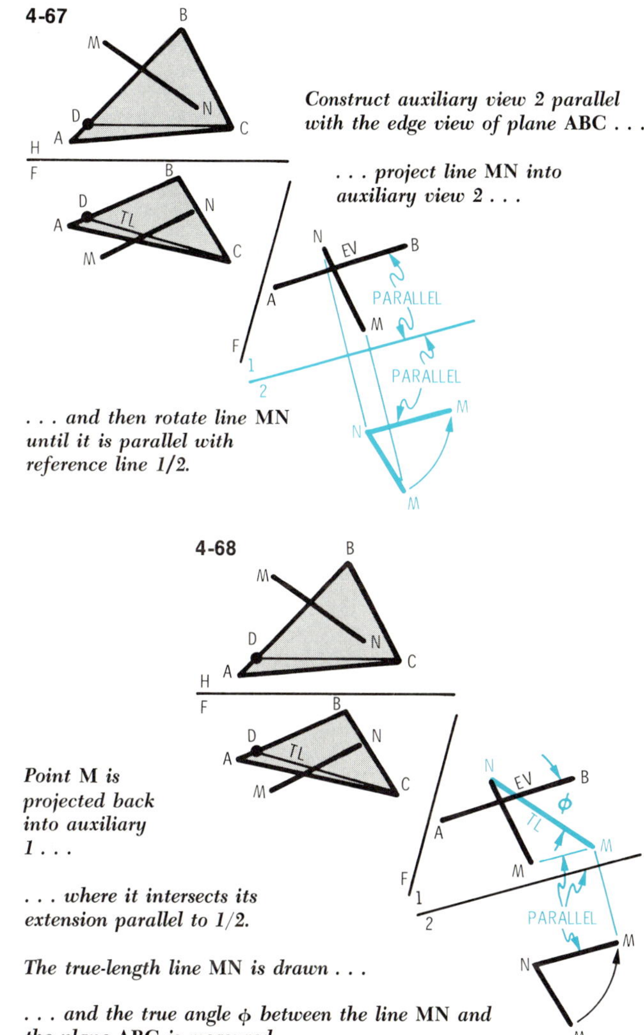

Construct auxiliary view 2 parallel with the edge view of plane **ABC** *. . .*

. . . project line **MN** *into auxiliary view 2 . . .*

. . . and then rotate line **MN** *until it is parallel with reference line 1/2.*

4-68

Point **M** *is projected back into auxiliary 1 . . .*

. . . where it intersects its extension parallel to 1/2.

The true-length line **MN** *is drawn . . .*

. . . and the true angle ϕ *between the line* **MN** *and the plane* **ABC** *is measured.*

4-69

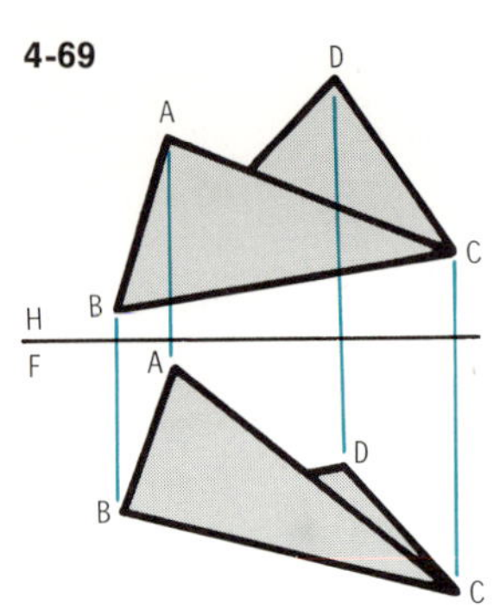

To find the true angle between planes **ABC** *and* **BCD** . . .

. . . *it is necessary to have an edge view of both planes in the same view.*

4-70

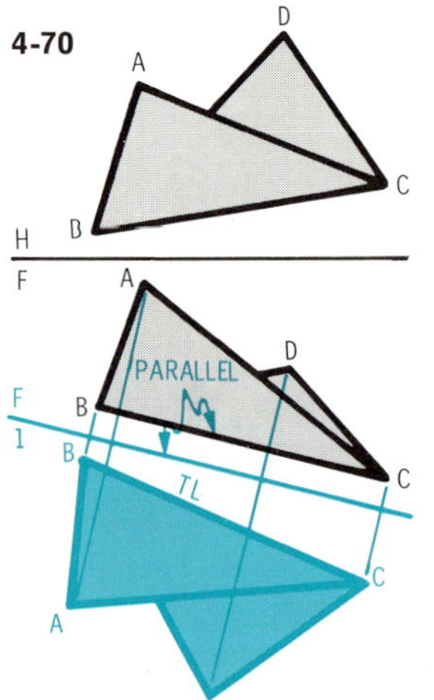

The edge view of both planes is achieved by establishing a point view of the mutually shared line of intersection **BC.**

Construct auxiliary view 1 parallel with the frontal view of line **BC** . . .

. . . *and project* **BC** *as true length into view 1.*

Project all the remaining lines into auxiliary view 1.

4-71

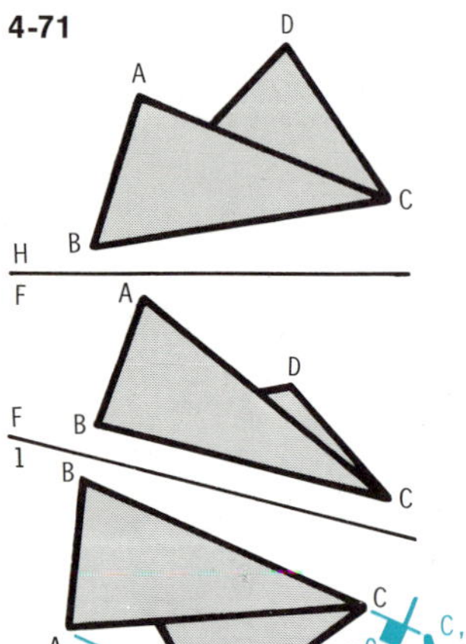

The point view of line **BC** *is found by constructing the second auxiliary view perpendicular to the line.*

Project the boundary lines of planes **ABC** *and* **BCD** *into view 3 as edge views* . . .

. . . *and measure the true angle* ϕ *between them.*

It is important to note that auxiliary view 2 was created to establish a path of rotation for point *M* in auxiliary view 1 that is parallel to the edge view of plane *ABC*, thus producing a true angle with *ABC*.

True Angle Between a Plane and a Plane

The true angle between two planes appears when both planes are edge views in the same view. The true angle between planes is often referred to as the *dihedral angle*. Two possible conditions confront the engineer: (1) the line of intersection is known; and (2) the line of intersection is *not* known. Of these, the first is the more common.

To find the true angle between planes *ABC* and *BCD* [4-69], all that is necessary is to obtain a point view of the line of intersection *BC*. Since this line is common to both planes, a point view of it will produce edge views of both planes in the same view. The point view of line *BC* is obtained by constructing auxiliary plane 1 parallel to this line of intersection. Project line *BC* in its true length into this newly constructed view. The other boundary lines of the planes are projected into auxiliary view 1 as well [4-70]. From there it is a simple matter to obtain the point view of the common line of intersection and the associated edge views of the planes by constructing auxiliary view 2 perpendicular to the true-length view of line *BC*. With the edge views of the planes complete, one can now measure the true angle ϕ between the planes—the dihedral angle [4-71].

If no common line of intersection is available, the basic principle of acquiring edge views of the planes in the

4-72

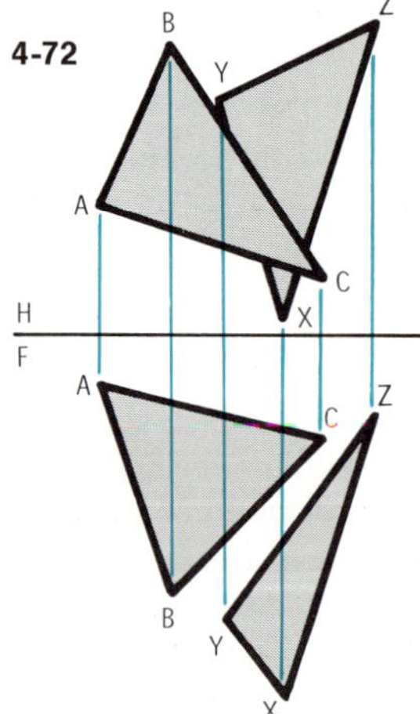

To find the true angle between planes **ABC** *and* **XYZ** *when there is no intersecting line given* . . .

. . . *it is necessary to show both planes as edge views in the same view.*

4-73

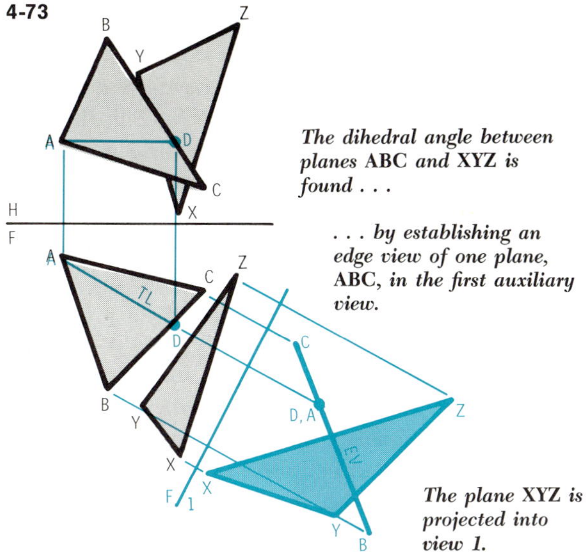

The dihedral angle between planes ABC and XYZ is found . . .

. . . by establishing an edge view of one plane, ABC, in the first auxiliary view.

The plane XYZ is projected into view 1.

same view still applies, but the procedure used is somewhat different.

To find the dihedral angle between the given planes ABC and XYZ [4-72], it is necessary first to establish the edge view of one of the planes. In this example, the edge view of plane ABC is projected into auxiliary view 1 using a true-length line AD [4-73]. The plane XYZ is also projected into this view. Auxiliary view 2 is constructed parallel with the edge view of plane ABC, and at the same time a line OY parallel to reference line 1/2 is constructed within the plane XYZ. The two planes are projected into auxiliary view 2 with plane ABC being in true size and the line OY in XYZ being in true length [4-74].

One additional auxiliary view is now constructed that is perpendicular to the true-length line in plane XYZ. The two planes are projected into auxiliary view 3 as edge views. Since we now have edge views of two planes in the same view, we can measure the true angle ϕ between them [4-75].

specifications of a line

The relationship between lines and principal reference planes has such common usage in engineering practice

4-74

Plane XYZ must become an edge view also.

Construct auxiliary view 2 parallel with edge view ABC . . .

. . . and construct a line OY parallel with the reference line 1/2.

Project the plane ABC as true size in auxiliary view 2 . . .

. . . and project plane XYZ with the true length line OY.

4-75

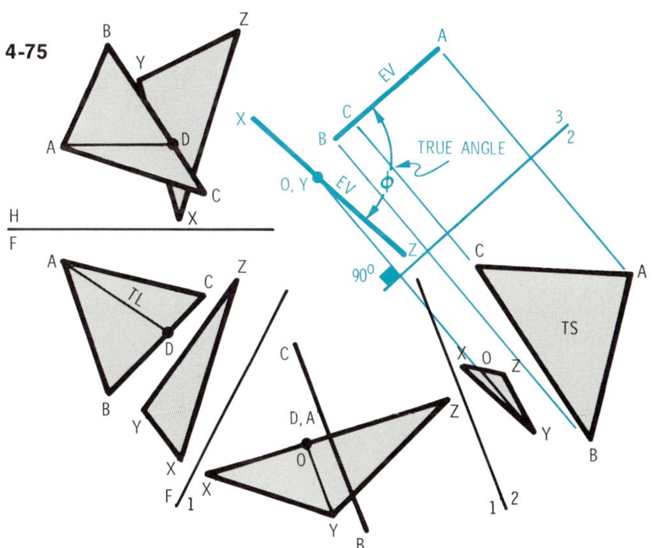

Construct auxiliary view 3 perpendicular to line OY . . .

. . . and project planes ABC and XYZ into the view as edge views.

Measure the true angle ϕ between the two planes.

4-76

The bearing of a line is designated by the angle between the line and the north-south axis of the earth.

that names for specific relationships have evolved. The most familiar names are *bearing* and *slope.*

The *bearing* of a line is used to specify the line's direction with respect to the earth's axis. Specifically, the bearing designates the angle between the line in question and the north–south axis of the earth [4-76]. Although the earth is basically a sphere, its surface is so great that a small segment of the surface can be regarded as a planar surface—the horizontal plane. The horizontal view that shows this is used to measure the bearing of a line. By convention, the top of the horizontal view in an engineering drawing is assumed to be north unless some other direction in the horizontal view is specified [4-77]. Since direction is the only consideration in a bearing, the angle can be measured directly from the horizontal view without having to establish the true length of the line. The bearing of a line is always less than 90° and is read as a deviation from north or south in the direction that the line is pointing, utilizing the quadrants of the compass [4-78].

The other important specification of a line's orientation in space is the *slope.* Slope is defined as the true angle between a line and the horizontal plane. Slope can therefore be measured in an auxiliary view oriented so that both the true length of the line and the edge view of the horizontal plane are seen [4-79]. Slope can also be found by the rotation method, as shown in [4-64].

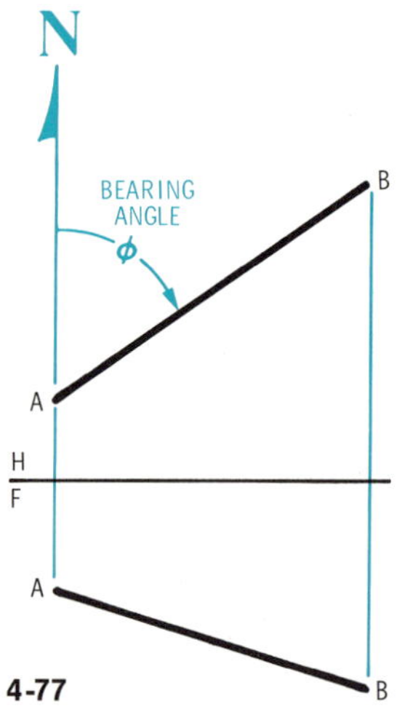

4-77

The bearing of a line is measured in the horizontal view.

4-78
Various bearing angles with the compass quadrants.

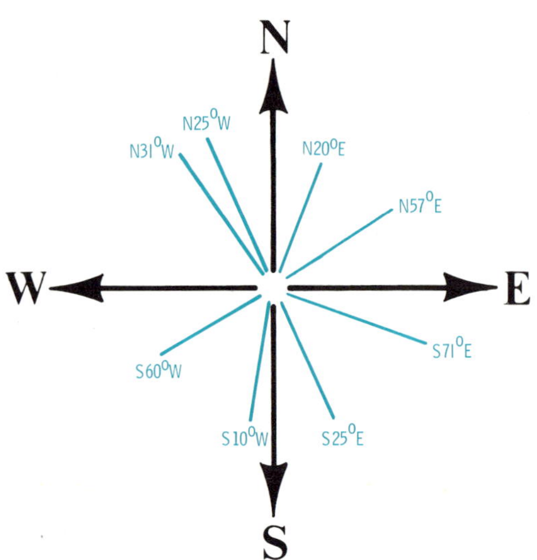

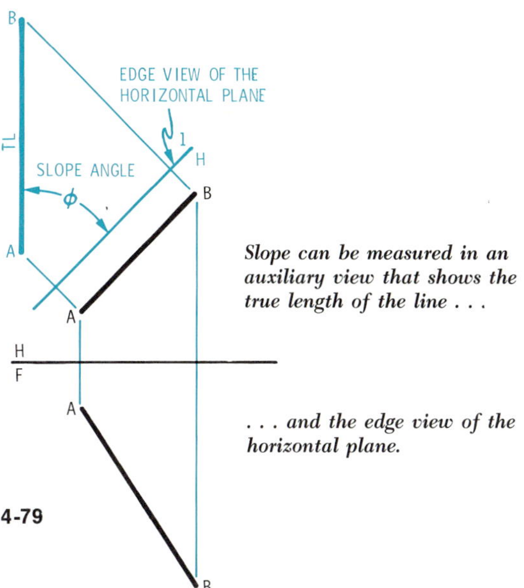

Slope can be measured in an auxiliary view that shows the true length of the line . . .

. . . and the edge view of the horizontal plane.

4-79

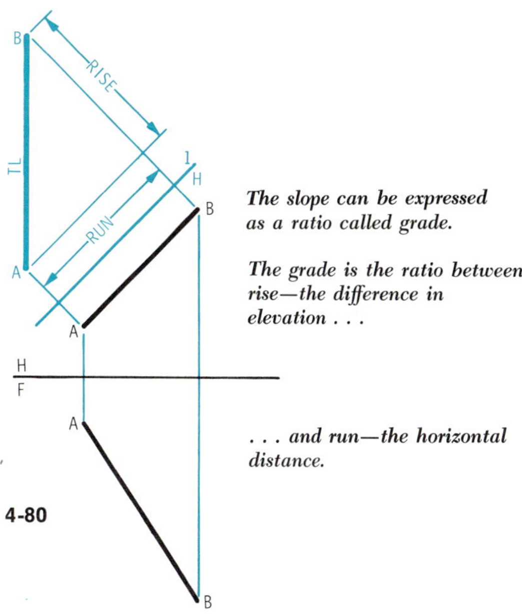

The slope can be expressed as a ratio called grade.

The grade is the ratio between rise—the difference in elevation . . .

. . . and run—the horizontal distance.

4-80

Instead of stating an angle in degrees, slope may also be stated as a ratio called *grade*. Grade is the ratio of the vertical difference in elevation of the end points of the line (*rise*) to the horizontal distance between the end points (*run*) [4-80]. This numerical ratio is multiplied by 100 to express the answer as a percentage. Grade can also be obtained by finding the tangent of the slope angle in a table of trigonometric functions and converting this to a percentage. The true values of rise and run can be measured readily on the same auxiliary view (or rotated view) of a line that shows the true slope angle. Grade is the term civil engineers use to designate the inclination of highways. A 6 per cent grade, for example, refers to a rise of 6 feet in every 100 feet of horizontal distance [4-81].

4-81
The grade of a highway is the rise for 100 feet of horizontal distance.

specifications of a plane

There is often a need to specify the location of a plane in space. Common terms used in such a specification are *strike* and *dip*. These terms come from the mining industry and describe the orientation of a vein of ore. The usefulness of the concepts has made them adaptable to other engineering applications.

The *strike* of a plane is the bearing of a horizontal line in the plane. It is found by constructing a horizontal line in the front view of the plane and projecting it into the

4-82

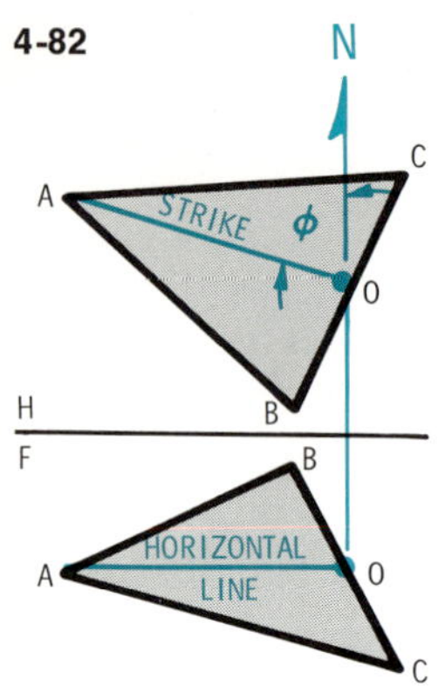

The strike of a plane is the direction that a horizontal line in the plane assumes when it is projected into the horizontal view.

The strike of plane **ABC** *is* **N φ° W.**

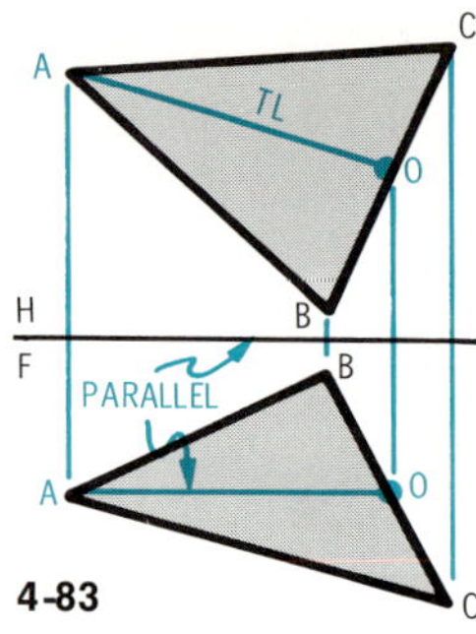

4-83

The dip of plane **ABC** *is found by constructing a horizontal line* **AO** *in the frontal view . . .*

. . . and projecting as true length into the horizontal view.

horizontal view. The angle between this projected line and a north–south line—the bearing—is the strike of the plane [4-82].

Dip is the true angle between a given plane and the horizontal plane. It is always an acute angle, and the compass direction of downward slope must also be stated. The dip of a plane, therefore, is specified by two characteristics—the angle of the downward slope and its direction.

The dip of a plane is found by constructing a line *AO* parallel with the horizontal plane in the front view [4-83]. By projecting this line and plane *ABC* as an edge view into an auxiliary view adjacent to the horizontal plane, the angle may be measured. This dip angle is the true angle between the plane and the horizontal plane [4-84].

The dip direction is the general compass direction that the plane slopes downhill. This direction is indicated in the horizontal view by constructing an arrow perpendicular to the true-length horizontal line and pointed towards the low side of the plane (North-East) [4-85]. There is no angular connotation to dip direction, since it is used merely to differentiate between two planes having the same strike and dip angle.

4-84

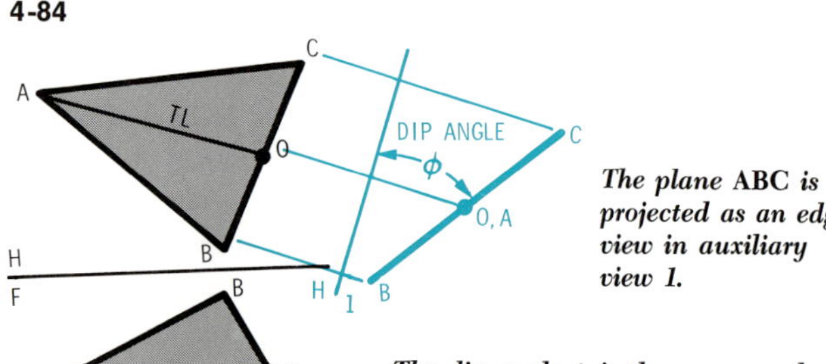

The plane **ABC** *is projected as an edge view in auxiliary view 1.*

The dip angle φ is then measured between the edge view of the plane and the reference line **H/1.**

4-85

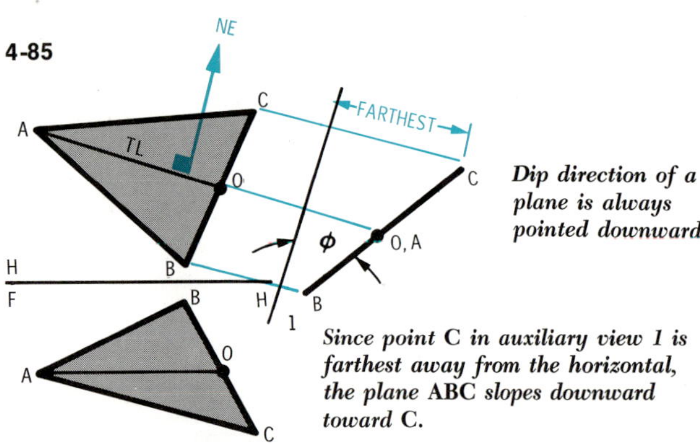

Dip direction of a plane is always pointed downward.

Since point **C** *in auxiliary view 1 is farthest away from the horizontal, the plane* **ABC** *slopes downward toward* **C.**

The dip direction is shown in the horizontal view as an arrow perpendicular to the strike line and pointing downward to the point **C** *side of the plane.*

true distances

Another important concept in engineering graphics is that of *true distances* between geometric elements. The shortest distance (often called *clearance*) between the various elements is most commonly used, but distances having a specific angle of inclination are not unusual.

Two lines are perpendicular if the angle between them is 90° in any view where either line appears as true length.

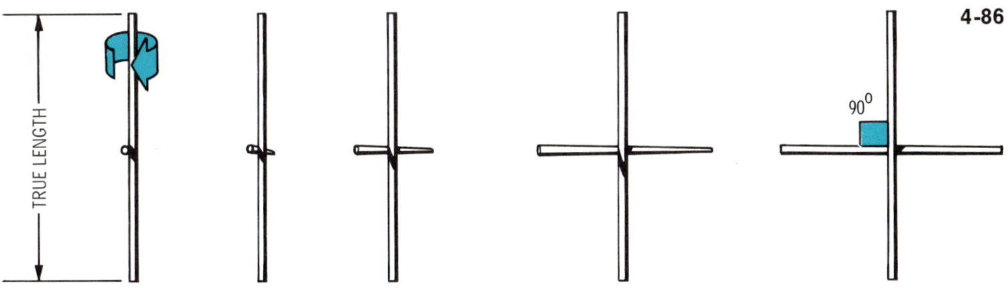

4-86

The shortest distance or clearance between geometric elements is the true length of a perpendicular line constructed between these elements—whether they are points, lines, or planes. *Two lines are perpendicular if the angle between them is 90° in any view where either line appears as true length* [4-86]. The importance of this rule will become evident in the following discussions of true distances.

Shortest Distance Between a Point and a Line

The shortest (perpendicular) distance between a point and a line appears as true length when the line is seen as a point view. To find the shortest distance between point *X* and line *AB*, it is first necessary to establish

The shortest distance between a point and a line is the true perpendicular distance.

4-87

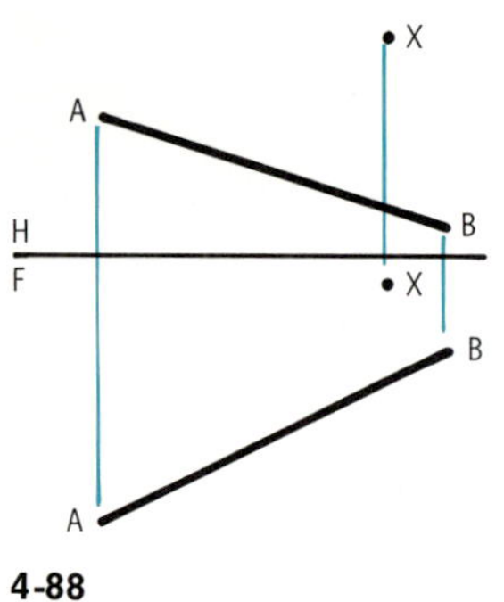

4-88

To find the shortest distance between point **X** *and line* **AB** *. . .*

. . . it is necessary to show the point and a point view of line **AB** *in the same view.*

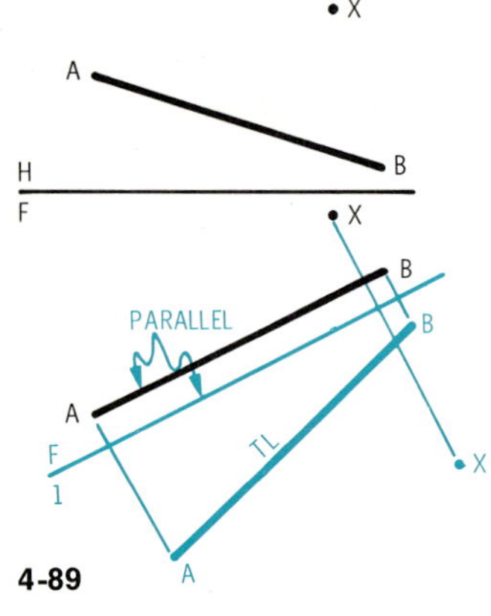

4-89

Construct auxiliary view 1 parallel with line **AB** *in the front view . . .*

. . . and project line **AB** *in true length into the auxiliary view 1.*

Project point **X** *into the auxiliary view.*

the true length of line *AB* [4-88]. This is done by constructing an auxiliary view parallel with the front view of line *AB* and projecting the line and the point into this auxiliary view 1 [4-89]. Once the true-length line has been established, it must be projected as a point view into a second auxiliary view. This view is constructed perpendicular to the line *AB* [4-90]. The point *X* and the point view of line *AB* are then projected into auxiliary view 2. A line drawn connecting these points represents a perpendicular line between point *X* and line *AB*. As such, it is the true length of the shortest distance—the clearance—between the point and the line.

If it is necessary to find the point of intersection of the clearance line with line *AB*, we must return to auxil-

4-90

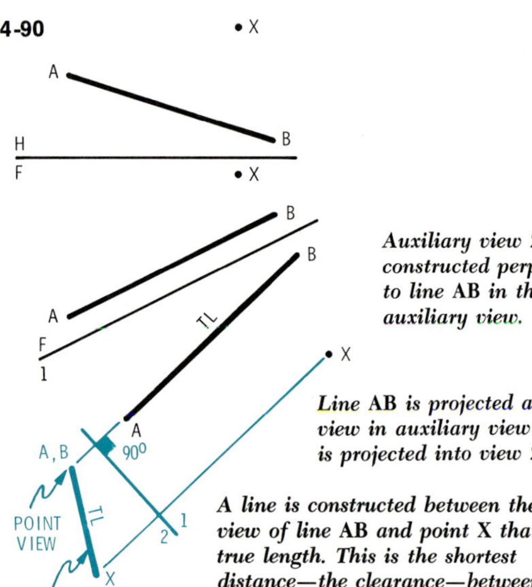

Auxiliary view 2 is constructed perpendicular to line **AB** *in the first auxiliary view.*

Line **AB** *is projected as a point view in auxiliary view 2. Point* **X** *is projected into view 2.*

A line is constructed between the point view of line **AB** *and point* **X** *that is in true length. This is the shortest distance—the clearance—between the point* **X** *and line* **AB**.

4-91

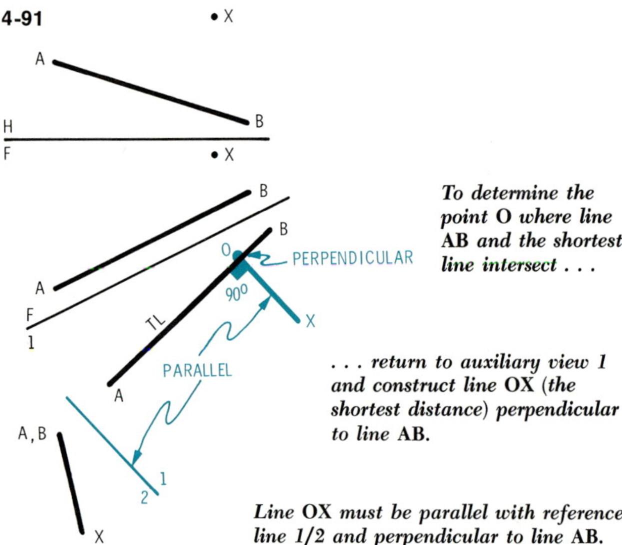

To determine the point **O** *where line* **AB** *and the shortest line intersect . . .*

. . . return to auxiliary view 1 and construct line **OX** *(the shortest distance) perpendicular to line* **AB**.

Line **OX** *must be parallel with reference line 1/2 and perpendicular to line* **AB**.

iary view 1. Recalling that two lines are perpendicular when the angle between them is 90° in any view where either is true length, construct a line from point *X* that is perpendicular to line *AB* [4-91]. Note that this line is also parallel with the reference line 1/2, thus verifying the statement that the same line in auxiliary view 2 is in true length. This perpendicular construction locates the point of intersection *O* on line *AB*, which can be projected to the preceding views of line *AB* and point *X* [4-92].

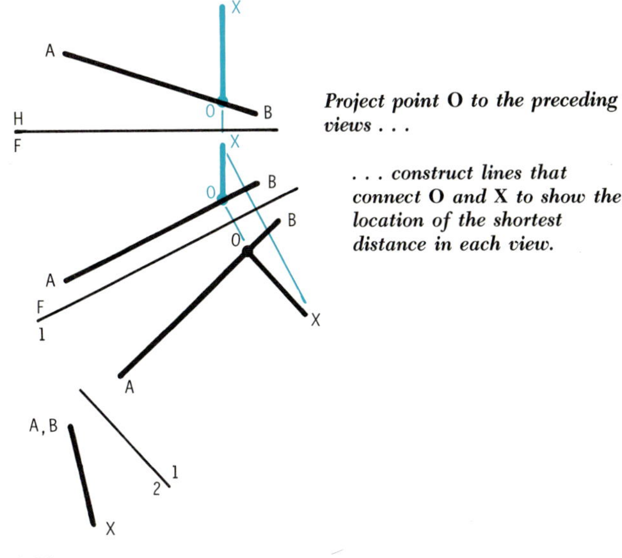

Project point **O** *to the preceding views . . .*

. . . construct lines that connect **O** *and* **X** *to show the location of the shortest distance in each view.*

4-92

Shortest Distance Between Two Lines

The shortest (perpendicular) distance between two lines appears as true length when either line is seen as a point view. The true distance in this definition is the common perpendicular between two skew lines.

To find the shortest distance between the given skew lines *AB* and *XY* [4-93], we must establish the true length of one of the lines in auxiliary view 1 [4-94]. Although we now have established a true-length view of one of the lines, our definition indicates the need to develop a point view of a line to determine the common perpendicular. A second auxiliary view is constructed to obtain a point view of line *XY* along with a view of *AB*. Now construct a line *PO* from the point view *XY* perpendicular to line *AB* [4-95]. This line is the common perpendicular between *AB* and *XY* and is true length, as in the case of the shortest distance between a point and a line. As such it represents the shortest distance (clearance) between the skew lines.

Often it is desirable to find the points on each line where the shortest distance line intersects. This is accom-

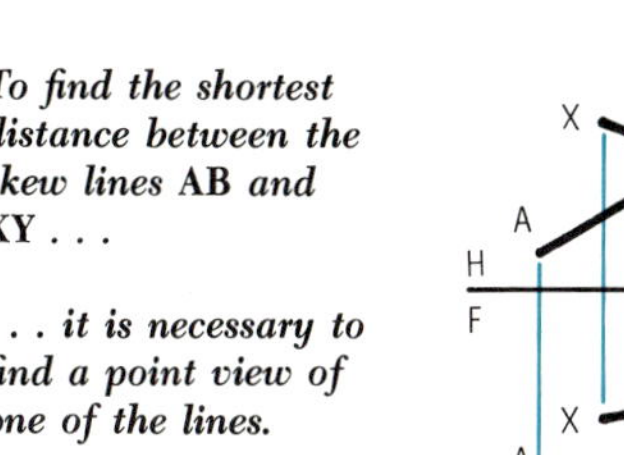

To find the shortest distance between the skew lines **AB** *and* **XY** *. . .*

. . . it is necessary to find a point view of one of the lines.

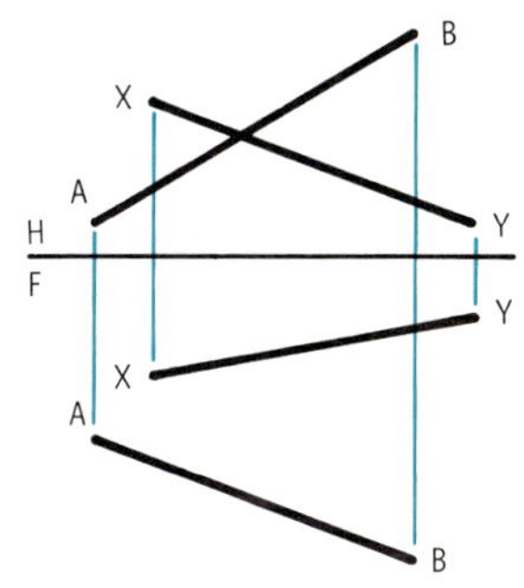

4-93

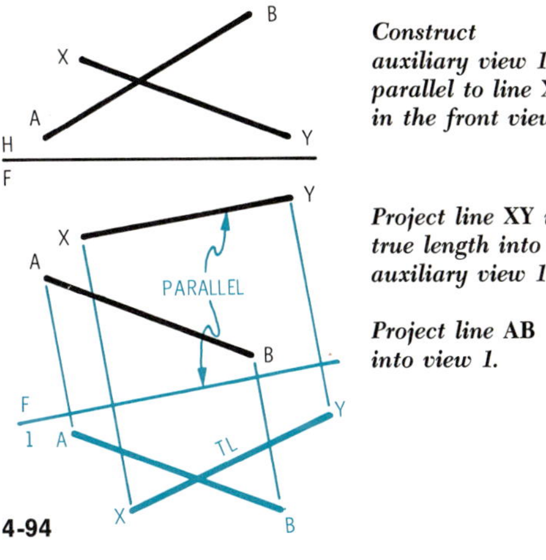

Construct auxiliary view 1 parallel to line **XY** *in the front view.*

Project line **XY** *in true length into auxiliary view 1.*

Project line **AB** *into view 1.*

4-94

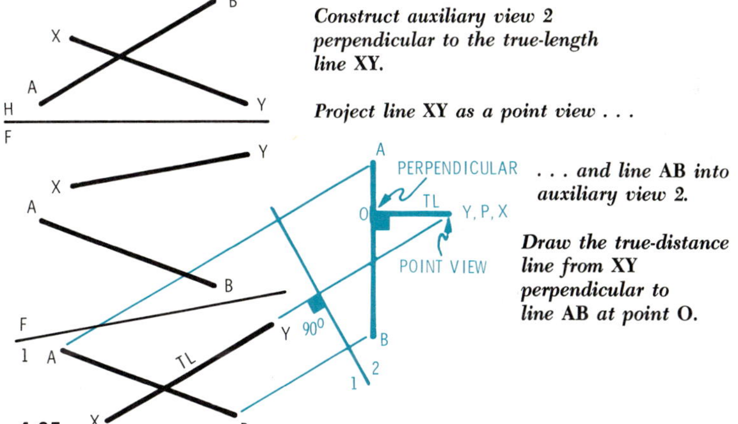

Construct auxiliary view 2 perpendicular to the true-length line **XY.**

Project line **XY** *as a point view . . .*

. . . and line **AB** *into auxiliary view 2.*

Draw the true-distance line from **XY** *perpendicular to line* **AB** *at point* **O.**

4-95

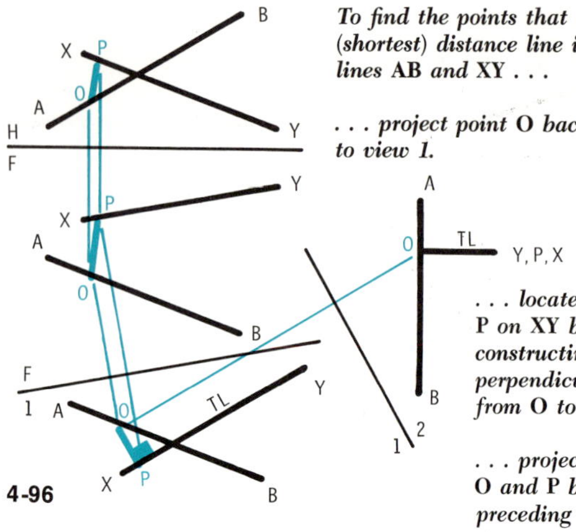

4-96

To find the points that the true (shortest) distance line intersects lines **AB** *and* **XY** . . .

. . . *project point* **O** *back to view 1.*

. . . *locate point* **P** *on* **XY** *by constructing a line perpendicular from* **O** *to line* **XY.**

. . . *project points* **O** *and* **P** *back to preceding views.*

4-98

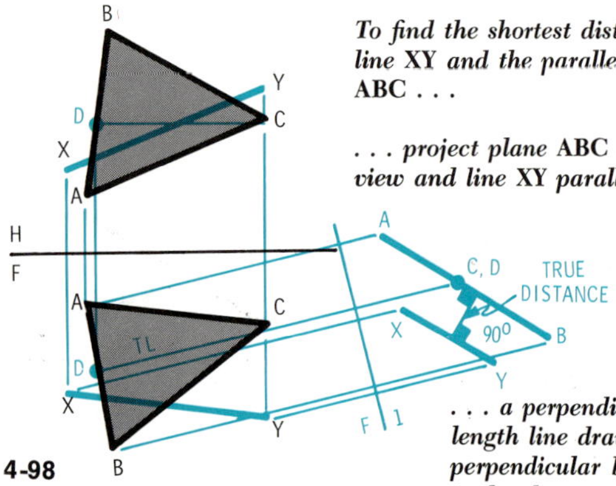

To find the shortest distance between line **XY** *and the parallel plane* **ABC** . . .

. . . *project plane* **ABC** *as an edge view and line* **XY** *parallel to it* . . .

. . . *a perpendicular true-length line drawn perpendicular between them is the shortest distance.*

4-99

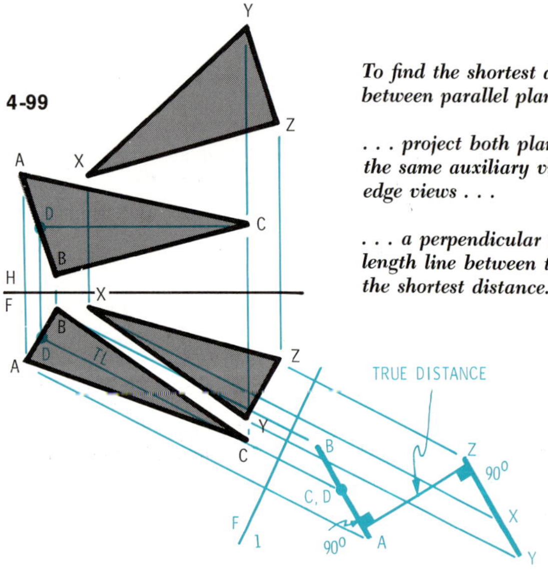

To find the shortest distance between parallel planes . . .

. . . *project both planes into the same auxiliary views as edge views* . . .

. . . *a perpendicular true-length line between them is the shortest distance.*

4-97

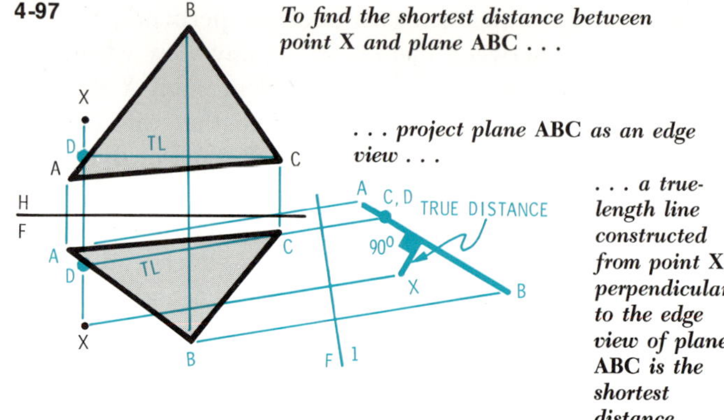

To find the shortest distance between point **X** *and plane* **ABC** . . .

. . . *project plane* **ABC** *as an edge view* . . .

. . . *a true-length line constructed from point* **X** *perpendicular to the edge view of plane* **ABC** *is the shortest distance.*

plished by projecting the true-distance line *PO* back into the preceding views of the line, just as in the example of the shortest distance between a point and a line [4-96]. (Note that line *PO* must be drawn perpendicular to line *XY* in the first auxiliary view, since *XY* appears as true length in this view.) In the first auxiliary view, the true-distance line is drawn from point *O* on line *AB* perpendicular to line *XY* at point *B*.

The true distances between other geometric elements are simply extensions of the analytical methods just discussed. As an example, to find the shortest distance between a point and plane, we must first obtain the edge view of the plane. Since the edge view is in effect a line, the method of finding the shortest distance between a point and a line applies [4-97]. Similarly, if an analysis requires the shortest distance between a line and a parallel plane, the edge view of the plane is once again needed. After this is accomplished, a perpendicular true- (shortest) length line can be drawn between them [4-98]. The distance between parallel planes is found through the technique of viewing both planes as edge views in the same view [4-99].

summary

Analysis of visual communication systems begins with the four fundamental views of geometric elements—lines and planes. These fundamental views are strongly interrelated and dependent on each other. The *true-length* view of a line is basic and is used to construct the point view of the line. The *point view* is then needed to construct an edge view of a plane that contains the line. This

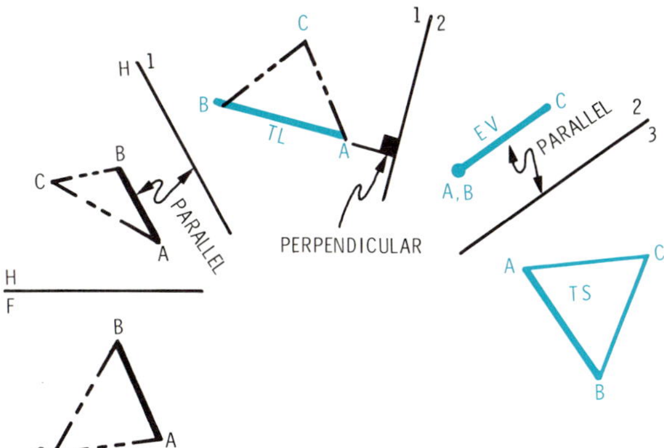

4-100
The four fundamental views of geometric elements.

edge view of a plane in turn leads to the development of the *true size* of the plane.

These fundamental views are then the means by which the two general classes of geometric relationships—*true angles* and *true distances*—are analyzed. True angles between lines require that lines be seen as true length in the same view. True angles between lines and planes require that the lines be true length and that the planes be edge views in the same view. True angles between planes require that the planes be edge views.

Certain special true-angle relationships, important to the engineer, are the *bearing* and *slope* of lines and the *strike* and *dip* of planes. Bearing describes the direction of the line relative to the north–south axis of the earth. The slope of a line is the true angle between the line and the horizontal plane. The strike of a plane is the bearing of a horizontal line in that plane. Dip defines the true angular relationship of the plane with the horizontal plane.

The true distances between various geometric elements are found by using the fundamental views to obtain true lengths and point views of a line and edge views of planes. Perpendicular lines are constructed between these elements to define the most important true distance . . . the shortest or clearance distance.

problems

4-1. Complete the missing third view in the three figures [4-101, 4-102, and 4-103] and identify every true-length line with the letters TL in each of the principal views. If a true-length line does not exist in a figure, print NO TL in the upper-right-hand corner of the view.

4-101

4-102

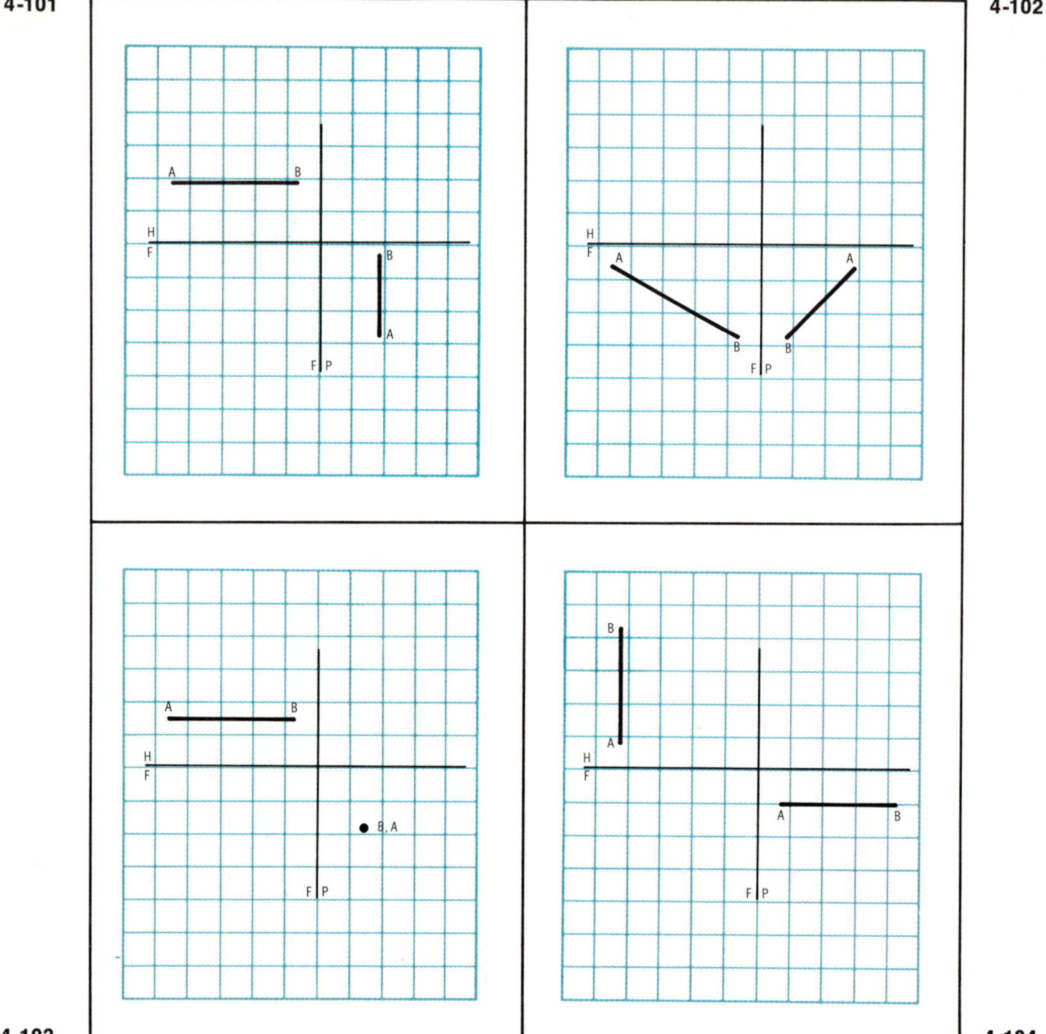

4-103

4-104

4-2. Complete the missing third view in the three figures [4-104, 4-105, and 4-106] and identify every point view of a line with the letters PV. If a point view of a line does not exist in a figure, print NO PV in the upper-right-hand corner of the view.

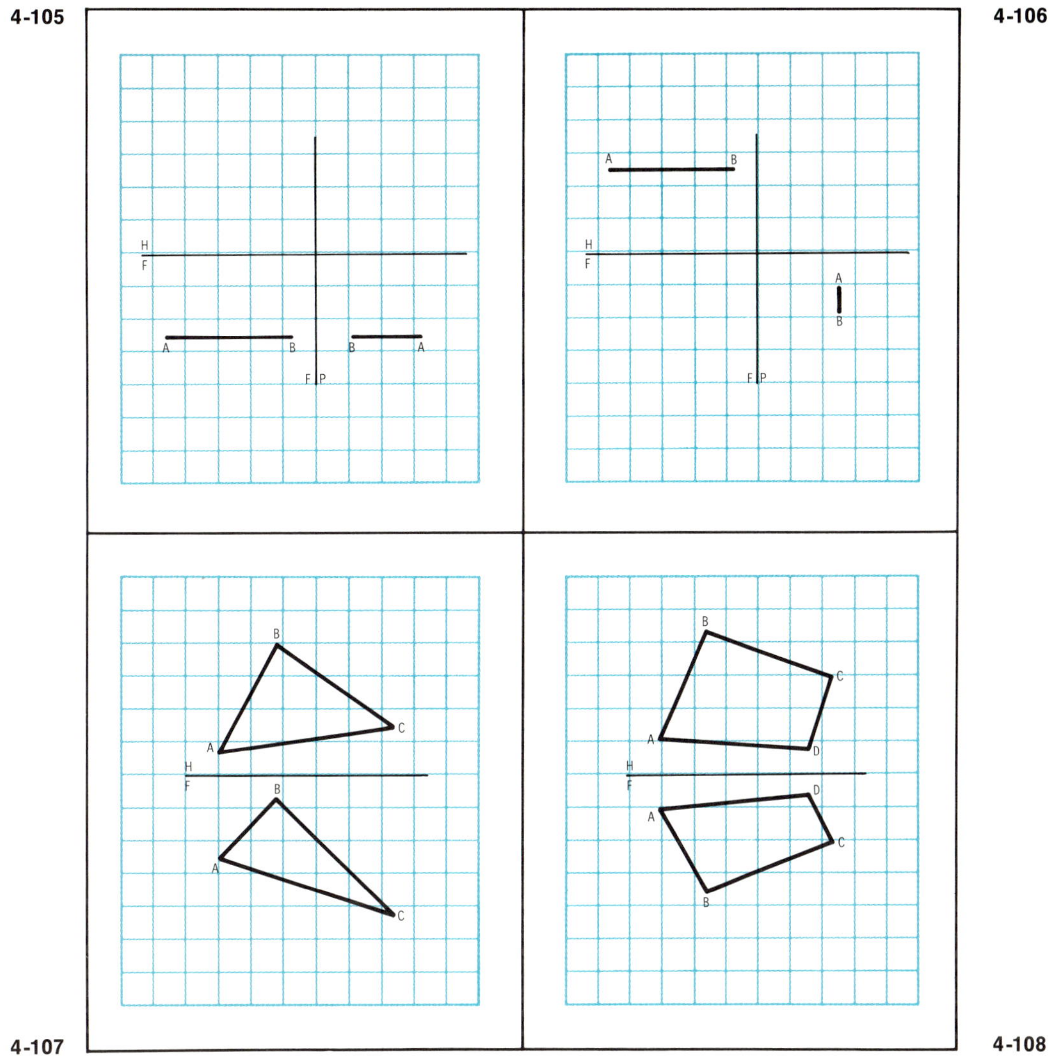

4-3. Construct the edge view of the planes shown in [4-107 and 4-108]. Using measuring and marking tools, transfer each figure to an $8\frac{1}{2}$- by 11-inch sheet of paper and increase the scale to double size before constructing the edge view. Plan the layout so that the views are neatly arranged on the paper while it is vertically oriented.

4-4. Construct the true-size view of the planes shown in [4-109 and 4-110]. Using measuring and marking tools, transfer each figure in double size to an $8\frac{1}{2}$- by 11-inch sheet of paper before completing the true-size view. Plan the layout so that the views are neatly arranged on the paper.

4-5. Determine if the lines given in [4-111 to 4-116] are intersecting. Measure and transfer each figure at double size to an $8\frac{1}{2}$- by 11-inch sheet of paper and complete the necessary views to determine the intersection. If the lines intersect, letter INT in the upper-right-hand corner; if not, letter SKEW.

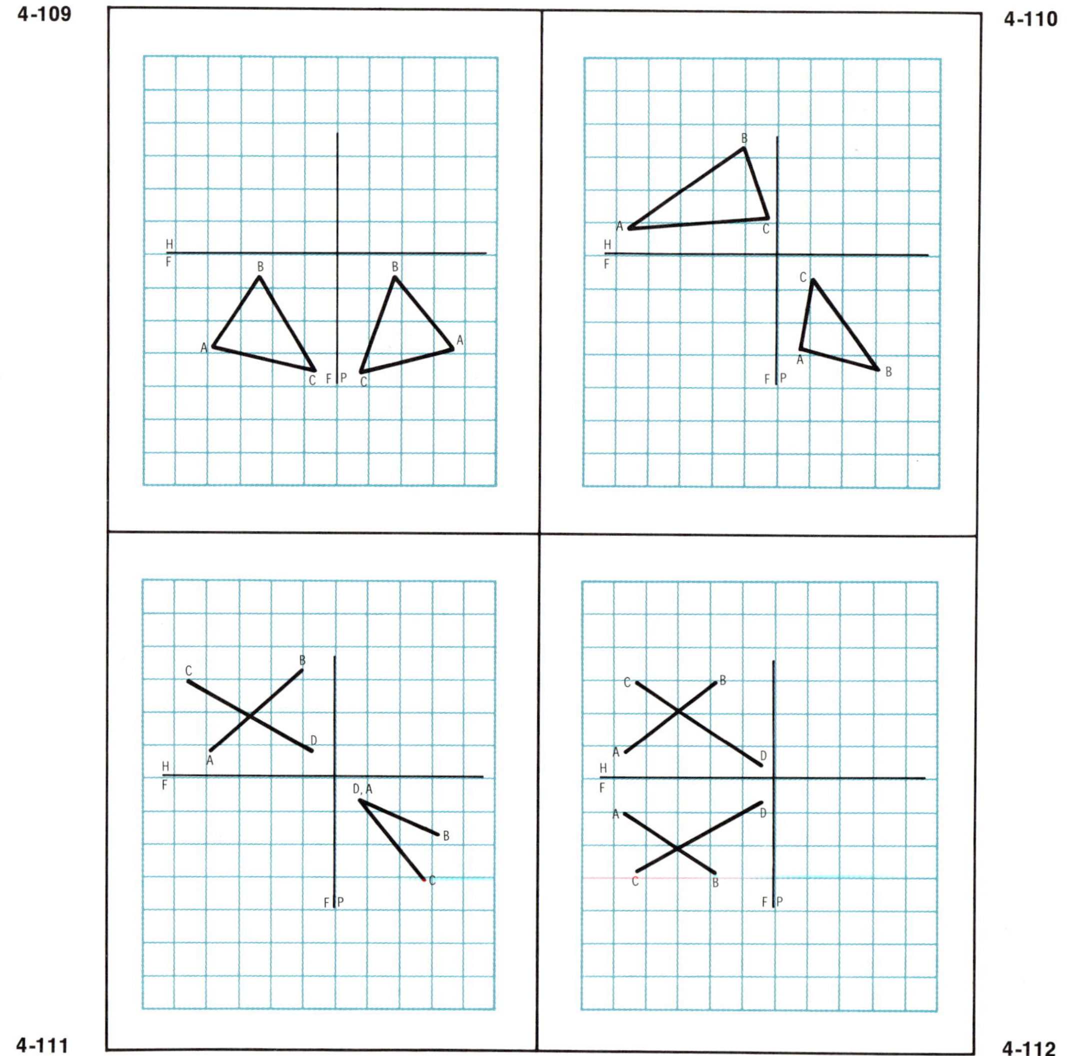

4-109

4-110

4-111

4-112

4-6. Determine if the lines given in [4-117 to 4-122] are parallel. Measure and transfer each figure at double size to an $8\frac{1}{2}$- by 11-inch sheet of unruled paper and complete the necessary views to determine parallelism. If the lines are parallel, letter PAR in the upper-right-hand corner; if not, letter NON.

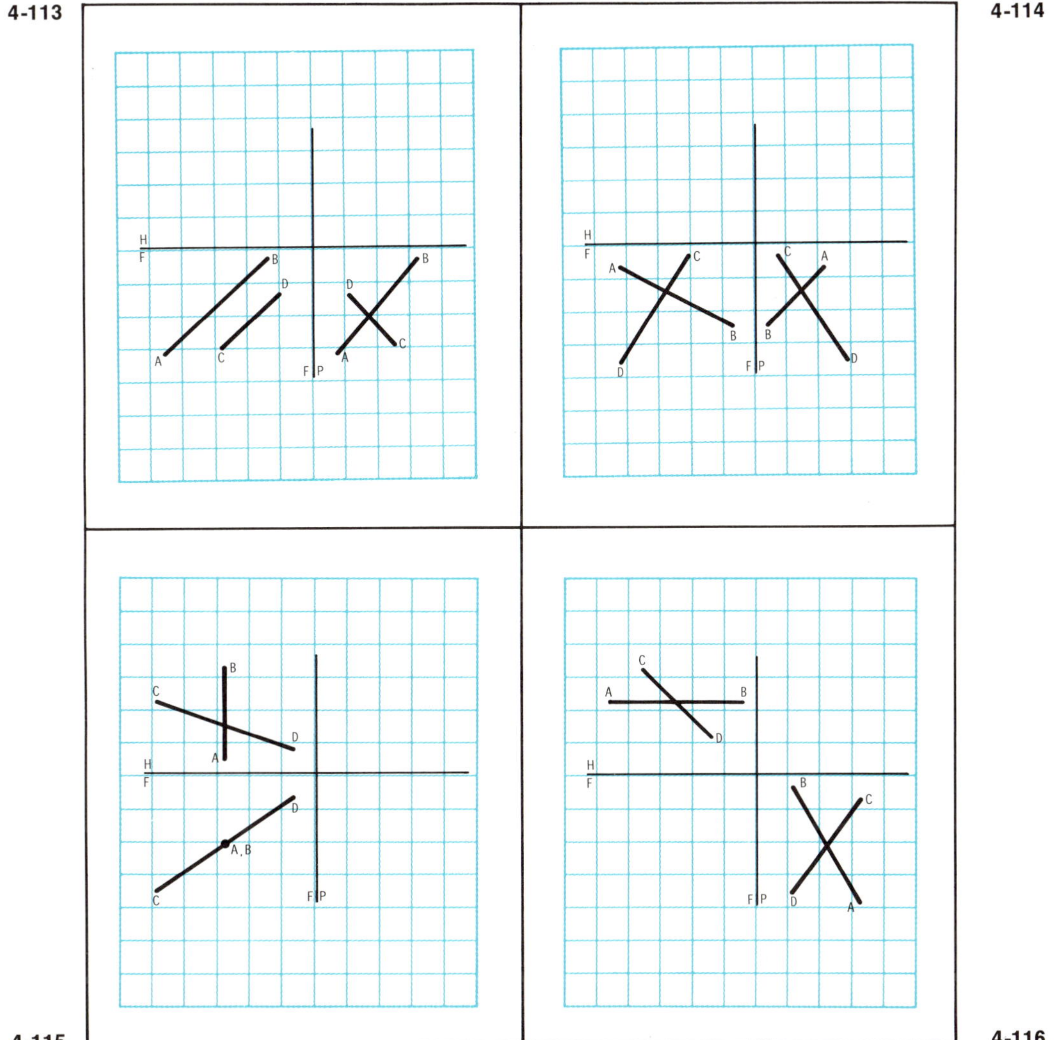

4-7. Determine the true angles between the pairs of lines given in [4-123 to 4-128]. Measure and transfer each figure at double size to an $8\frac{1}{2}$- by 11-inch sheet of paper and complete the necessary views to determine the true angle. Letter the angular value between the lines in the true-angle view.

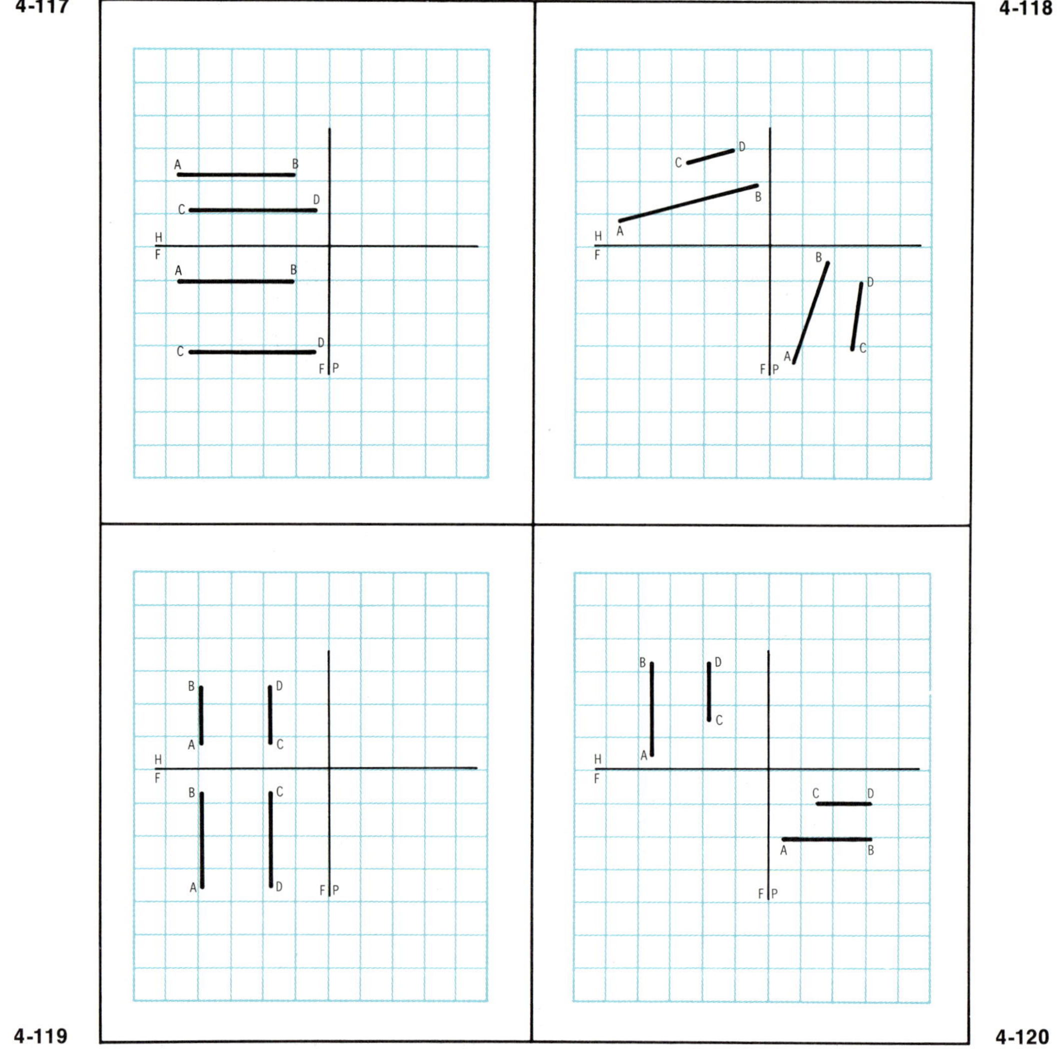

4-8. Using the auxiliary-view method, determine the true angle between the line and the plane in [4-129, 4-130, and 4-131]. Measure and transfer each figure at full size to an $8\frac{1}{2}$- by 11-inch sheet of paper and complete any necessary views to determine the true angle. Letter the angular value on the true-angle view.

4-9. Using the rotation method, determine the true angle between the line and the plane in [4-132, 4-133, and 4-134]. Measure and transfer each figure at full size to an $8\frac{1}{2}$- by 11-inch sheet of paper and complete any necessary views to determine the true angles. Letter the angular value on the true-angle view.

4-121 **4-122** **4-123** **4-124**

4-10. Specify the bearing of the lines *AB* in [4-129, 4-130, and 4-131]. Measure and transfer each figure at double size to an $8\frac{1}{2}$- by 11-inch sheet of paper and complete any necessary views to determine the bearing of the line. Letter the angular value on the proper view. Read the bearing along the line from point *A* to point *B*.

4-11. Specify the slope of lines *AB* in [4-132, 4-133, and 4-134]. Measure and transfer each figure at double size to an $8\frac{1}{2}$- by 11-inch sheet of paper and complete any necessary views to determine the slope of the line. Letter the angular value on the proper view.

4-12. For Problem 4-11 specify the grade of the lines *AB* instead of the slope.

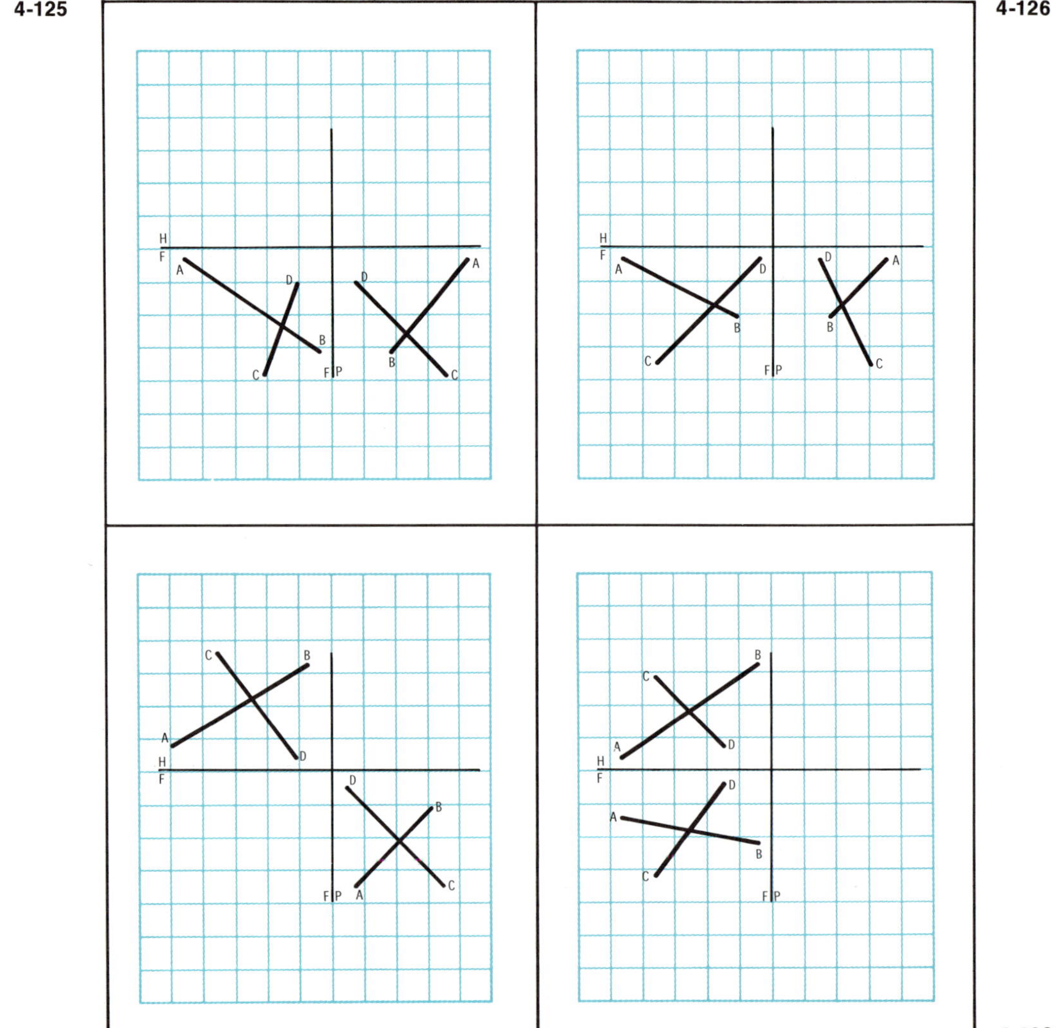

4-13. Determine the true angles between the intersecting planes in [4-135, 4-136, and 4-137]. Measure and transfer each figure at full size to an $8\frac{1}{2}$- by 11-inch sheet of paper and complete the necessary views to determine the true angles. Letter the angular value on the true-angle view.

4-14. Determine the true angles between the nonintersecting planes in [4-138, 4-139, and 4-140] as instructed in Problem 4-13.

4-15. Specify the strike of the planes *ABC* in [4-135, 4-136, and 4-137]. Measure and transfer *only* the plane *ABC* for each figure at double size to an $8\frac{1}{2}$- by 11-inch sheet of paper and complete the necessary views to specify the strike. Letter the value of the strike angle in the appropriate view.

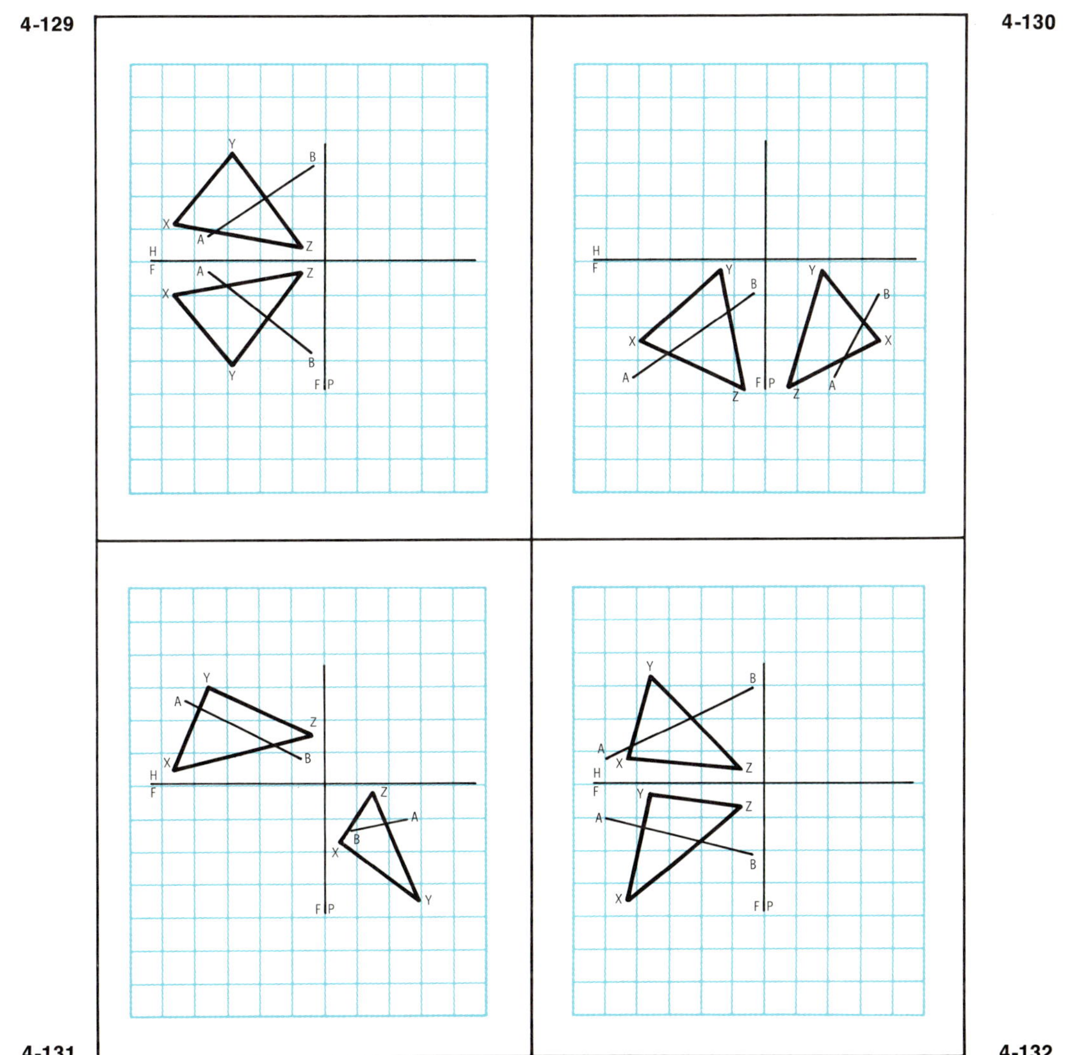

4-16. Specify the dip of the plane *XYZ* in [4-129, 4-130, and 4-131]. Measure and transfer *only* the plane *XYZ* for each figure at double size to an $8\frac{1}{2}$- by 11-inch sheet of paper and complete the necessary views to specify the dip. Letter the value of the dip angle in the appropriate view. Also, show the dip direction in the appropriate view.

4-17. Determine the shortest distance between the point *X* and the line *AB* in [4-141, 4-142, and 4-143]. Measure and transfer each figure at full size to an $8\frac{1}{2}$- by 11-inch sheet of paper and complete the necessary views to determine the true distance. Measure and letter the true distance on the appropriate view.

4-18. Determine the shortest distance between the two lines *AB* and *XY* in [4-144, 4-145, and 4-146] as instructed in Problem 4-17.

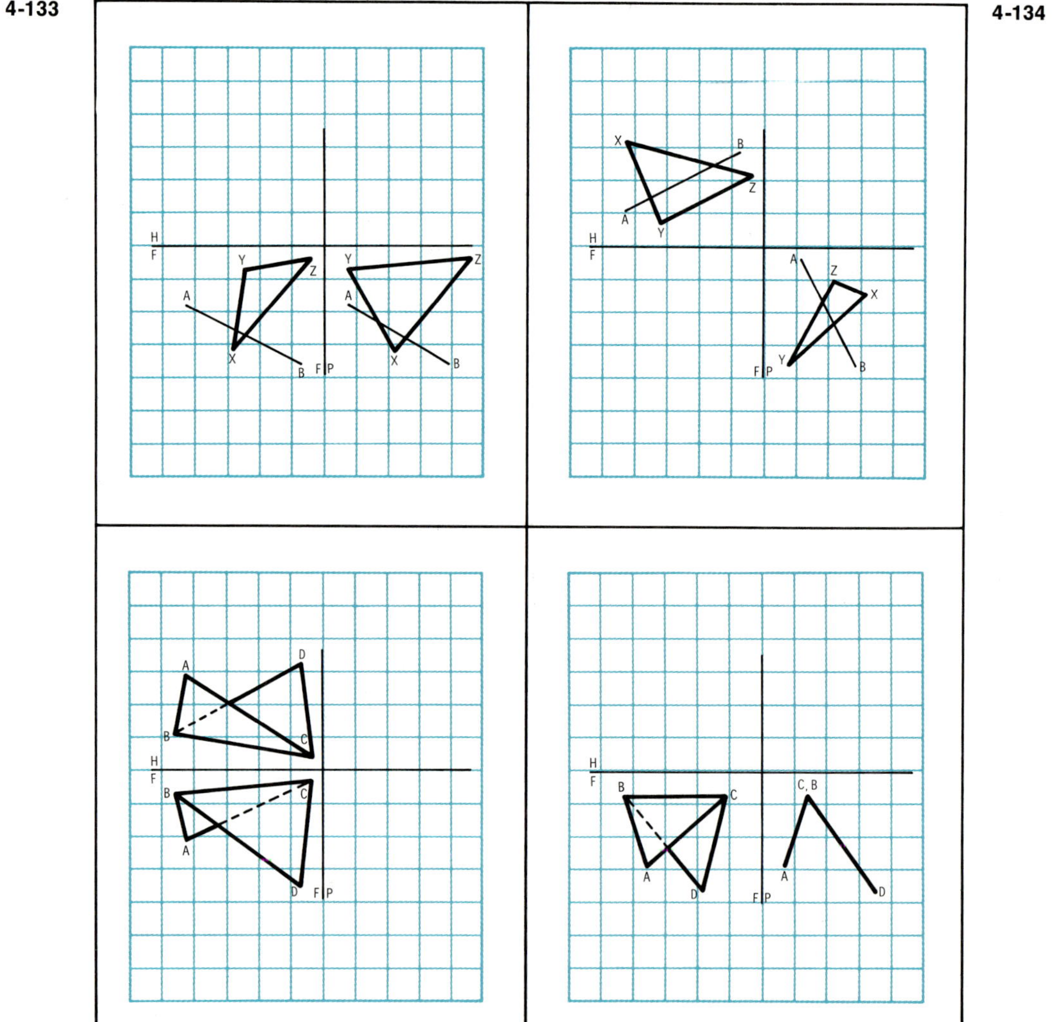

4-19. A policeman and a dangerous criminal are involved in a gun battle on Main Street (which is an east-west street). The policeman is shooting from a street position on the south side of the street at the corner of the Commercial Building. The criminal is across the street and is crouched on the third-story landing of a fire escape 30 feet above street level, firing back. The fire escape is on the south wall of the Enterprise Building, 90 feet from the corner that is opposite the policeman. Main Street, which is 65 feet wide including the sidewalks, separates the Commercial and Enterprise Buildings. Using graphical techniques, determine the distance the policeman is shooting and the angle from the street level.

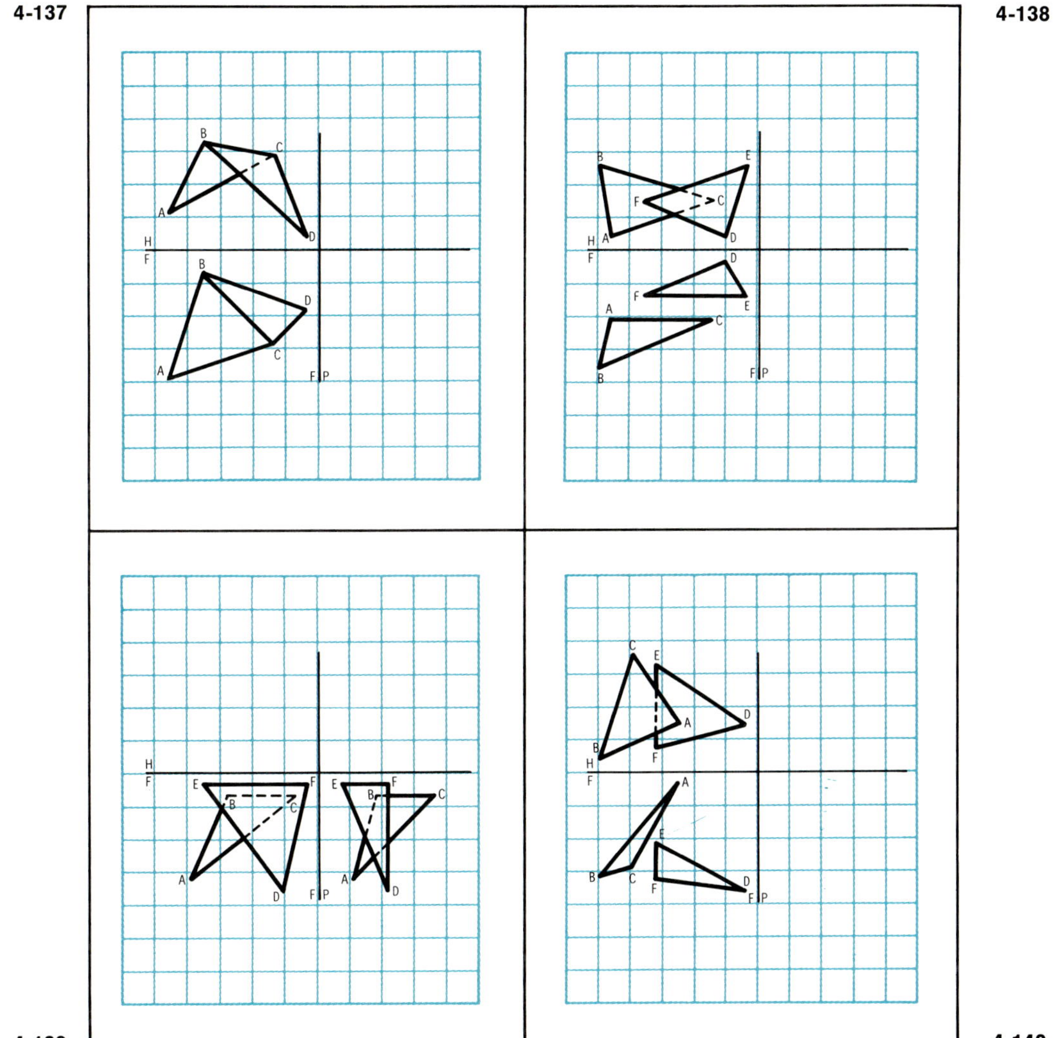

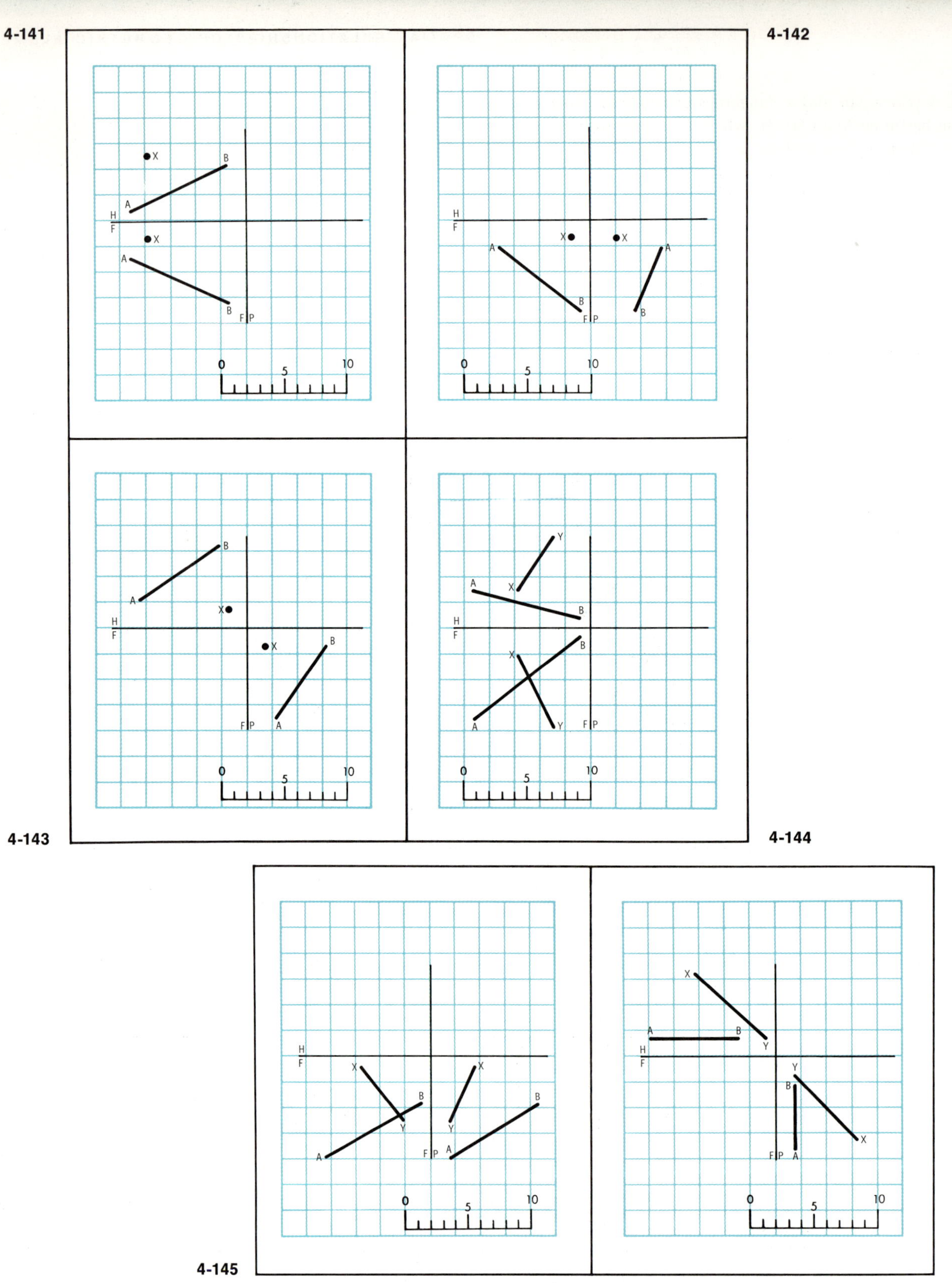
4-141
4-142
4-143
4-144
4-145
4-146
H
F
A
B
X
Y
F P
0
5
10

applications and conventions 5

This ornament's decorative shape arose from the rectilinear limitation of spheroids.

Most of us have an awareness of the value of technical sketches and engineering drawings. However, many of us may not have realized that competent sketches and drawings must provide much more information than the general shape of the object. Many symbols and standardized procedures are needed to convey precise information. Dimensioning, tolerancing, sectioning, threaded fasteners, etc., all require specialized but uniform treatment. Also, in designing it is frequently necessary to expand the principles discussed in Chapters 3 and 4 to include the precise location of intersecting prisms or planes and the development of surface areas. This chapter is devoted to an explanation of the general principles involved for these specialized conventions and applications.

5-1

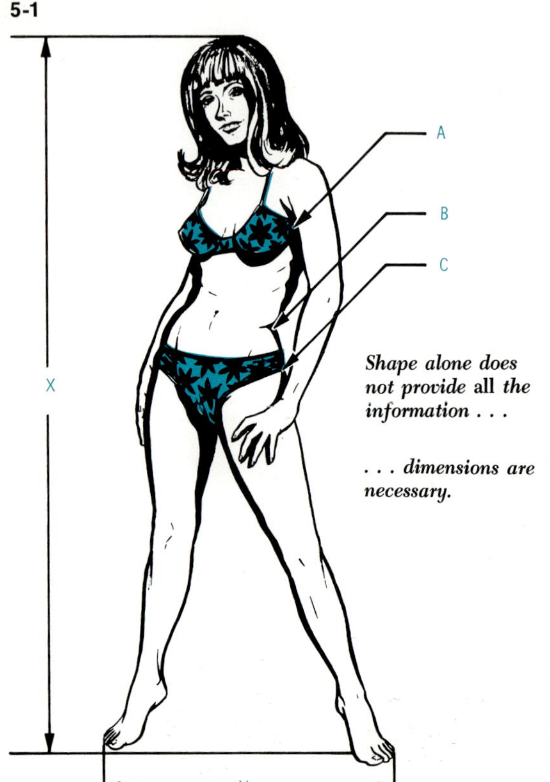

Shape alone does not provide all *the information . . .*

. . . dimensions are necessary.

dimensioning

Although the graphic image of a design is very important, it alone is not sufficient. If we wish to manufacture the design, dimensions are needed to specify sizes, locations, and other pertinent information so that it can be constructed accurately and repeatedly.

Dimensioning refers to the numbers and notes on a drawing that define the size of the object, the methods of manufacture, or the relation of the object to other objects. To produce correctly dimensioned sketches and drawings, the engineer must be able to visualize the final form of the product, the function it is intended to per-

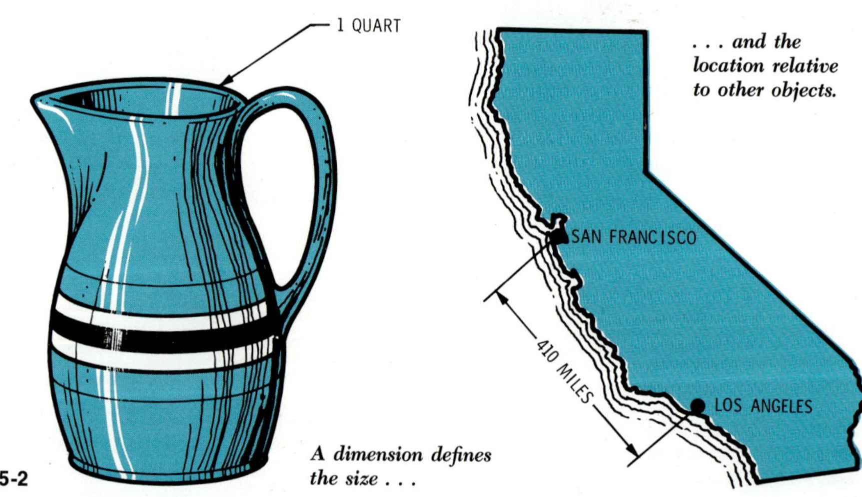

5-2

A dimension defines the size . . .

. . . and the location relative to other objects.

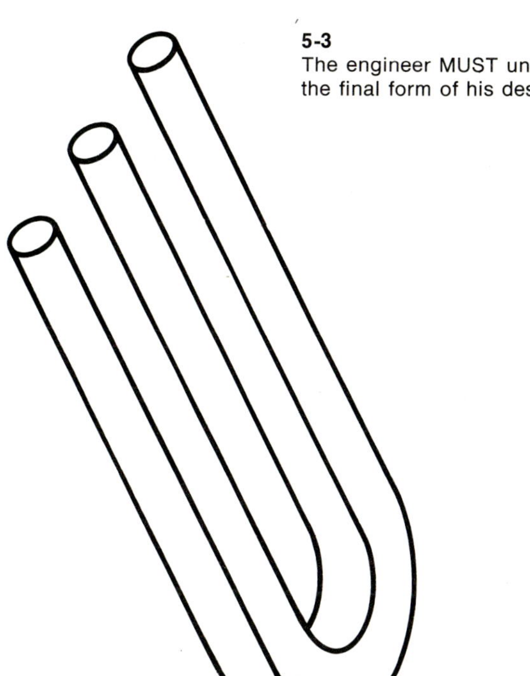

5-3
The engineer MUST understand the final form of his design.

Do not provide confusing dimensions.

5-4

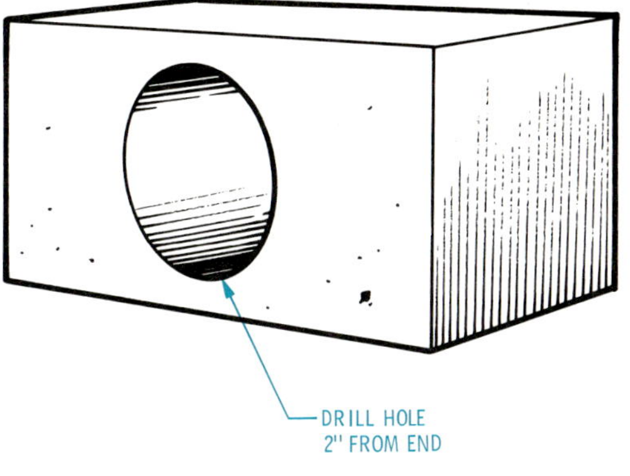

form, how it will be manufactured, and how it will be checked for conformity with the specifications.

Dimensions enable craftsmen to build the design. Remember, it is the craftsman's responsibility to build the part, product, or device—not to design it. If he must stop the manufacturing process and calculate or scale a missing dimension, or reset his machine tools to accommodate an inaccessible reference point, the dimensions provided him have been poorly conceived and inadequately presented.

Practically all man-made objects can be reduced to various combinations of four geometric shapes: rectangular prisms, cylinders, cones, and spheres. The primary function of dimensioning is to specify the *size* of the geometric elements and their relative *location* one to another.

Dimensioning Conventions

Dimensioning is a language, a way of communicating information from the engineer to the craftsman. Often the design engineer and the craftsmen never meet, so the drawing must convey all the information needed. Therefore, dimensions used must transmit consistent and mean-

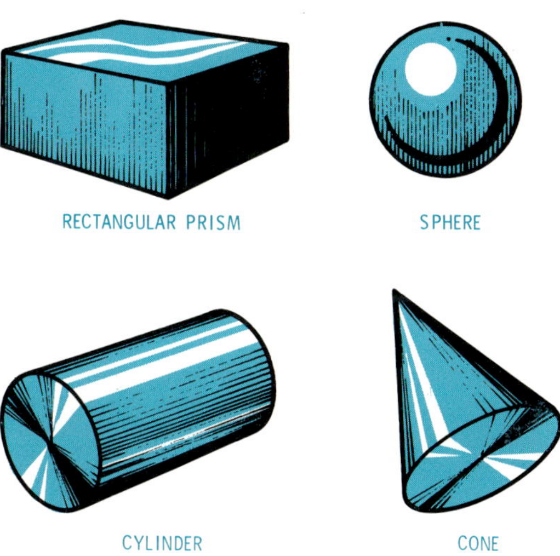

5-5
Four basic solids of man-made objects.

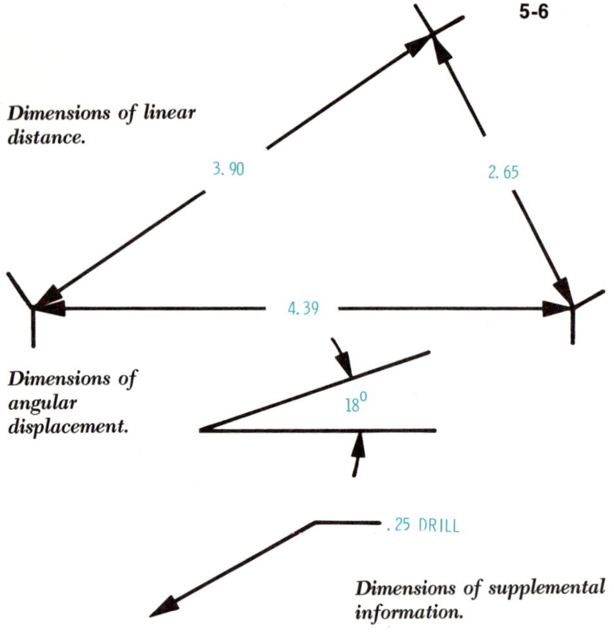

FULL SIZE	or	1.00 = 1.00	or	1 = 1	or	1/1
HALF SIZE	or	.50 = 1.00	or	$\frac{1}{2} = 1$	or	1/2
QUARTER SIZE	or	.25 = 1.00	or	$\frac{1}{4} = 1$	or	1/4
EIGHTH SIZE	or	.125 = 1.00	or	$\frac{1}{8} = 1$	or	1/8
TWICE SIZE	or	2.00 = 1.00	or	2 = 1	or	2/1
TEN TIMES SIZE	or	10.00 = 1.00	or	10 = 1	or	10/1

5-7
Specification of scales.

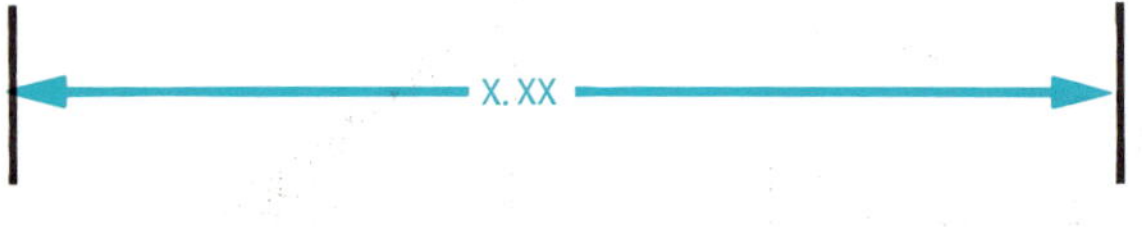

5-8
Dimension line.

ingful information. To accomplish this, certain conventions and standard procedures have been developed.

Scale

Unlike sketches, drawings must be drawn to *scale*, that is, all dimensions on the drawing made in the same proportion (or scale) to their counterpart on the real object. Selection of an appropriate scale is usually determined by the size of the object and the size of the paper on which it is to be drawn. Large objects, such as buildings, ships, and bridges, must be drawn at a scale less than full size. On the other hand, very small objects (such as the parts of a watch) usually are drawn larger than actual size so that the detail and dimensional notation may be seen and understood. Whatever scale is selected must be clearly indicated on the drawing [5-7]. In some cases it is advantageous to include on the drawing a measured scale length with appropriate divisions (similar to a map scale). Then, if the drawing is photographically reduced, the scale length will be consistent with the new size.

Dimensioning Symbology

Dimensioning uses many symbols, of which lines are an important part. Four lines are used extensively: *dimension lines*, *extension lines*, *center lines*, and *leaders*. Each of these lines has a specific meaning and rules that govern its use.

Dimension Lines. *Dimension lines* are thin solid lines with arrows at each end [5-8]. In most applications there is a space at or near the midpoint of dimension lines to accommodate the dimension numerals. Dimension lines are drawn parallel to the feature being dimensioned, and they are placed a minimum of $\frac{3}{8}$ inch away from the object [5-9]. Succeeding dimension lines should be placed at constant intervals of at least $\frac{1}{4}$ inch with shorter

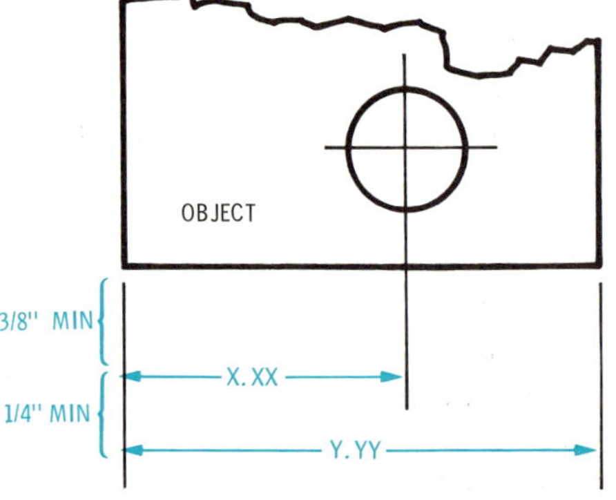

5-9
Placement of dimension lines.

dimensions placed inside longer ones. Dimension lines must *never* be crossed by *any* type of line. Remember that dimensions, although important, must never interfere with a clear view of the object.

Extension Lines. *Extension lines* are thin solid lines that extend the edge of the object or the location of an important feature to a place that permits convenient and clear dimensioning [5-10]. The extension line should leave a small visible gap (about $\frac{1}{32}$ inch) at the object and terminate approximately $\frac{1}{8}$ inch past the last dimension line. Extension lines may cross object lines and other extension lines.

Center Lines. *Center lines* are thin, alternately long and short dashed lines and are used to indicate the center of a cylindrical form or the axis of symmetry of an object. When center lines cross, as at the center of a circle, there must be a definite crossing of either two short or two long dashes. Center lines are used also as extension lines when locating cylindrical forms.

Leaders. *Leaders* are thin solid lines that have an arrow or a small dot at one end. They are used to show where a dimension or a note applies [5-12]. When an arrow is used, the leader must aim at the center of the feature with its point just touching the feature. When a dot is used, the note usually relates to a general area on the object, and the dot is placed inside the area. Leaders are never drawn horizontally or vertically and should terminate with a short horizontal line meeting the center of the first line of the note.

Arrowheads. Well-executed arrowheads on leaders and dimension lines do much to improve the clarity and attractiveness of a drawing. Arrowheads should be about $\frac{1}{8}$ inch long in a ratio of three units long to one unit wide [5-13]. If the space between the extension lines is very small, the arrows should be placed outside the extension lines, pointing inward [5-14].

5-10
Extension lines.

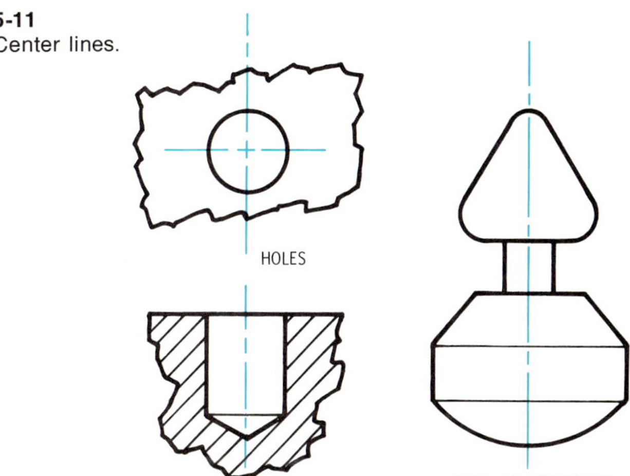

5-11
Center lines.

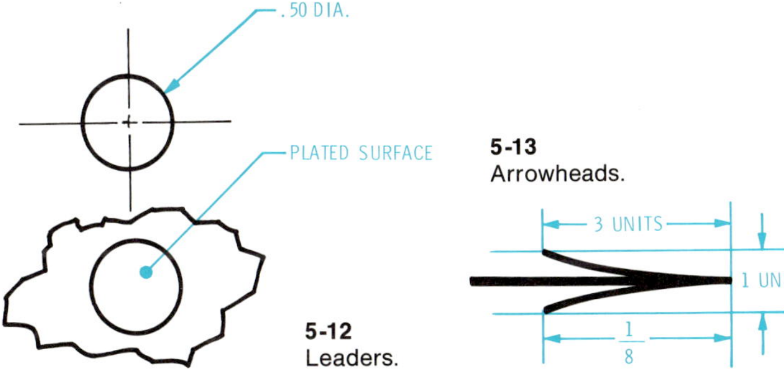

5-12
Leaders.

5-13
Arrowheads.

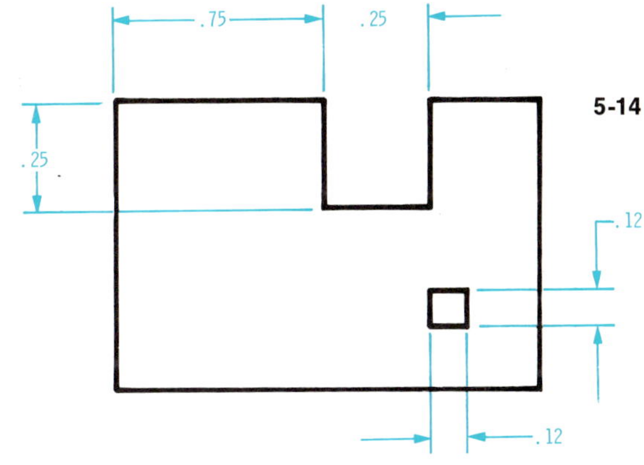

5-14

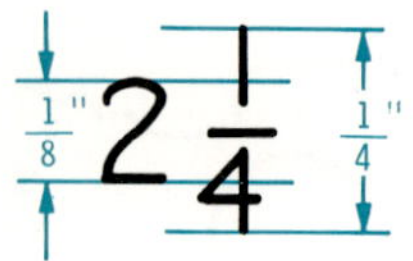

5-15
Lettering.

Dimensional Lettering

Except for the case of notes, the majority of dimensional lettering consists of numbers and associated symbols, such as degree, inch, and foot marks. The style of lettering used should be either vertical or inclined single-stroke Gothic, as discussed in Chapter 1. Recall that most letters and numbers are about $\frac{1}{8}$ inch high with fractions being $\frac{1}{4}$ inch [5-15]. Use guidelines to assist in producing clear and attractive lettering.

Units. Always identify the units that apply to a dimensional figure unless all or nearly all are identical. In the United States, where dimensions on mechanical engineering drawings normally are stated in inches, it is conventional practice to omit the inch marks. However, to avoid the possibility of confusion, it is wise to use a general note, such as ALL DIMENSIONS IN INCHES UNLESS OTHERWISE SPECIFIED. A good rule of thumb for showing units might be: if their use eliminates confusion—use them. For example, the notation, 1 VALVE, is decidedly different from 1″ VALVE.

Dimensional Number Systems. Historically, dimensions on drawings in the United States have appeared as combinations of whole and fractional numbers such as $10\frac{7}{16}$, $3\frac{1}{4}$, $7\frac{13}{32}$, and so on. As a result of this practice, many standard parts are known by their fractional dimensions, such as a $\frac{5}{8}$ inch bolt, $\frac{3}{4}$ inch board, $1\frac{1}{4}$ inch pipe, and $\frac{1}{4}$ inch plate. Although cumbersome, it is quite likely that these names and the procedures for specifying them will remain unchanged for some time. However, with the increasing requirements of mass production, the need for greater accuracy, and convenience of base-ten mathematical systems, more and more industries are adopting decimal measurement. Preferably, most decimal measurements are written with two significant figures to the right of the decimal point; however, three significant figures are quite common because of increased accuracy requirements. For numbers less than 1, the zero before the decimal point is omitted on drawings.

Dimensional Notes. When there is a need to specify or describe a feature on a drawing other than a length or an angle, a *note* is often used. Notes usually specify such things as the dimensions of a hole, the size and type of screw thread, or a machine process, such as drilling, countersinking, or reaming [5-16]. Notes should always be oriented horizontally with the drawing and usually are connected to the feature that they describe by leaders.

5-16
Typical uses of notes.

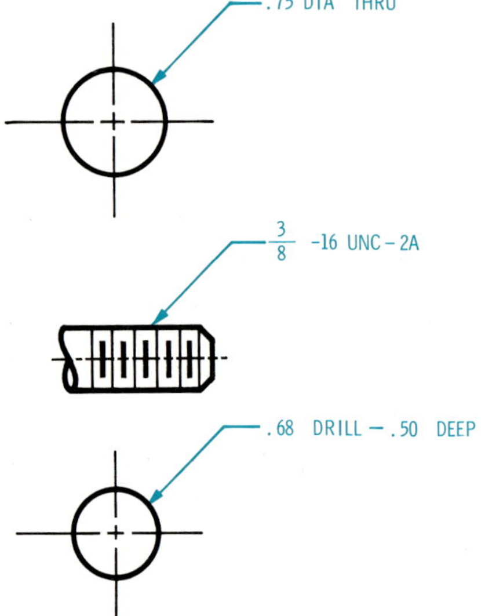

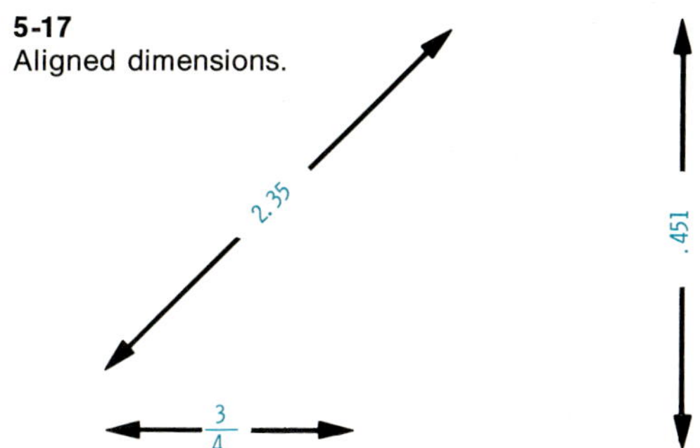

5-17
Aligned dimensions.

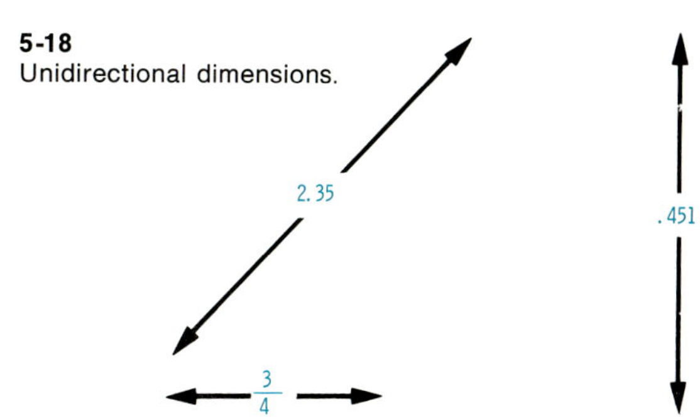

5-18
Unidirectional dimensions.

Orientation of Numbers. Dimension numbers either may be *aligned* with the corresponding dimension line, or they may be *unidirectional.* If they are aligned, the numbers are oriented so that they are either read from left to right or from bottom to top (that is, from the bottom or right side of the drawing) [5-17]. Unidirectional dimensions are becoming the more popular. They are always horizontal and are read from the bottom of the drawing [5-18].

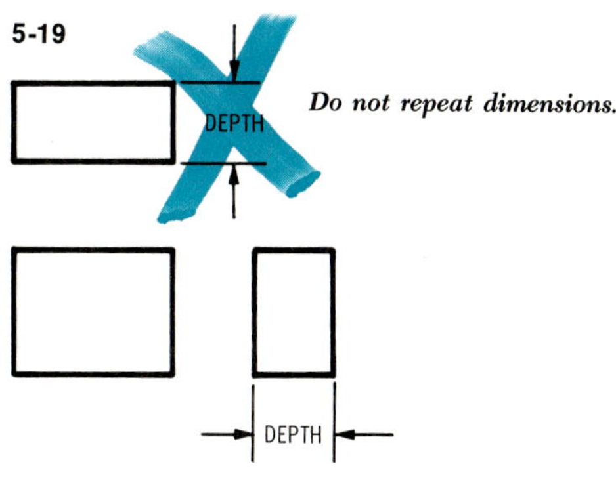

5-19

Do not repeat dimensions.

Placement of Dimensions

Dimensions must define precisely the shape of the object and be placed so as not to detract from the views of the object. Each feature of the object should be defined by one and only one dimension. For example, if the depth of an object is given in a right side view, it should not be repeated in the top or in any other view [5-19]. Dimensions are normally placed outside the outline of the object and positioned so as to maintain at least a $\frac{3}{8}$-inch open space around the object [5-20].

Overall dimensions are normally desirable, because they are needed for raw material requisitions and other facets of the manufacturing process [5-21]. If the indi-

5-20

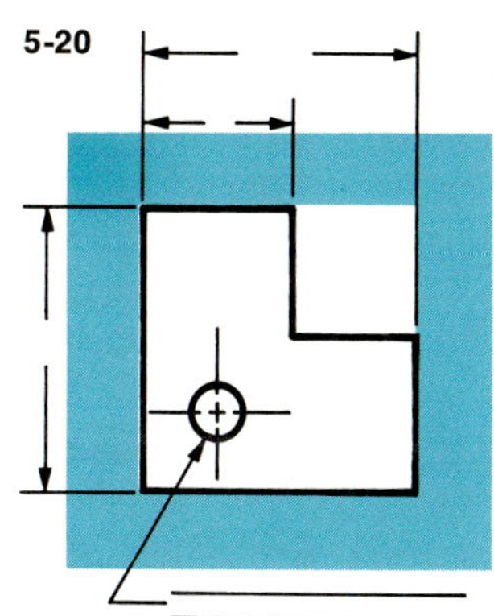

Leave $\frac{3}{8}''$ minimum open space around object . . .

5-21

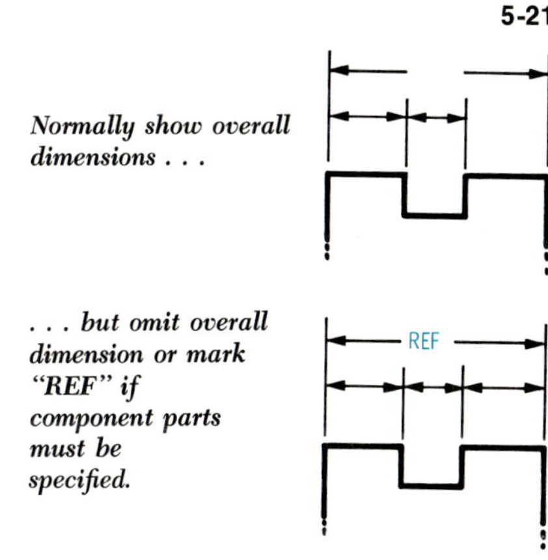

Normally show overall dimensions . . .

. . . but omit overall dimension or mark "REF" if component parts must be specified.

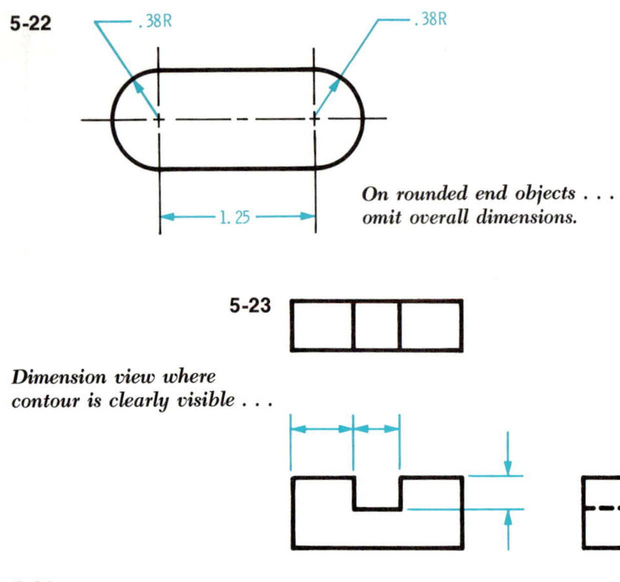

5-22

On rounded end objects . . . omit overall dimensions.

5-23

Dimension view where contour is clearly visible . . .

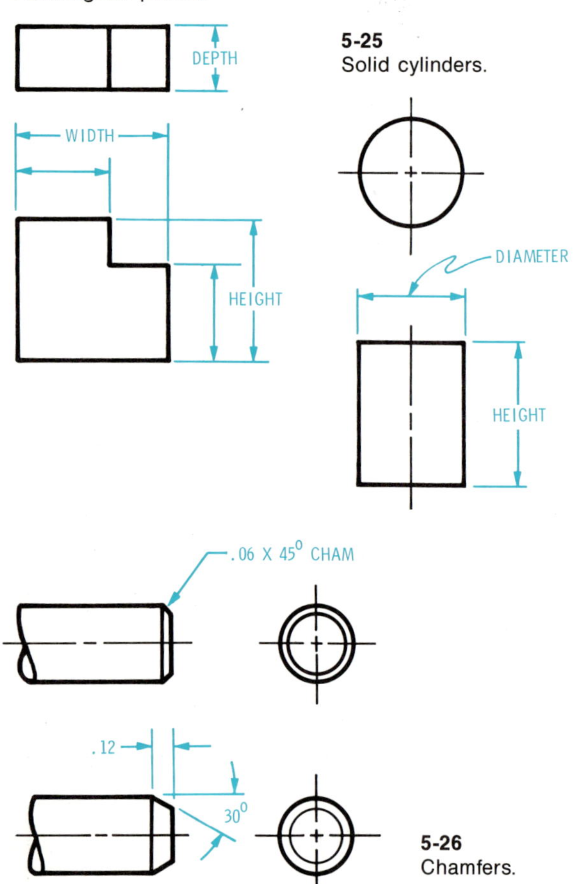

5-24
Rectangular prisms.

5-25
Solid cylinders.

5-26
Chamfers.

vidual dimensions making up an overall dimension are more important, the overall dimension is accompanied by the note REF. This indicates that the dimension is to be used for reference only and not for manufacturing. One situation in which overall dimensions are seldom used (unless REF is added) is in the case of rounded-end objects, where both the distance between radius centers and the size of the radii must be specified [5-22].

Dimensions can be understood best when directly related to that view which pictures the shape or contour most clearly [5-23]. Avoid dimensioning to hidden lines. Extension lines are directed to only one view and must extend fully to the feature being dimensioned.

Size Dimensions

Preferred methods for specifying the size and location of the four basic geometric shapes commonly found in manufactured objects have been developed through the years. The rectangular prism is the most common of these.

Rectangular Prisms. When dimensioning a prism, the height and the width are usually specified in the front view [5-24]. The depth is shown in either the top or side views. If a dimension (such as width) applies to more than one view, it is good practice to place it between the two views. If possible, related dimensions are grouped together for clarity.

Cylinders. Next to the prism, the cylinder is the most commonly used geometric shape. The cylinder can be a bolt, a drive shaft, a wheel, or a round hole. The size (diameter and height) of a solid cylinder is preferably indicated in the view that shows the cylinder as a rectangle [5-25]. Full circles are dimensioned by their diameters, but partial circles (arcs) are dimensioned by their radii [5-28].

It is much more graphic and understandable to provide two views of a cylindrical form, although not entirely necessary. When two views are given, place the diameter dimension between the rectangular and circular views. If only one view (the rectangular view) is given, it is necessary to follow the diameter dimension with the note DIA. Chamfers are dimensioned by their angle and width [5-26].

Cylindrical holes are normally dimensioned in the circular view with a leader and a note specifying the diameter and the depth of the hole [5-27]. Manufacturing-process information required to make the hole is

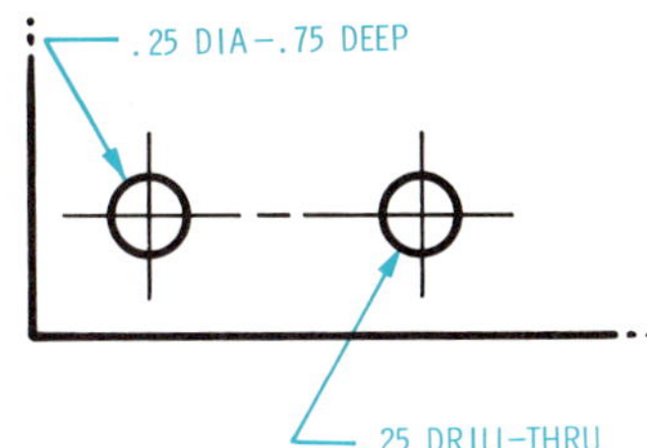

5-27
Holes (hollow cylinders).

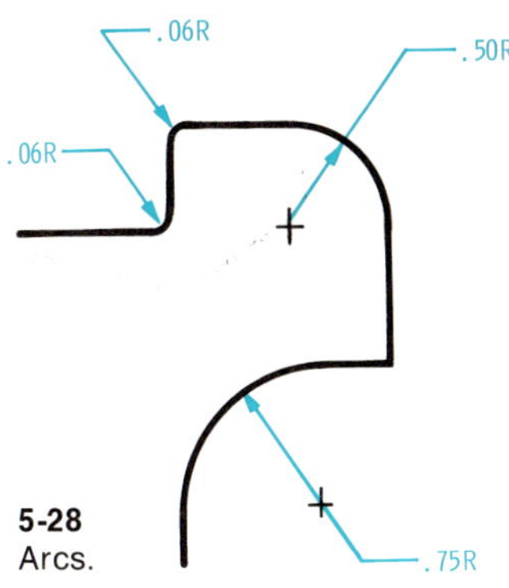

5-28
Arcs.

often included in the note. The depth is not specified if the hole goes through the object. If this is not readily apparent from the drawing, the note THRU follows the diameter specification. When more than one hole has exactly the same size, only a single leader pointing to one of the holes is used with the number of holes added to the note [5-39].

Arcs and Irregular Curves. Arcs are dimensioned in the circular view using a leader [5-28]. The size is specified as a radius and is always followed by the symbol R. If the object has many fillets and rounds of the same radius, a general note such as, ALL FILLETS AND ROUNDS .06 R UNLESS OTHERWISE SPECIFIED, is used. An alternative method is to use a leader with a note .06R TYP (typical) pointing to one radius of the group. Irregular curves are best dimensioned using a coordinate system bounded by two datum lines that have been established at right angles [5-29].

Sphere and Cones. Spheres are dimensioned by giving their diameter. Partial spheres are dimensioned by their radius [5-30]. For cones or portions of conical shapes, it is common practice to dimension the height and the diameter in the triangular or trapezoidal view [5-31]. Occasionally, it may be useful to dimension a cone using an angular measurement with the angle being specified in the triangular view.

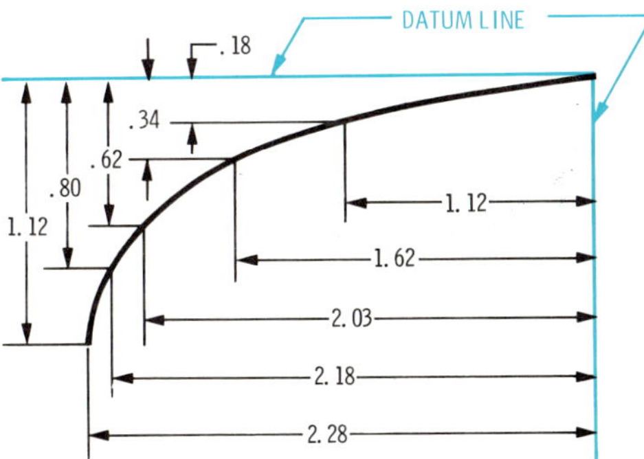

5-29
Noncircular curves.

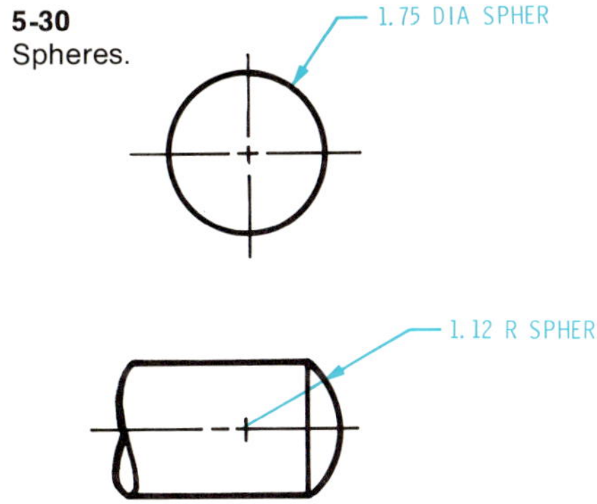

5-30
Spheres.

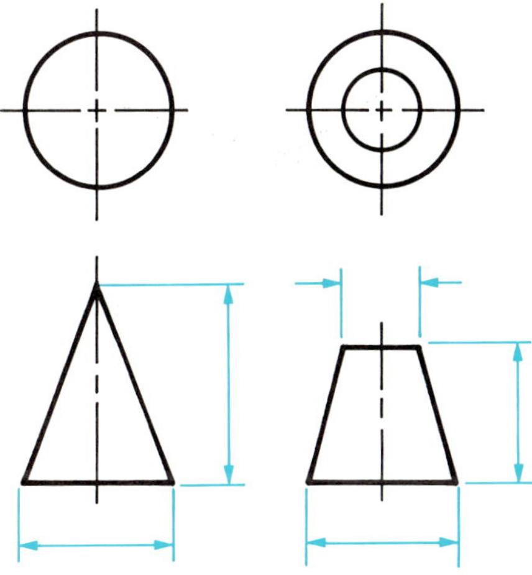

5-31
Cones.

Location Dimensions

In dimensioning, the location as well as the size of each geometric shape must be specified. The first step in establishing location dimensions is to select a single appropriate set of planes of the object to serve as reference planes or datums from which to initiate all location dimensions [5-32]. Often the choice of these datums is obvious from the shape of the object, its intended use, or its relationship to adjacent parts. In general, the largest or most prominent flat surface in each of the three basic directions (height, width, and depth) is selected. In the case of a symmetrical object, the center line (or lines) is normally taken as a datum [5-33].

Prisms must be located as well as sized in the view that clearly shows the contour [5-23]. Round features—

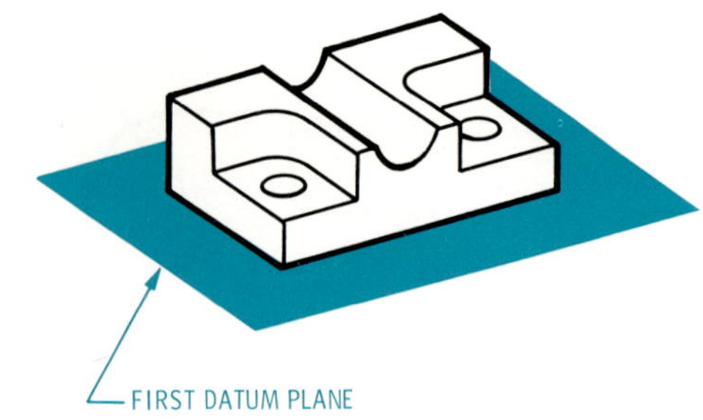

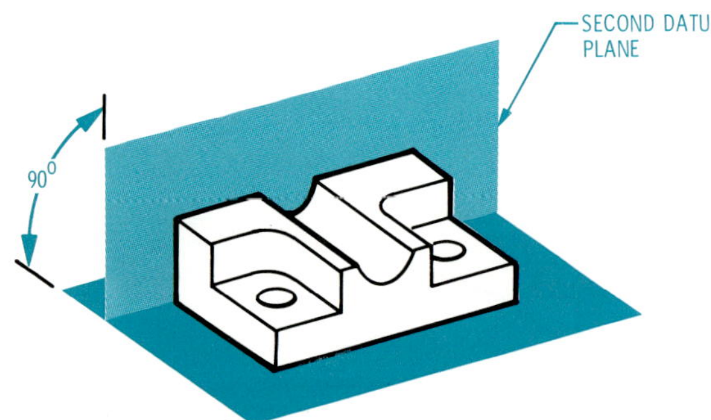

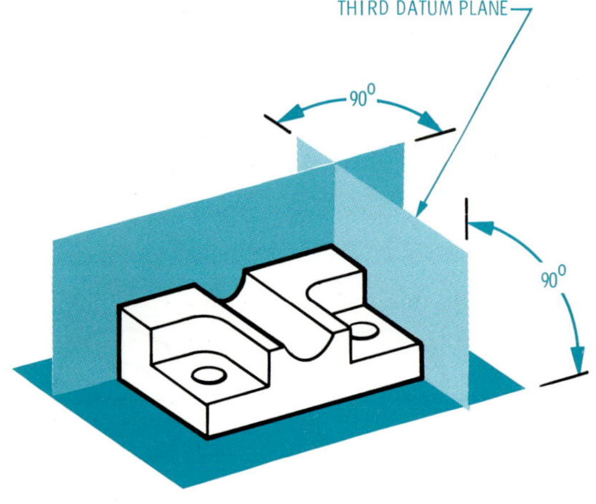

5-32
The use of datum planes in dimensioning.

5-33

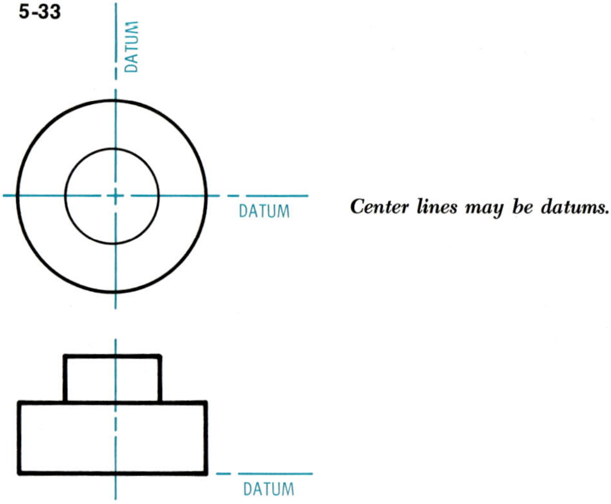

Center lines may be datums.

5-34

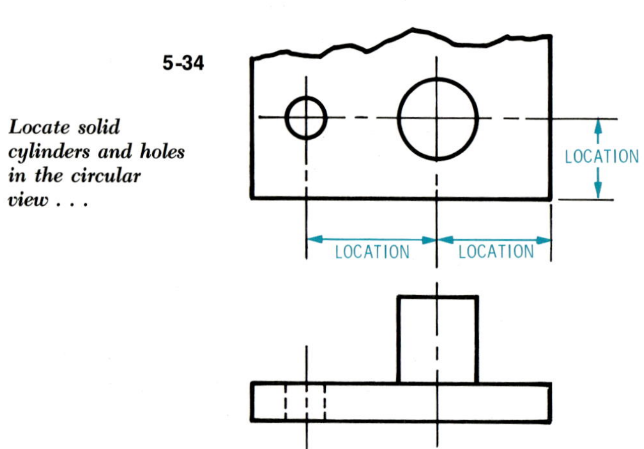

Locate solid cylinders and holes in the circular view . . .

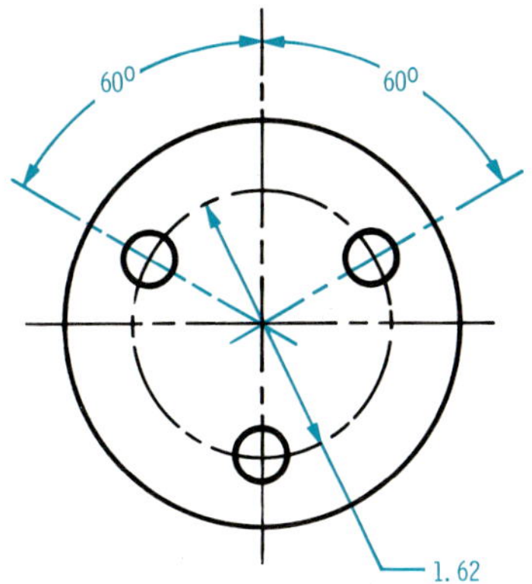

5-35
Angular dimensioning.

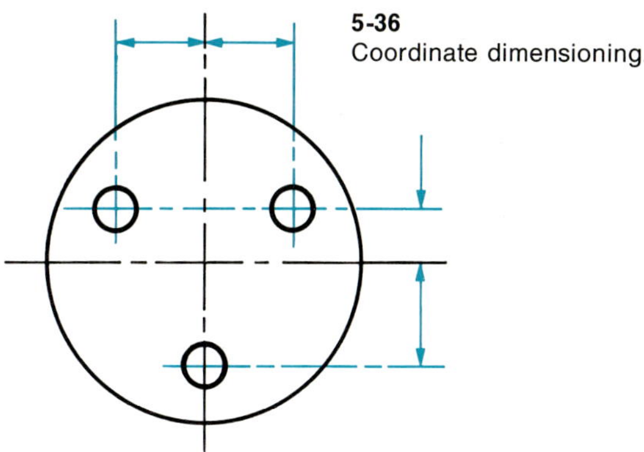

5-36
Coordinate dimensioning.

both solid cylinders and holes—must be located by their center lines in the circular view [5-34]. Holes arranged in a circular pattern called a *bolt circle* may be located by their angular separation and the diameter of the circle [5-35]. For more accurate location, coordinate dimensions are usually used [5-36].

Threaded Fasteners

Fasteners are perhaps the most common features found in manufactured objects. Special methods have been established for drawing and dimensioning threaded fasteners and the holes associated with them. Threads may be drawn as they actually appear, but since this is very time consuming, either *schematic* or *simplified* thread symbols are usually used [5-37]. Schematic symbols use short thick lines to represent the roots of the thread with thin longer lines representing the crests. Simplified symbols use hidden lines to represent the root circle of the thread.

Threads are usually dimensioned with a note specifying in sequence the diameter, pitch (number of threads per inch), thread series, class of fit, and the letter A for external thread or B for internal thread (if a Unified thread series) [5-38]. If the thread is not a standard right-hand thread, the letters LH must be added to indicate left-hand thread.

The leader for an external thread note is usually directed to the rectangular view [5-38]. The leader for an internal thread note points to the circular view, but the thread size in the note must be preceded by a specification for the size of the hole that must be made prior to cutting the threads [5-39]. The diameter of this hole, called the *tap drill size*, is slightly larger than the root

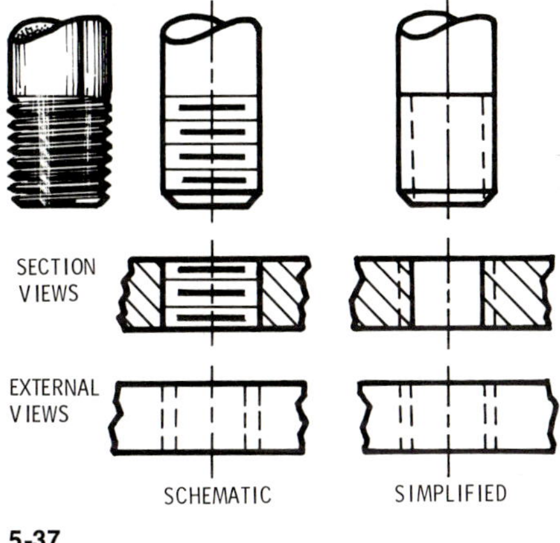

5-37
Thread symbols.

5-38
Dimensioning external threads.

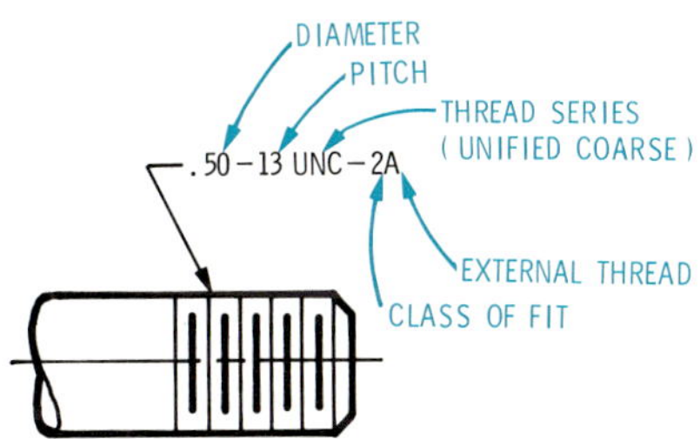

5-39
Dimensioning internal threads.

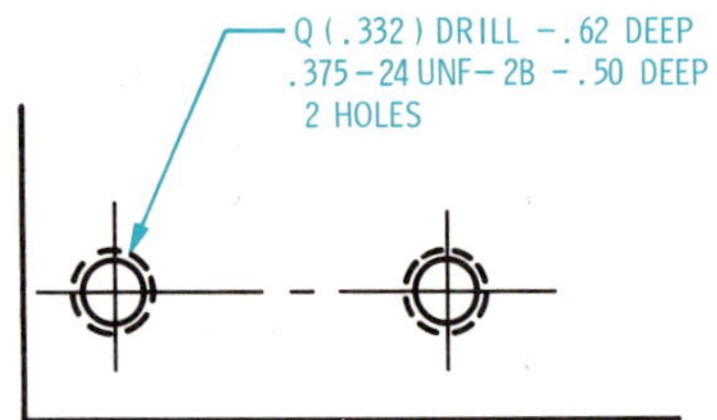

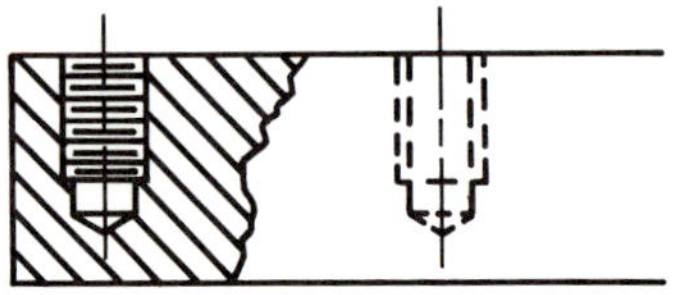

diameter of the thread and may be found in standard thread tables.

The method for representing bolts and screws on drawings is shown in [5-40]. In addition to thread size, fastener specifications usually include such things as length, finish, material, type of drive, head shape, and type of fastener. See Appendix IV for detailed information on types of threads and fasteners.

5-40
Threaded fasteners.

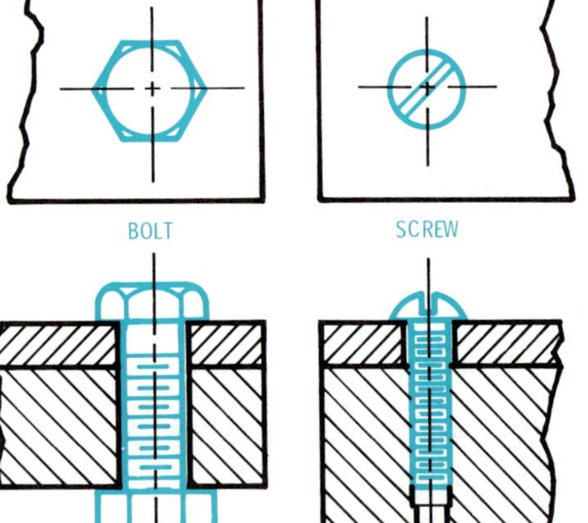

Special Holes

The most common type of hole associated with fasteners is a drilled hole [5-41]. Note that the specified depth does not include the cone point left by the drill. (The point angle of the drill is standard and need not be specified. It is usually drawn at 30°.)

Other common types of holes and the tools that produce them are shown in [5-41]. Typical dimension notes are also shown. Reamers are used to bring an existing hole (usually drilled) to exact size. Counterbores create an enlarged hole to permit recessing the head of a fastener. The spotfacing tool is used only to produce a smooth surface for the fastener to bear upon and requires no

5-41

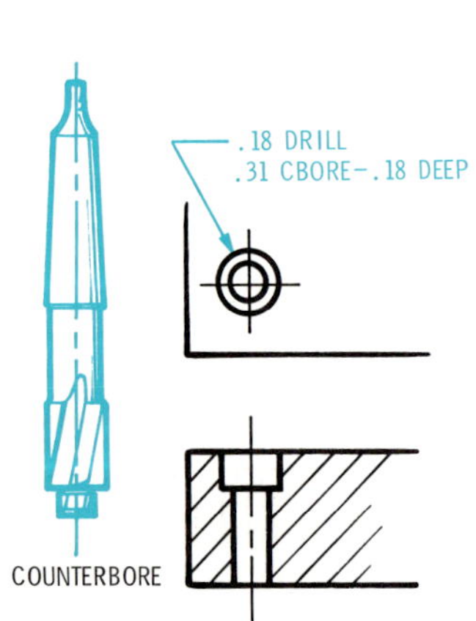

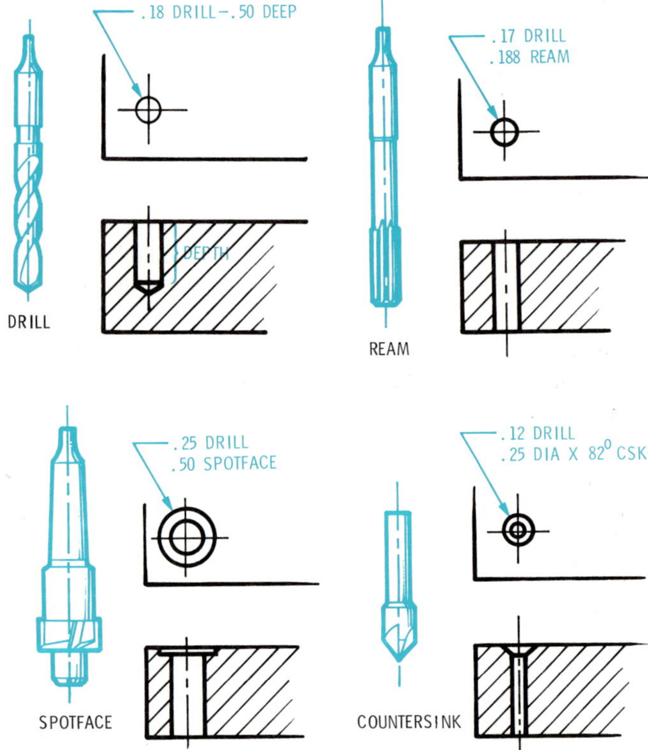

depth specification. A countersink creates a conically shaped depression to permit recessing the head of a flat or oval head screw. Note that the size of a countersink is the included angle of the point rather than the diameter, and that the diameter required at the surface is specified instead of depth.

5-42
A dispersed production system needs a method to make things fit together . . . tolerancing.

tolerancing

Modern technology and the increased production capability of industry allow a product to be designed in one place, the component parts to be manufactured in another (or even in numerous plants throughout the world), and the final assembly to take place in yet another location. How can we assure that all parts, components, and subassemblies fit together?

The engineer may have accurately specified the size, location, and form of all parts of a design. But what do we mean by "accurate"? May an actual dimension of the final part differ from the one specified by $\frac{1}{4}$ inch? by $\frac{1}{16}$ inch? by 0.001 inch? Variations in the raw materials, in the adjustment of machine tools, and in the errors of the gages used to measure the parts all contribute to inaccuracies. These can be reduced but not eliminated. *Tolerances* are the maximum inaccuracies that can be permitted if the finished object is to be assembled and operate as originally designed.

The function of a design determines the level of tolerance needed. For example, the function of a skate board allows greater component tolerance than does the function of a lunar-excursion module. As a general rule, the smaller the tolerance the higher will be the production cost of the design. Thus, if skate boards were built to the same tolerances as lunar-excursion modules, their cost would be such that few people would consider buying them.

Terms of Tolerancing

Tolerance is the total allowable variation of a dimension. The dimensions involved may relate to size, position, and form of an object.

The *nominal* size is an approximate size that has been assigned for easy identification. Examples are $\frac{1}{2}$-inch pipe, $\frac{3}{4}$-inch board, and $\frac{1}{2}$- by 1-inch bar stock. The nominal size may vary considerably from the actual size. For example, while the outside diameter of $\frac{1}{2}$-inch pipe re-

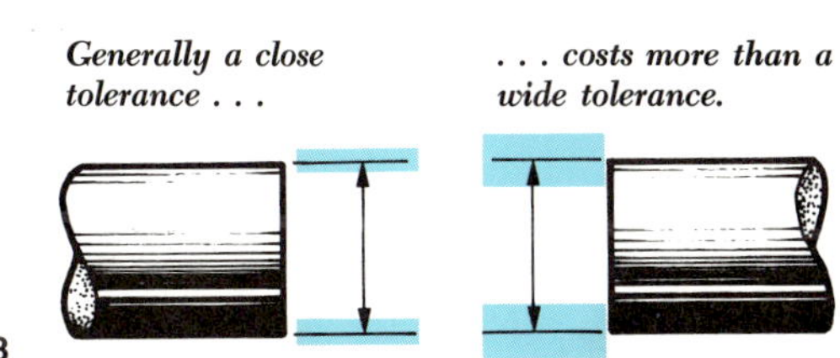

5-43

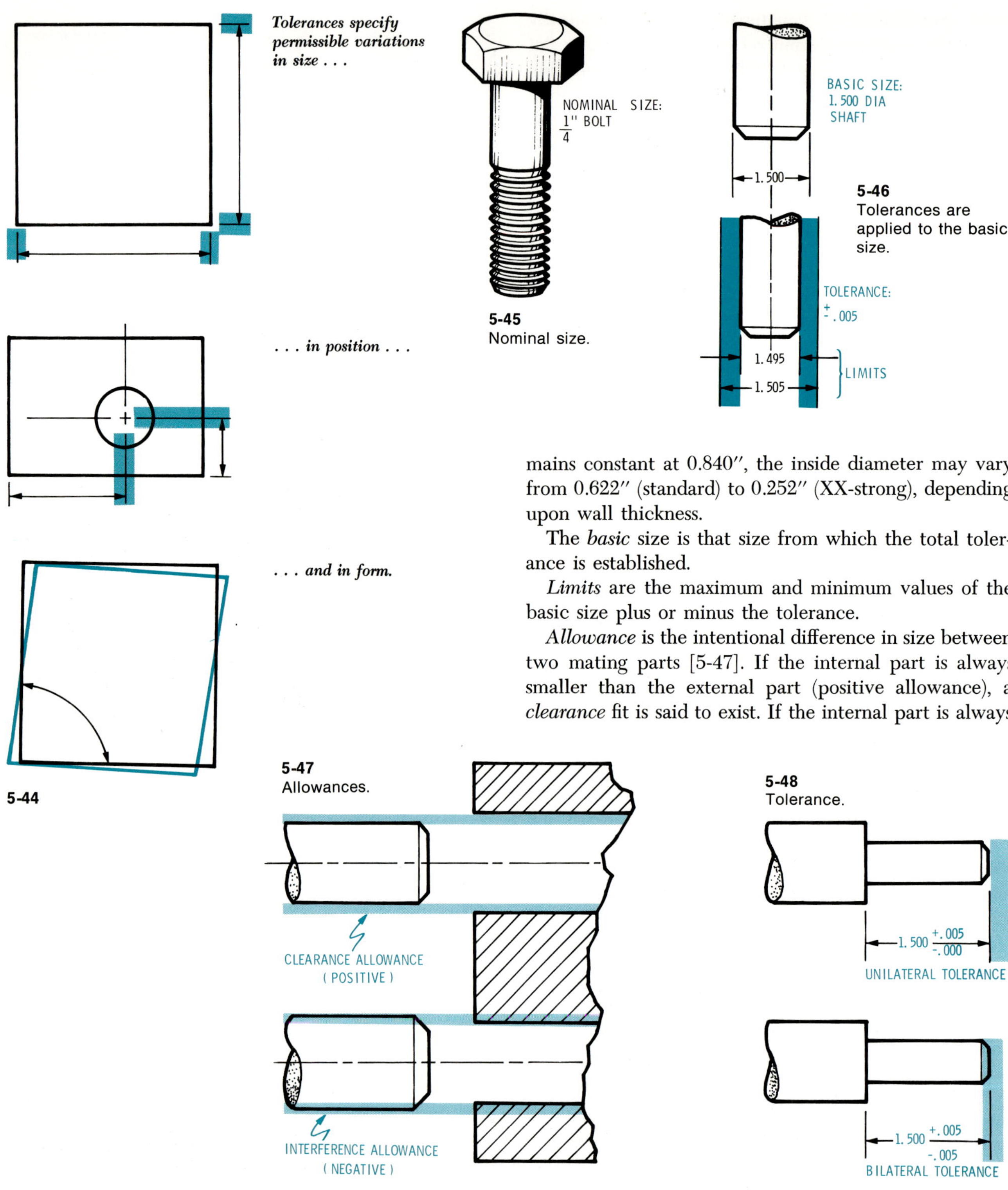

5-44

5-45
Nominal size.

5-46
Tolerances are applied to the basic size.

mains constant at 0.840″, the inside diameter may vary from 0.622″ (standard) to 0.252″ (XX-strong), depending upon wall thickness.

The *basic* size is that size from which the total tolerance is established.

Limits are the maximum and minimum values of the basic size plus or minus the tolerance.

Allowance is the intentional difference in size between two mating parts [5-47]. If the internal part is always smaller than the external part (positive allowance), a *clearance* fit is said to exist. If the internal part is always

5-47
Allowances.

5-48
Tolerance.

larger than the external part (negative allowance), an *interference* fit is said to exist. If the maximum and minimum limits on both parts are established such that either clearance or interference could exist, the result is a *transition* fit.

A tolerance established such that variation is permitted in only one direction from the basic size is called a *unilateral* tolerance (for example, $1.500^{+.005}_{-.000}$) [5-48]. When variation is permitted in both directions, this is a *bilateral* tolerance (for example, $1.500^{+.005}_{-.005}$). The tolerance on a dimension may be specified on a drawing by stating the basic size followed by the plus and minus tolerance ($1.500 \pm .005$, $1.500^{+.005}_{-.005}$) or by stating the limits directly ($\frac{1.505}{1.495}$). To save having to state a tolerance on each dimension, a general note is very often used on the drawing to cover the majority of dimensions. A typical note might be ALL LINEAR DIMENSIONS ±.015; ALL ANGULAR DIMENSIONS ±.5°; UNLESS OTHERWISE SPECIFIED.

Tolerances of Position

Tolerances of position are the basis for locating precisely one part in relation to another. For example, if gears mounted on parallel shafts are to run together smoothly, the distance between the shafts must be precisely controlled by the tolerances on the hole locations in the shaft mounting plates [5-49]. Positional tolerances are established on the basic dimensions that locate a feature from specific *datum* (reference) points, lines, or planes. These datums must be used for all related features on an object and on all mating parts [5-50].

Tolerancing Systems

There are two systems for specifying location tolerances—*coordinate* and *true position*. The coordinate system is the older of the two and is based on the use of coordinate axes. Since most manufacturing machinery is also constructed to operate on the coordinate system (to move either the work piece or the working tool in coordinate directions), this system of tolerancing is very commonly used.

The coordinate system of tolerancing first specifies a particular dimension and then specifies the allowable tolerance for that dimension. Because of the rectangular nature of the coordinate system, the tolerance zone (variance from the exact location) is also rectangular [5-52]. The diagonal of this rectangle permits consid-

5-49

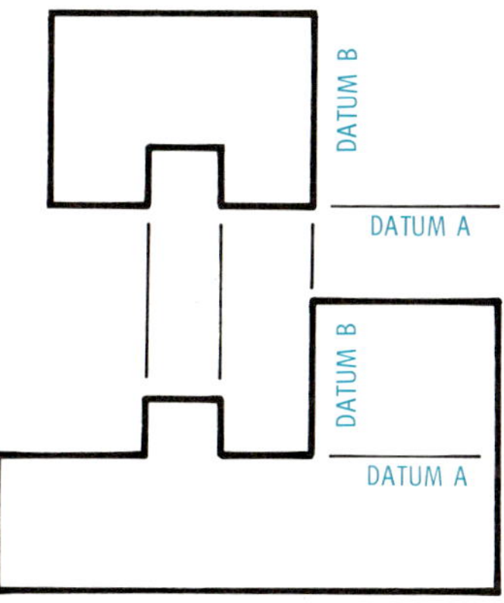

5-50
Mating parts use the same datum lines.

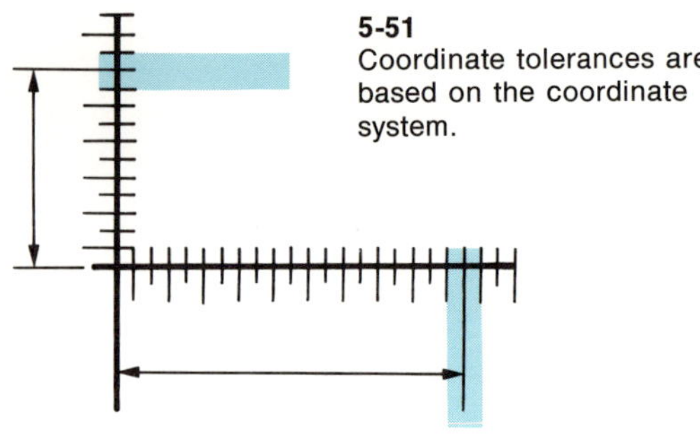

5-51
Coordinate tolerances are based on the coordinate system.

TOLERANCE ZONE
TOLERANCE ZONE
EXACT POSITION

5-52
The tolerance zone is rectangular in the coordinate system.

erably larger variation than either of the coordinate dimensions [5-53].

The true-position system of tolerancing overcomes this limitation by first specifying an exact location and then applying a tolerance relative to that location in the form of a note or a symbol. This system develops a circular tolerance zone, where the allowable variance from the exact dimension is the same in all directions [5-54]. The manner of specifying this on a drawing is shown in [5-55].

In using a particular system, the engineer must be aware of the capabilities of a machine shop so that he can specify tolerances that are economically attainable and still allow the design to function. Often an incorrectly specified tolerance will lead to misinterpretation by a craftsman or inspector. When such misinterpretation occurs, the designer is at fault for not specifying the tolerances clearly.

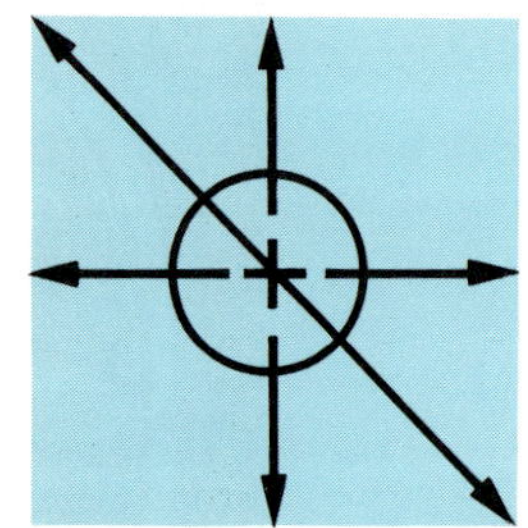

5-53
The range of variance in the coordinate system.

Tolerances of Form

Tolerances of form specify the permissible variation in the geometric shape or contour of an object. Examples

The basic position is exactly located . . .

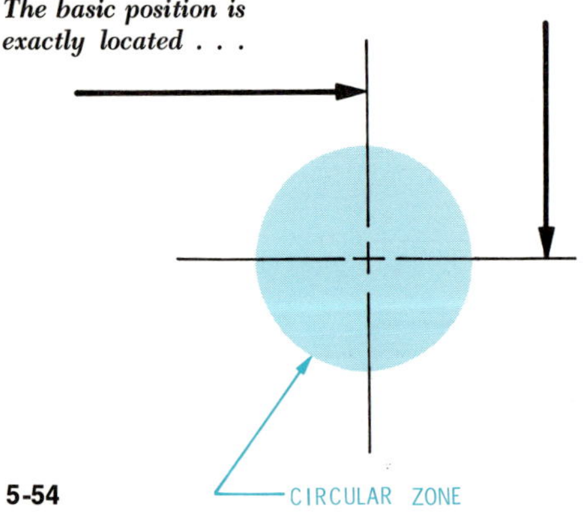

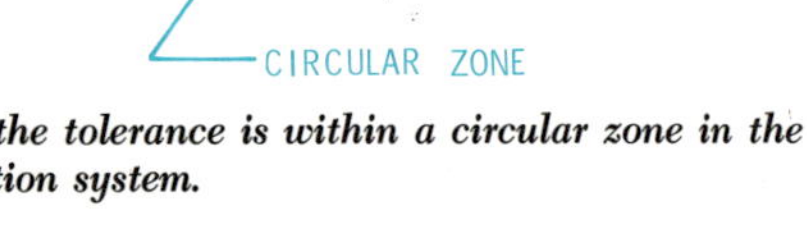

5-54
. . . and the tolerance is within a circular zone in the true position system.

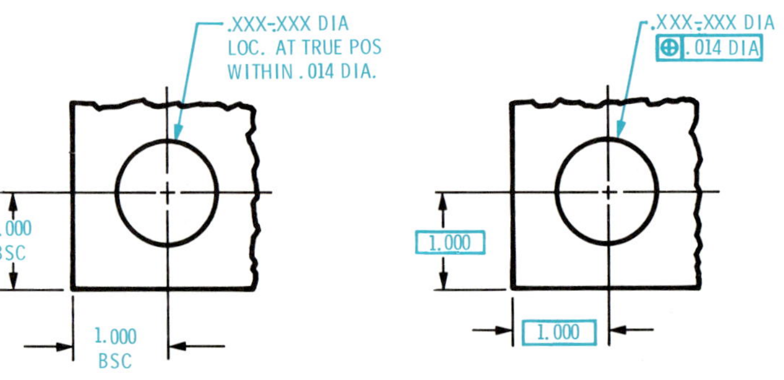

5-55
True-position tolerancing.

5-56
Specifying flatness.

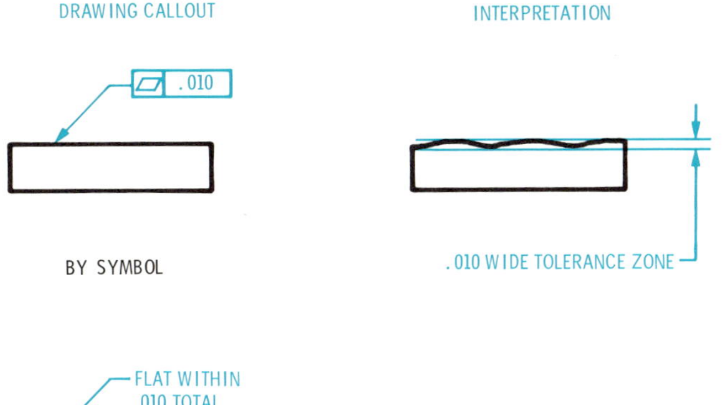

The surface must be within the specified tolerance of size and must lie between two parallel planes (.010 apart).

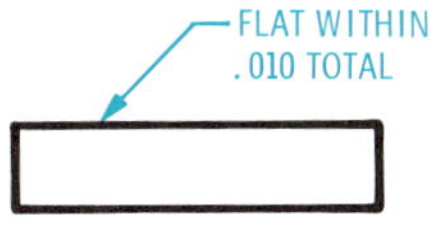

BY NOTE

5-57
Specifying the profile of a surface.

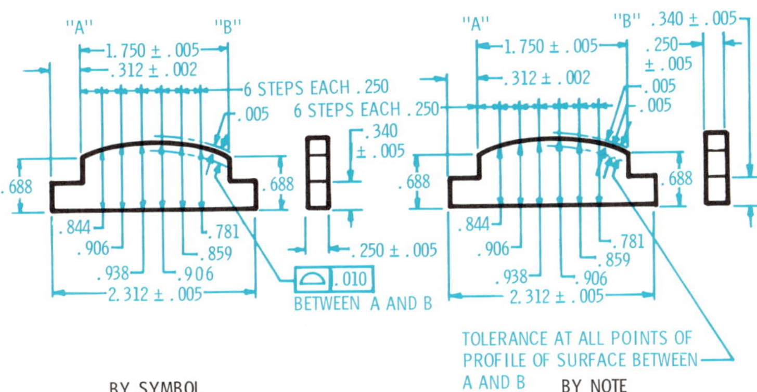

of geometric tolerances are flatness, straightness, roundness, parallelism, and perpendicularity. Such characteristics can be toleranced using symbols or notes on the drawing [5-56]. Contour tolerances specify the variation that is acceptable in profile elements, particularly curvilinear forms—regular and irregular [5-57].

section views

Many objects have complex internal features. When orthographic (multiview) drawings or sketches are made of such objects, the resulting multiplicity of hidden lines (necessary to depict the internal shape of the object) frequently is confusing. *Section views* can be used to reduce the confusion and clarify the design.

5-58
The cutting plane.

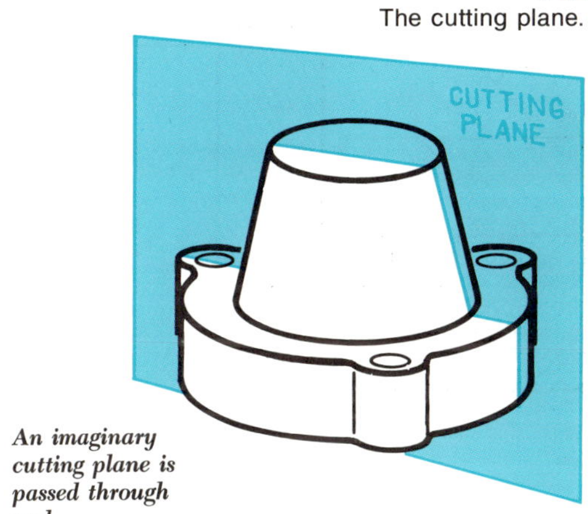

An imaginary cutting plane is passed through and . . .

. . . a section view results.

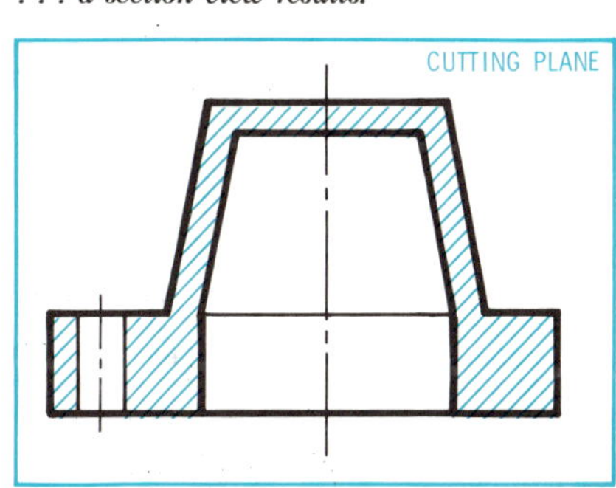

5-59
The cutting-plane line.

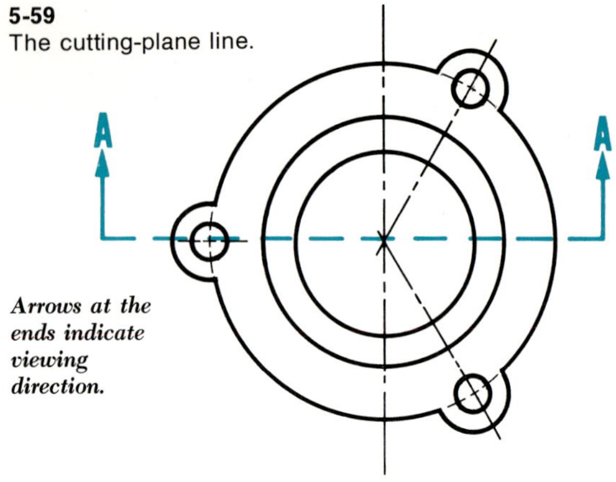

Cutting Plane

A section view is created by drawing the object as if part of it had been removed—cut away. An imaginary plane or combination of planes (known as *cutting planes*) is passed through selected parts of the object. The image that would be projected onto the cutting plane is the section view. It is drawn in accordance with the rules of orthographic projection.

Cutting-plane Line

The edge view of the cutting plane is represented by a *cutting-plane line*. This line is composed of a series of thick dashed lines. An arrow is placed at each end of the cutting-plane line to indicate the direction that the viewer is looking at the cut surface of the object. When more than one section view is cut through a drawing, each pair of arrows of the individual cutting-plane lines is lettered successively using capital letters, such as *AA*, *BB*, etc.

5-60
Sectioning of adjacent parts.

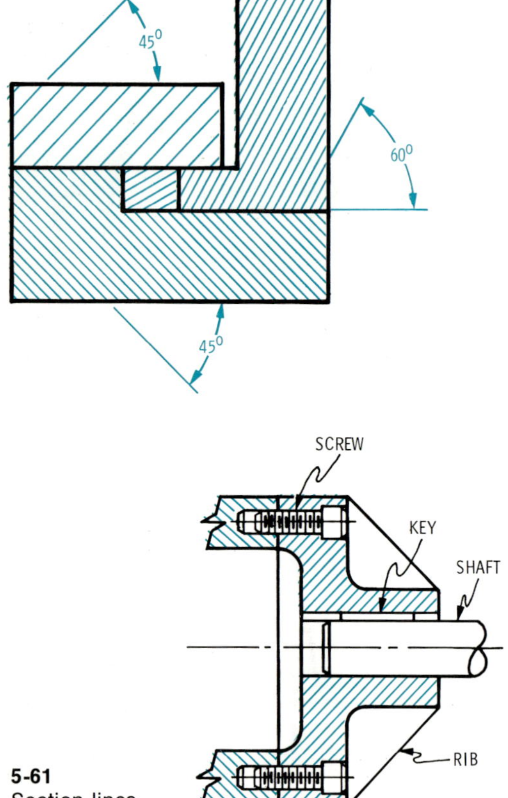

5-61
Section lines.

Section Lines

To distinguish between material that has been cut by the cutting plane and the remainder of the object, the surface of the cut material is crosshatched with thin, solid, parallel lines called *section lines*. These lines are drawn at some constant angle relative to the object, usually 45° with the horizontal, and are spaced $\frac{1}{16}$ to $\frac{1}{8}$ inch apart. Other angles (especially 30° and 60°) are used to avoid having the section lines parallel with a principal line of the object and for additional contrast between adjacent parts [5-60]. Cut portions of the same part are designated throughout the drawing with section lines that have the *same* slant and spacing. Cut portions of different pieces in the same view are differentiated by section lines at different angles and spacing. If any of the mating parts at the cutting plane are shafts, ribs, or standard parts (such as screws, pins, rivets, or keys) these parts are *not* sectioned [5-61]. Hidden lines behind the cut surface are omitted unless absolutely necessary for clarity. Specific patterns of section lines are sometimes used to represent different types of materials (see Appendix I).

Types of Sections

A cutting plane may cut through an object in a variety of ways, resulting in different portions of the object being removed.

5-62
A half-section.

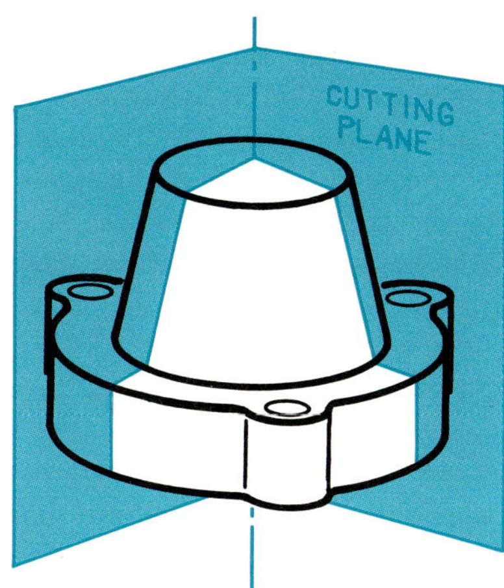

A half-section is developed by passing the cutting plane through the object to the center line.

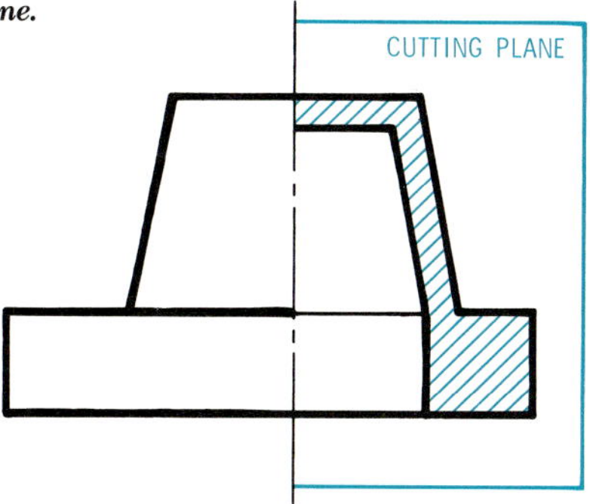

Full Section. A full section is generated by passing a cutting plane through the entire object [5-58]. This is generally done so that it divides the object into halves. By removing the front half, a full view of all the internal features can be seen. If the object is symmetrical and the cutting plane passes through the axis of symmetry, it is not necessary to draw the cutting-plane line.

Half Section. Many times it is desirable to show both external and internal features of an object in the same view. This is especially useful with symmetrical objects. When such a view is needed, a cutting plane can be passed through the object up to the midpoint (normally the center line). The resulting view is a *half-section* showing both external and internal features [5-62]. In this type of view, hidden lines are normally omitted from the entire drawing. However, the center line or a visible line representing the end point of the cutting plane must be shown.

Broken-out Section. A *broken-out section* is formed by passing a cutting plane through only a small selected

5-63
Broken-out sections.

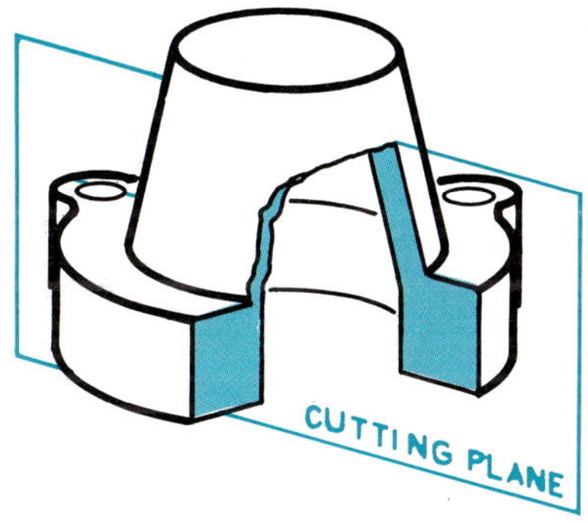

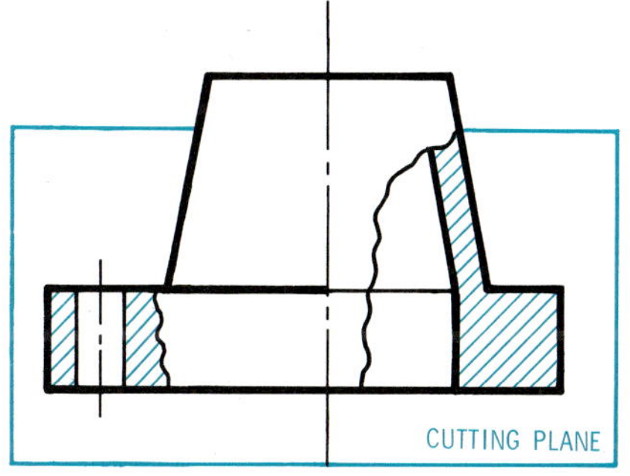

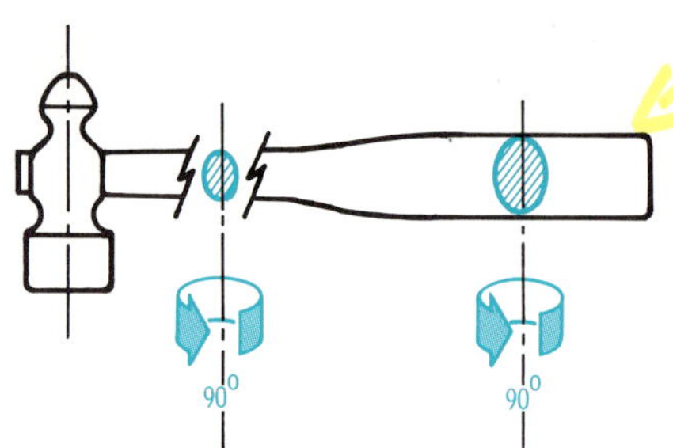

5-64
Revolved sections.

5-65
Removed sections.

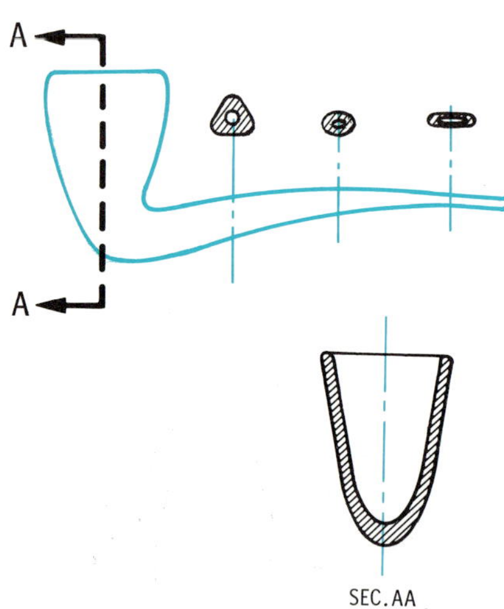

portion of the object. This is often useful in pictorial drawings as well as in orthographic drawings.

Revolved Section. At times it is desirable to draw a section view of a long object, such as a pipe, beam, or handle. A *revolved section* can be used to show the section contour at desired locations. Here the cutting plane is passed perpendicular to the object's long axis. The resulting section is then revolved 90° in place to reveal the cut section in true size. Break lines may be used if desired to separate the section from the view.

Removed Section. A *removed section* is similar to a revolved section, except that it is drawn at some removed location, preferably in line with the axis of the cut. If the removed section must be placed at some other location on the sheet, it must be carefully labeled to correspond with the appropriate cutting-plane line.

Offset Section. The *offset section* is similar to the full section, except that the path of the cutting plane is not straight. Rather, it conforms (using 90° turns) to the significant features of the object. Note that there are no vertical lines in the section view or changes in the sec-

5-66

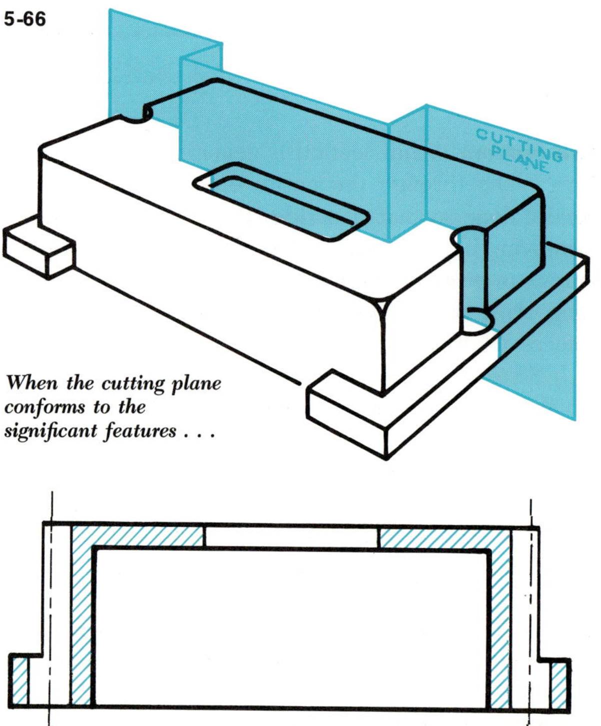

When the cutting plane conforms to the significant features . . .

. . . an offset section is developed.

tion-line pattern where the cutting plane has changed direction.

Aligned Section. Some objects do not lend themselves readily to section views. In these cases it may be necessary to rotate a portion of the object to a more suitable position before sectioning. This practice is particularly common in circular objects with odd numbers of radial features, such as a spoked hand wheel. When this situation arises, the element of concern is rotated to a new position that lends itself to conventional section practice. If rotation were not used, a standard section view would make the object appear unsymmetrical and might be misleading.

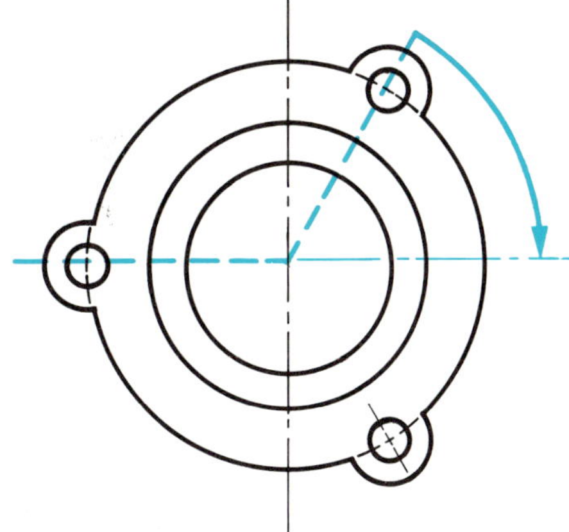

Some features are rotated to get a conventional view.

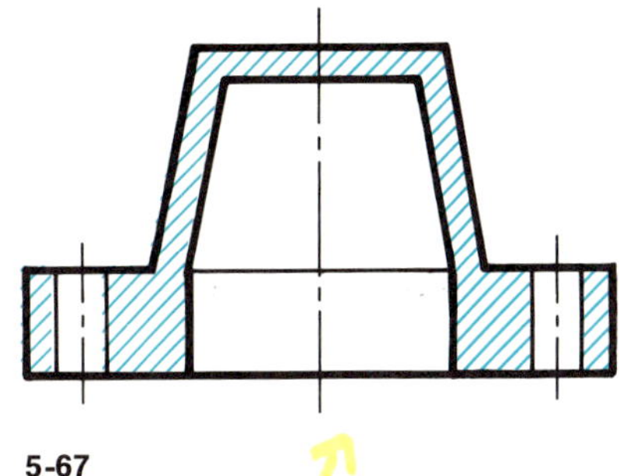

5-67
An aligned section.

intersections

Most manufactured objects are composed of simple geometric shapes. The line at which any two of these shapes meet is called their *intersection*. This intersection may be a straight or a curved line.

The engineer often must find intersection lines on his drawings so that he can specify such things as the location of fasteners, the shape of the intersecting forms, and the quantity of materials required. Since any line is defined by a series of points, the process for defining an intersection involves establishing the necessary intersection points. To illustrate the techniques for establishing an intersection, we shall use the simplest and most general intersection—a line intersecting a plane. Two methods are commonly used to determine this point, the *edge-view* method and the *cutting-plane* method.

Intersection of a Line and a Plane

Edge View Method. Standard orthographic views of common objects very often show edge views that give intersection points directly [5-68]. If not, the intersection point of a line oblique to a plane is found by establishing the edge view of the plane in an auxiliary view [5-69]. In the example, line *AB* and plane *XYZ* are given, and the intersection point between the line and the plane is desired. The line *AB* is left incomplete, since the intersection point is not known at this time.

A line *XO* is constructed in the horizontal view parallel to the reference line *H/F* and projected into the frontal

5-68
The intersection point often can be found directly from the edge view.

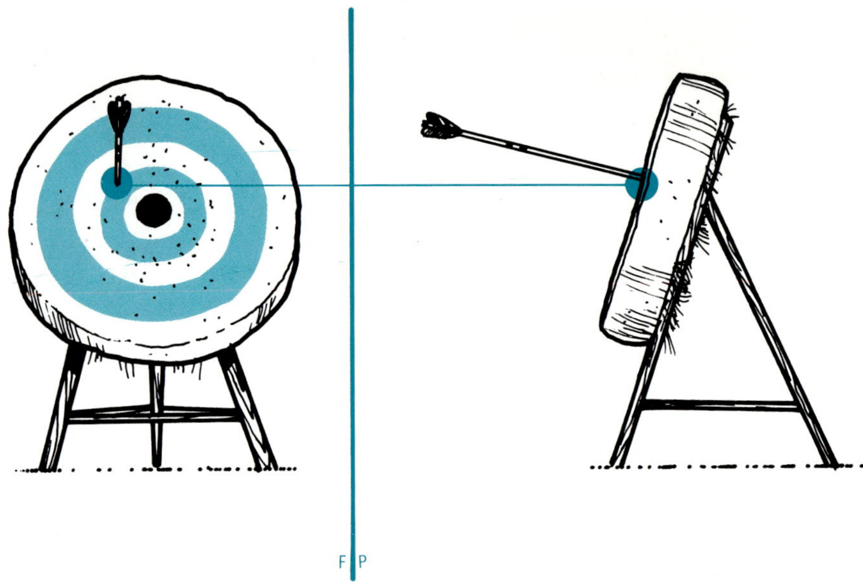

The intersection point of line AB and plane XYZ is found . . .

. . . by constructing the horizontal line XO . . .

. . . and projecting it as true length.

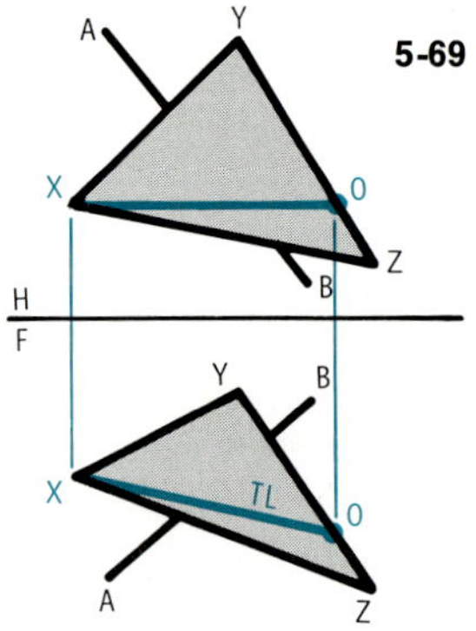

5-69

view as a true-length line. The edge view of the plane is found in auxiliary view 1 by projecting a point view of the true-length line *XO*. The line *AB* also is projected into the auxiliary view and the intersection point with the plane is established [5-70].

The original views can be completed to show the intersection point of line *AB* with the plane *XYZ* by projecting the intersection point found in the auxiliary view back into the frontal and then into the horizontal view. These views of line *AB* are then completed by extension to the intersection point as visible or hidden lines [5-71].

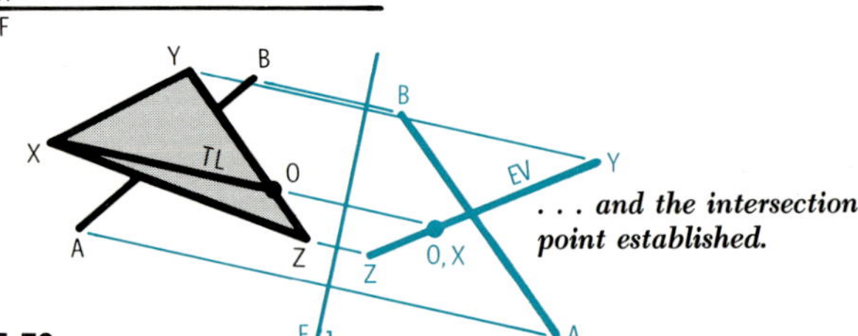

The plane XYZ is projected as an edge view in the auxiliary view . . .

. . . along with line AB . . .

. . . and the intersection point established.

5-70

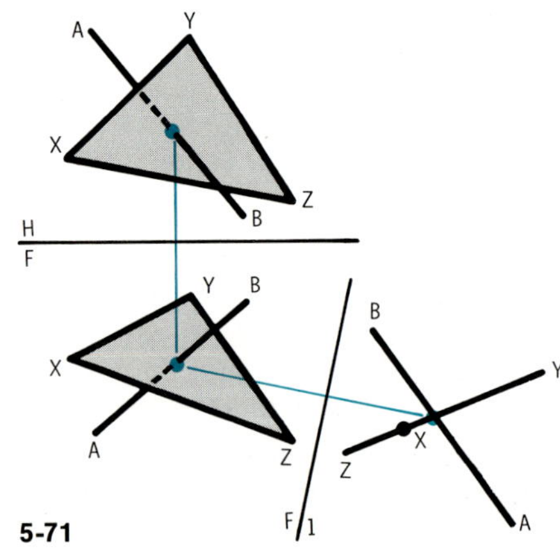

5-71

The intersection point is projected back to the preceding views.

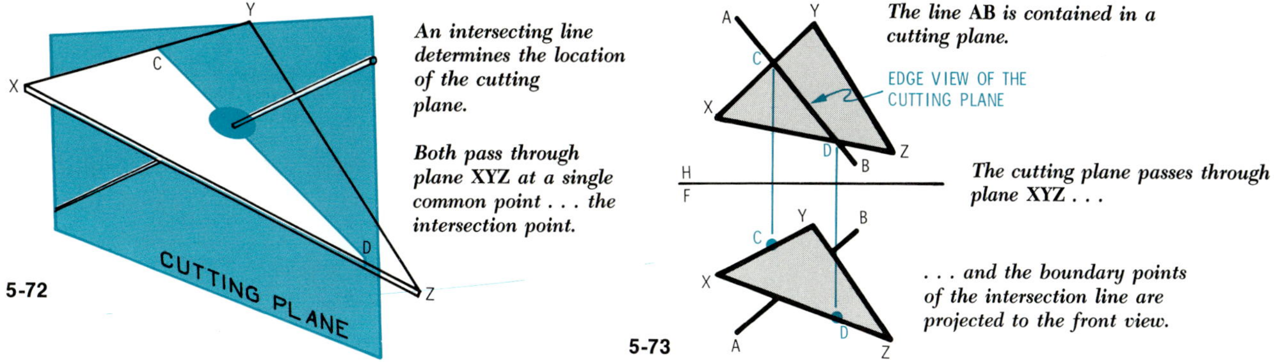

5-72 *An intersecting line determines the location of the cutting plane.*

Both pass through plane **XYZ** *at a single common point . . . the intersection point.*

5-73 *The line* **AB** *is contained in a cutting plane.*

The cutting plane passes through plane **XYZ** *. . .*

. . . and the boundary points of the intersection line are projected to the front view.

Cutting-plane Method. Using the same line *AB* and plane *XYZ* as in the previous example, the intersection point also can be found by the cutting-plane method. The intersecting line is considered as lying within a plane that cuts through the subject plane [5-72]. The edge view of a plane is passed through the horizontal view of line *AB* and cutting plane *XYZ*. The points *C* and *D*, where the cutting plane crosses the boundaries of the plane *XYZ*, are projected into the front view [5-73]. These projected points are connected with a straight line, which represents the intersection line of the cutting plane with the plane *XYZ*. The point at which the line *AB* crosses the intersection line in the front view is the intersection point [5-74]. The intersection point in the horizontal view is found by projection from the front view [5-75]. The number of geometric shapes used in design creates a large number of possible intersection conditions. Most intersection problems involve more complex geometric elements than just a line and a plane, but the methods are the same. Because of the complexity, it is a good practice to number the intersection points as they are developed and to pay careful attention to the relative visibility of the major lines.

The basic types of intersections that will be discussed are the intersection of a plane and a prism, of a plane and a cylinder, of two prisms, and of two cylinders.

Intersection of a Plane and a Prism

Since the intersection of a plane and a prism [5-76] is a basic and common type of intersection, we shall discuss the use of both the edge-view method and the cutting-plane method.

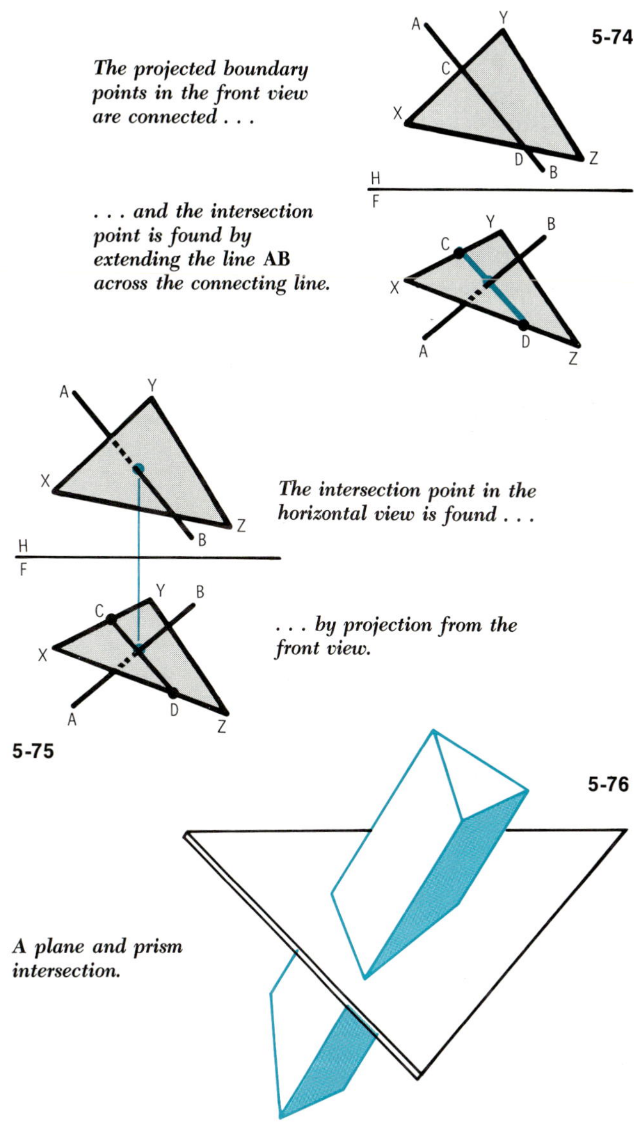

5-74 *The projected boundary points in the front view are connected . . .*

. . . and the intersection point is found by extending the line **AB** *across the connecting line.*

5-75 *The intersection point in the horizontal view is found . . .*

. . . by projection from the front view.

5-76 *A plane and prism intersection.*

5-77

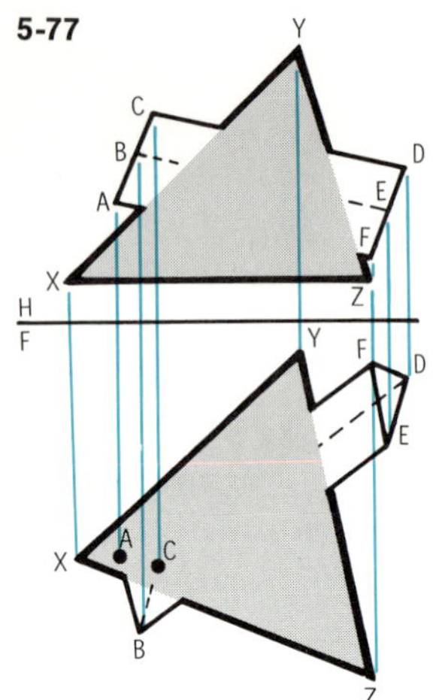

Using the triangular prism **ABCDEF** *and the plane* **XYZ** *. . .*

. . . find the intersection lines.

5-78

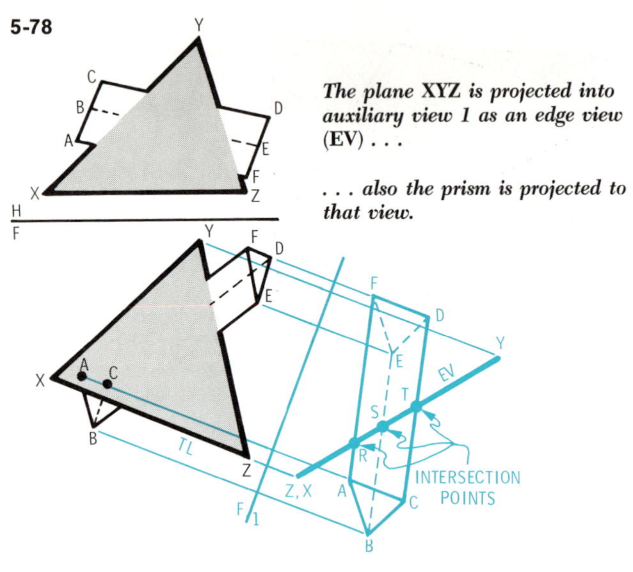

The plane **XYZ** *is projected into auxiliary view 1 as an edge view* **(EV)** *. . .*

. . . also the prism is projected to that view.

5-79

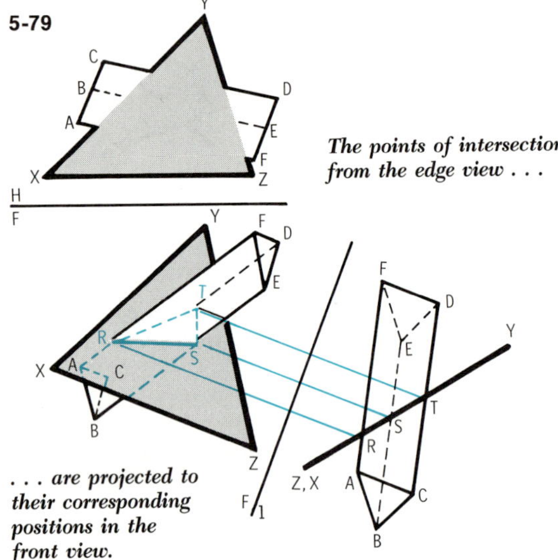

The points of intersection from the edge view . . .

. . . are projected to their corresponding positions in the front view.

The points are connected to define the shape of the intersection.

5-80

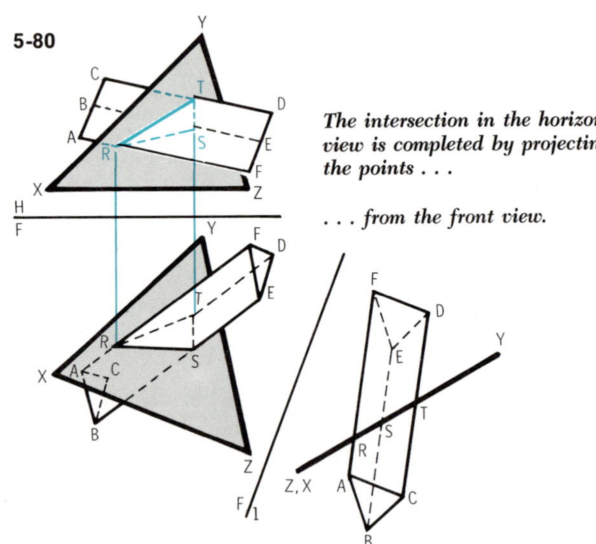

The intersection in the horizontal view is completed by projecting the points . . .

. . . from the front view.

Edge-view Method. The first example is the intersection of the triangular prism *ABCDEF* with the plane *XYZ* [5-77]. Note that the views are incomplete at this time, and the plane is not seen as an edge view. We must first construct the auxiliary view 1, which shows the plane *XYZ* as an edge view passing through the triangular prism [5-78]. The intersection points *R*, *S*, and *T* of the edge view of the plane and the vertex lines of the prism are then projected back from the auxiliary view to the front view and to the horizontal view. These points are con-

5-81

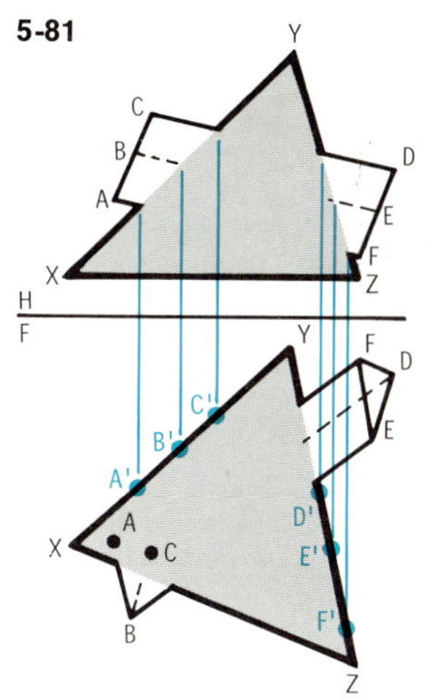

Finding the intersection by the cutting plane method . . .

. . . cutting planes **AF, BE** *and* **CD** *are established.*

Their points of intersection with plane **XYZ** *are projected to the front view.*

5-82

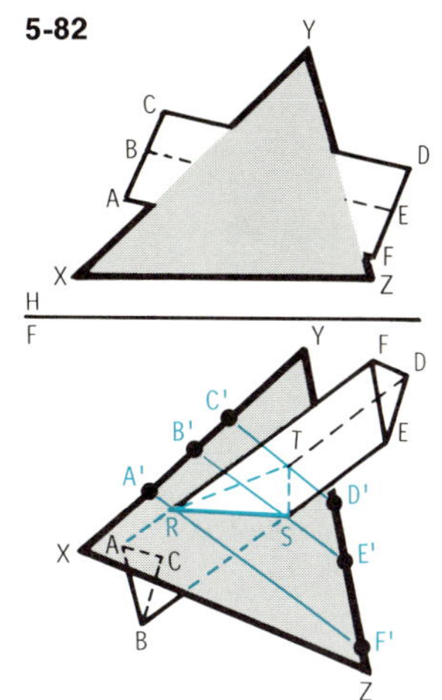

The cutting plane intersection lines **A′F′, B′E′,** *and* **C′D′** *cross the appropriate vertex lines to develop the intersection points.*

nected by straight lines to establish the shape of the intersection in both views [5-79 and 5-80].

Cutting-plane Method. The cutting-plane method is generally preferred to the edge-view method because it requires less construction. Using the same example of the triangular prism and the plane *XYZ*, cutting planes are passed through the vertex lines of the triangular prism in the horizontal view. The points at which these cutting planes pass through the boundaries of the plane *XYZ* are found and projected to their appropriate positions in the front view [5-81]. These points are then connected with lines that represent the intersection of the respective cutting planes with the plane *XYZ*. Where these lines cross vertex lines of the prism, intersection points *R*, *S*, and *T* are defined. These intersection points are then connected with straight lines, and the shape of the intersection in the front view is established [5-82]. By projection, the points are then transferred to the horizontal view and connected with straight lines [5-83].

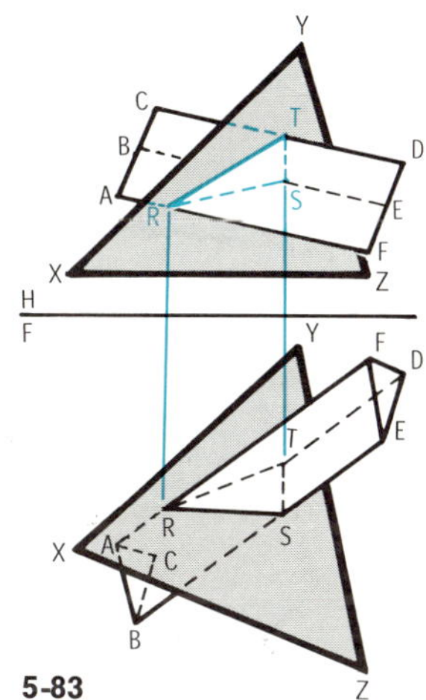

5-83

The intersection points are projected to the horizontal view to establish the shape of the intersection.

Intersection of a Plane and a Cylinder

The intersection of a plane and a cylinder is another common type of intersection problem [5-84]. To find the intersection between the cylinder and the plane *ABCD* [5-85], we shall use the cutting-plane method. The views are incomplete because the intersection has not yet been defined.

Using light construction lines, complete both views of the intersecting shapes. Since the cylinder does not have

5-84

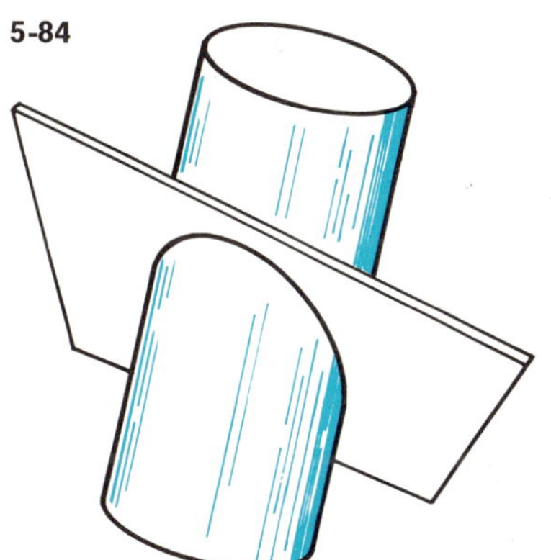

The plane and cylinder intersection.

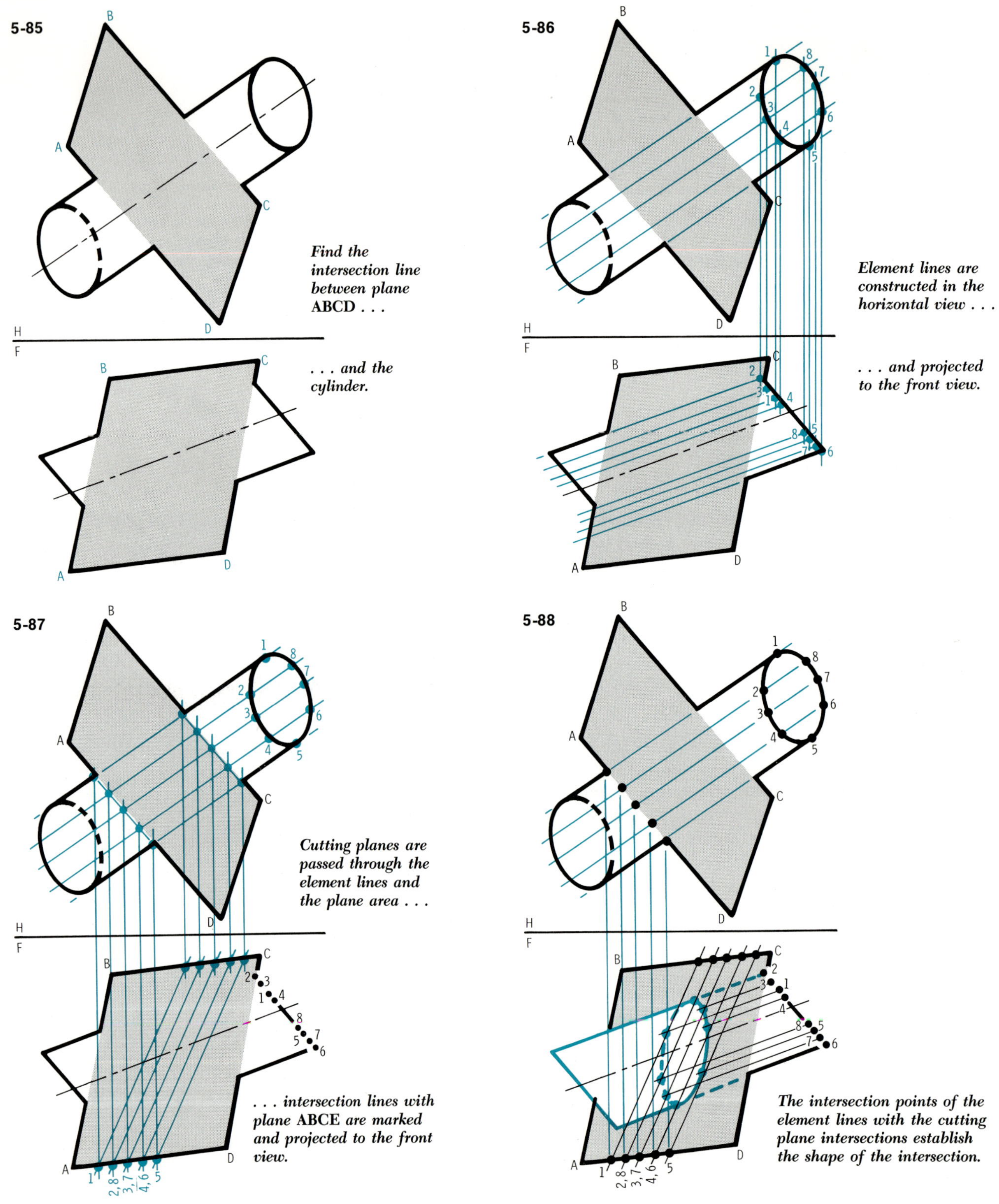

Find the intersection line between plane **ABCD** . . .

. . . *and the cylinder.*

Element lines are constructed in the horizontal view . . .

. . . *and projected to the front view.*

Cutting planes are passed through the element lines and the plane area . . .

. . . *intersection lines with plane* **ABCE** *are marked and projected to the front view.*

The intersection points of the element lines with the cutting plane intersections establish the shape of the intersection.

lines corresponding to the vertex lines of the prism on which intersection points can be located, it is first necessary to construct parallel element lines on the surface of the cylinder at regular intervals in the horizontal view [5-86]. The points where these lines meet the end of the cylinder are numbered and projected to the front view. The element lines are then extended from these points parallel to the axes of the cylinder in the front views.

Cutting planes are now passed through the element lines in the horizontal view [5-87]. The lines of intersection of the cutting planes with plane *ABCD* are marked and projected to the front view. The intersection points of the element lines are now defined in the front view, where these lines cross the cutting-plane intersection lines. These points must be connected to complete the description of the shape of the intersection [5-88]. The shape of the intersection in the horizontal view is found by projecting the points from the front view to the appropriate element lines in the horizontal view and drawing the intersection line [5-89].

Intersection of Two Prisms

The intersection of two prisms is essentially the problem of multiple planes intersecting each other. The lines of intersection of the various planes are determined by the intersection points of the lines of one prism piercing the planes of the other and vice versa. The edge-view and the cutting-plane methods of establishing the intersection points may be used independently or cooperatively. The example shows a rectangular prism and a triangular prism intersecting [5-90]. Inspection of the given views reveals that the planes of the rectangular prism appear as edges in the horizontal view. The intersection points of the vertex lines of the triangular prism with the planes of the rectangular prism are therefore immediately available and may be projected directly to their appropriate locations in the front view [5-91].

This, however, does not complete the shape of the intersection because the points for the intersection of the vertical corner line of the rectangular prism with the planes of the triangular prism are not defined yet. To establish the location of these points, the cutting-plane method is applied. In the horizontal view the cutting plane (CP) is drawn through the corner line and parallel with the vertex lines of the triangular prism [5-92]. The cutting-plane intersection lines are projected to the front

5-89

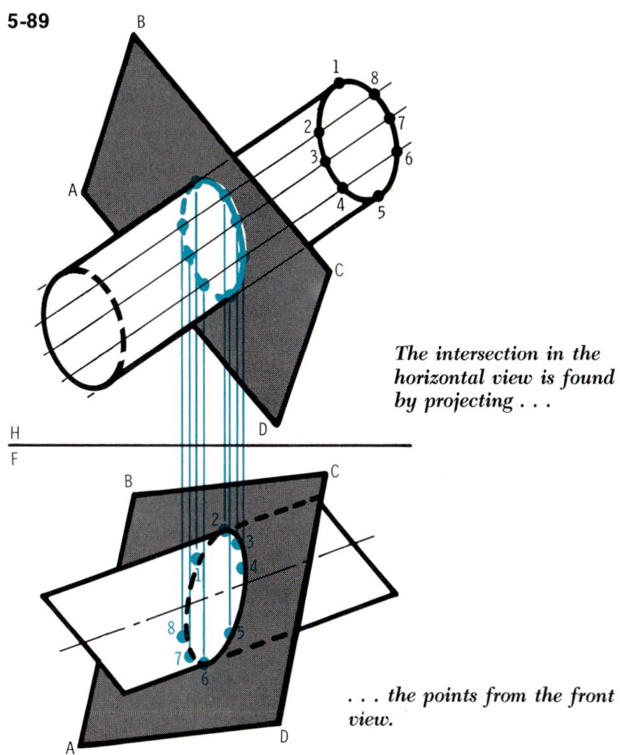

The intersection in the horizontal view is found by projecting . . .

. . . the points from the front view.

5-90

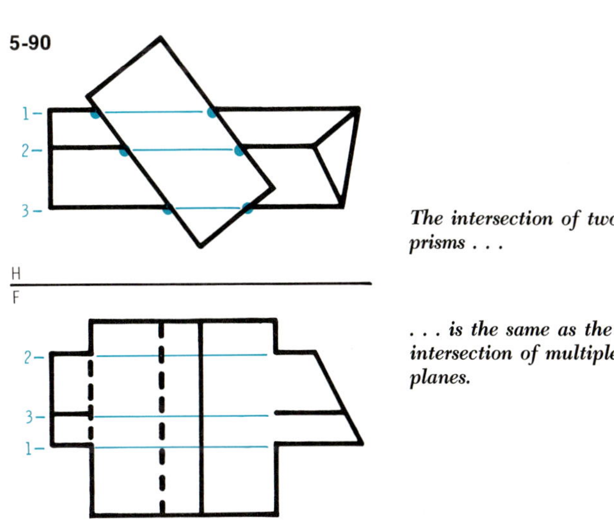

The intersection of two prisms . . .

. . . is the same as the intersection of multiple planes.

5-91

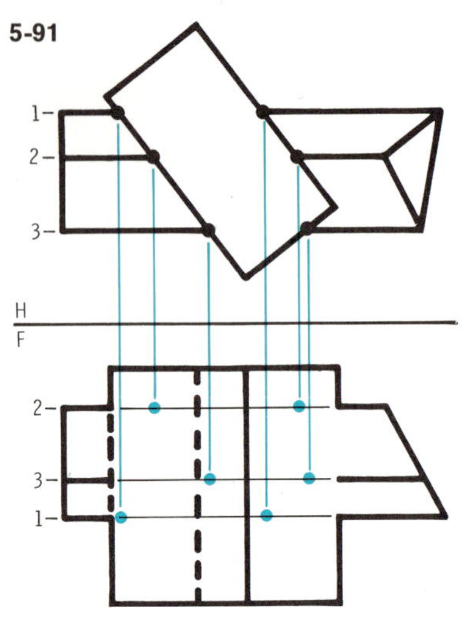

The intersection points along the edge views . . .

. . . are projected directly to the front view.

5-92

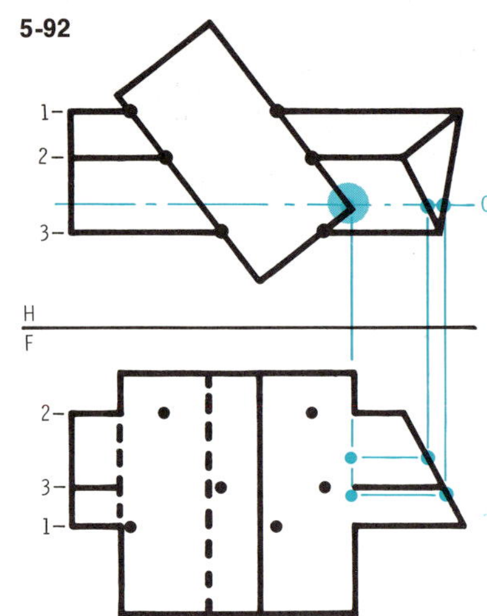

A cutting plane (CP) is passed through the corner of the rectangular prism . . .

. . . intersection lines are projected to the front view . . .

. . . and intersection points located.

view and extended parallel with the vertex lines of the triangular prism. (Since the corner is the point of interest, there is no need to extend the lines beyond the projection of the corner.) Note that the cutting plane passes through two faces of the triangular prism and that both intersection lines are projected to the front view. The intersection points are now connected with straight lines to define the shape of the intersection [5-93].

5-93

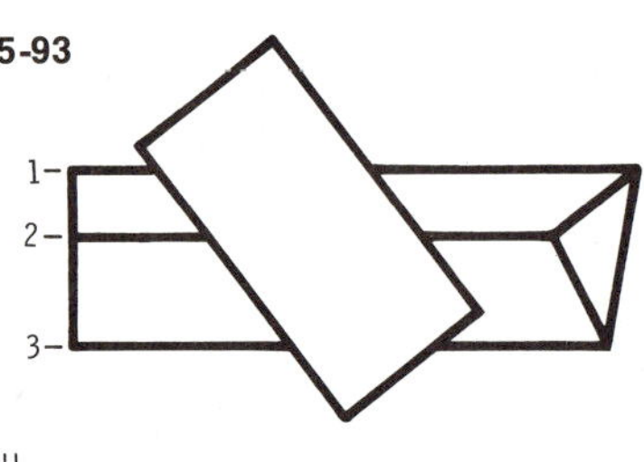

2
3
1

The points are connected to define the shape of the intersection.

5-94
Some designs require an understanding of how two cylinders can be joined together.

Often it is difficult to visualize the final shape of the intersection, so careful analysis of the various intersecting planes and lines is necessary. This analysis is performed with both views of the prisms to define their basic relationships. Each line of each object must be studied to determine its possible intersection with each plane . . . if it does and where.

Intersection of Two Cylinders

The intersection of two cylinders is quite similar in principle to the intersection of two prisms. The primary difference is that cylinders are continuous surfaces; they have no vertex lines, as did the prisms. It is therefore necessary to construct parallel element lines on the surface of the small cylinder as in the case of the intersection of a cylinder and a plane. In the example [5-95], eight linear elements are drawn on the surface of the smaller cylinder parallel with its axis. These lines are then projected to the front view. The number and location of the cylinder elements are selected to give the best definition to the curved intersection line that will result from their intersection with the large cylinder. For accuracy more than eight cylinder elements are frequently necessary.

The intersection points of the element lines of the small cylinder with the large cylinder can be seen in the horizontal view, where the large cylinder appears as an edge view. These intersection points can be projected to the front view of the element lines and a smooth curve drawn through them [5-96 and 5-97].

5-95

To find the intersection of two cylinders, element lines are constructed on the small cylinder.

5-96

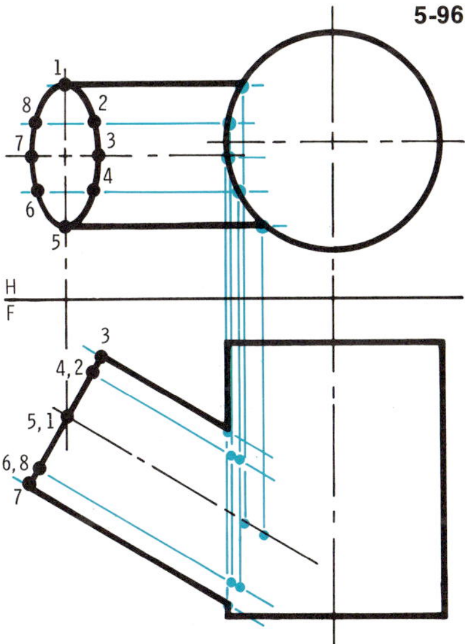

The intersection points of the element lines with the edge view of the larger cylinder . . .

. . . are projected to the element lines in the front view.

5-97

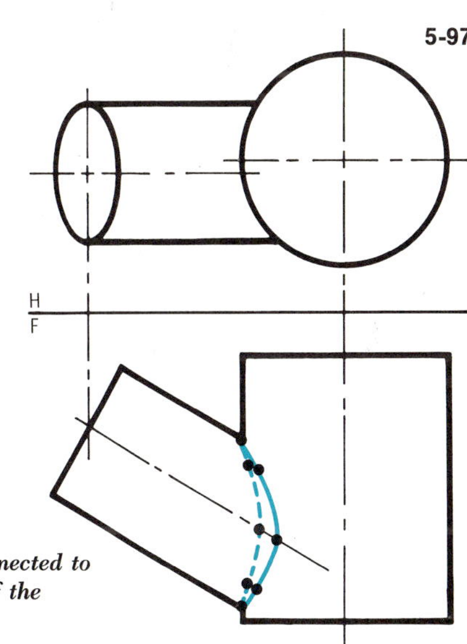

The points are connected to define the shape of the intersection.

A tomato juice can . . .

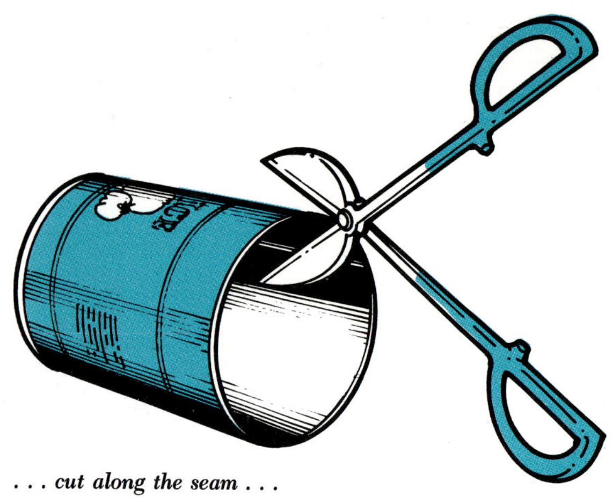

. . . cut along the seam . . .

. . . and rolled flat is a development of a cylinder.

5-98

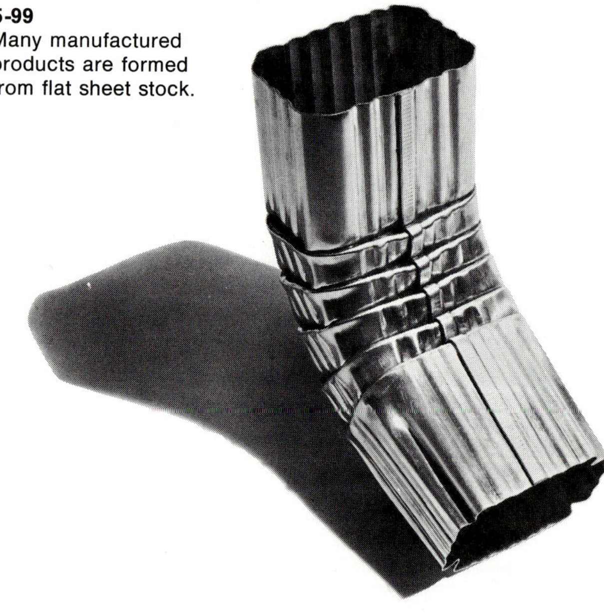

5-99
Many manufactured products are formed from flat sheet stock.

developments

A development is the flattened-out surface of a hollow geometric shape. Developments are used mostly in industries that manufacture sheet metal parts. Air conditioning and heating ducts, metal furniture, and the skin of airplane wings are a few examples.

If we were to take an old tomato juice can, remove the top and bottom, cut it along its seam, and roll it out into a flat sheet, we would have the development of a cylinder [5-98]. Similarly, the orthographic box in Chapter 3 that was unfolded to show the six principal views is a development of a rectangular prism [3-29]. The surface of the orthographic box is composed of a series of flat planes; the tomato juice can is a continuous single-curved surface. Any geometric shape that is composed of several plane or single-curved surfaces can be developed. Complex surfaces such as double-curved and warped surfaces cannot be developed except by special techniques that provide reasonable approximations of a developable form.

The principle that governs most developable surfaces is that the surface must be definable with straight lines. Each of these lines must be a true-length line on the developed surface.

Developments are classified according to the orientation of the linear elements that compose the geometric forms. Our discussion will be limited to the two most common classes—*parallel-line developments* and *radial-line developments*.

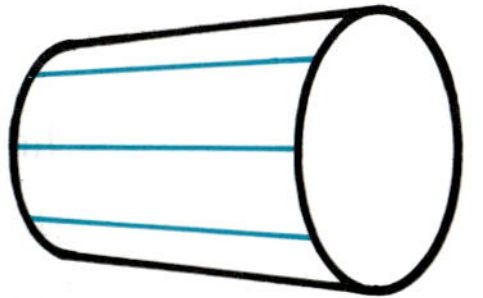

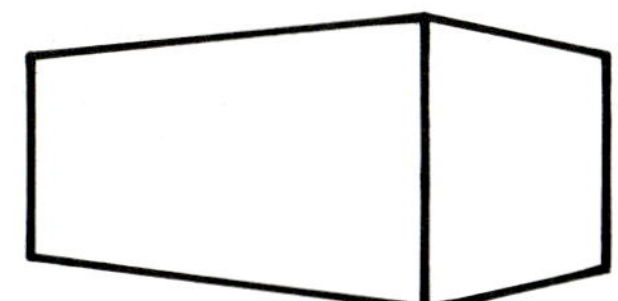

5-100
The cylinder and the rectangular prism are examples of parallel-line geometric shapes.

Parallel-line developments apply to all geometric shapes whose linear elements are parallel with each other. The cylinder and the rectangular prism are good examples of this group [5-100].

The linear elements of a radial-line development converge to a point. The cone and the pyramid are good examples of this type [5-101].

Since a development represents only the surface of a developable geometric form, this surface can be viewed from either the inside or the outside. In the case of an inside development, the layout lines will appear on the inside of the object after it is formed. In an outside development the layout lines appear on the exterior surface after forming [5-102].

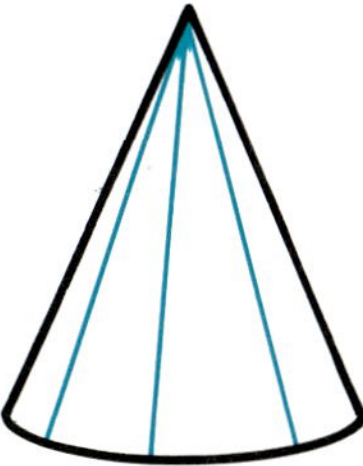

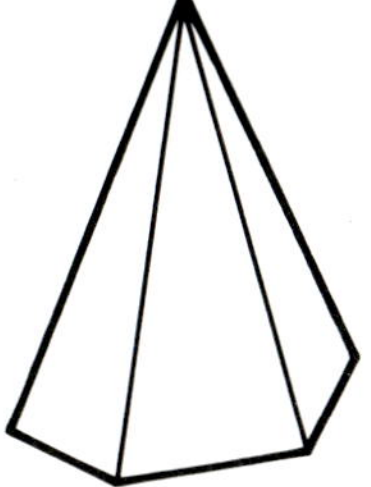

5-101
The cone and the pyramid are examples of radial-line geometric shapes.

5-102

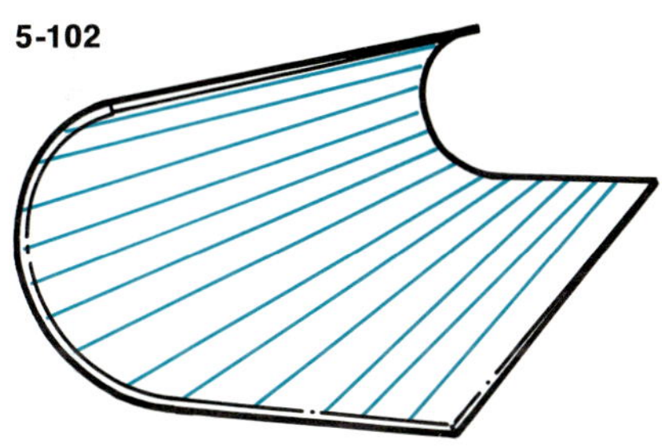

The layout marks of an inside pattern are on the interior surface.

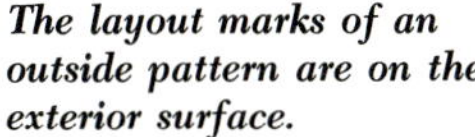

The layout marks of an outside pattern are on the exterior surface.

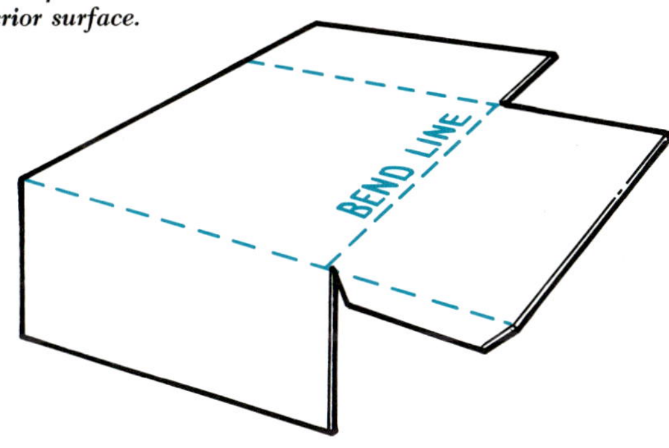

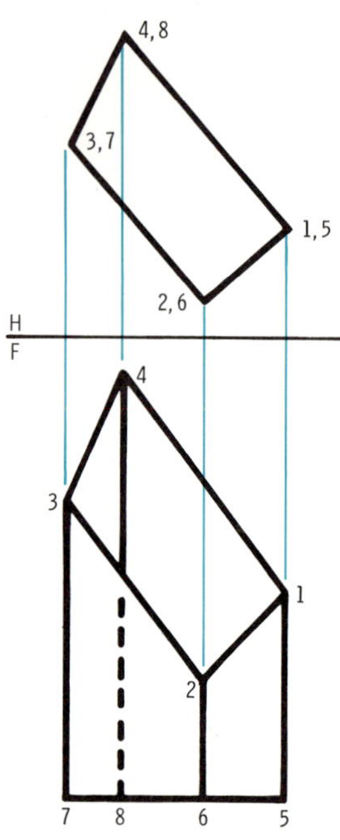

Construct a development for a hollow, rectilinear prism.

5-103

Development of a Prism

All parallel-line geometric shapes are developed on the same basic principles. We shall develop a hollow prism to demonstrate these basic principles [5-103]. Remember that every line on the surface of the form must be seen as *true length* in the development. In the example, every plane that encloses the prism will have to be shown as true size. This could be achieved by constructing four auxiliary views around the horizontal view of the prism [5-104]. With these true-size planes we could construct the rectilinear solid that they define. In this configuration, however, it would be a cumbersome process requiring cutting out each of the planes, positioning them properly relative to each other, and then fastening them together.

This is far too time consuming and unacceptable to industry because of the many seams involved. A better approach is to unfold the prism sequentially, thus retaining the common lines between the true-size planes as

5-104

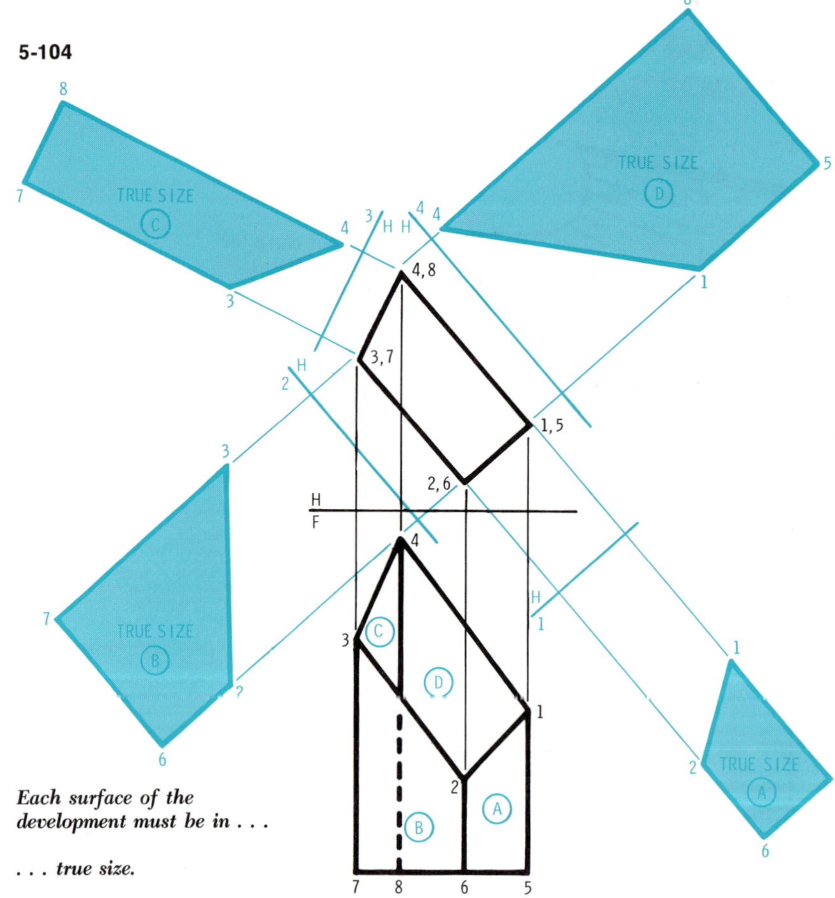

Each surface of the development must be in . . .

. . . true size.

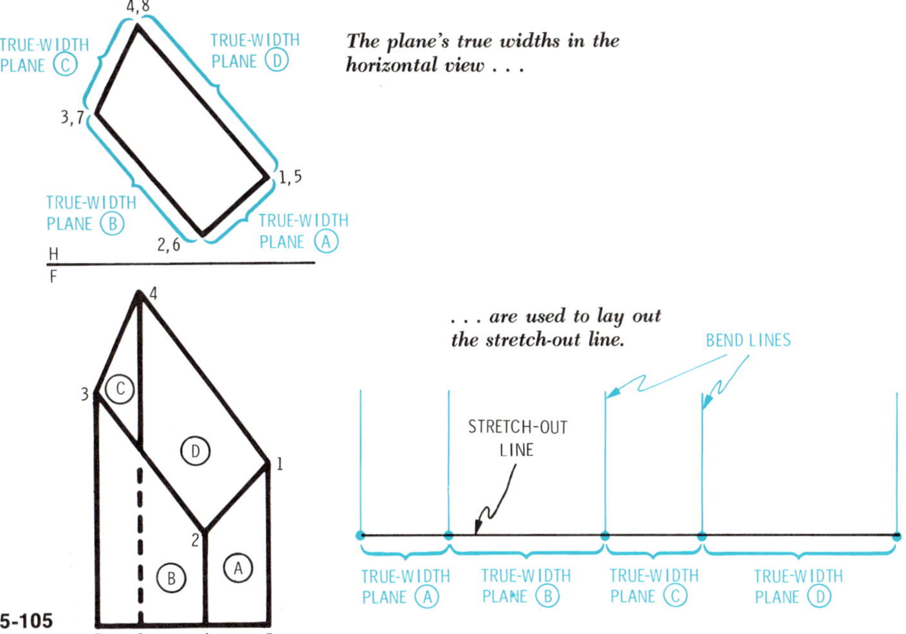

5-105

bend lines. This can be done with a *stretch-out line* constructed perpendicular to the axis of the prism and having a length equal to the perimeter of the prism. Examination of [5-105] shows that the true length of the perimeter can be found in the horizontal view, where the edge

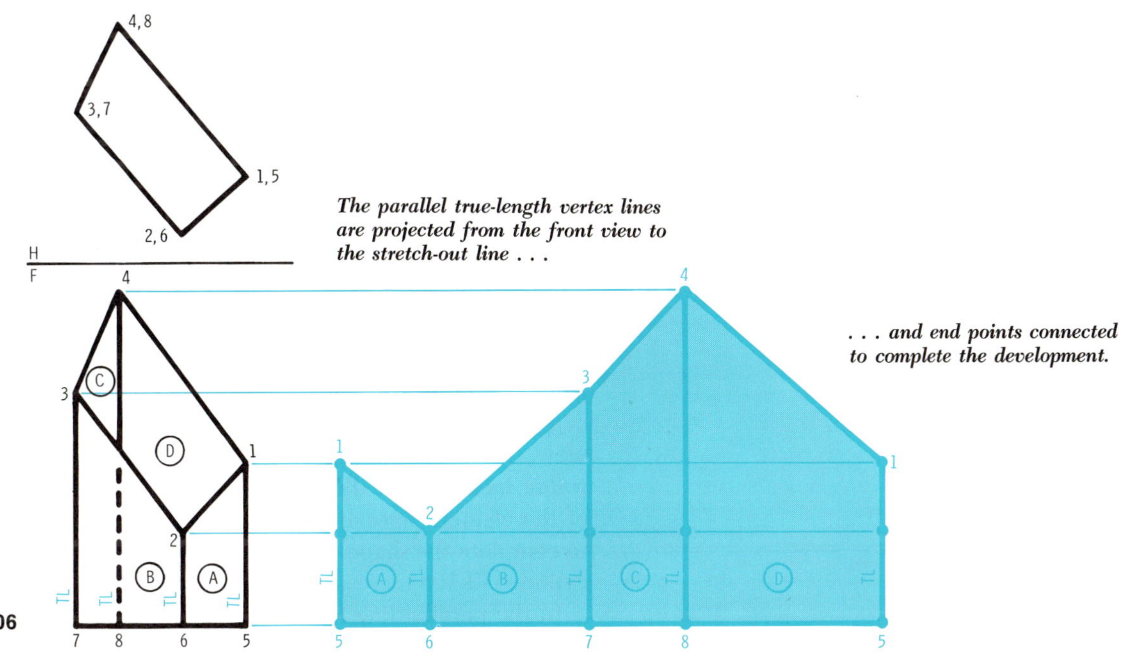

5-106

views of the prism's planes can be seen. The width of each plane is transferred to the stretch-out line in sequence Ⓐ Ⓑ Ⓒ Ⓓ, starting with the 1-5 line. Since the vertex lines of the prism are all true length in the front view, they can now be transferred parallel to the appropriate locations on the stretch-out line to indicate where the material must later be bent [5-106].

The projected end points of the vertex lines are connected by straight lines to define the shape of the development. Compare the true-size planes that comprise this development with the true-size planes made earlier [5-104]—they are the same. This method is not only quicker but also joins the planes in their proper relationship for easy fabrication by bending and fastening along only one seam. Note that this example shows the interior surface of the prism, so this is an inside development. To produce an outside development, the same procedure can be used, except that the planes must be layed out in reverse sequence—Ⓓ Ⓒ Ⓑ Ⓐ.

Development of a Cylinder

The methods for constructing the development of a cylinder are basically the same as those for a prism. A stretch-out line is constructed perpendicular to the axis of the cylinder with a length equal to its circumference. Although there are methods for graphically determining this circumference, the simplest method is to calculate it mathematically.

Since a cylinder is a continuous single-curved surface, no edges or corners are available. It is therefore necessary to construct a representative group of parallel linear elements on the surface of the cylinder. The number of lines is arbitrary, but the greater the number, the greater the accuracy, just as in the case of intersections. Eight equal segments were chosen in the example [5-107]. The stretch-out line is similarly segmented into eight equal parts and numbered to correspond with the segments on the circumference of the circle in the horizontal view. The segment points on the edge of the cylinder are projected into the front view, and parallel elements drawn.

Since the element lines are true length in the front view, they may now be projected directly over to appropriate locations on the stretch-out line. The end points of the element lines are connected with a smooth curve to complete the shape of the development for the hollow cylinder [5-108].

In the parallel-line forms developed above, the length

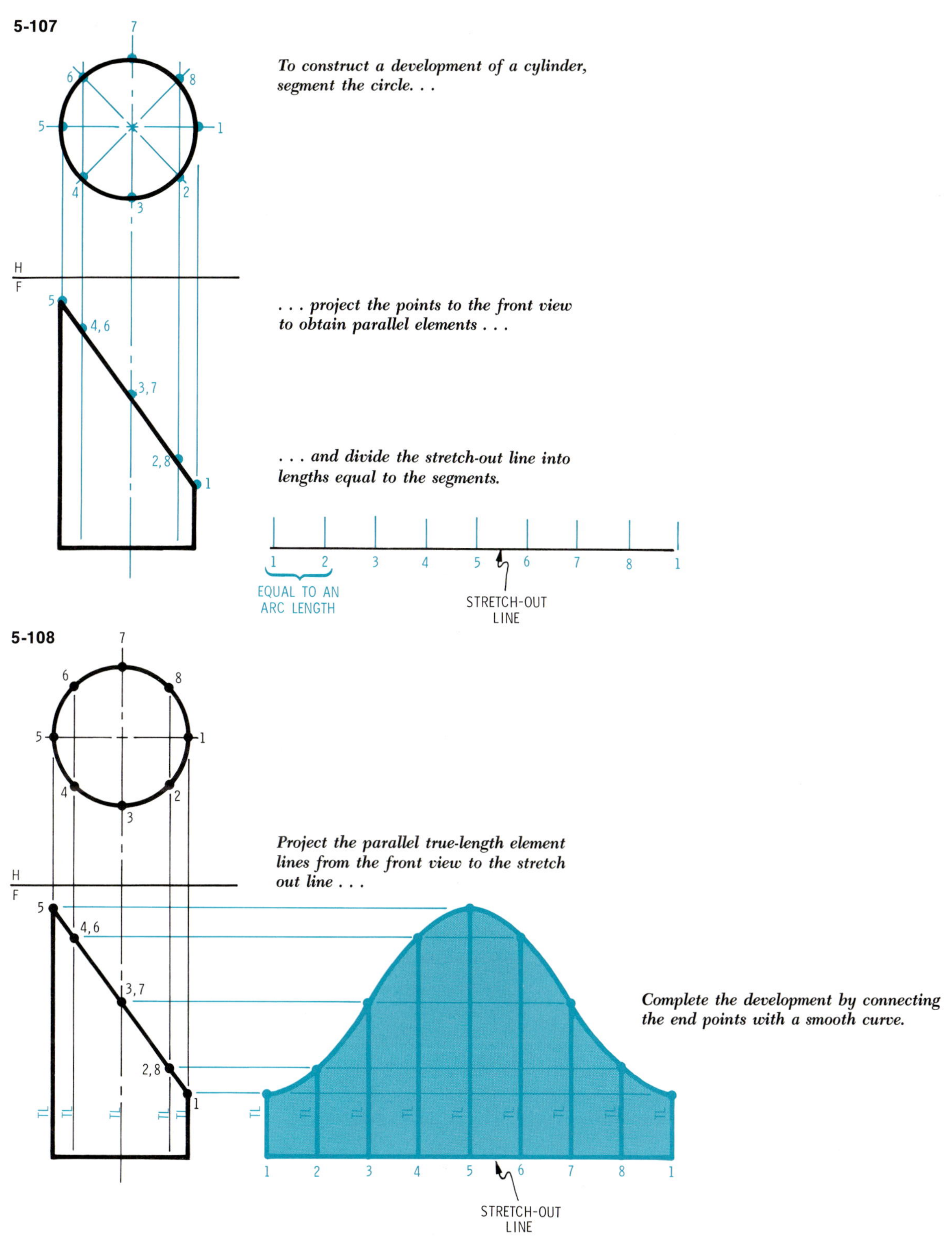

To construct a development of a cylinder, segment the circle. . .

. . . project the points to the front view to obtain parallel elements . . .

. . . and divide the stretch-out line into lengths equal to the segments.

Project the parallel true-length element lines from the front view to the stretch out line . . .

Complete the development by connecting the end points with a smooth curve.

of the stretch-out line was found by using the true widths from the edge views (the horizontal views on our examples). When the edge view of a geometric shape is not available, one must establish its edge view in an auxiliary view and transfer its dimensions to the stretch-out line.

5-109

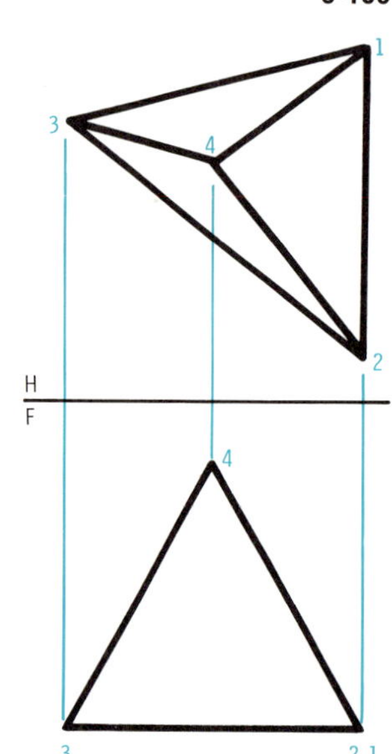

To develop a radial line form, all the lines of the form must be true length.

Development of a Pyramid

The triangular pyramid [5-109] is an excellent example of a geometric form whose linear elements have a radial orientation. Again, all lines that comprise the surface of the development must be in true length, but a different method of layout must be used. Following the example of the prism, the development of the pyramid could be obtained by rotation of the planes or by constructing the necessary auxiliary views to show each plane in true size [5-110]. Note that two auxiliary views would be needed for each plane, since the edge views are not given. The planes then could be cut out, properly aligned, and fastened together again. This procedure is just as unacceptable here as it was before. The better approach is to sequentially unfold the triangular planes using the radial boundaries as bend lines. Unfortunately, radial-line forms cannot be developed with a stretch-out line because the linear elements are not parallel. Instead, the true lengths of the boundary lines of each triangular plane

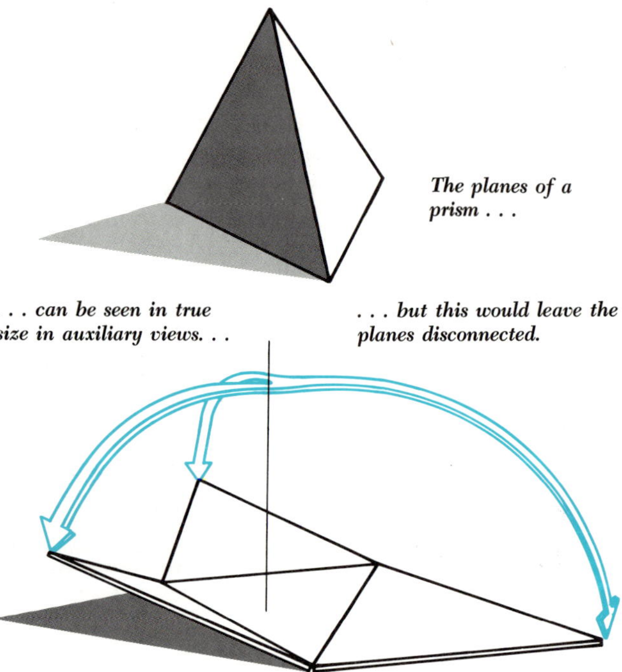

The planes of a prism . . .

. . . can be seen in true size in auxiliary views. . .

. . . but this would leave the planes disconnected.

5-110

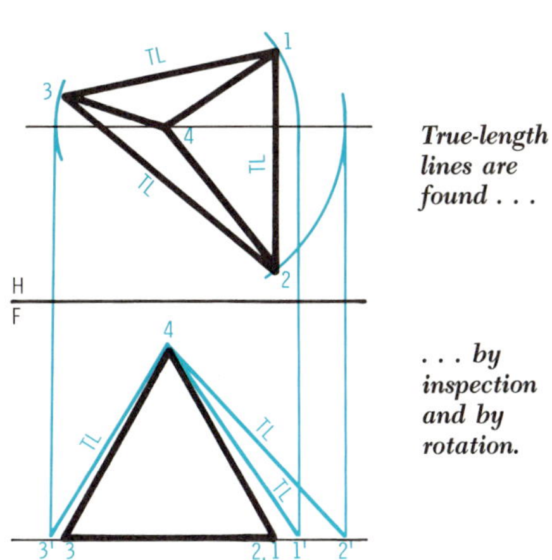

True-length lines are found . . .

. . . by inspection and by rotation.

5-111

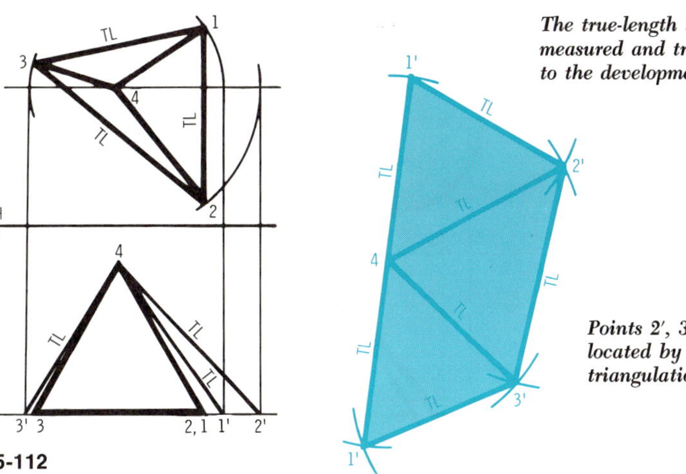

5-112

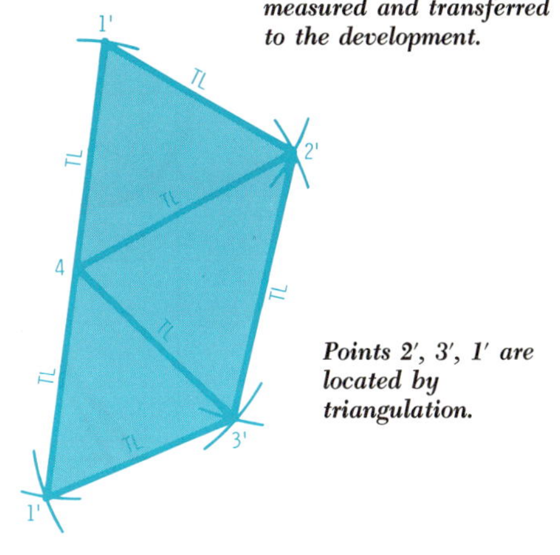

The true-length lines are measured and transferred to the development.

Points 2′, 3′, 1′ are located by triangulation.

must be found and transferred to the development. If the lines do not appear true length in one of the given views, the rotation method is often very convenient for determining the required true lengths [5-111].

The development is started with a convenient line 4-1′ adjacent to the given views [5-112]. After that, arcs equal to the true lengths of sides 1-2 and 4-2 are swung from points 1′ and 4, respectively, on the development. The intersection of these two arcs will determine the location of point 2′. This procedure, called *triangulation*, is continued until all the points of the development are completed. The points are joined with straight lines to complete the outline of the development. (Note that in this case we have produced an outside development. If we reverse our procedure and locate point 3 first, instead of point 2, an inside development will be produced.)

The triangulation procedure discussed above is basic to all radial-line forms. Since radial lines converge at some point, the surfaces of the object always are based on true-size triangles in the development.

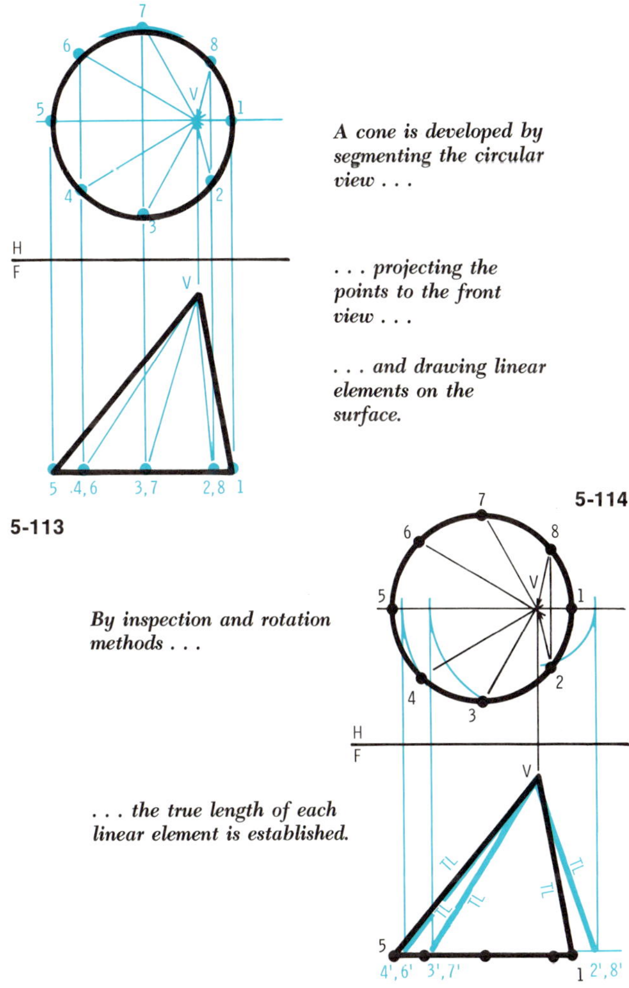

A cone is developed by segmenting the circular view . . .

. . . projecting the points to the front view . . .

. . . and drawing linear elements on the surface.

5-113

By inspection and rotation methods . . .

. . . the true length of each linear element is established.

5-114

Development of a Cone

The cone is related to the pyramid as the cylinder is related to the prism. As with the cylinder, the surface of the cone is continuous, so no corners or edges exist. It is therefore necessary to create radial linear elements in order to develop the cone. The circular view of the cone is segmented into an arbitrary number of equally spaced sections. These are projected to the front view to establish the linear elements of the cone [5-113]. The true length of each of the linear elements is found by inspection and by rotation [5-114], and transferred to

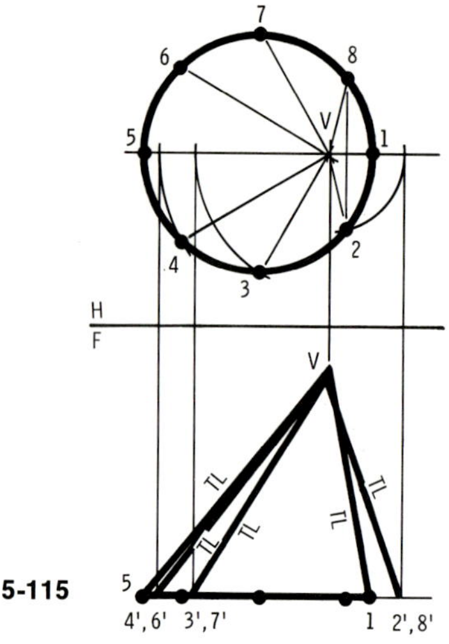

5-115

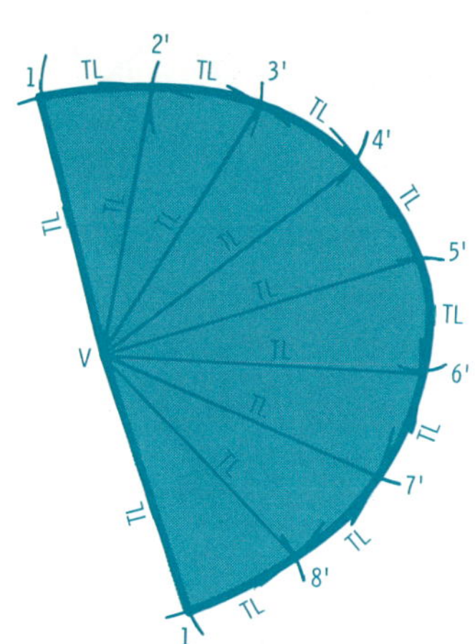

The true-length elements are measured and transferred to the development.

The final shape is developed using the triangulation technique.

the development using the triangulation method [5-115]. After all the points have been established, the shape of the development is completed with a smooth curve.

problems

5-1. Select one of the objects listed below. Measure the object and make a drawing of the three principal views. Dimension the drawing using the scale indicated (unidirectional dimensions). Include all appropriate dimensional symbols.

(a) desk chair (eighth size)
(b) four-drawer file cabinet (eighth size)
(c) chalkboard eraser (full size)
(d) door knob (full size)
(e) waste paper basket (quarter size)

5-2. Using a table of decimal equivalents, convert the following fractional numbers to decimal numbers of three significant figures. Carefully letter the numerals using single-stroke vertical Gothic style.

$10\frac{3}{4}$	$12\frac{1}{16}$	$7\frac{3}{4}$	$2\frac{9}{16}$	$6\frac{19}{24}$
$1\frac{7}{8}$	$\frac{5}{8}$	$41\frac{1}{4}$	$7\frac{27}{32}$	$9\frac{1}{8}$
$3\frac{3}{32}$	$5\frac{13}{16}$	$\frac{25}{64}$	$4\frac{13}{16}$	$\frac{15}{16}$

5-3. Make a list of ten items with which you are familiar that are identified by their *nominal size*.

5-4. Using the list established in Problem 5-3, determine the *basic size* of each of the ten items using decimal notation.

5-5. Give five examples in which dimensions in decimal notation are preferable to fractional notation.

5-6. Dimension [5-116 and 5-117] using aligned dimensions. Add extension, dimension, and center lines necessary to complete the drawings. Scale the drawings to establish the dimensions. Redraw the figures at double size on an $8\frac{1}{2}$- by 11-inch sheet of paper.

5-7. Dimension [5-118 and 5-119] using unidirectional dimensions. Prepare the drawing as described in Problem 5-6.

5-8. Using the dimension 0.750 inch as the basic size of a shaft, write the specifications for a hole and a shaft using a tolerance of ± 0.002 to always accomplish an interference fit. Make a drawing in full size, similar to [5-47], and apply those specifications.

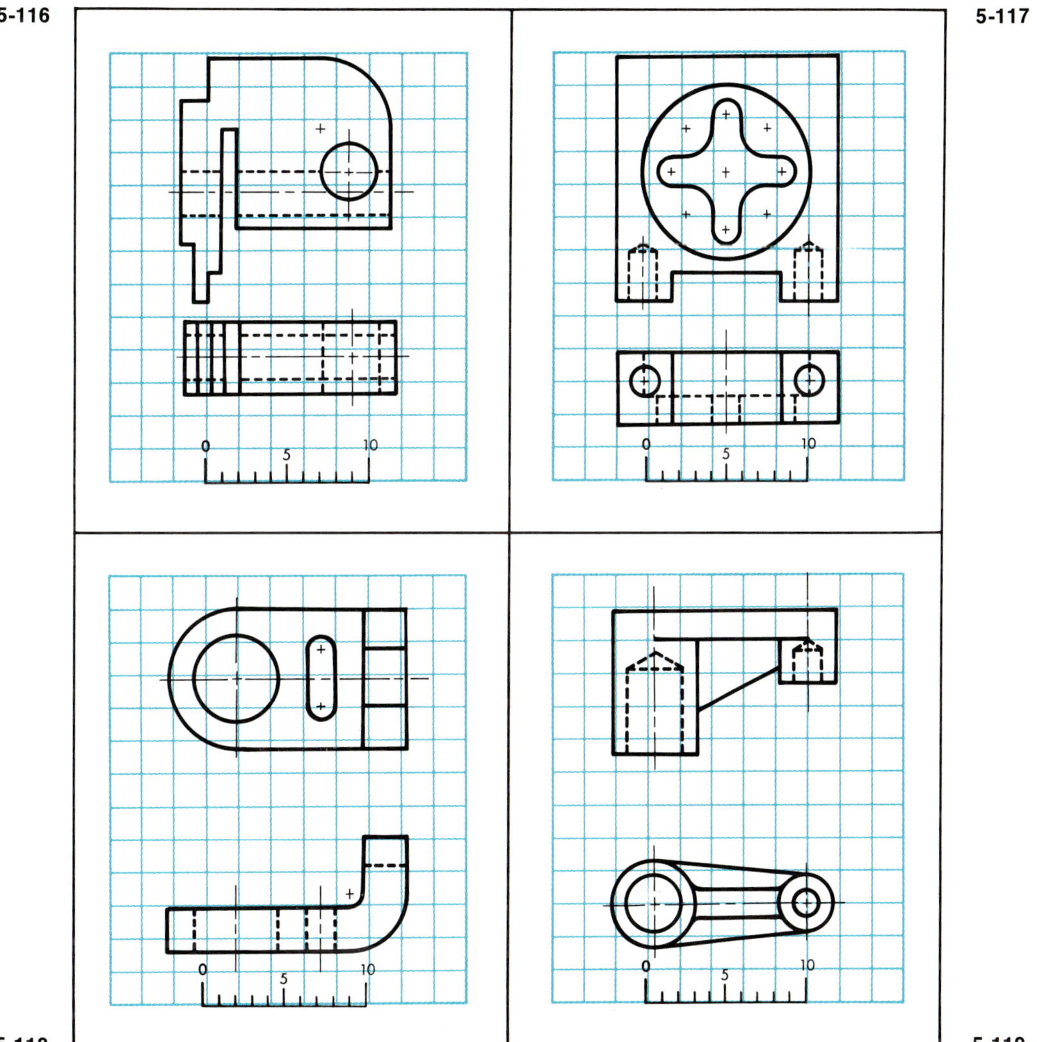

5-9. Using the dimension 1.250 inches as the basic size with a maximum limit of 0.005 inch and minimum limit of 0.002 inch, prepare the following:

(**a**) Make a drawing in full size of a cylinder that has a height and diameter equal to the basic size.

(**b**) Show the unilateral tolerances for the height and the bilateral tolerances for the diameter on the drawing.

5-10. In the problems in [5-120], the views indicated by the circles and arrows are to be changed to full sectional views taken along the center line in the direction indicated by the arrows in the remaining view. For each, select the correct answer from the 24 proposed views shown at the right, and place its number in the appropriate circle.

5-120

5-11. In the problems in [5-121], the views indicated by the circles and arrows are to be changed to half sectional views as indicated by the letters A–A taken along the center line in the remaining view. For each, select the correct answer from the 24 proposed views shown at the right, and place its number in the appropriate circle.

A 11
B 21
C 16
D 13
E 24

5-121

5-12. From the 30 section views shown in [5-122], select one to complete each problem and place its number in the appropriate circle.

5-122

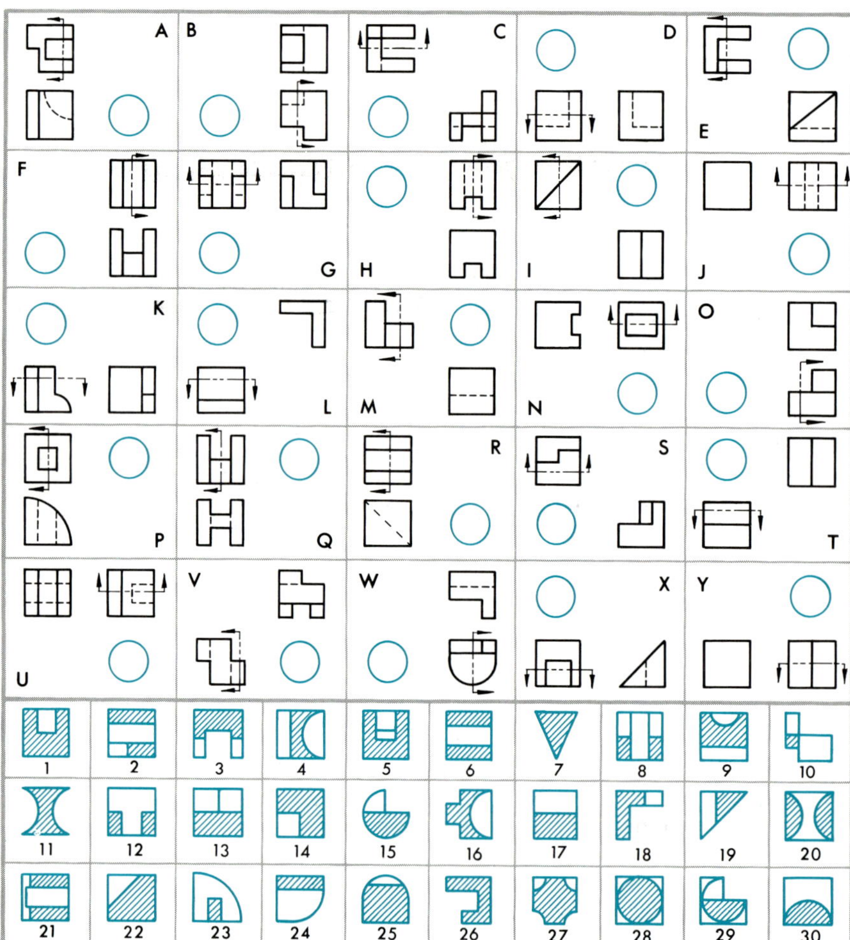

5-13. From the 30 section views shown in [5-123], select one to complete each problem and place its number in the appropriate circle.

5-123

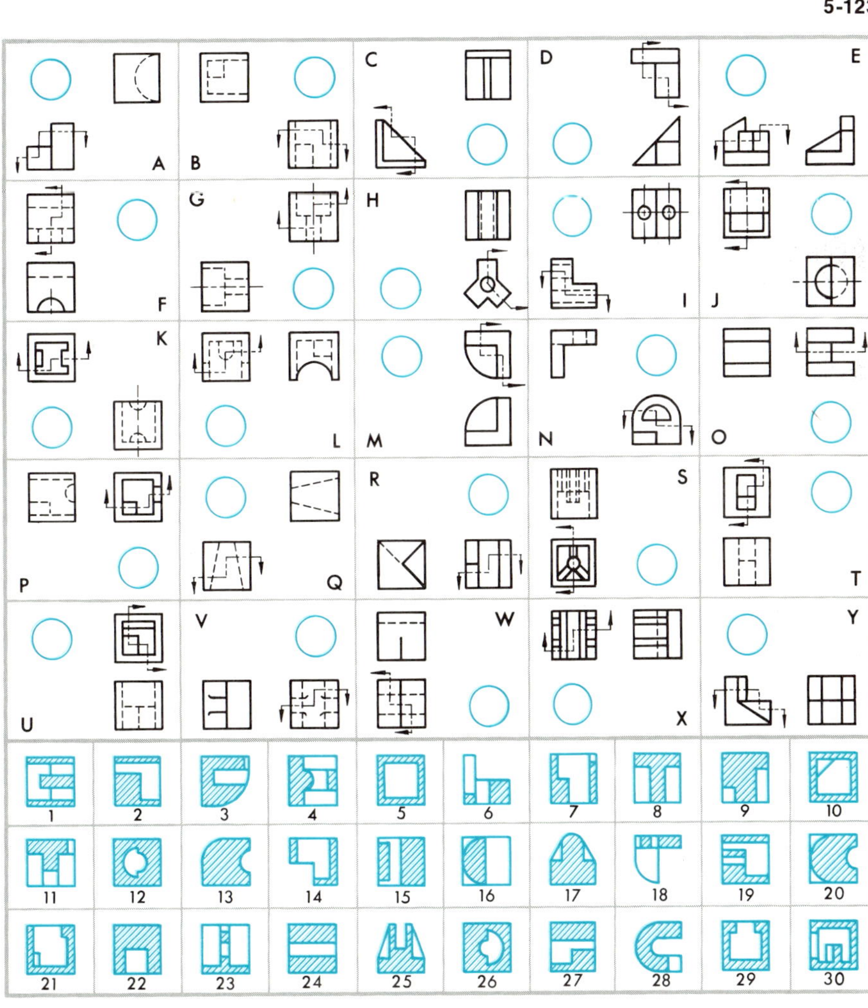

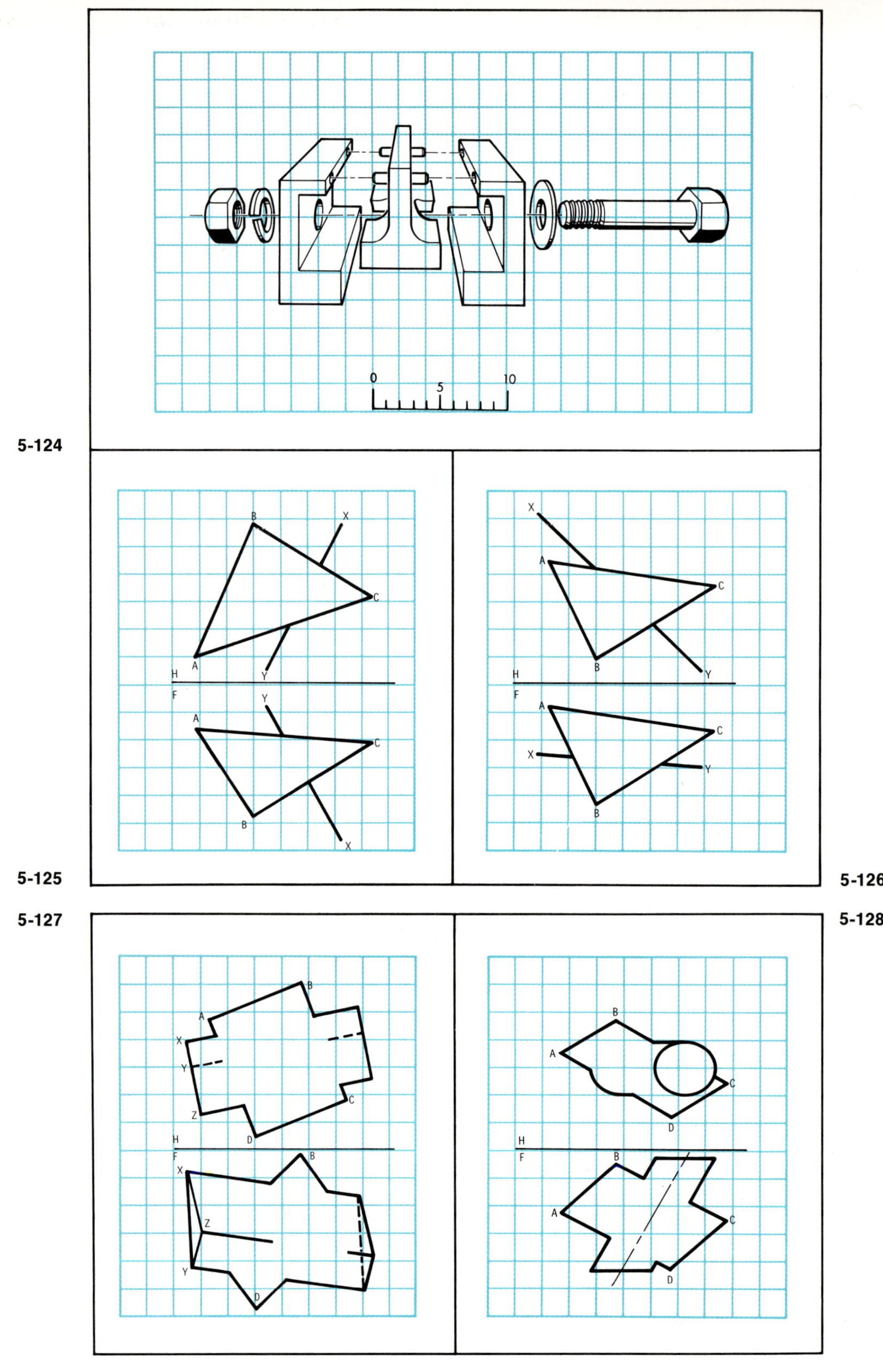

5-124

5-125

5-126

5-127

5-128

5-14. Figure [5-124] is an exploded drawing of a *whidjet.* Make a full-section drawing at double scale of the whidjet using a cutting plane that slices along the vertical center line of the bolt. Draw the full section as if the whidjet were assembled and use appropriate section lines to distinguish the various components of the assembly.

5-15. Make a half-section drawing of a coffee mug. Orient the side view so that a section view of the handle will result. Draw the mug in full size and dimension it appropriately.

5-16. Study several issues of current technical and trade magazines. Using a copying machine, reproduce three examples of different types of section views, such as full sections, half-sections, revolved, and removed sections. On each reproduction, write the type of section shown and explain why you think it was used as an illustration.

5-17. In [5-125 and 5-126], determine if the lines *XY* intersect the planes *ABC*. Use the edge-view method to find the intersection point (if any) and complete the views accordingly, showing the line *XY* where it is visible. Draw the views on $8\frac{1}{2}$- by 11-inch paper, one problem to a page at double size.

5-18. Find the intersection point of lines *XY* and planes *ABC* in [5-125 and 5-126], using the cutting-plane method. Prepare the drawings double size, as described in Problem 5-16. Show the line *XY* where it is visible.

5-19. In [5-127] establish the intersection lines between the plane and the triangular prism using the cutting-plane method. Scale the dimensions in the text and reproduce the drawing double size on $8\frac{1}{2}$- by 11-inch paper. Indicate all visible lines.

5-20. Make an isometric sketch of [5-127]. Orient the sketch such that the intersection between the plane and the prism is clearly visible and the end of the prism with the letters *XYZ* is on the left. (To work this problem, it may be necessary to complete Problem 5-19 in order to visualize the shape and general location of the intersection.)

5-21. Find the intersection between the plane and the cylinder in [5-128]. Use the cutting-plane method and make the necessary measurements from the text. Produce the drawing at double size on $8\frac{1}{2}$- by 11-inch paper. Show all lines where they are visible.

5-22. Make an isometric sketch of [5-128].

5-23. Draw a development of the portion of the triangular prism that lies to the left of the intersection line in Problem 5-20. Use the lettered (*XYZ*) end of the prism as the base. Assume plane *ABCD* passes entirely through the prism.

5-24. Find an unusual example of intersecting prisms that exists on your campus. (Some examples to investigate might be heating and air conditioning ducts, hallways and buildings, and storage bins.) Make the necessary measurements to draw the prisms and then establish the shape of the intersection using the cutting-plane method. Select a scale such that the drawing can be presented on an $8\frac{1}{2}$- by 11-inch sheet of paper.

5-25. Following the general procedure outlined in Problem 5-24, construct a drawing of cylinders intersecting at other than a right angle.

graphic presentation and computation 6

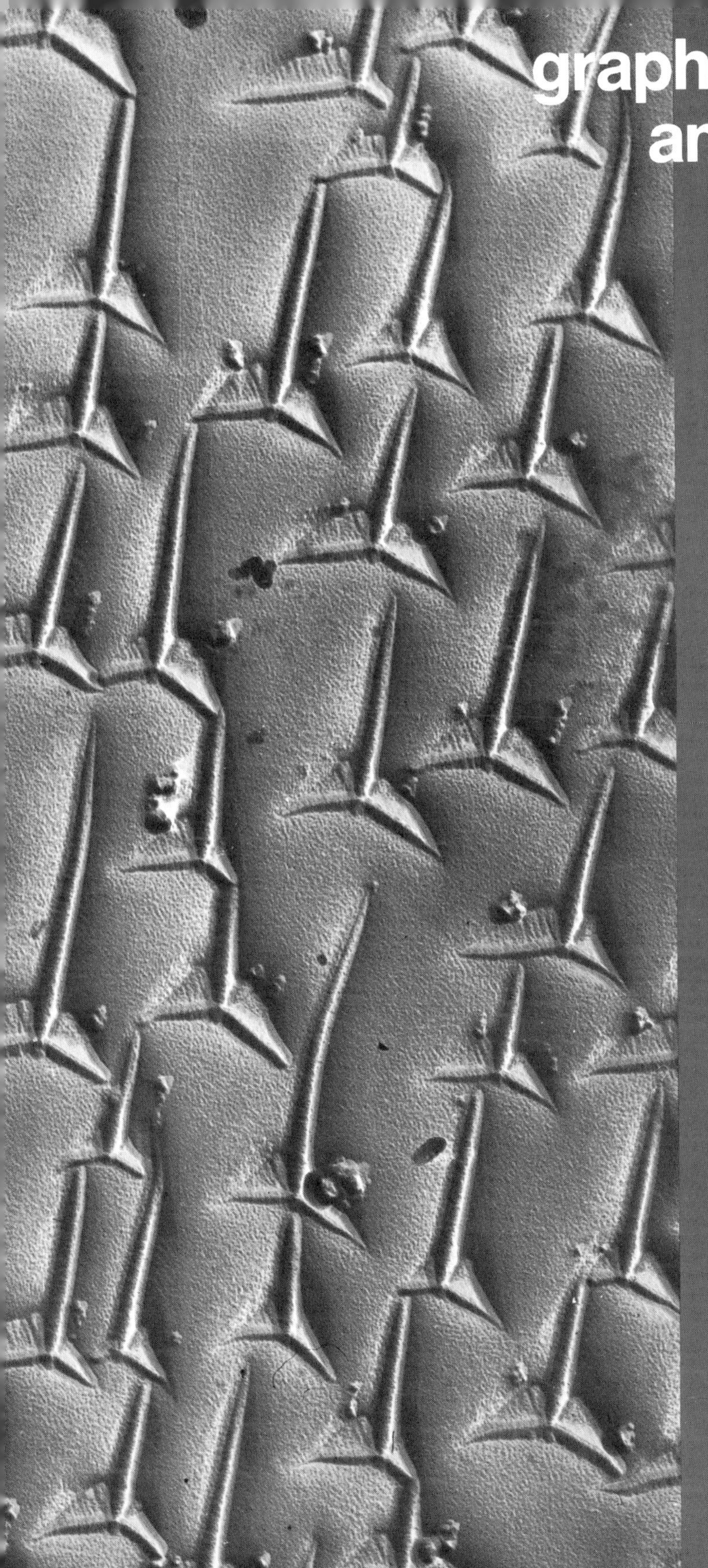

Manganese precipitates.

6-1

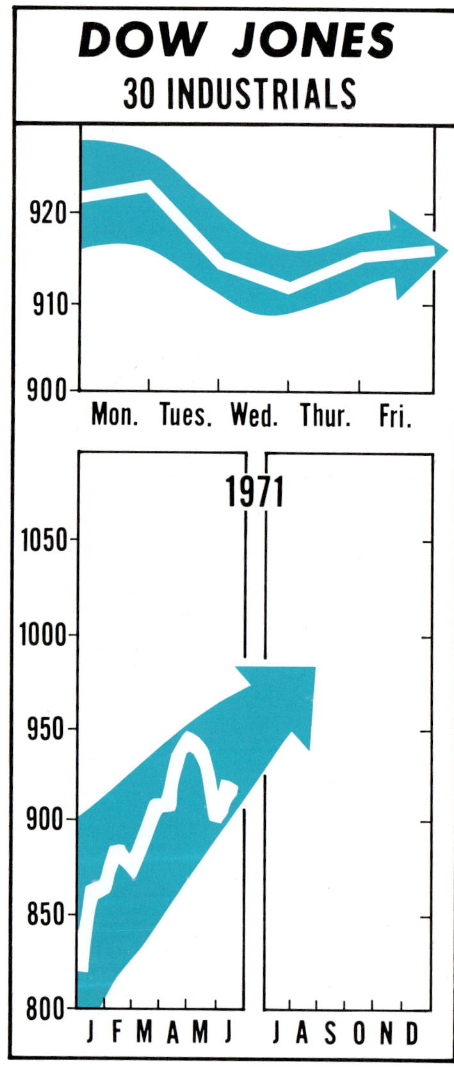

Although the engineer may complete a superb design, a valid test, or a study of great importance, the significance of his efforts may be ignored or misunderstood if he is unable to communicate properly the results of his work. For this reason it is very important that he understand the most effective ways to present technical data. This chapter deals with the presentation of numerical data—such as computed performance figures, the results of experiments, and survey data. The use of design drawings and diagrams and the layout of a written or oral report are discussed elsewhere in this book.

The engineer must write for two different audiences: the general public and technically trained people. Certain techniques and types of graphic format lend themselves more readily to general understanding and popular appeal; others are more useful for technical audiences. Simplicity of format and emphasis on gross dynamic changes between variables are more important factors for the general public [6-1]; accuracy and preciseness of the presentation of the data are paramount considerations in engineering and science [6-2].

Data may be presented in several different formats or styles as follows:

1. Tables
2. Charts
    - Pictorial charts
    - Bar charts
    - Pie charts
    - Segment charts
    - Organizational charts
    - Flow charts
    - Alignment charts
3. Graphs
    - Rectilinear graphs
    - Semilogarithmic graphs
    - Logarithmic graphs
    - Polar graphs

The progression of data presentations shown in [6-3] illustrates the use of tables, graphs, and charts to communicate the same information.

tables

Frequently, raw data are presented in tabular format. Tables are useful for detail and exactness, but it is recommended that some other graphic form be used for a final presentation. In a table, data are arrayed in rows (horizontally) and columns (vertically), usually in ascending or descending order. Each row and column should be labeled clearly and completely, and the table should be assigned a title that effectively describes its function or purpose. For examples of tables, see Appendix V.

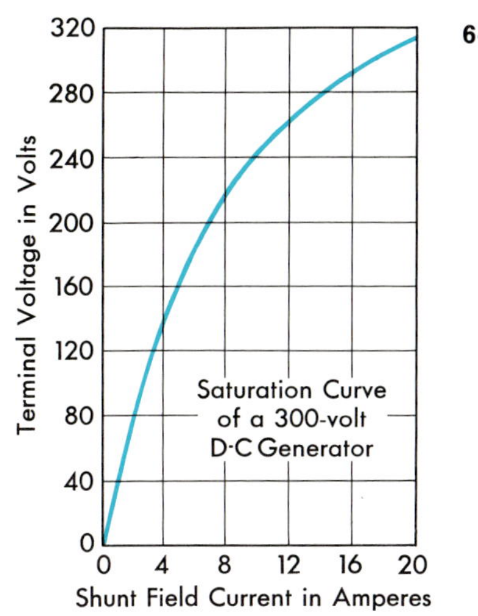

6-2

charts

Charts are most useful in communicating with nontechnical audiences. They are particularly effective when comparing several dependent variables with a single independent variable. For charts, it is usually best to use commercially prepared chart board or paper. If a preprinted grid does not exist, one should be prepared to accommodate the ranges of each variable.

Pictorial Charts

The average layman has little interest in evaluating technical information. Therefore, the simplest form of

6-3
Growth in product sales.

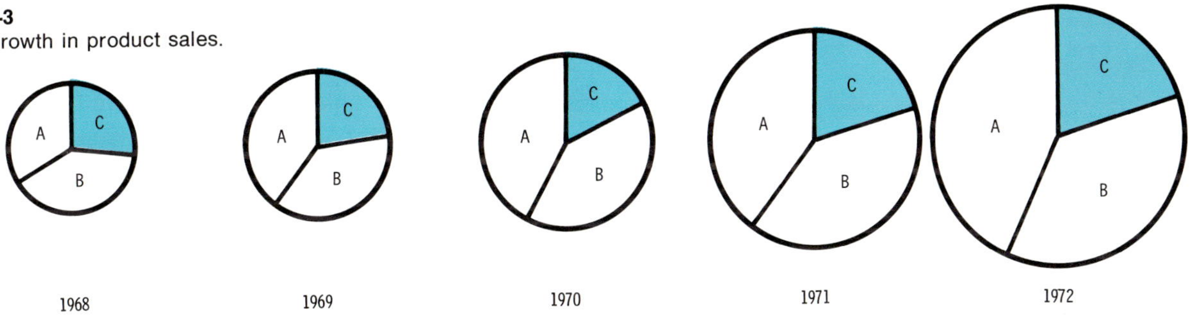

PRODUCT SALES (x 100)				
YEAR	A	B	C	TOTAL
1968	10	12	8	30
1969	14	13	8	35
1970	17	16	7	40
1971	20	20	10	50
1972	26	22	12	60

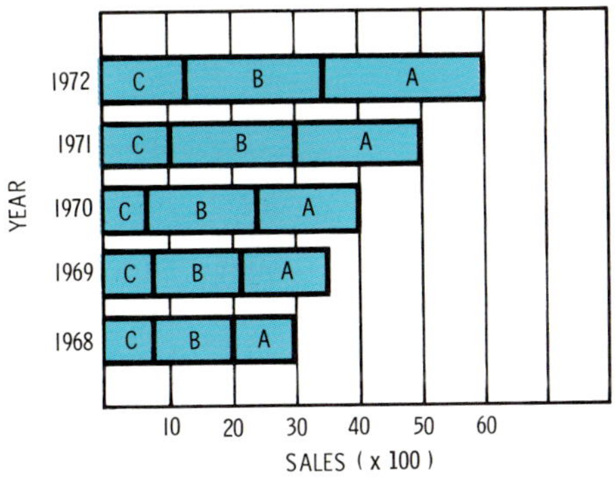

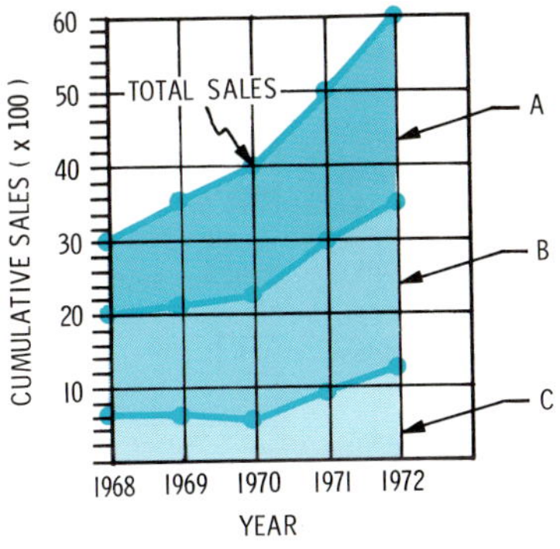

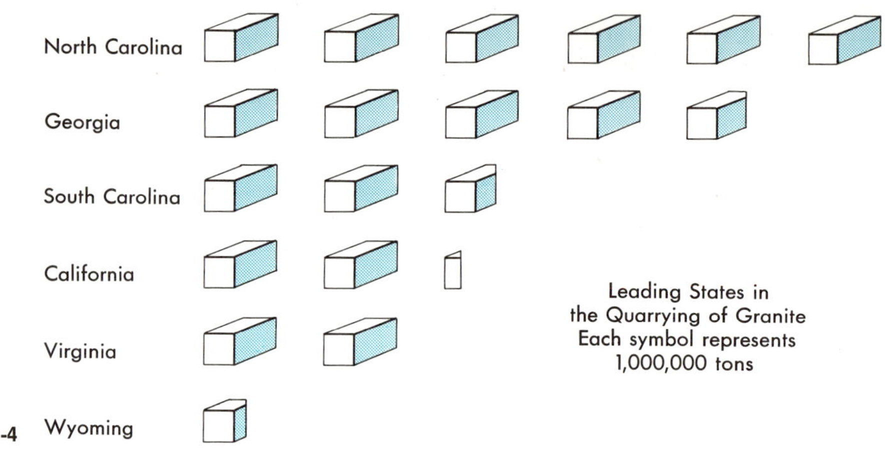

6-4

chart should be used for presentations to the general public. The pictorial chart or map-type chart is most commonly used for such purposes [6-4].

Bar Charts

Bar charts are commonly used in popular literature, particularly when the results of economic or industrial surveys are reported. The bars may be drawn horizontally or vertically, but they should extend from a common base (axis). Horizontal bars are preferable for most applications because their respective positioning facilitates length comparison [6-5]. Vertical bars are particularly useful when time is the independent variable [6-6]. For emphasis and clarity the exact value represented by each bar may be printed within the body of the bar or adjacent to it.

Custom more than anything else dictates which variable will be plotted on the horizontal axis. (The horizontal X axis is called the *abscissa* and the vertical, or Y axis, is called the *ordinate*.) Generally, we would expect to find the controlled or *independent-variable* scale values plotted horizontally from left to right and the *dependent-variable* scale values plotted vertically from bottom to top.

The choice of scales is perhaps the most important step in bar-chart construction, because the selections have a controlling influence on the viewer's interpretation of the relationships that are presented. It would be a breach of professional ethics, therefore, for the engineer to relatively stretch a horizontal or vertical scale to create an erroneous impression. The same care must be exercised when using volume or area relationships. For example, in [6-7] the relative comparison of A, B, and C is easy,

6-5

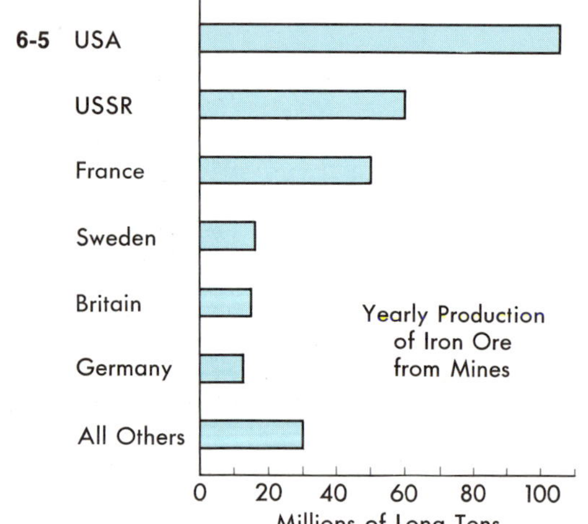

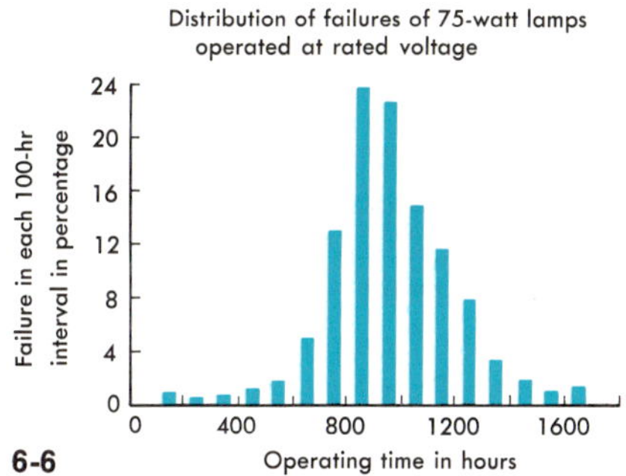

6-6

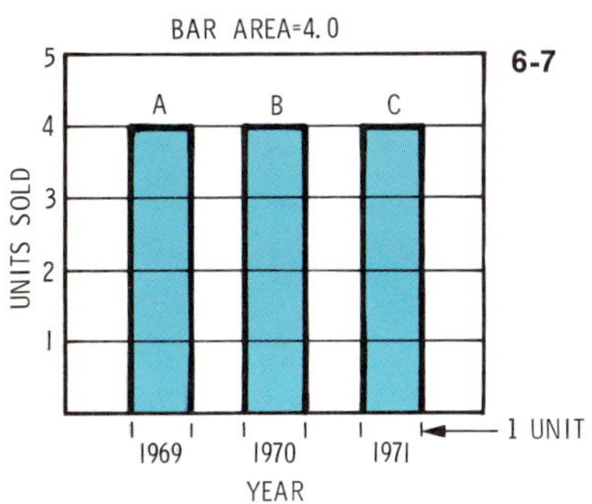

6-7

The engineer should not stretch a horizontal or vertical scale that might create an erroneous impression.

but in [6-8] (where areas are also used), the comparison is likely to be misunderstood. If a visual comparison is desired, the zero point should be clearly indicated. The charts [6-9 and 6-10] demonstrate the differences in interpretation that may result.

It is also important that each scale or part of the chart be labeled clearly and neatly. The printed information should be readable from the bottom and right side of the paper. Each chart should have a title that is clear, concise, and complete. Legends or keys are sometimes useful in conveying necessary explanatory information.

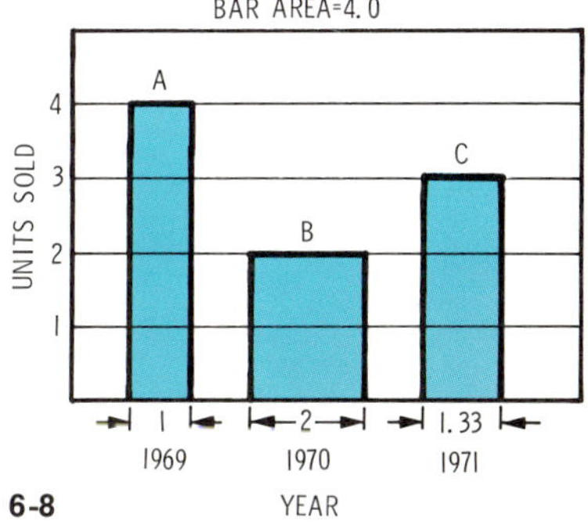

6-8

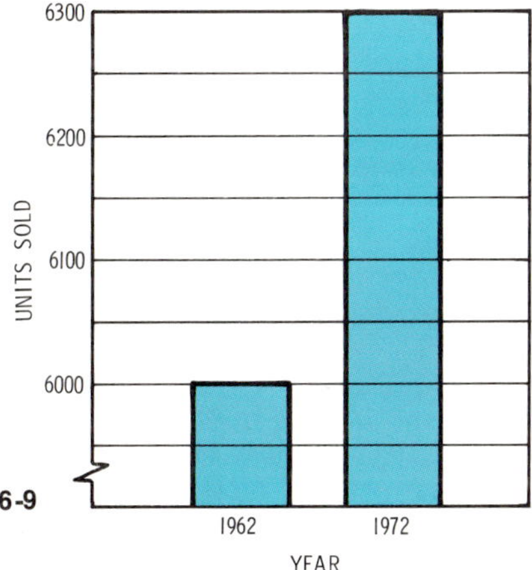

6-9

The zero point should be clearly indicated.

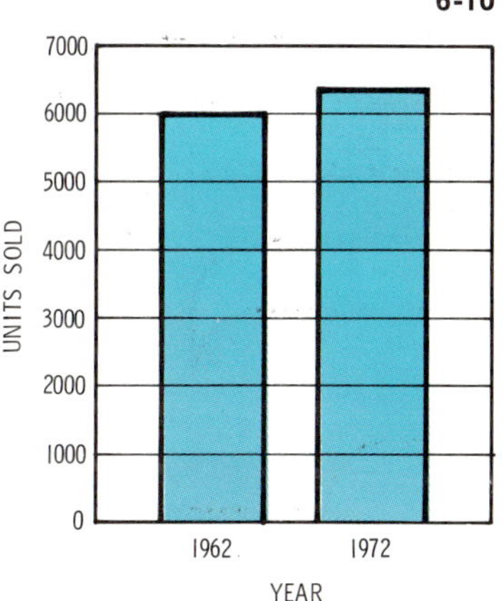

6-10

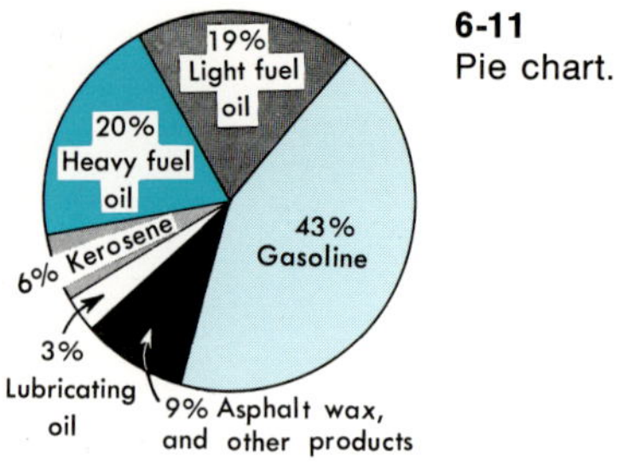

6-11
Pie chart.

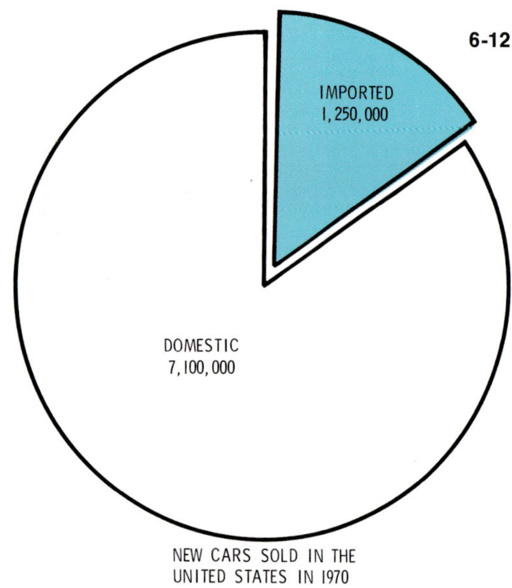

6-12

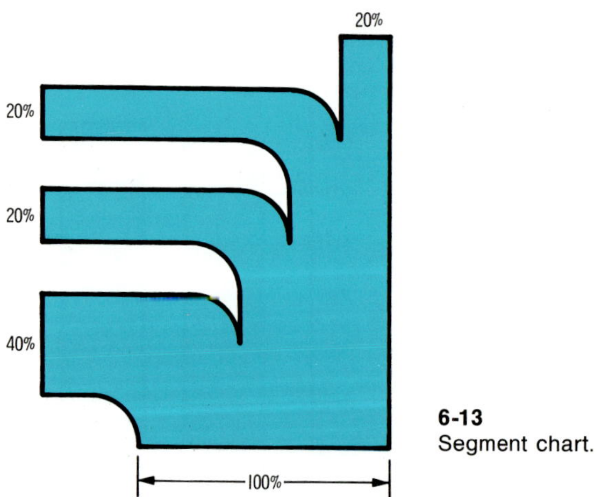

6-13
Segment chart.

Pie Charts

One of the most effective ways to show the percentage or distribution of related parts of a unit is to make a pie chart [6-11]. The total area of the circle represents 100 per cent of the sample and the sectors represent percentages of the total. Each sector should be identified, and the lettering should be completed prior to adding the shading or coloring. For added emphasis one or more sectors can be removed slightly from their normal position [6-12].

Segment Charts

Segment charts are useful to show how some total quantity is divided into individual components—somewhat similar in this respect to the pie chart. The base of the chart (assumed to be 100 per cent) should be wide enough to provide an adequate area for the smallest width limb [6-13].

Organizational Charts

From time to time management studies, computer analyses, and material or energy flows (for example, in the process industries) all require charts that identify and clarify the sequence of the operations involved [3-1].

Alignment Charts

The purpose of alignment charts (frequently called *nomographs*) is to reduce laborious and repetitive calculations of mathematical relationships and to study the interrelationships among the variables. A more complete explanation of this type of chart begins on page 206.

graphs

For the technical audience, graphs are the most useful form of graphical presentation and are especially advantageous when technical facts must be grasped readily. They also aid in the analysis of engineering data and facilitate the presentation of statistical and test information. Graphs also offer an opportunity to interpolate or extrapolate values, and to draw conclusions as to the behavior of the variable quantities involved. In general, they are composed of a grid background, calibrated scales, and one or more trend lines. Some basic procedures in the preparation of rectangular coordinate graphs are noted here.

1. Graphs are most often drawn on special preprinted paper and, of those available, rectangular-coordinate paper is the most common. Its grid is formed by horizontally and vertically equally spaced lines that form small rectangles. Other types of grid rulings will be discussed below. Popular types are those whose squares are 1 millimeter, $\frac{1}{10}$ inch, or $\frac{1}{20}$ inch on a side. In each case the ruling of every fifth or tenth line is emphasized by being slightly wider or darker, or both.
2. As with charts, the dependent variable is plotted on the vertical axis and the independent variable on the horizontal. The axes are drawn as heavy lines and should intersect near the lower-left corner of the paper[1] about 1 inch inside the grid field to allow for lettering. The sheet may be turned so that the abscissa is along either the short or long side of the paper. If the graph is prepared for a report, the margin for the binding should be either to the left or at the top of the sheet.
3. Scale selection is a function of the engineer's sublety and discernment. An appropriate scale selection will do much to place the viewer in a proper mental attitude to evaluate the significance of the orientation of the trend line (curve). A poor choice might lead to a misunderstanding of the data [6-14]. Also, the scales should be consistent with the precision of the data. A trial computation to select the scale on each axis can be made with the following expression:

$$\text{scale} = \frac{\text{range in the variable}}{\text{scale length available}}$$

4. Scale divisions should start at zero, unless the curve would be compressed unnecessarily, and each unit (small square) on the scale should represent 1, 2, 5, or 10 units of measurement, multiplied by 1, 10, 100, 10^{-3}, 10^{-8}, etc., for very large or very small numbers. Do not use a scale that will require awkward fractions in the smallest calibration on the paper. Normally, only the accented lines of the printed grid are designated by graduations.

[1] When both positive and negative values of a function are to be plotted, the origin should be located so that all desired values can be shown.

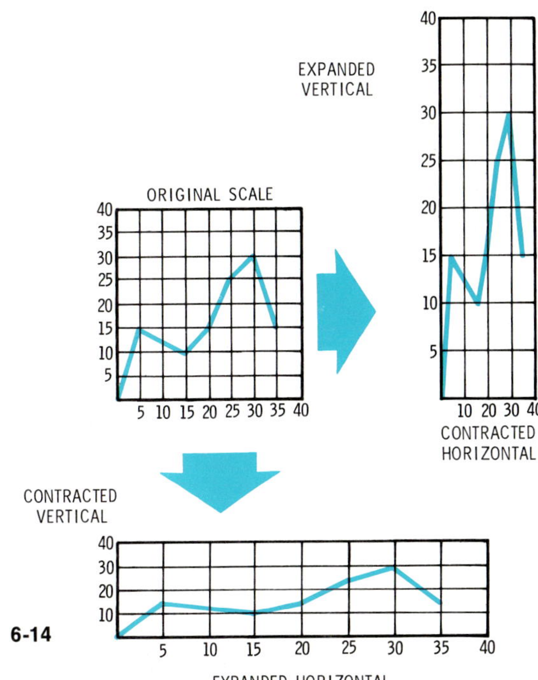

6-14

5. The ordinate and abscissa should be labeled with the names and dimensional units of the variable it represents. For example, WEIGHT IN POUNDS. Considerable care should be taken to ensure that the lettering is neat, uncrowded, and well proportioned. It should be about $\frac{3}{16}$ inch high and may be either vertical or slant lettering. Always place a cipher to the left of

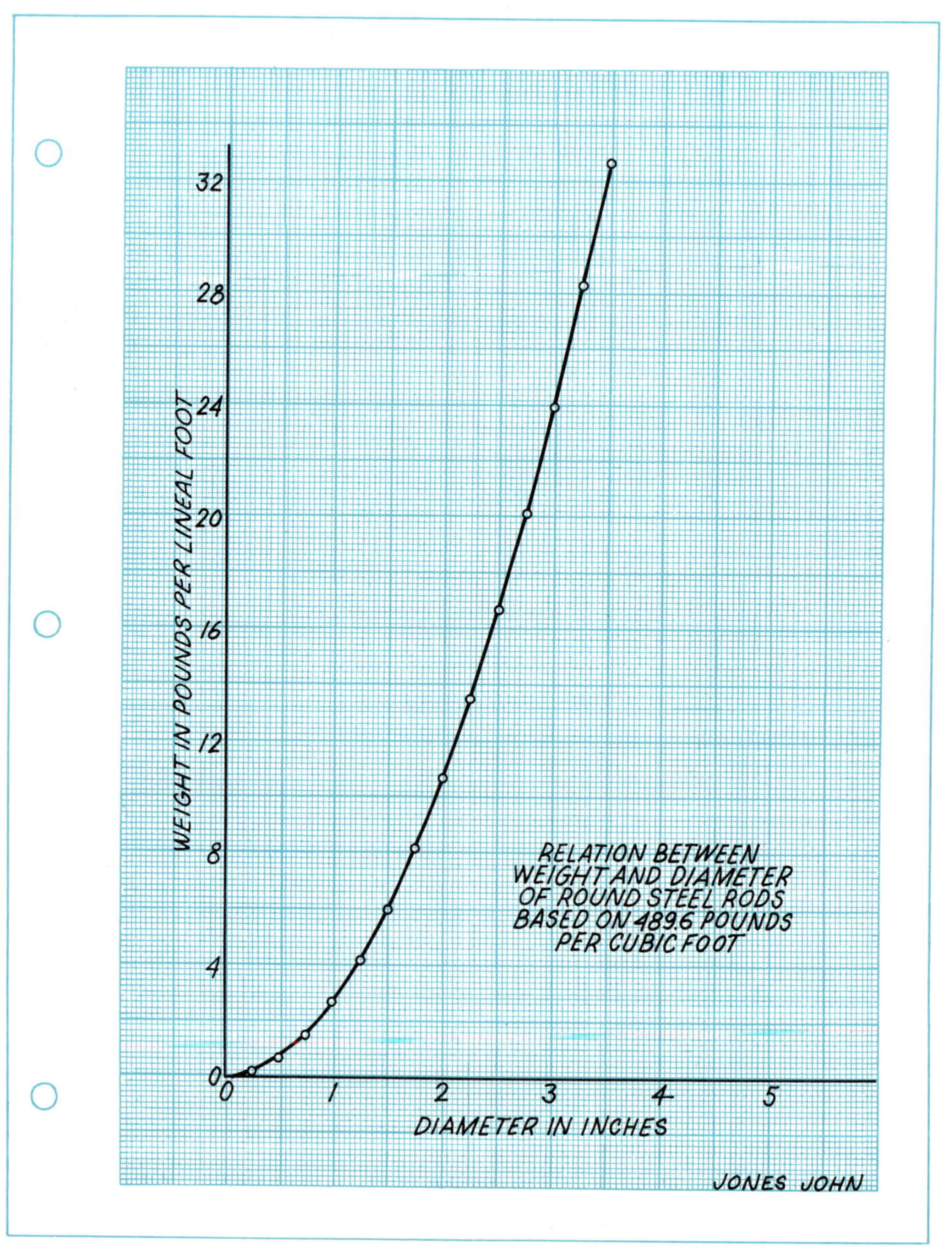

6-15

the decimal point if decimal fractions are used. When space is limited, standard abbreviations such as those in Appendix II should be used, particularly for designating the unit of measurement.

6. If the data represent a set of experimental observations, the plotted points of a single trend line should be marked by small circles having a diameter of $\frac{1}{10}$ inch or less. Where several trend lines are to be indicated on a single graph, partially filled in circles, open squares and triangles, and filled in squares and triangles may be used to locate the plotted points of the various curves. Filled-in symbols should be drawn $\frac{1}{16}$ inch or less in diameter. Do not use crosses or X's.

6-16

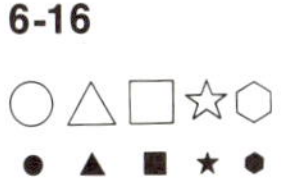

7. Most frequently, the engineer will be concerned with plotting data that are related to physical phenomena for which the relationships are apt to be direct and continuous—resulting in a smooth trend line. When such a line approaches one of the plotted points, it will touch but not pass through the small circle circumscribing the plotted point [6-15]. Where multiple trend lines are shown on the same graph, identification can be enhanced by using different lines, such as solid lines, dashed lines, or long dash–short dash lines. Trend lines that have been determined theoretically will not normally have point designations. Data points in measured relationships not supported by mathematical theory or empirical relationships should be connected by straight lines drawn directly from point to point [6-14].

When plotted, much experimentally determined data will not fall on a smooth trend line. This is due to experimental errors and other factors. When this happens draw a smooth trend line (straight or curved) as the data indicate, which as accurately as possible (by eye) will average the plotted points. A light freehand line will aid in locating the average, but the final line should be drawn with the assistance of mechanical aids (irregular curve, French curve, etc.). The graph [6-17] displays data from an actual test. Note the dispersion of the data points.

6-17

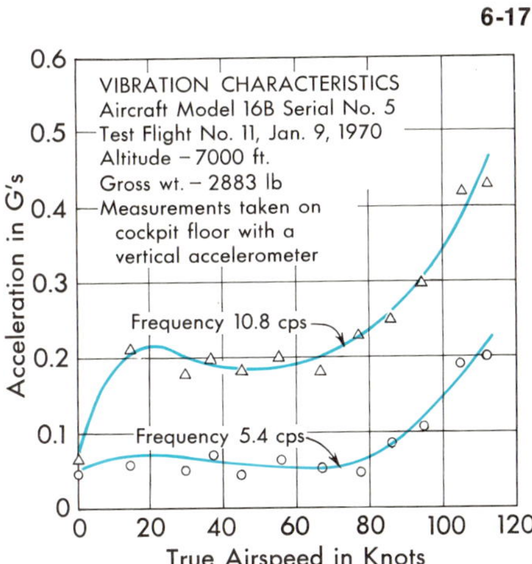

8. The title of the graph may include the names of the plotted quantities and other descriptive information, such as sizes, weights, names of equipment, date that the data were obtained, serial numbers of apparatus, name of manufacturer, and any other information that would help describe the graph. Usually, graphs are designated by naming ordinate values first, then

abscissa quantities. The title should be placed on the sheet where it will not interfere with the graph. The title section of display graphs is usually placed across the top of the sheet. Simple graphs that comprise parts of reports frequently have the title in either the lower-right or upper-left quadrant. The name of the person preparing the graph may be placed in the lower-right corner of the sheet [6-15].

Plotting on Semilogarithmic Graph Paper

The preceding discussion has concerned the graphing of data on rectangular-coordinate paper. There are cases when the variation of the data is such that it may be desirable to compress a variable. To do this, semilogarithmic graph paper may be used. Semilog paper, as it is usually called, is graph paper that has one coordinate ruled in equal increments and the other coordinate ruled in logarithmic increments. When plotting on this type of paper, it can be turned so that either the horizontal coordinate or the vertical coordinate will have the logarithmic divisions. Semilog paper is available in either one-cycle, two-cycle, three-cycle, four-cycle, or five-cycle ruling.

A semilog grid is especially useful in the derivation of relationships where it is difficult to analyze the rate of change or trend as depicted on rectangular-coordinate paper. Data that will plot as a curve on rectangular-coordinate paper may plot as a straight line on semilog paper. If this happens the rate of change of the variable is constant. Where straight lines do not occur on a semilog grid, the rate of change is varying.

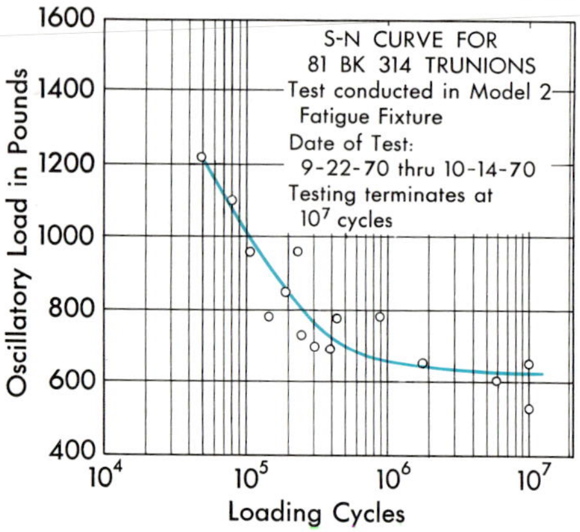

The same rules apply for plotting on semilog paper as were given for rectangular-coordinate paper, except that the numbering of the logarithmic divisions cannot begin with zero. Each cycle on the paper represents a multiple of 10 in value, and the graduations may begin with any power of 10. When reading from a logarithmic graph, interpolations should be made logarithmically rather than arithmetically. An example of data plotted on semilog paper is shown in [6-18].

6-18

Plotting on Log–Log Graph Paper

Log–log graph paper, as its name indicates, has both coordinate divisions expressed as logarithmic functions. This serves to compress data. In addition, data that plotted as a curve on rectangular-coordinate paper may plot as a straight line on log–log paper.

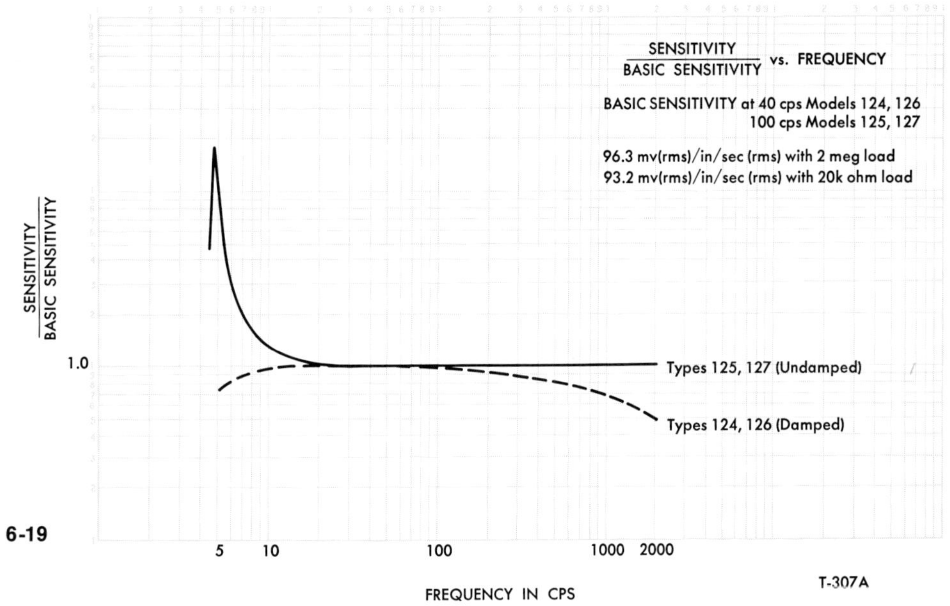

6-19

As an example, the plot of the algebraic expression

$$X = Y^2$$

on rectangular-coordinate paper is a parabola. However, if its values are plotted on log–log paper, it forms the expression

$$\log X = 2(\log Y)$$

This is a linear equation having a slope of 2. Thus, a relationship of variable quantities that may be expressed as $X = Y^2$ when plotted on log–log paper will be a straight line with a slope of 2.

Log–log paper may be secured with one or more cycles for each coordinate direction. The axis lines are drawn on the sheet in a manner similar to the procedure described for plotting on rectangular-coordinate paper. However, the beginning values for the axes will never be zero but will always be a power of 10.

An example of data plotted on log–log paper is shown in [6-19].

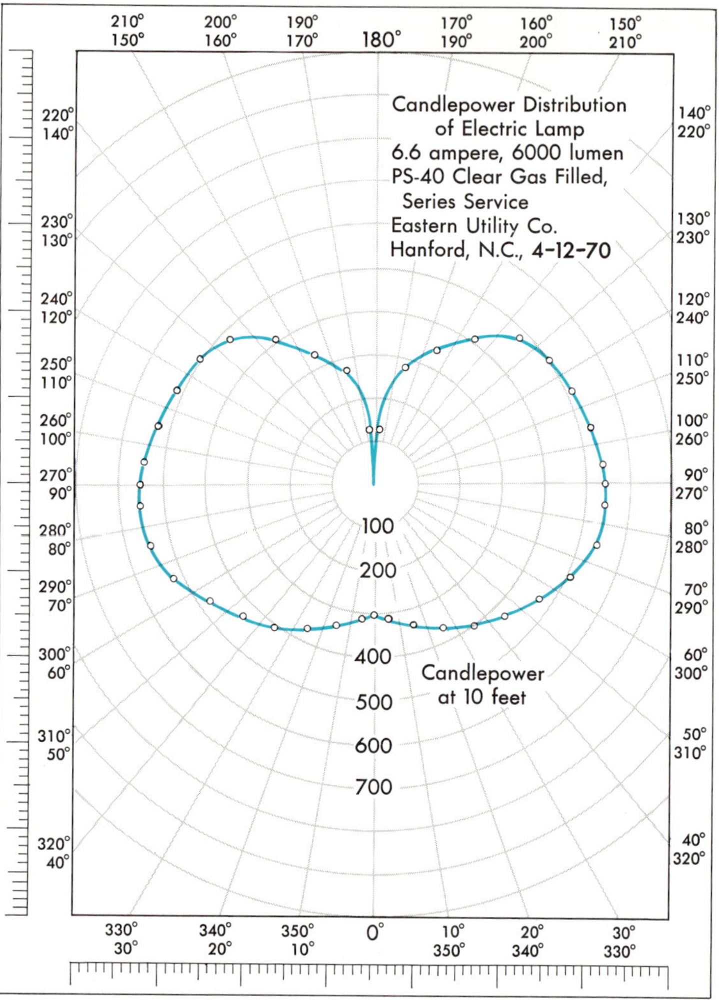

6-20

Plotting on Polar Graph Paper

Polar graphs are sometimes used when a variable quantity is to be examined with respect to various angular positions. The same general principles of plotting apply as were outlined for rectangular plots, except that the outer border is marked off in degrees for the independent variable, and either the horizontal or vertical radial line is marked off for the dependent variable.

Polar graphs frequently are used to display the light output of luminous sources, the response of microphone pickups, and the behavior of rotating objects at various angular positions. An example of a graph plotted on polar-coordinate paper is shown in [6-20].

Determining Empirical Equations from Curves

Experimentally determined data when plotted often will approximate a straight line or a simple curve. By plotting experimentally determined data, it is frequently possible to obtain a mathematical equation that closely expressed the relations of the variables.

Many equations encountered in engineering work have the form

$$y - k = m(x - h)^n$$

where n may have either positive or negative values. If the exponent n is 1, the equation reduces to the familiar straight-line slope-intercept form. If the value of n is other than 1 but positive, the equation is parabolic; if the value of n is negative, the equation is hyperbolic. This expression affords a means of securing empirical equations from experimental data.

If experimental data are to be plotted and an empirical equation is to be determined, it is advisable first to plot the test data on rectangular-coordinate paper in order to gain some idea of the shape of the graph. If the locus approximates a straight line, the general equation $y = mx + b$ may be assumed. The Y intercept b and the slope m may be measured by taking a straight line drawn so as to average the plotted points.

A plot of data taken in the laboratory for a test involving the magnitude of frictional forces is shown in [6-21]. A straight line is drawn to average the plotted points, and the slope of the line is found by taking any two points along the line and determining the X component and the Y component between the two points according to the plotted scales. In this example, the slope is approximately $\frac{65}{500}$, or 0.13. If the line is projected to the Y axis, corresponding to a value of $x = 0$, the Y intercept is seen to correspond approximately to 14 pounds. An approximate equation of these data would be $y = 0.13x + 14$.

If a plot of experimental data on rectangular-coordinate paper should appear to be approximately parabolic or hyperbolic in shape, an empirical equation may be obtained by plotting the data points on log–log paper. The slope of the line determines the exponent of the independent variable, and the Y intercept, when $x = 1$, defines the coefficient of the independent variable. For example, a plot of data taken in the laboratory is shown in [6-22].

A straight line is drawn to average the plotted points.

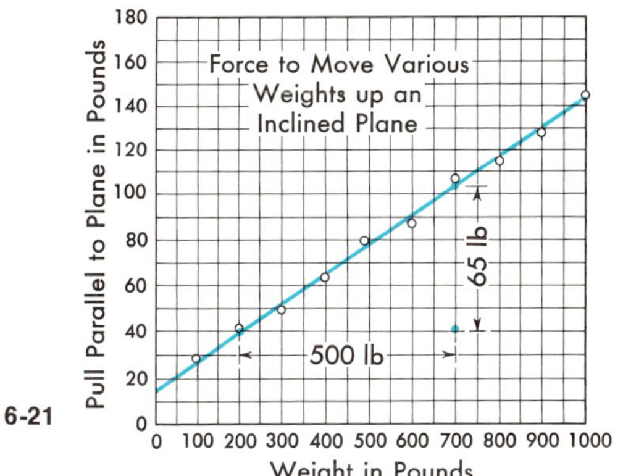

6-21

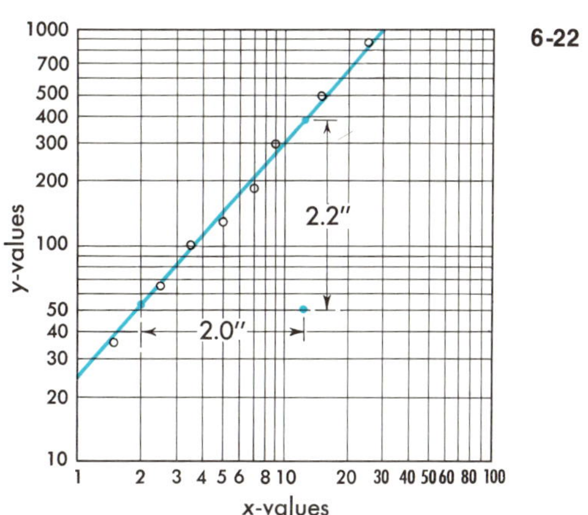

6-22

Using a linear scale, measure the X component and Y component values for two points on the plotted line. The slope of the graph in [6-22] is 2.2/2.0, or 1.1, and the Y intercept is 23.3. Substituting these values in the basic equation of a parabola gives $y = 23.3(x^{1.1})$ for the approximate equation.

In case the plotted points on log–log paper curve upward as x increases, the expression may approximate the form $y = ax^n + k$, or $y - k = ax^n$. To straighten the curve, try subtracting a constant from the y values. By trial and error, a value of k may be found that will cause the plot to follow a straight line. If this is done, the approximate equation may be determined.

There are other methods of determining empirical equations, such as the method of least squares, but a complete discussion of such techniques is beyond the scope of this book. Also, data that plot into curves following harmonic laws or exponential laws are not discussed here.

calculation and analysis

Graphical methods of analysis can be used as aids in determining functions, calculating areas, or developing procedures for simple repetitive calculation. Graphical techniques can solve arithmetic, algebraic, and calculus problems. They are particularly advantageous when one type of equation or formula must be solved repeatedly for different conditions or variables. For example, a chemical engineer who designs heat exchangers has to deal repeatedly with tubes of various length, diameter, and wall thickness. He may need to know surface area, volume, or weight for all types of tubes again and again. Although he could calculate these data easily, he is better off with a quick way of looking them up—and graphs or nomographs are useful aids.

6-23
Heat exchanger tube.

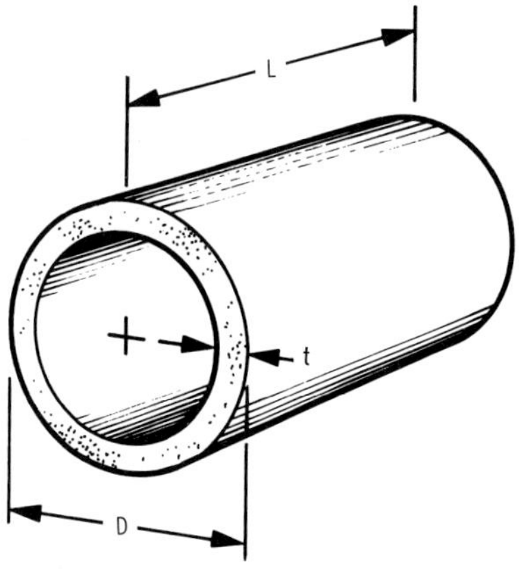

In addition, certain types of equations cannot be solved easily by purely mathematical methods but are quickly handled graphically.

We shall illustrate some of these techniques next. One word of caution: graphical solutions are inherently less accurate than numerical ones. How accurate they are depends, of course, on the competence of the draftsman and the size of paper and scale he uses. One per cent accuracies are not too hard to achieve usually, but when much greater accuracy is needed, the cost of achieving it graphically tends to become prohibitive.

Besides the construction of graphical equations and nomographs, we shall deal briefly with vectors.

Graphical Arithmetic

Graphical-arithmetic charts consist of a plot of three or more variables on a coordinate grid such that the value of any one variable can be found quickly if the others are known. For example, the chemical engineer needs the outside surface area of his heat-exchanger tubes and is tired of calculating it every time from the equation

$$\text{area } A \text{ (in.}^2) = \pi \times \text{diameter } D \text{ (in.)} \times \text{length } L \text{ (ft)} \times 12 \text{ (in./ft)}$$

or

$$A = 37.7DL$$

He knows he will have to deal with tubes of all lengths up to 20 feet, but only with four different standard diameter sizes: $\frac{1}{2}$, 1, $1\frac{1}{2}$, and 2 inch. He first selects his scales; he decides to plot length on the abscissa or horizontal axis; it will go from 0 to 20 feet. He also decides to read his answer, the surface area, on the ordinate, or vertical axis. To find his vertical scale, he computes the largest possible area: 1,508 square inches for a 2-inch diameter, 20-foot-long tube. This establishes his scales [6-24].

Now he wants one line for each tube diameter that will permit him to find the surface area for any length of that size tube. For example, he computes that 15 feet of $1\frac{1}{2}$-inch-diameter tube has a surface area of 849 square inches (see point *A* in [6-25]). He also realizes that 0 length of any size tube has zero surface area; that is, each line goes through the origin. So he draws a straight line between 0 and *A*,[2] and he labels the line with the pipe diameter [6-25].

For the other pipe diameters he produces similar lines [6-26]. Such lines are often called *parametric* lines, and the variable that fixes them—in this case the pipe diameter—is called the *parameter*. His graphical equation is now completed. Any time he has to find a surface area he goes from the proper length on the abscissa vertically upward to the appropriate diameter line. When he meets

[2] The relation between pipe length and surface area is linear. If it had not been linear, the line would have to be curved.

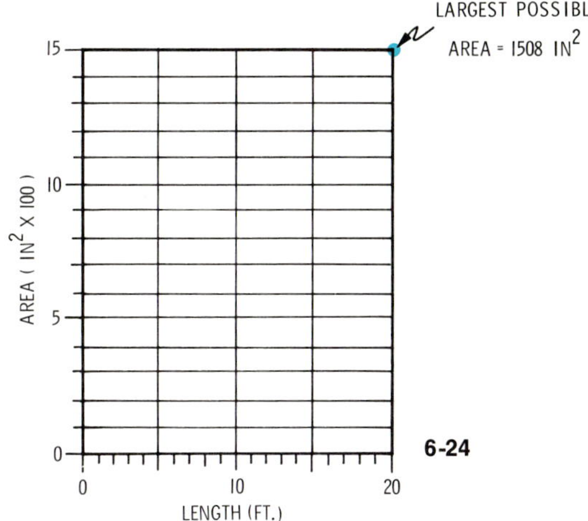

6-24

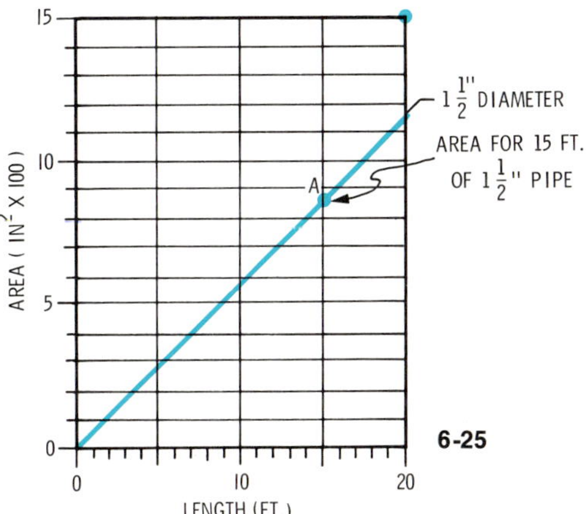

6-25

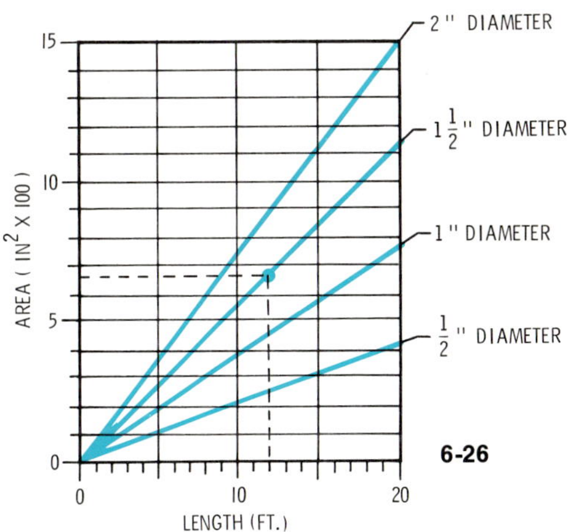

6-26

that line he "turns left" until he meets the ordinate, where he reads the area he wants. (Example: A 12-foot-long, $1\frac{1}{2}$-inch-diameter tube has a surface area of 680 square inches—as shown with the dotted line in [6-26].)

The graphical method can be used for more than one parameter by putting several graphs back to back. If, for example, the engineer is interested in the total amount of heat transferred from the tube, this is determined by the equation

$$Q = A \times \Delta T \times k$$

where Q is the heat transferred (Btu/hour)
A is the surface area, as before (ft^2)
ΔT is the temperature difference between the inside and the outside of the tube (°F)
k is the heat transfer coefficient (Btu/hour-ft^2-°F)

This equation has two new parameters, ΔT and k. How these are used graphically will become self-evident when you study [6-27].

6-27

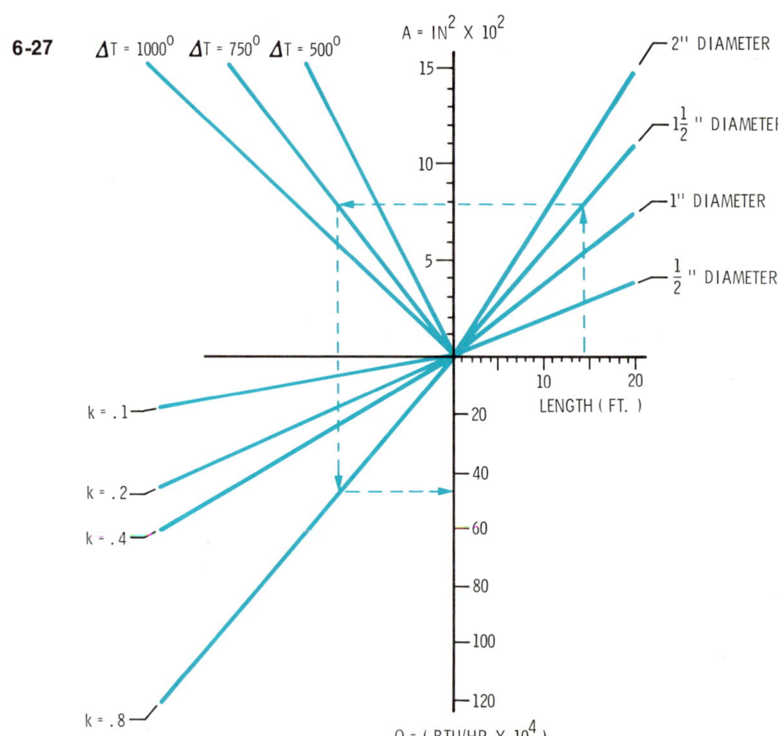

Graphical Algebra

Graphics can be used to solve equations in one or two variables. This is particularly worthwhile when the equation does not have a simple algebraic solution. For example, the equation

$$\tan X = 1 + \cos X$$

can be solved for X only by trial and error—or graphically, as follows:

1. Make yourself a table that contains the values of the right-hand side (RHS) and left-hand side (LHS) of the equation for various values of X.
2. Plot RHS and LHS versus X separately [6-28].
3. Find the intersection. It gives the desired value of X (57.2°).

Solve the equation
$\tan X = 1 + \cos X$

X	LHS	RHS
45°	1.000	1.707
50°	1.192	1.645
55°	1.429	1.573
60°	1.731	1.500
65°	2.141	1.423

6-28

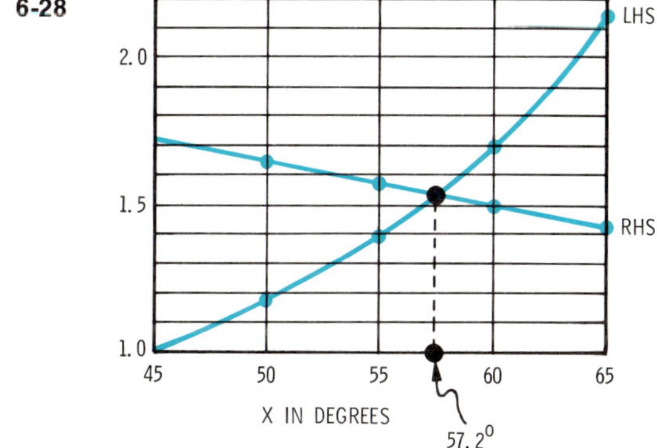

Another useful place for graphical solutions is in problems where it is easy to plot the relations but difficult to express them in algebraic form. For example:

A car leaves Chicago traveling toward Atlanta at an average speed of 55 miles per hour and increases this speed to 65 miles per hour after 3 hours. A 1-hour lunch stop is made after traveling for 5 hours, and then an average speed of 55 miles per hour is maintained. A bus leaves Atlanta toward Chicago 2 hours after the car leaves Chicago and averages 70 miles per hour. How many hours after the car left Chicago will the bus and car meet, assuming that they both travel the same route, the bus makes no stops, and the distance between cities is 700 miles? How far from Atlanta will they meet?

The solution to this problem would be quite complex algebraically but can be solved graphically by following these steps.

1. Develop a coordinate grid with miles on the horizontal axis (to 700) and hours on the vertical (to 10).
2. Plot the path of the car on the grid [6-29]. The slope of the line is determined by the car's velocity: 55 miles per hour for 3 hours; then 65 miles per hour for 2 hours. The 1-hour rest stop produces no change on the horizontal axis because no distance is traveled. The velocity then returns to 55 miles per hour.
3. Plot the path of the bus on the same grid [6-30]. The plot is started from the 700 value of the horizontal

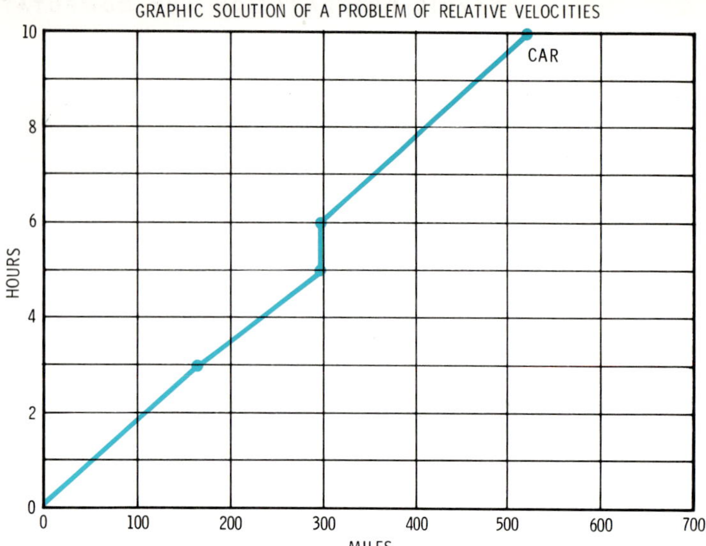

6-29
Car travel.

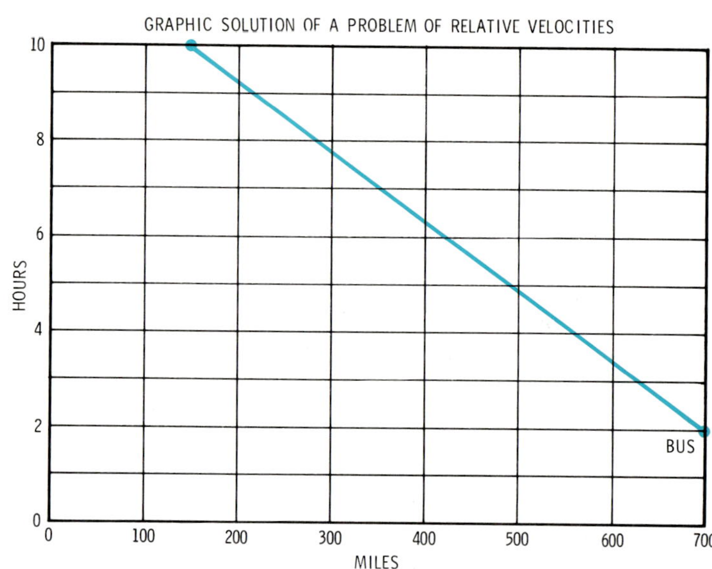

6-30
Bus travel.

6-31
Combined car and bus travel.

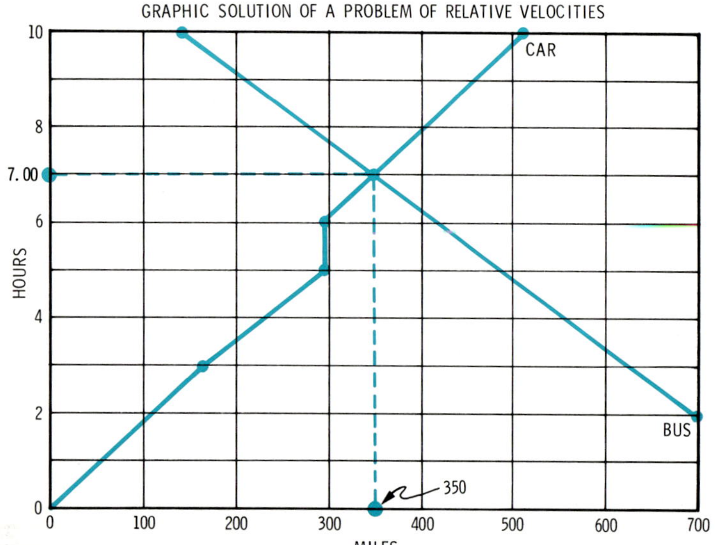

scale, since the bus starts in Atlanta. Its slope is toward the 0 value since it is traveling toward Chicago.

4. The coordinates of the intersection of these lines are 7.0 and 350, indicating the bus will pass the car 7 hours after the car leaves Chicago and 350 miles from Atlanta [6-31].

Graphical Calculus

Graphical techniques are powerful tools to differentiate or integrate a function that cannot be expressed easily in algebraic form, such as the readings of an instrument during most any experiment [6-32].

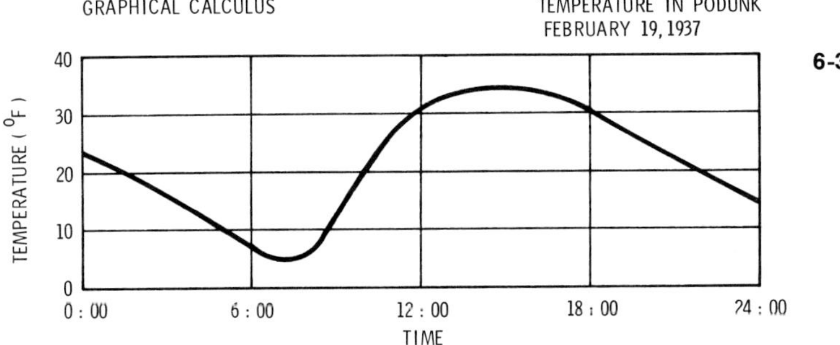

6-32

We shall illustrate one method of graphical integration,[3] and use it to find the average temperature from the graph shown. Mathematically, the average is the integral of the curve with respect to time, divided by the integration period. In other words, it is the area under the curve from midnight to midnight (in degrees Fahrenheit-hours) divided by 24 hours.

To determine the area beneath a curve, select several intervals along the *X* axis (the time axis) and construct lines parallel to the *Y* axis (the temperature axis) through the original curve. Each interval selected is chosen to intersect a relatively straight portion of the original curve. This will necessitate selecting smaller intervals in those areas where curve change is great [6-33].

Rectangles are formed by drawing a horizontal line through each interval so as to strike the curve where the area improperly excluded from the rectangle approximates the area improperly included. The area of this rectangle is a close approximation of the area beneath that section of the curve, and the summation of the areas of all these rectangles approximates the area beneath the whole curve. The approximation is good because small

[3]For other integration methods and for graphical differentiation, see A. S. Levens, *Graphics*, New York: Wiley, 1968, p. 287ff.

6-33

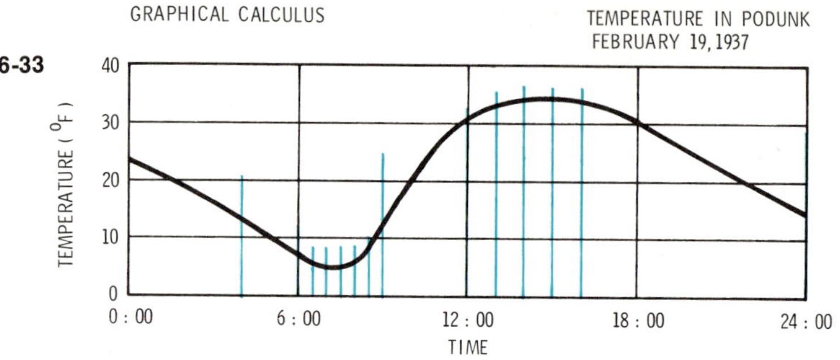

6-34

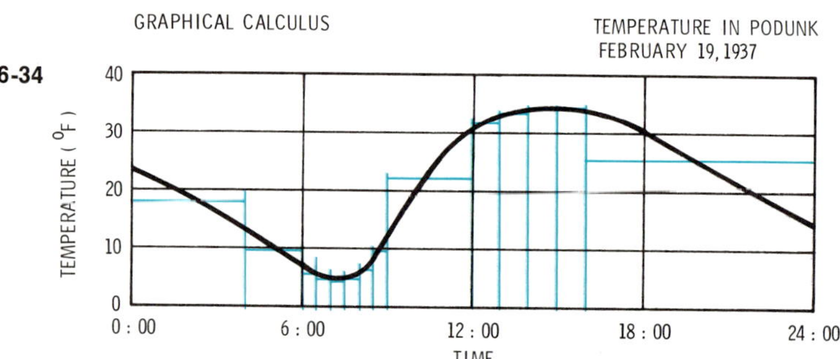

Time	Δt	Temp.	°F × Hr
0:00	—	—	
4:00	4.0	19°	76.0
6:00	2.0	10.5°	21.0
6:30	.5	6°	3.0
7:00	.5	5°	2.5
7:30	.5	5°	2.5
8:00	.5	6°	3.0
8:30	.5	7°	3.5
9:00	.5	10°	5.0
12:00	3.0	22°	66.0
13:00	1.0	32°	32.0
14:00	1.0	33°	33.0
15:00	1.0	34°	34.0
16:00	1.0	34°	34.0
24:00	8.0	25°	200.0
	24.0		515.5

The average temperature is

$$\frac{515.5°\text{F} - \text{Hr}}{24.0\ \text{Hr}}$$

errors of line location are rarely cumulative, and most often offset one another [6-34]. A table helps in accumulating the data.

Nomography

The *nomograph* or alignment chart is another method for depicting a mathematical relationship so that it can be solved quickly and easily. It is an elaboration and refinement of the graphical methods described earlier, and it is used for much the same purpose, that is, to simplify solving equations that are used frequently. The nomograph always uses a linear (straight-line) relationship between scales. To take care of nonlinear equations, the scales can be generated nonlinearly and they can be inclined to one another. (In some cases they can even be curved.)

To construct a scale you need to know the values of its end points, that is, the biggest and the smallest value to be marked on the scale. The difference between the two is the scale's *range*. When you have decided how long the scale is to be and divide the length by the range, you obtain the *scale modulus* (M_S).

If the scale is linear, one can lay it out by a convenient graphical technique. This technique consists of placing a calibrated ruler at an arbitrary angle with the new scale making the origins coincide [6-35]. A line is drawn from the other end of the new scale to that number on the ruler which corresponds to the largest value on the scale. A line is drawn along the edge of the ruler and the desired graduations are marked along it [6-36]. Lines are then drawn through each of these graduations parallel to the line between the terminal values. The intersection of these lines on the new scale forms the graduations desired [6-37].

For example, let us construct a temperature conversion scale, to convert degrees Celsius to degrees Fahrenheit. It is to go from −60°C to +140°C, a range of 200°C. Let us assume we wish to make the Celsius scale 9 inches long. Then the scale modulus is 9/200 = 0.045 inch/°C. The corresponding end points of the Fahrenheit scale are −76°F and +284°F, a range of 360°F. If we want to make the Fahrenheit scale equal in length to the Celsius scale, its modulus will be 9/360 or 0.025 inch/°F. Both scales are linear, so they are easy to graduate graphically. Since we made them the same length, we can put them back to back for easy reading [6-38]. Such scales are called *adjacent conversion* scales.

Another type of nomograph, which might be described as nonadjacent conversion scales, uses separate scales for each variable and places the scales parallel to each other. The scales are constructed inversely so that the minimum value of one scale is opposite the maximum value of the other. A pivot point is determined by the intersection of lines connecting the minimum and maximum values of each scale. Figure [6-39] shows such a nomograph used to convert inches of liquid in a 55-gallon drum to gallons.

One advantage of this type of nomograph is that the length of both scales is arbitrary. A more important ad-

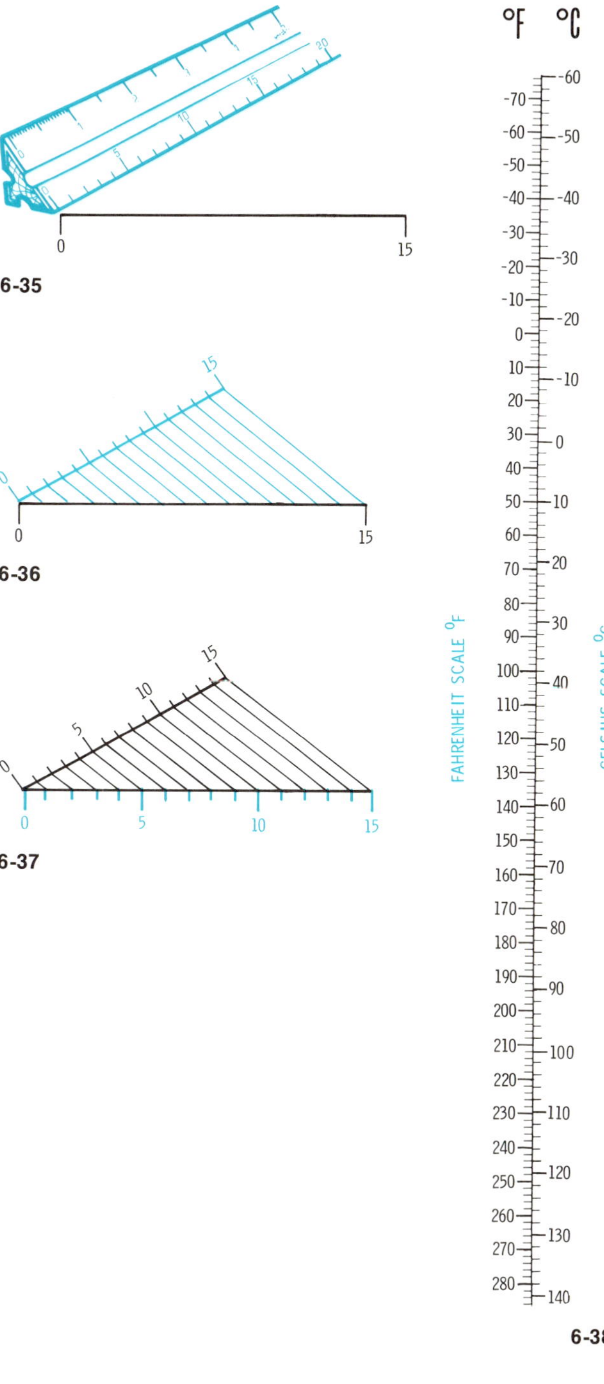

6-35

6-36

6-37

6-38

6-39

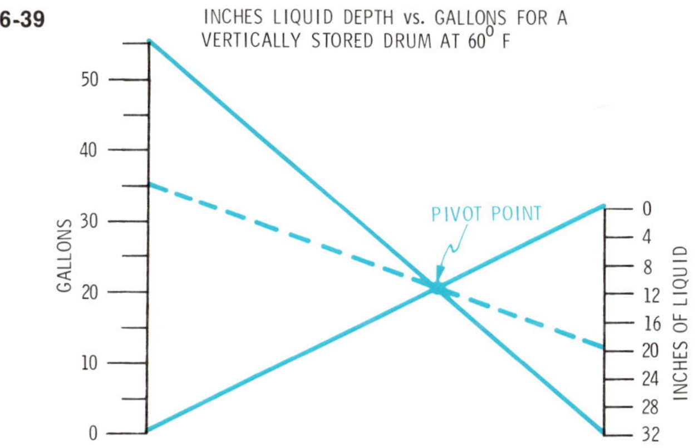

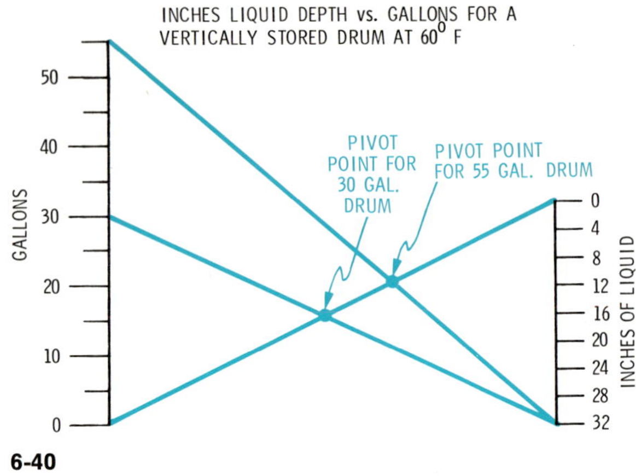

6-40

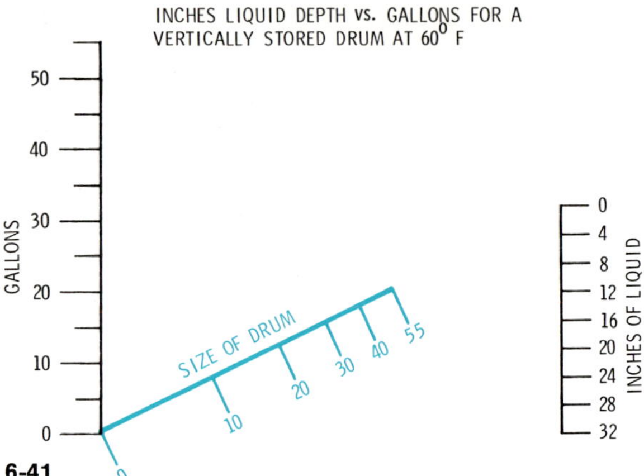

6-41

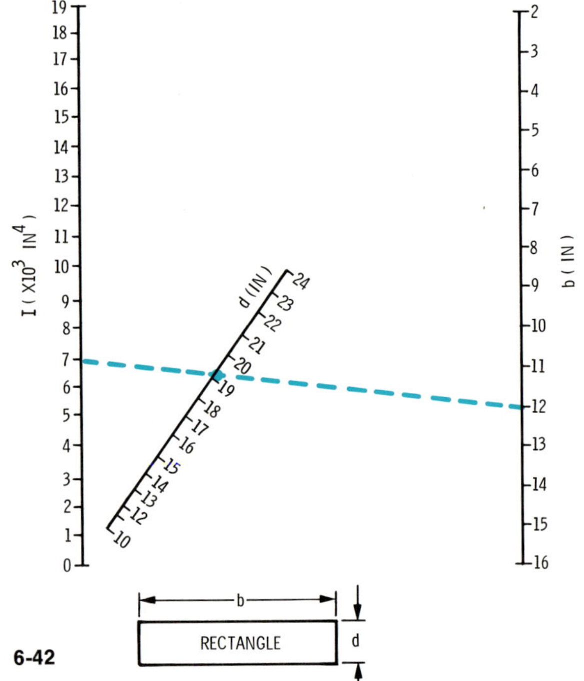

6-42

vantage is that it is possible to use the same nomograph for different equations of the same type. If, for example, there were another, smaller diameter drum that contains only 30 gallons when the depth of liquid is 32 inches, its equation can be added easily by finding the appropriate pivot point [6-40].

For a whole series of different-diameter drums, the line on which the pivot points lie could become another scale [6-41]. A nomograph of this type is called an N or Z alignment chart (depending on whether the parallel scales are vertical or horizontal). Two useful N or Z charts follow. The first chart [6-42] is used to find the moment of inertia of a rectangle about a horizontal axis through its midpoint. (The moment of inertia is important when calculating stresses and deflections of a beam with that cross section.) Note that the d scale is not linear, since the dimension d enters the equation as the cube.

The second [6-43], though not strictly an N or Z chart (for none of the lines are parallel), is of a type that represents the equation $1/u + 1/v = 1/f$; u, v, and f may be the actual variables as in the example, in which case the scales are linear; or they may be functions of some other basic variables, in which case the scales may not be linear. In any event, for this type of relation the scales—if extended—will converge at a common point.

Nomographs may be used to solve equations with more than three variables. For example, assume that the chemical engineer designing heat exchangers (p. 201) must often find the combined surface area (inside plus outside) of his tubes. The equation he would have to solve is

$$A = [\pi D + \pi(D - 2t)] \times 12L = 75.4(D - t)L$$

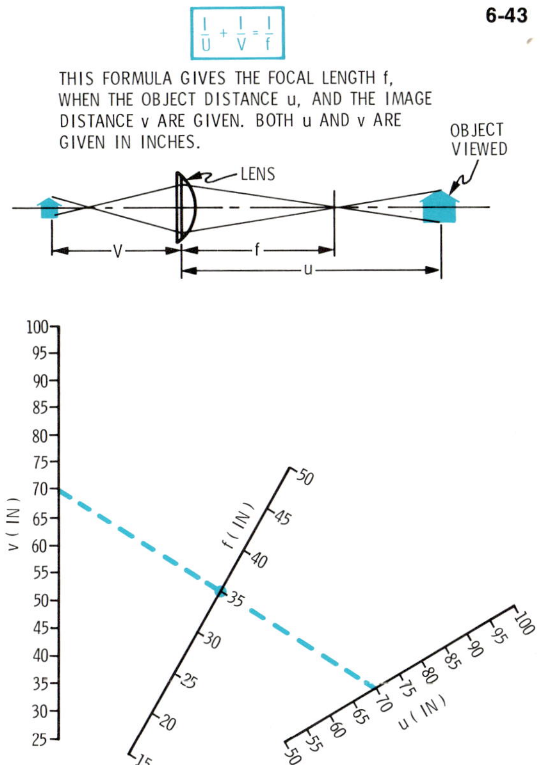

6-43

where t is the wall thickness in inches, D is the outside diameter, and L is the length. The three leftmost vertical lines are used to find $D - t$, whose value is then used in the N chart to find A [6-44].

Vectors

Certain physical units, such as velocities and forces, must be defined by direction as well as magnitude. It is not enough to say that two forces of 100 pounds push on a cart, unless we also say in which direction each one pushes. Graphically, a force can be represented by a *vector*, a line with an arrowhead on one end. The length of the line represents the magnitude of the force, and the arrowhead points in the direction in which it acts.

When determining what forces act on a body, it is necessary to mentally (or on a drawing) remove the body from its surroundings and *replace its environment with forces* that are equivalent to the action of the surroundings. When this is completed we have a *free-body diagram* [6-46].

6-44 MULTIPLE VARIABLE NOMOGRAM $A = [\pi D + \pi(D-2t)] \times 12L$

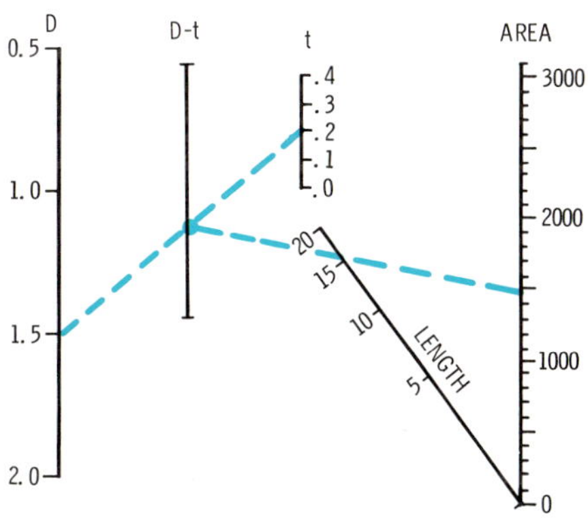

Often we may not know the magnitude or direction of all the forces and may have to find them. A graphical method, called the *polygon of forces,* is very useful for this. What is it and how is it used? Let us illustrate by some very simple examples:

A ball resting on a horizontal floor [6-47] is easy to turn into a free-body diagram [6-48]. Also, it is obvious that the push from the floor is just equal in size and opposite in direction to the pull of gravity (the weight) of the ball. No graphical construction is needed here.

If the same ball is resting on an incline [6-49], we have a different free-body diagram [6-50]. Now the size of the "pushes" from the floor and the stop are not obvious, although we do know their direction. Their size can be found by drawing the force vectors to scale and in the proper direction. After the weight vector has been drawn, we trace two lines of action in the direction of the two pushing forces, one from each end of the weight vector. The point at which the two lines meet must be the head of one push vector and the tail of the other [6-51]. A closed triangle like the one shown implies that the body is in equilibrium (i.e., all forces that act on the body act with such magnitudes and directions that no net external force results). For three forces the *force polygon* is a triangle. If there are more than three forces, the *polygon* will have more sides.

6-45

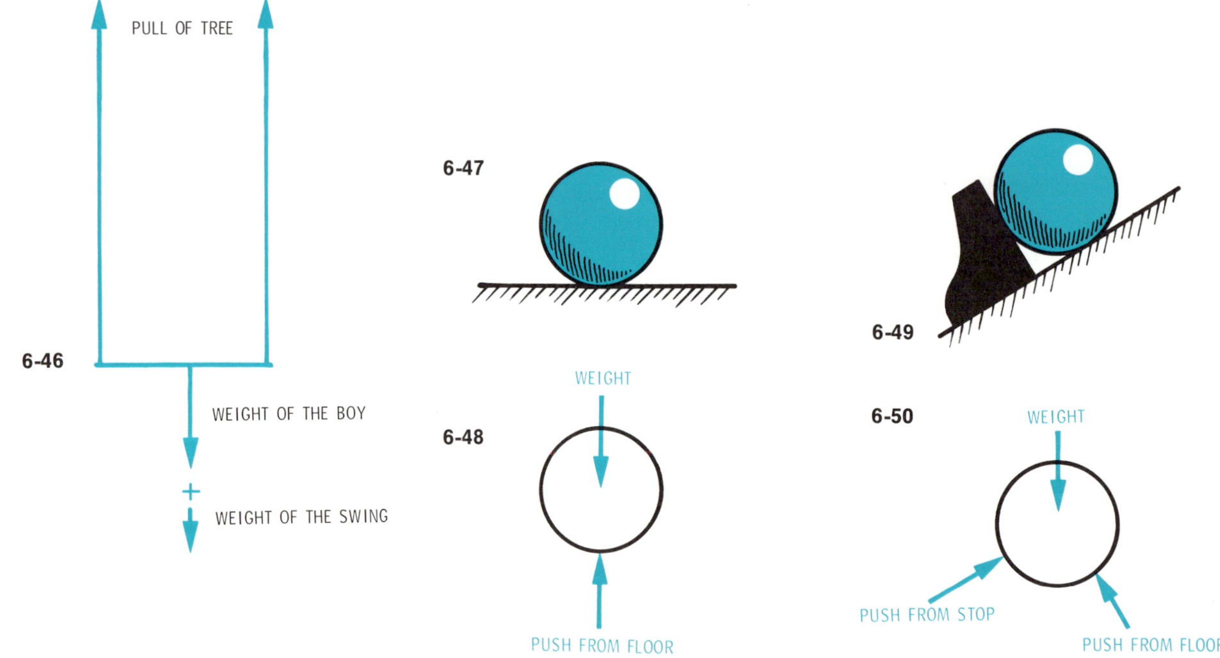

6-46

6-47

6-48

6-49

6-50

One can look at [6-50] in another equally valid way. If the two pushing forces were replaced by a single equivalent force, this force would have to be equal in size and opposite in direction to the weight, just as the push from the horizontal floor in [6-48]. This equivalent force is often called the *resultant*. The resultant is a force equivalent to two or more individual forces[4] [6-52].

So far we have talked only of forces in one plane in which each of the force vectors was drawn full length. The method is still useful when forces are not coplanar. In that case, the *projections* of all forces on every one of the three perpendicular projection planes sum up to the projection of the resultant (which must be zero if the body is in equilibrium). We shall not attempt to prove this statement. Rotation methods for finding true lengths of lines are convenient in some non-coplanar situations [6-53].

[4]The resultant of *all* forces acting on a body in equilibrium is always zero.

6-51

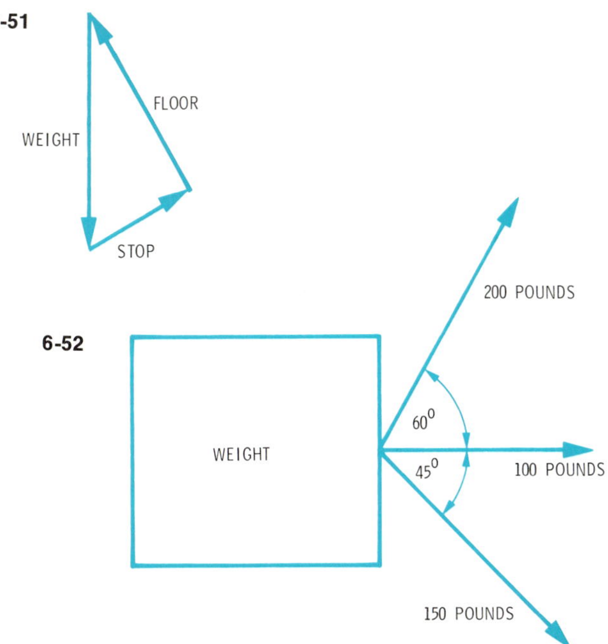

6-52

In the situation shown the three forces are not in equilibrium. However their resultant can be determined graphically by adding the force vectors.

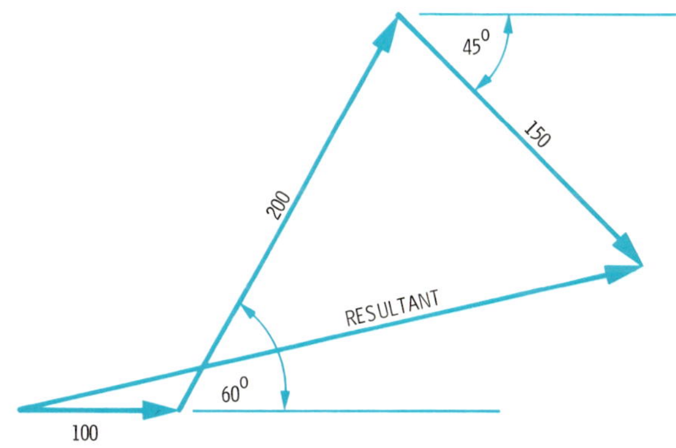

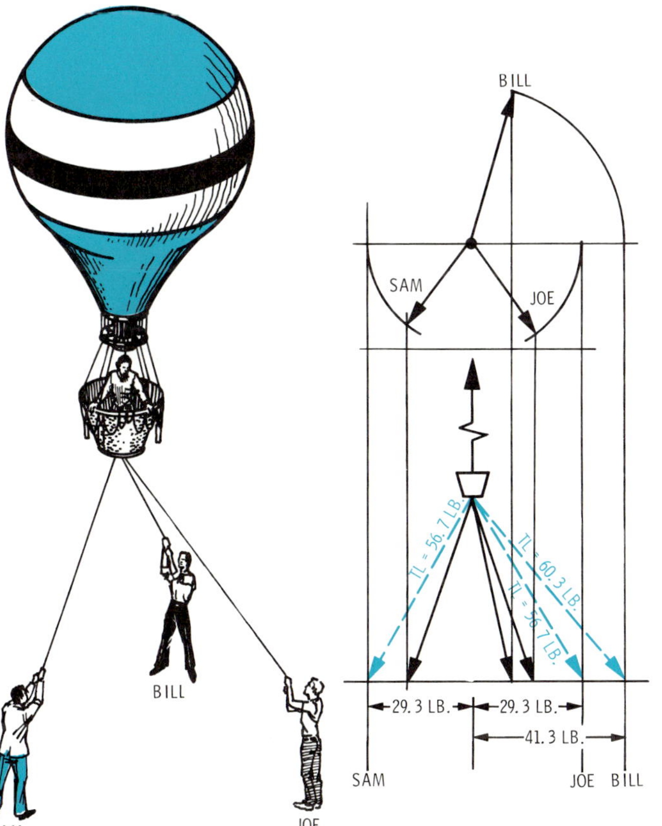

6-53

6-54
The computer can reassemble these symbols and density values . . .

If forces are not concurrent, that is, if they do not meet in one point, their resultant must still be zero if the body is to be in equilibrium; but this condition, though necessary, is not sufficient. Other conditions which assure that the body does not rotate must also be fulfilled. For more information on these conditions, see any fundamental text on static mechanics.[5]

[5]For example, F. P. Beer and E. R. Johnston, Jr., *Statics*, New York: McGraw-Hill, 1972.

computer-aided graphics

As in so many other areas of engineering, the digital computer has found a place in graphics. Here its major function is to relieve the designer of drudgery and repetitive tasks and to free his time for creative design and experimentation with new ideas. To do this, techniques have been developed to have the computer "read" a drawing, to have it display what it has read on a big TV screen, and to permit the designer to alter the drawing stored in the computer easily and quickly.

Just how the computer reads a drawing is not important here and will vary from one computer installation to another. Basically, a grid is placed over the drawing and the coordinates of every point and line are entered into the computer's memory. Originally, this required manual reading and transferring onto punched cards or paper tape. More recently, optical readers have been

6-55
. . . into a meaningful pattern. (Place the book about 10 feet away and observe how the small details of the micropatterns fuse together, and the overall picture takes on the quality of a continuous-tone photograph.)

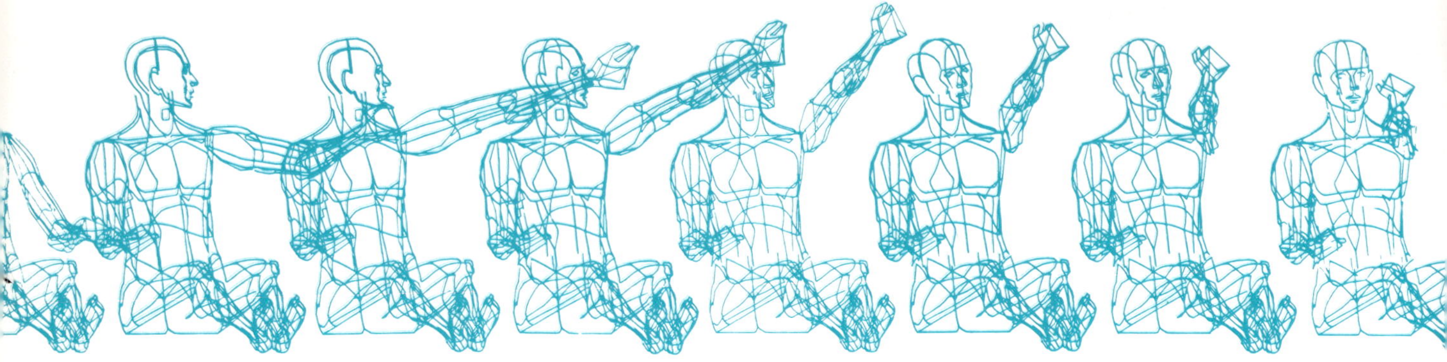

6-56
It is sometimes advantageous to direct the computer to "rotate the drawing," as shown here in considering a design for maximum pilot comfort.

6-57
Working with a light pen.

developed that can do this tedious transfer automatically.

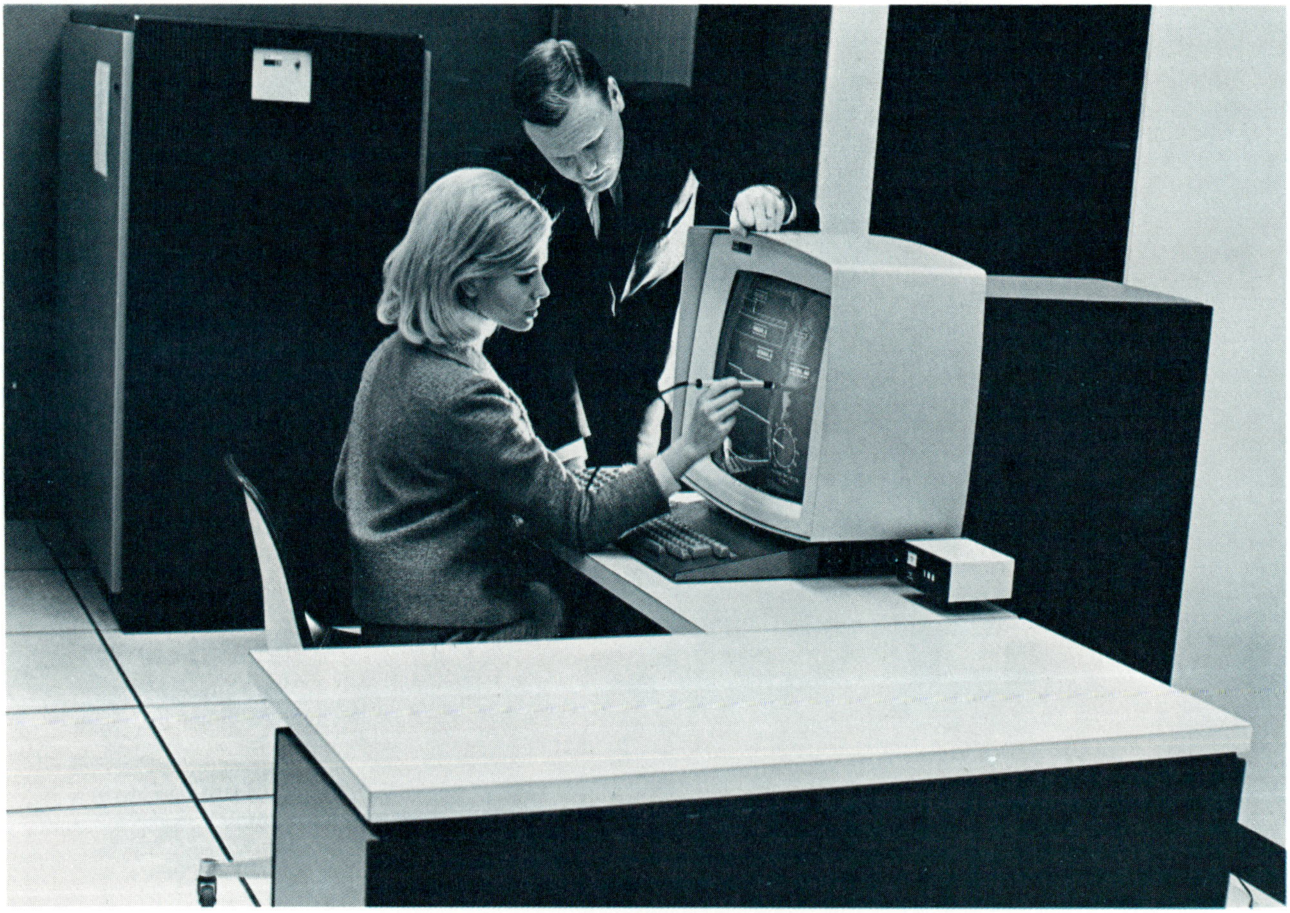

Once in the computer memory, the drawing can be displayed visually on the screen of a cathode-ray tube (TV screen). More important, computer programs are now available to magnify portions of the drawing for closer analysis, to rotate the drawing and present different views from those originally read in, and to combine the drawing with *stored* drawings of other parts to produce assemblies [6-56]. Most important, however, is the ability to modify stored drawings at the TV screen, to erase lines, and to add others quickly and easily. This is done with a light pen, much like a pocket-size flashlight with a narrow light beam at the tip [6-57]. By playing this light on the TV screen and pressing appropriate switches on the console, one can erase or add at the places where the light shines. The computer automatically smoothes out small wiggles in the new lines.

Many mathematical analyses can be performed by the computer on the part whose drawing is stored in its memory. It can be loaded, heated, vibrated, impacted, and the test results printed out. Of course, every important change in the design can be retested. The extent of the program is limited only by the ingenuity of the designer—and the cost of operating the computer. A point of caution: although the results of such computer tests have all the appearance of authentic experiments, they are still subject to the validity of the assumptions made in the selection of the model. They are not equivalent to and should never be confused with prototype testing.

Once a computer-stored drawing has been approved for production, the computer can be used to compute raw-material requirements and to program the machine tools that are to make the part. Many modern machine tools can be operated directly by programs produced in a computer and stored on punched cards, punched paper tape, or on magnetic tape.

problems

6-1. Investigate several issues of a scientific or technical journal that are available in the library. Select five different methods of graphically portraying technical data and make copying machine reproductions of each type. Write a brief discussion describing the advantages and limitations of each type.

6-2. From the want ads of a daily newspaper, select your favorite brand of American-made automobile. Find five listings of the two-door models for this year and for each of the preceding 4 years. Prepare a table that shows the price of the 25 cars. Use a separate column to indicate each year. Title and label the table.

6-3. Using the raw data developed in Problem 6-2, calculate the average yearly price for the two-door sedans. Construct a vertical bar chart that shows these yearly averages. Construct the chart on an $8\frac{1}{2}$- by 11-inch sheet of paper and complete it with title and proper labeling.

6-4. Using last month as a basis, calculate your total living expenses. Include your rent (with utilities), food, school supplies including prorated book costs, entertainment costs, transportation (such as bus fares or car expenses), and other miscellaneous expenses. Construct a pie chart showing these six costs and label it appropriately.

6-5. Construct an organization chart of the management structure of your university or college. Start at the level of the president, chancellor, or provost, and continue to the student body of your discipline.

6-6. From the admissions office or other appropriate office at your school, determine the number of students that were enrolled as engineering students for each of the past 10 years, including this year. Collect similar data for at least two other major disciplines at your school. Plot a graph of the data collected on rectangular-coordinate graph paper. Use different symbols for each group of data. Title and label the graph.

6-7. Analyze the trend lines developed in Problem 6-6. Write a brief summary of your conclusions about the growth or decline of engineering enrollment relative to the other two disciplines at your school. Assuming a continuation of the indicated trends, what will be the likely enrollments of these disciplines 5 years in the future?

6-8. Plot a graph showing the relation of weight to diameter for round steel rods. Plot values for every quarter-inch to and including $3\frac{1}{2}$ inches in diameter.

Weight of Round Steel Rods in Pounds Per Lineal Foot (Based on 489.6 lb/ft^3)

Size, in	Weight, lb/ft	Size, in	Weight, lb/ft
$\frac{1}{4}$	0.167	2	10.66
$\frac{1}{2}$	0.668	$2\frac{1}{4}$	13.50
$\frac{3}{4}$	1.50	$2\frac{1}{2}$	16.64
1	2.68	$2\frac{3}{4}$	20.20
$1\frac{1}{4}$	4.17	3	24.00
$1\frac{1}{2}$	6.00	$3\frac{1}{4}$	28.30
$1\frac{3}{4}$	8.18	$3\frac{1}{2}$	32.70

6-9. Plot a graph for the following experimental data showing the relation between the period in seconds and the mass of a vibrating spiral spring.

Period, Sec	Mass, G	Period, Sec	Mass, G
0.246	10	0.650	70
0.348	20	0.740	90
0.430	30	0.810	110
0.495	40	0.900	130
0.570	50	0.950	150

6-10. Plot a graph of the variation of the boiling point of water with pressure.

Boiling Point, °C	Pressure, cm Mercury	Boiling Point, °C	Pressure, cm Mercury
33	3.8	98	72.9
44	5.3	102	85.8
63	17.2	105	93.7
79	34.0	107	102.2
87	48.1	110	113.5
94	69.1		

6-11. Plot the variation of pressure with volume, using data as obtained from a Boyle's-law apparatus.

Pressure, cm Mercury	Volume, cm^3	Pressure, cm Mercury	Volume, cm^3
50.3	23.2	76.8	15.1
52.5	22.4	79.7	14.7
54.5	21.5	82.7	14.1
56.9	20.9	84.2	13.6
59.4	19.6	87.9	13.2
63.0	18.5	90.6	12.8
65.3	17.8	93.5	12.5
67.2	17.3	95.7	12.3
72.6	16.1	101.9	11.4
74.5	15.6		

6-12. The formula for converting temperatures in degrees Fahrenheit to the equivalent reading in degrees Celsius is

$$C^\circ = \tfrac{5}{9}(F^\circ - 32^\circ)$$

Plot a graph so that by taking any given Fahrenheit reading between 0° and 220° and using the graph, the corresponding Celsius reading can be determined.

6-13. The following data were taken in the laboratory for a 60-watt, gas-filled, tungsten-filament light bulb. Plot a resistance–voltage curve.

Voltage, V	Resistance, Ω	Voltage, V	Resistance, Ω
10	47.5	70	160.2
20	77.5	80	170.0
30	100.3	90	178.3
40	119.0	100	189.0
50	132.6	110	200.1
60	144.2		

6-14. Plot the variations of efficiency with load for a $\frac{1}{4}$-horsepower, 110-volt, direct-current electric motor, using the following data taken in the laboratory.

Load Output, hp	Efficiency, %
0	0
0.019	24.0
0.050	42.0
0.084	44.9
0.135	50.7
0.175	56.5
0.195	58.0
0.248	59.1
0.306	58.0
0.326	56.2

6-15. Plot the values given in Problem 6-8 on semilog paper.

6-16. Plot the values given in Problem 6-8 on log–log paper.

6-17. Plot the values given in Problem 6-11 on log–log paper. Determine the slope of the line and give the approximate form of the equation shown by the plot.

6-18. The electrical frequency response of a Type X501 microphone is given below.

Frequency, Hz	Relative Response, dB
20	−40
40	−33
80	−22
100	−18
200	−11
400	−5
600	−2
1,000	+1
2,000	+2
4,000	−1
6,000	−4
10,000	−10

Plot a graph on semilog paper showing the decibel response with frequency.

6-19. Some test specimens of a crank arm, part No. 466-1, were tested for the number of cycles needed to produce fatigue failure at various loadings. The results of the tests are tabulated below.

Specimen Number	Oscillatory Load, lb	Operating Cycles to Produce Failure
1	960	1.1×10^5
2	960	2.2×10^5
3	850	1.5×10^5
4	850	2.4×10^5
5	800	4.2×10^5
6	800	6.0×10^5
7	700	2.4×10^5
8	700	3.1×10^5
9	700	5.1×10^5
10	650	1.8×10^6
11	650	2.6×10^6
12	600	7.7×10^6
13	550	1.0×10^7

Plot a graph of load against operating cycles (*S-N* curve) on semilog paper for the tests above.

6-20. A test on an acorn-type street lighting unit shows the mean vertical candlepower distribution to be as given in the table below.

Midzone Angle, Deg	Candlepower at 10 ft	Midzone Angle, Deg	Candlepower at 10 ft
180	0	85	156
175	0	75	1110
165	0	65	1050
155	1.5	55	710
145	3.5	45	575
135	5.5	35	500
125	8.5	25	520
115	13.5	15	470
105	22.0	5	370
95	40.0	0	370

Plot the data above. (Although data for only half the plot are given, the other half of the plot can be made from symmetry of the light pattern.)

6-21. From data determined by the student, draw a circle chart (pie chart) to show one of the following.

(a) Consumption of sulfur by various industries in the United States.

(**b**) Budget allocation of the tax dollar in your state.
(**c**) Chemical composition of bituminous coal.
(**d**) Production of aluminum ingots by various countries.

6-22. Using the data collected in Problem 6-6, plot the graph on one-cycle semilog graph paper. Compare the shape of the trend line with that in Problem 6-6.

6-23. Prepare a line chart that will permit converting readings from grams to ounces up to 64 ounces and a line chart that will convert readings from pounds to kilograms up to 10 pounds.

6-24. Using data and divisions described in Problem 6-4, construct a segment chart.

6-25. Prepare a graphical arithmetic chart to calculate the board feet of lumber in various lengths of standard widths of 2-inch-thick planks. The nominal widths are 4, 6, 8 and 10 inches, and the maximum length is 20 feet.

6-26. Graphical arithmetic charts can be combined to provide additional related information about a variable. Using the chart developed in Problem 6-25, combine another arithmetic chart that calculates the cost of four types of lumber based on the board feet. The woods used are oak ($1.10/board foot), pine ($0.85/board foot), maple ($1.35/board foot), and teak ($1.75/board foot).

6-27. Construct a concurrent scale that shows the conversion between English and metric systems of measuring length. Use the inch and the centimeter as the basic units and construct the scale at true size. Make the scale 10 inches long.

6-28. Construct a concurrent scale that shows the conversion between quarts and liters. One side of the scale should have a range of 0 to 100 quarts and be 8 inches long.

6-29. Construct a nomograph that will convert miles traveled by automobile to gallons of gasoline consumed. Use the average gasoline mileage of your car (or a car with which you are familiar) as the pivot point of the nomograph.

6-30. Construct a nomograph that converts gallons of root beer to number of mugs. Use a 12-ounce mug as the pivot point of the nomograph and a standard beer keg to establish the maximum gallons.

6-31. Construct a Z chart that will compute the average gasoline consumption in gallons per mile of your automobile. Use 400 miles and 30 gallons as the maximum values of each parallel scale.

6-32. Construct a Z chart that will calculate the cost of lumber as outlined in Problem 6-26. Design the chart to calculate the board-foot cost through the range of costs that are listed.

6-33. Draw a free-body diagram of this book resting on some inclined surface of your choice. From this free-body diagram, determine graphically the value of the force that prevents the book from sliding off the inclined surface. Use a scale of $\frac{1}{2}$ inch = 1 pound.

6-34. Imagine an old tire suspended by two ropes from a large limb of a tree. The limb grows at a 15° angle upward from the horizontal. One rope is 12 feet long and is attached to the tree 2 feet from the trunk (measured along the limb). The second rope, 17 feet long, is attached 10 feet from the trunk. Both ropes attach to the tire at the same point. If a 160-pound person were to sit in the tire and lift his feet from the ground, what would be the forces acting along each rope. Determine your answer by graphical analysis.

A cluster of cadmium-chromium-selenide crystals grown by a novel-liquid-phase transport technique is shown on a background of iron filings organized by a magnetic field. These crystals are unique in that they display hole-election and spin-wave phenomena as well as other exotic effects.

appendixes

SEQUENCE OF STROKES VERTICAL LETTERS

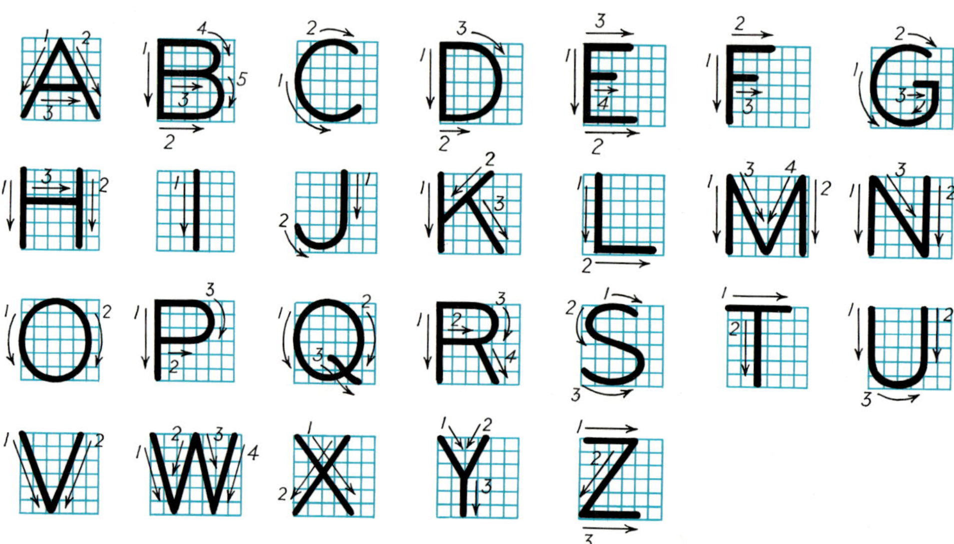

graphics construction and symbols

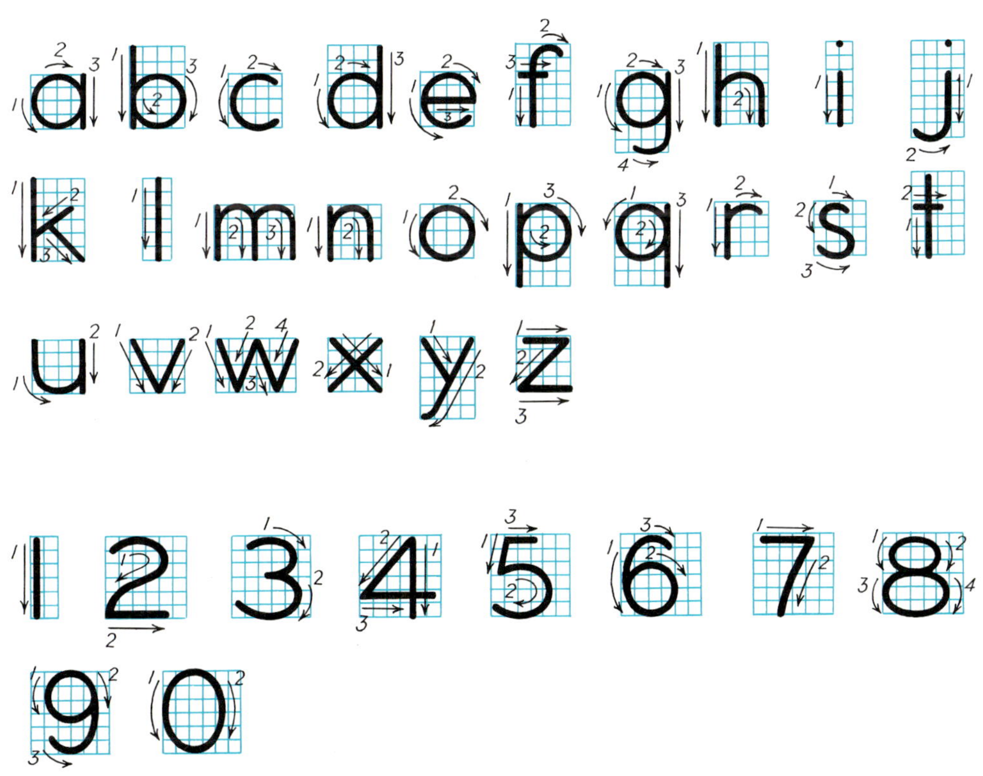

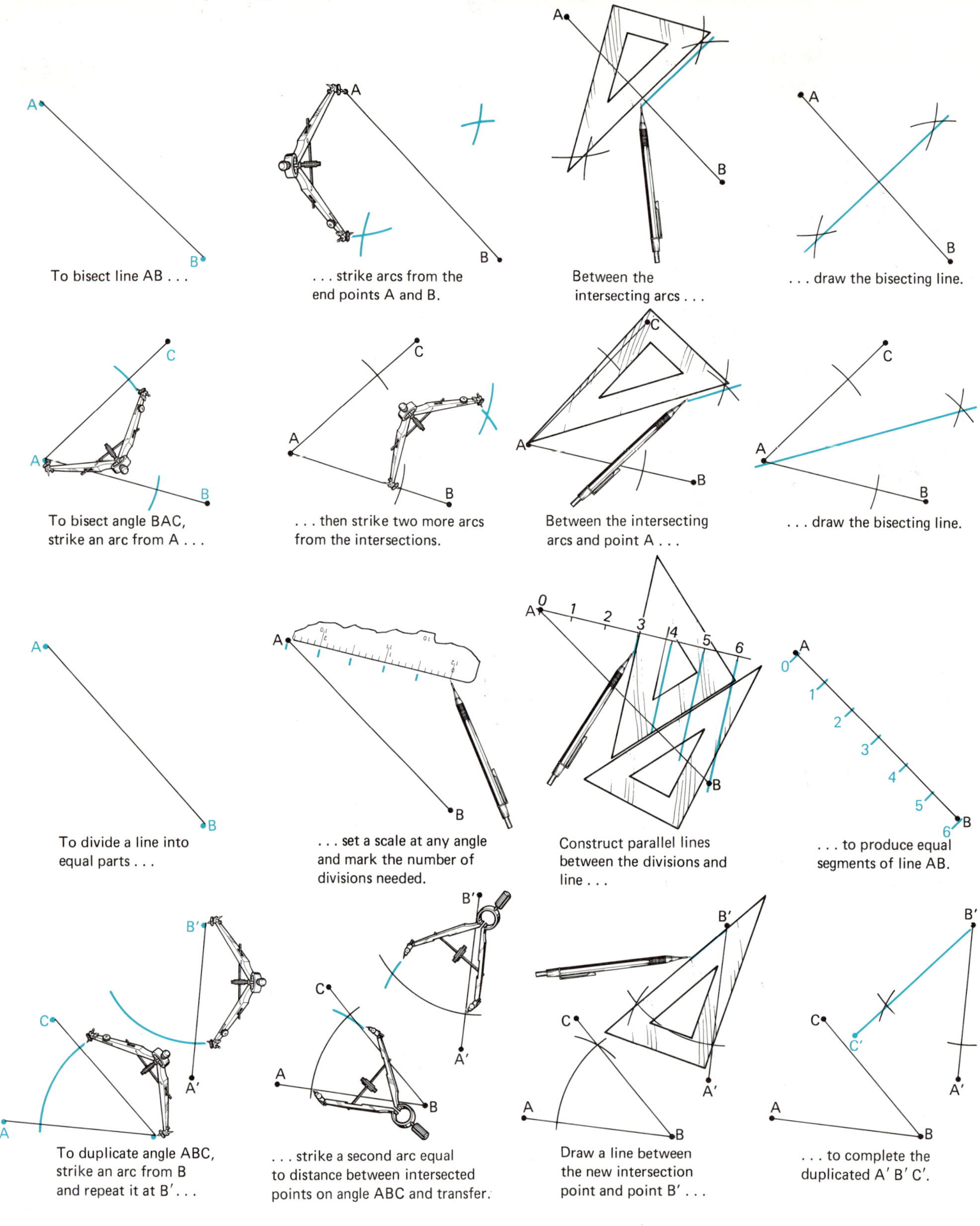

To bisect line AB . . .

. . . strike arcs from the end points A and B.

Between the intersecting arcs . . .

. . . draw the bisecting line.

To bisect angle BAC, strike an arc from A . . .

. . . then strike two more arcs from the intersections.

Between the intersecting arcs and point A . . .

. . . draw the bisecting line.

To divide a line into equal parts . . .

. . . set a scale at any angle and mark the number of divisions needed.

Construct parallel lines between the divisions and line . . .

. . . to produce equal segments of line AB.

To duplicate angle ABC, strike an arc from B and repeat it at B′ . . .

. . . strike a second arc equal to distance between intersected points on angle ABC and transfer.

Draw a line between the new intersection point and point B′ . . .

. . . to complete the duplicated A′ B′ C′.

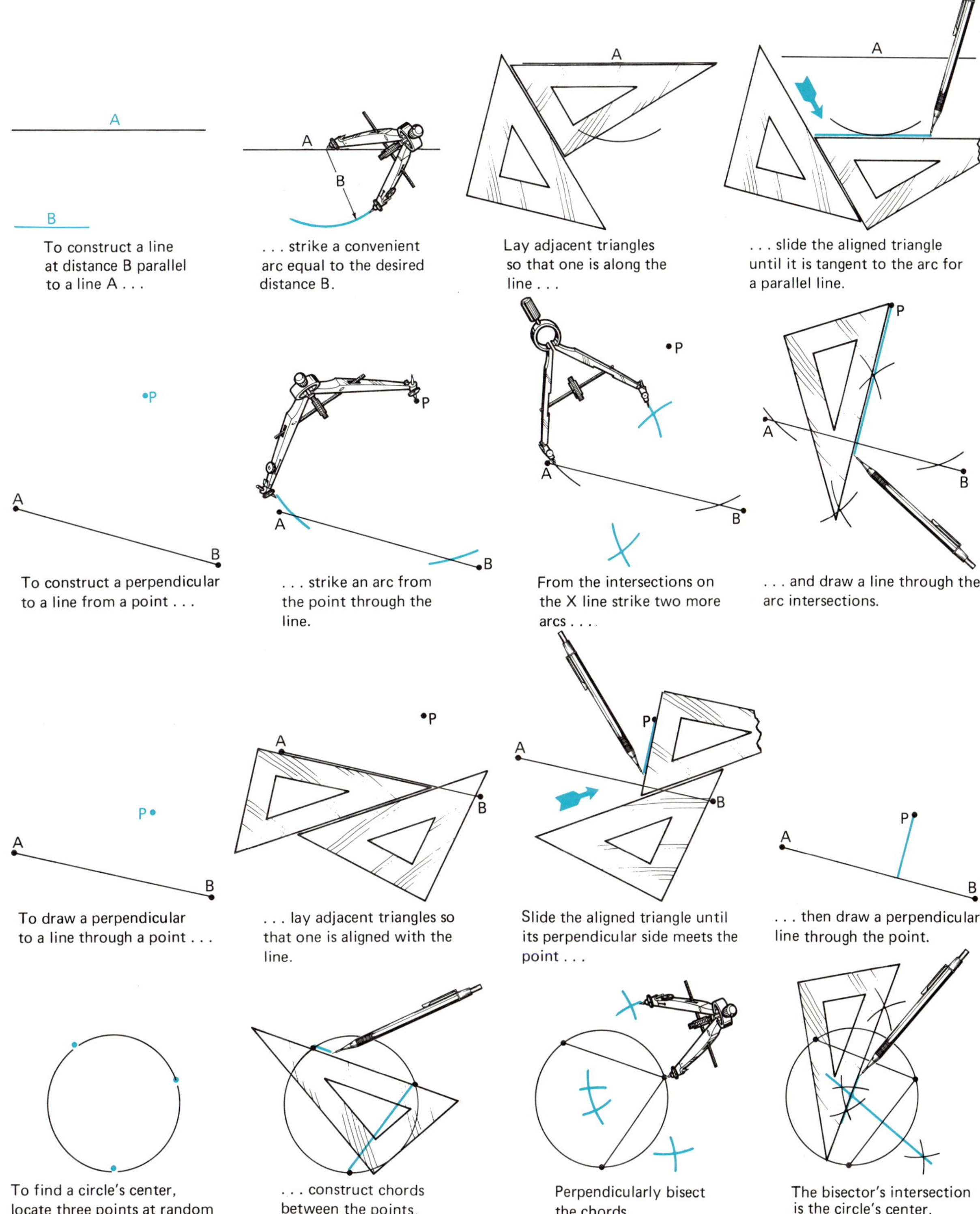

To construct a line at distance B parallel to a line A . . .

. . . strike a convenient arc equal to the desired distance B.

Lay adjacent triangles so that one is along the line . . .

. . . slide the aligned triangle until it is tangent to the arc for a parallel line.

To construct a perpendicular to a line from a point . . .

. . . strike an arc from the point through the line.

From the intersections on the X line strike two more arcs . . . .

. . . and draw a line through the arc intersections.

To draw a perpendicular to a line through a point . . .

. . . lay adjacent triangles so that one is aligned with the line.

Slide the aligned triangle until its perpendicular side meets the point . . .

. . . then draw a perpendicular line through the point.

To find a circle's center, locate three points at random along the circumference . . .

. . . construct chords between the points.

Perpendicularly bisect the chords.

The bisector's intersection is the circle's center.

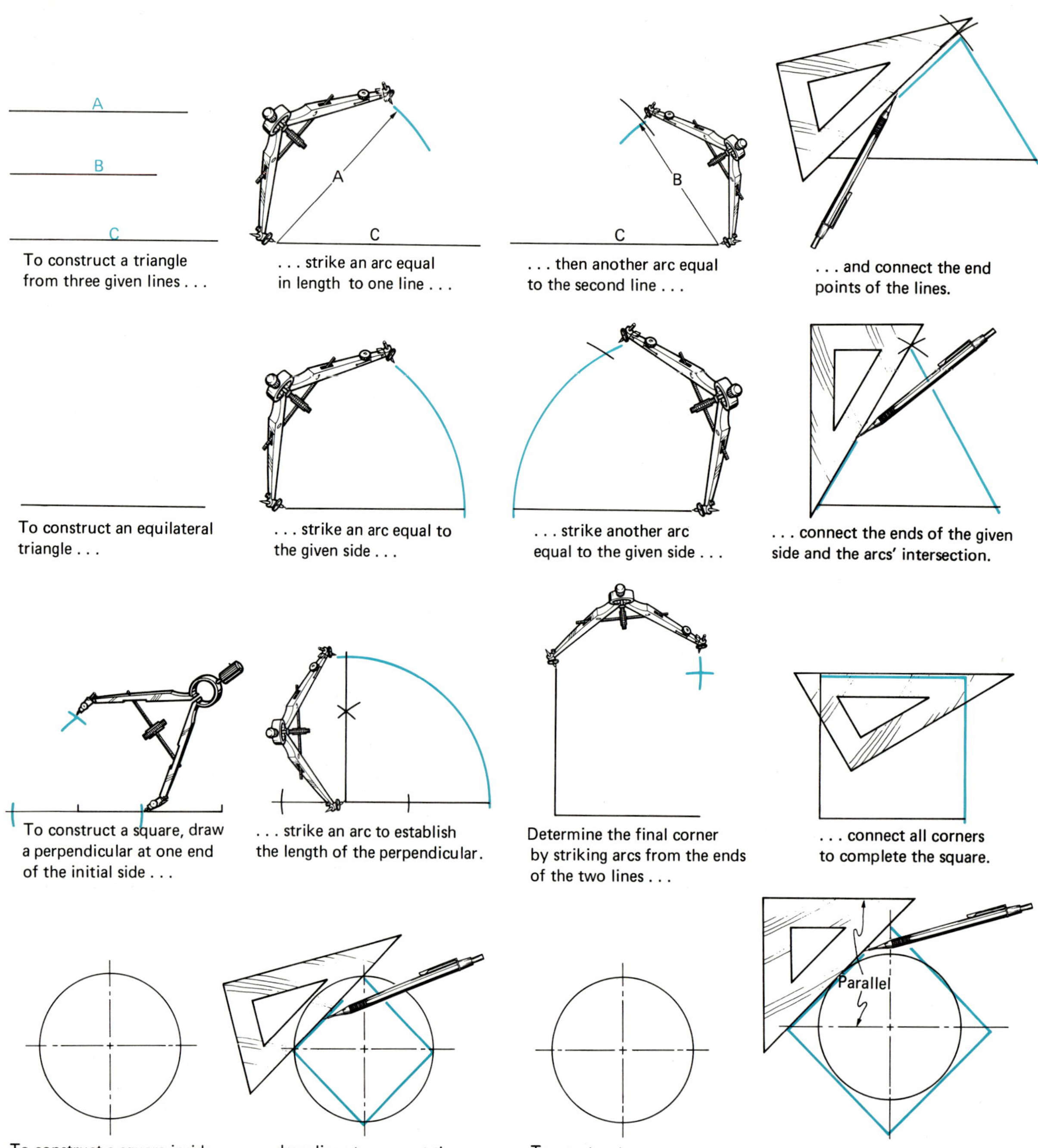

To construct a triangle from three given lines . . .

. . . strike an arc equal in length to one line . . .

. . . then another arc equal to the second line . . .

. . . and connect the end points of the lines.

To construct an equilateral triangle . . .

. . . strike an arc equal to the given side . . .

. . . strike another arc equal to the given side . . .

. . . connect the ends of the given side and the arcs' intersection.

To construct a square, draw a perpendicular at one end of the initial side . . .

. . . strike an arc to establish the length of the perpendicular.

Determine the final corner by striking arcs from the ends of the two lines . . .

. . . connect all corners to complete the square.

To construct a square inside a given circle . . .

. . . draw lines to connect the center line and circumference intersections.

To construct a square outside a given circle . . .

. . . connect center line extensions with 45° tangents.

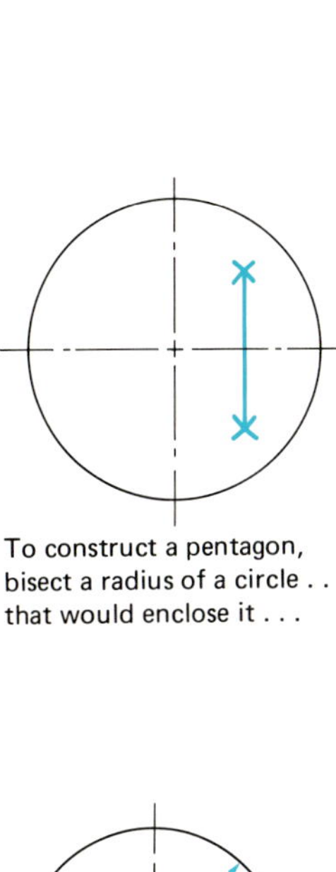

To construct a pentagon, bisect a radius of a circle . . . that would enclose it . . .

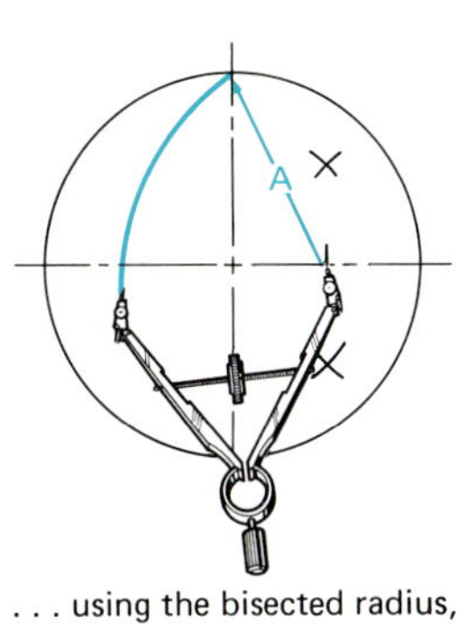
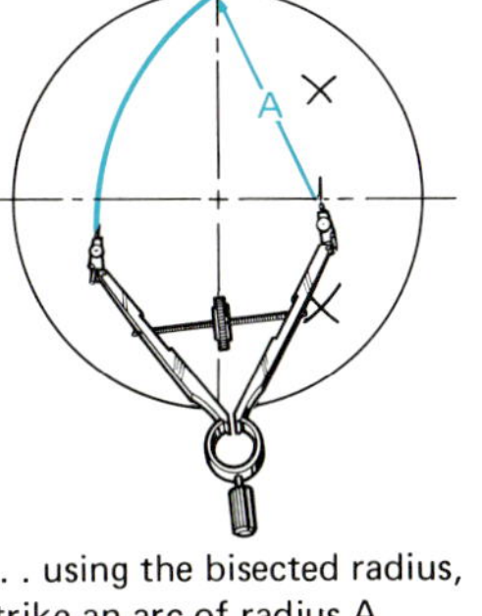

. . . using the bisected radius, strike an arc of radius A.

The resulting radius R is . . .

. . . used to establish the five points of a pentagon.

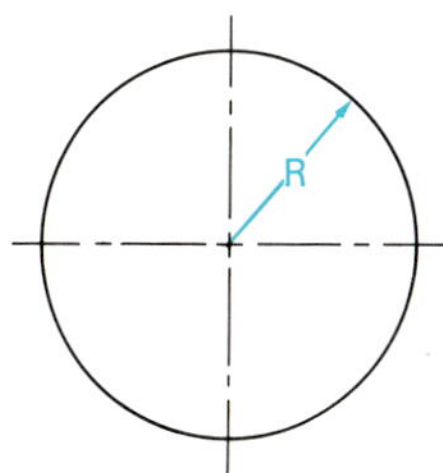

To construct a hexagon transfer the radius R of a circle that would enclose it . . .

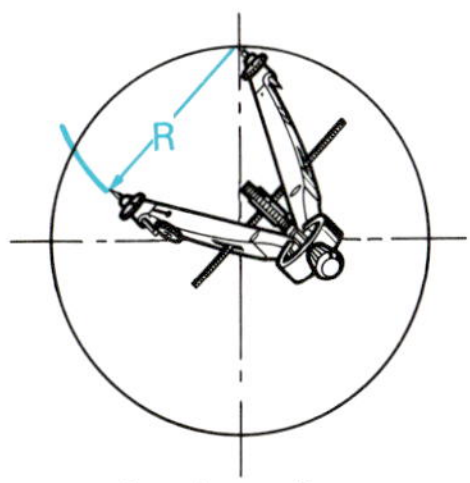

. . . to the circumference . . .

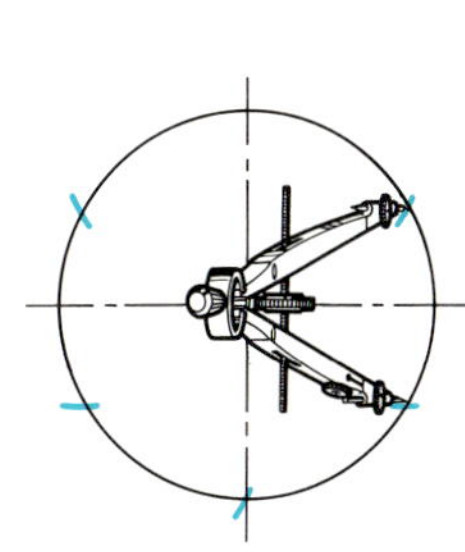

. . . and mark the six points . . .

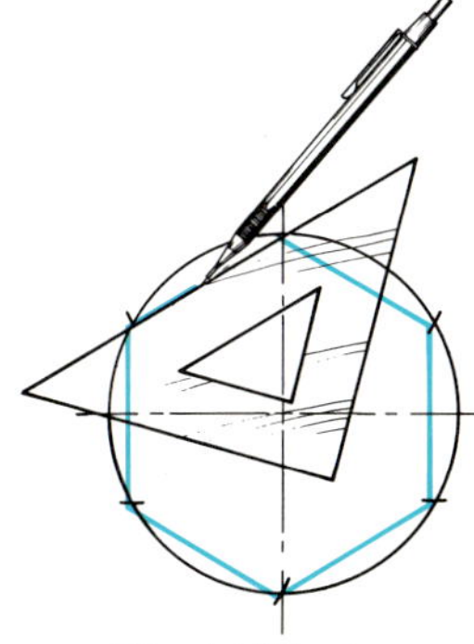

. . . used to establish a hexagon.

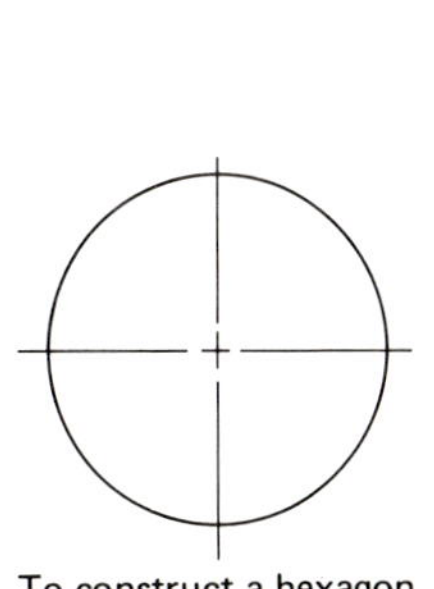

To construct a hexagon around a given circle . . .

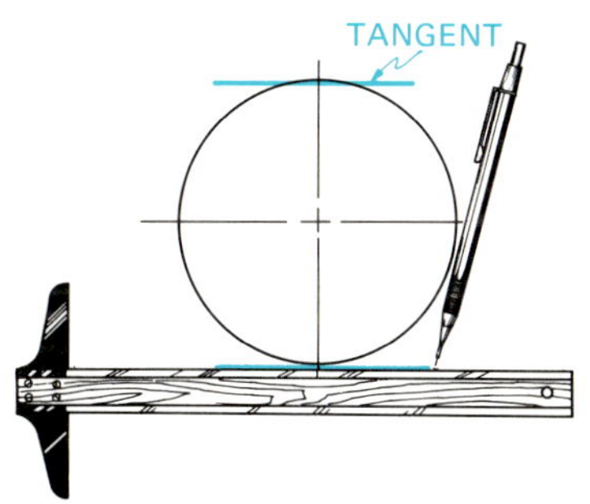

. . . draw parallel lines, using a "T" square, that are tangent to opposite sides of the circle.

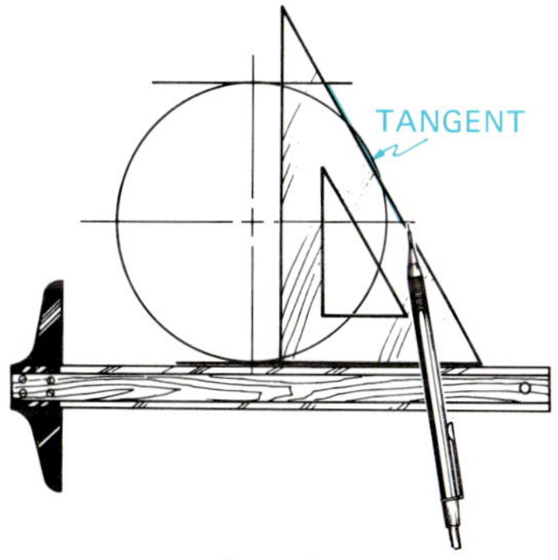
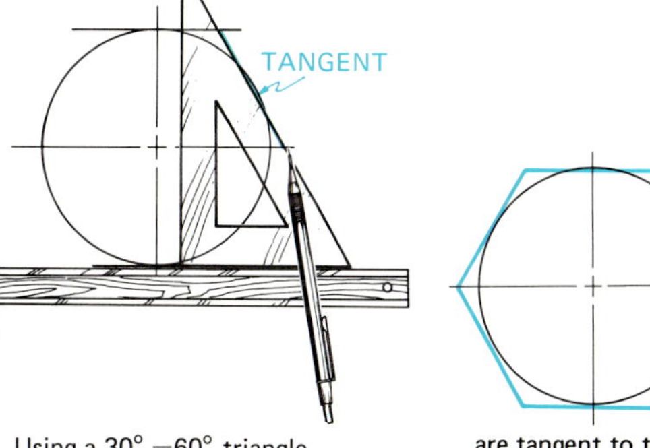

Using a 30° —60° triangle, construct two pairs of parallel lines that . . .

. . . are tangent to the circle to complete the hexagon.

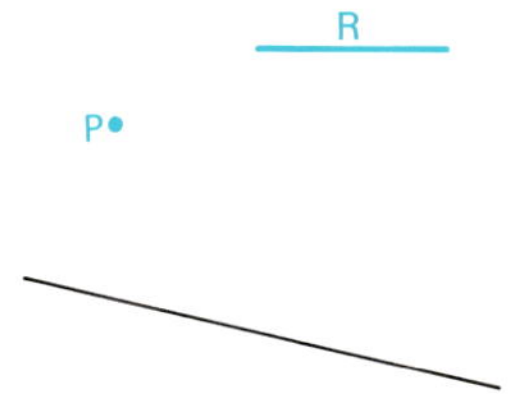

To draw an arc of radius R tangent to a line and through a point . . .

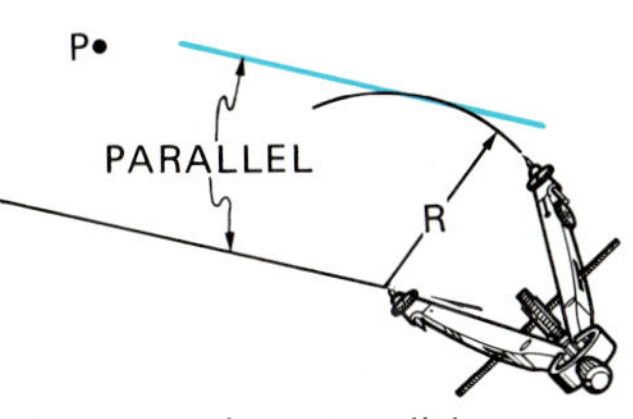

. . . draw a parallel at distance R.

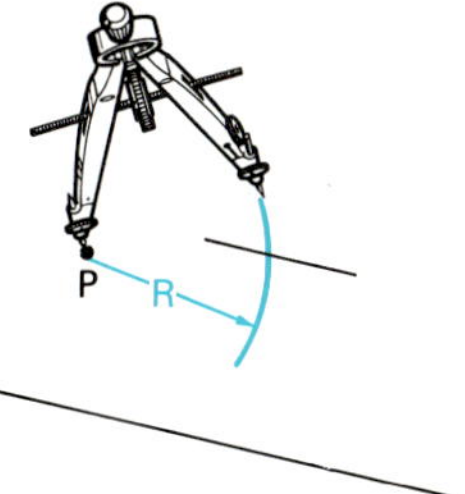

Intersect the parallel with an arc R from the point.

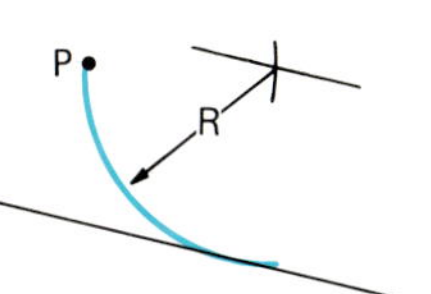

Using the intersection draw an arc tangent to the line through P.

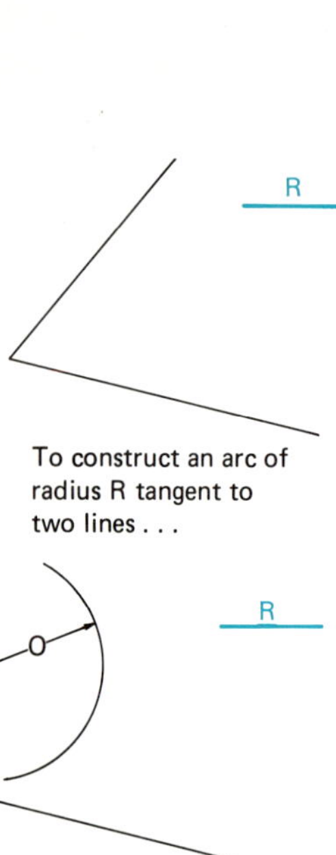

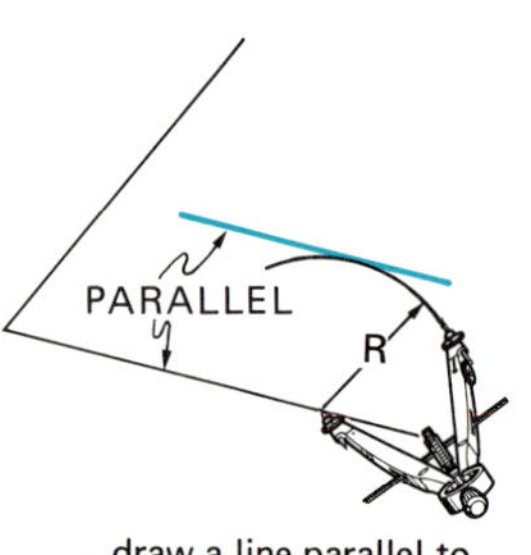

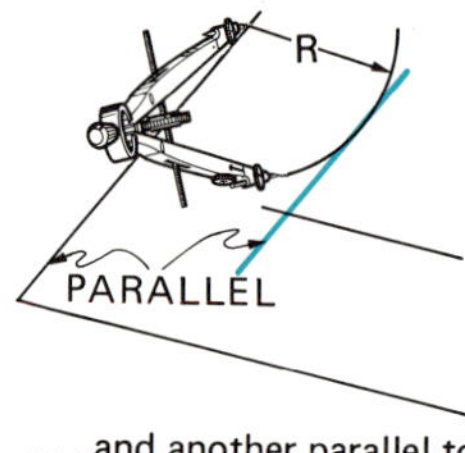

To construct an arc of radius R tangent to two lines . . .

. . . draw a line parallel to one line at distance R . . .

. . . and another parallel to the second line.

The intersection is the point from which to strike the tangent arc.

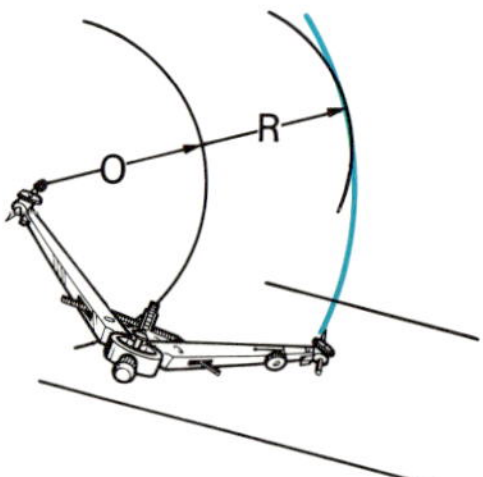

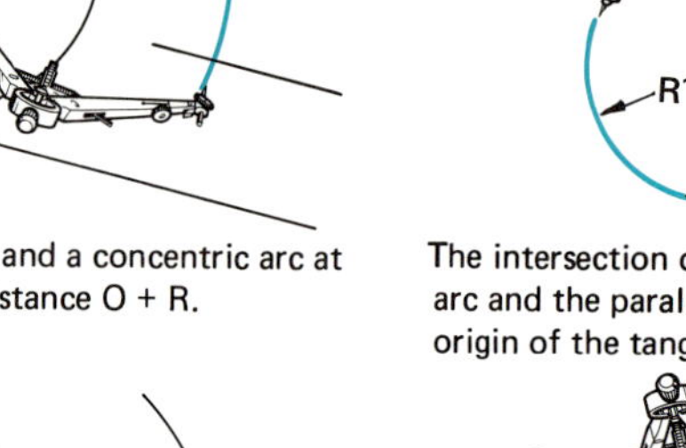

To construct an arc of radius R between an arc and a line . . .

. . . draw a line parallel to the line at distance R . . .

. . . and a concentric arc at a distance O + R.

The intersection of the large arc and the parallel line is the origin of the tangent arc.

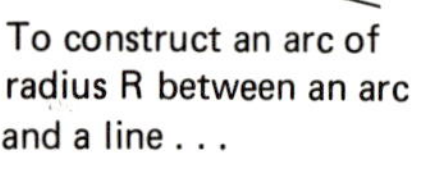

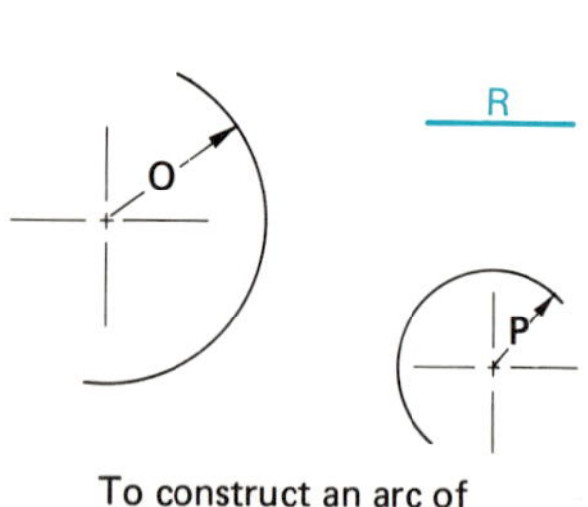

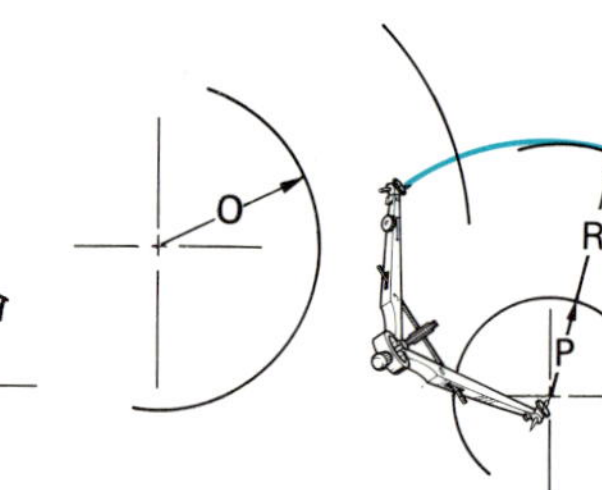

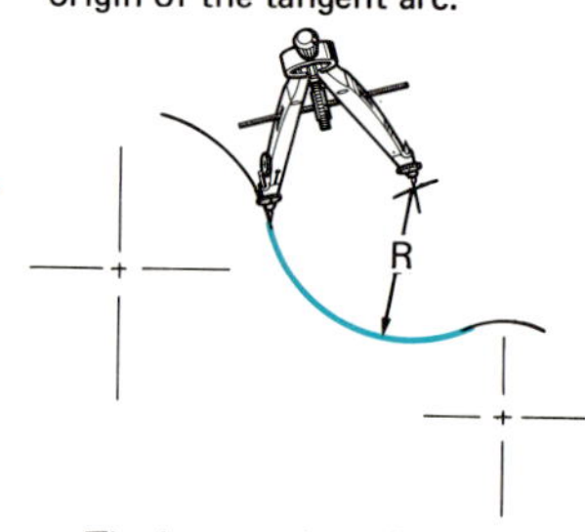

To construct an arc of radius R between two arcs . . .

. . . draw a concentric arc of length O + R . . .

. . . and another concentric arc of length P + R.

The intersection of arcs is the origin of the tangent arc R.

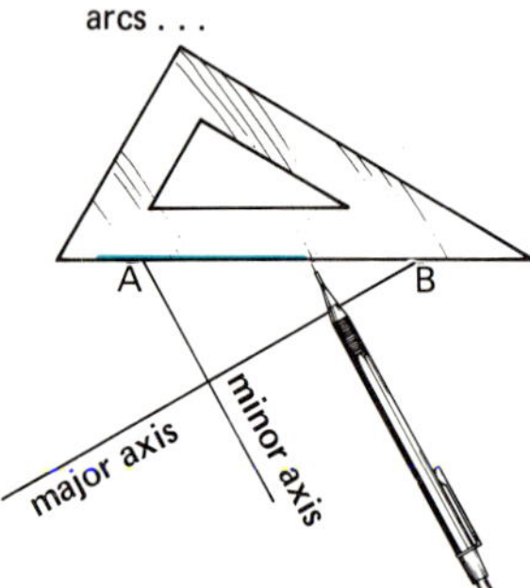

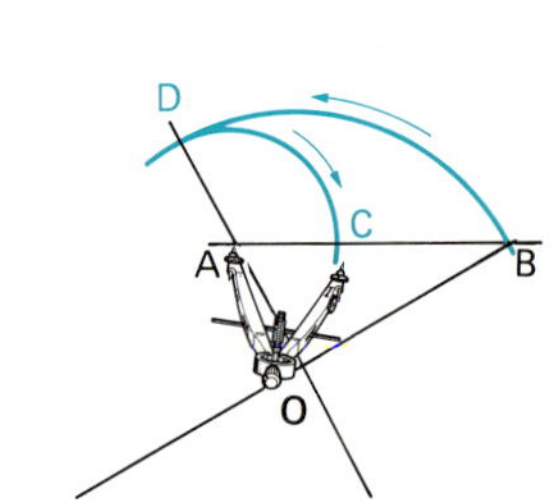

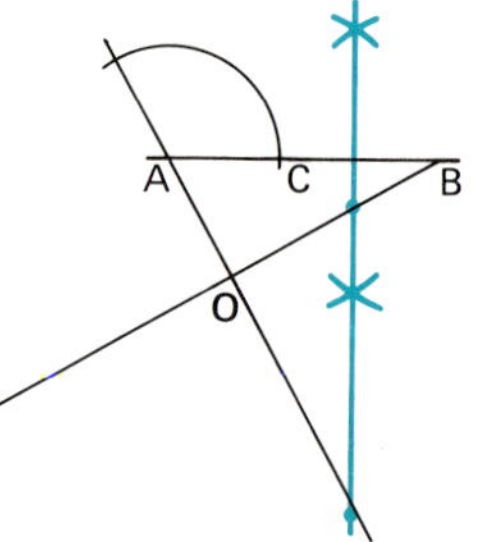

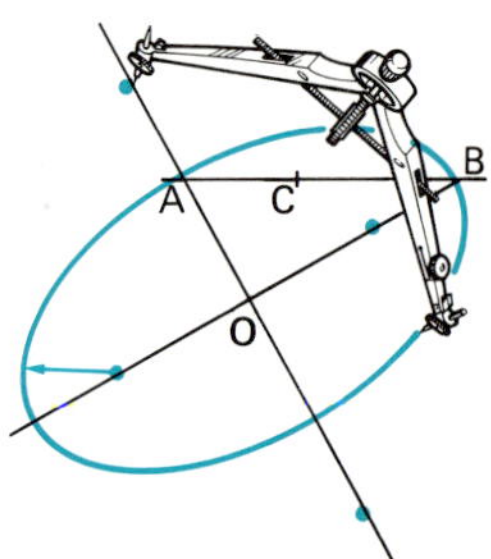

To construct an approximate ellipse, draw a line between the end points (A and B) of the major and minor axes.

Construct OD equal to OB and AC equal to AD.

Construct a perpendicular bisector of CB to obtain pivot points on the major and minor axes.

Using these pivot points, draw arcs to form the ellipse.

Abbreviations for Use on Drawings, Specifications, Standards, and in Technical Documents

Abbreviation	Meaning
A	
AB	anchor bolt
ABBR	abbreviate
ABN	airborne
ABNL	abnormal
ABRSV	abrasive
ABS	absolute
ABSORB	absorption
ABSTR	abstract
ABT	about
ABV	above
AC	alternating current
AC	asbestos cement
ACB	air circuit breaker
ACB	asbestos-cement board
ACC	accumulator
ACCEL	accelerate
ACCESS	accessory
ACCOM	accommodate
ACCT	account
ACCUM	accumulate
ACCUR	accurate
ACFT	aircraft
ACI	American Concrete Institute
ACLD	aircooled
ACPT	accept
ACR	across
ACS	access
ACS	alternating current synchronous
ACST	acoustic
ACTR	actuator
AD	ampere demand meter
AD	anode
ADD	addendum
ADD	addition
ADH	adhere
ADH	adhesive
ADJ	adjacent
ADJ	adjust
ADP	automatic data processing
ADPTR	adapter
ADV	advance
AERO	aeronautic
AERODYN	aerodynamic
AF	audio frequency
AFB	antifriction bearing
AFC	automatic frequency control
AFT	after
AFT	automatic fine tuning
AGA	American Gas Association
AGC	automatic gain control
AGDS	American Gage Design Standard
AGGR	aggregate
AGMA	American Gear Manufacturers Association
AHM	ampere-hour meter
AIL	aileron
AISC	American Institute of Steel Construction
AISI	American Iron and Steel Institute
AIR COND	air condition
AL	aluminum
AL	amber light
ALIGN	alignment
ALLOW	allowance
ALM	alarm
ALT	altitude
ALTN	alternate
ALY	alloy
AM	amplitude modulation
AMB	amber
AMB	ambient
AMER	American
AMLG	amalgam
AMM	ammeter
AMP	ampere
AMPH	amphibious
AMPL	amplifier
AMPTD	amplitude
AMS	aeronautical material specification
AMT	amount
AMT	armature
AN	alphanumeric
AN	Army-Navy
ANDZ	anodize
ANG	angle
ANHYD	anhydrous
ANL	anneal
ANLG	analog
ANLR	angular
ANLR	annular
ANN	annunciator
ANO	alphanumeric output
ANSI	American National Standards Institute
ANT	antenna
AO	access opening
AOQ	average outgoing quality
AOQL	average outgoing quality limit
AP	access panel
AP	acidproof
APC	automatic phase control
APERT	aperture
API	American Petroleum Institute
APL	airplane
APLQ	applique
APP	appearance
APPAR	apparatus
APPD	approved (altn abbr)
APPROX	approximate
APPV	approve
APPVL	approval
APPX	appendix
APR	April
APRT	airport
APS	auxiliary power supply
APT	apartment
APVD	approved
AQL	acceptable quality level
AQT	acceptable quality test
AR	acid resisting
AR	as required
ARB	arbitrary
ARCW	arc weld
ARL	acceptable reliability level
ARM	armature
ARMD	armored
ARR	arrange, arrangement
ARS	asbestos roof shingles
ARSR	arrestor
ART	article
ARTF	artificial
AS	airspeed
AS	automatic sprinkler
ASA	American Standards Association (now ANSI)
ASAP	as soon as possible
ASB	asbestos
ASGMT	assignment
ASGN	assign
ASME	American Society of Mechanical Engineers
ASPH	asphalt

Extracted from MIL-STD-12C, Supp. 1 (Feb. 1, 1971).

Abbreviation	Term
ASR	automatic sprinkler riser
ASRT	assort
ASSEM	assemble
ASSN	association
ASSOC	associate
ASST	assistant
ASSY	assembly
ASTM	American Society for Testing and Materials
ASV	angle stop valve
ASW	auxiliary switch
ASWG	American Steel Wire Gage
ASYM	asymmetric
ASYMP	asymptote
AT	airtight
AT	ampere turn
ATC	acoustic tile ceiling
ATCH	attach, attachment
ATM	atmosphere
ATMG	atomizing
ATTD	attitude
ATTN	attention
AUD	audible
AUD	audio
AUG	August
AUTH	authorize
AUTO	automatic
AUTO	automobile
AUTOM	automotive
AUX	auxiliary
AV	audio-visual
AVAIL	available
AVC	automatic volume control
AVDP	avoirdupois
AVE	avenue
AVG	average
AVN	aviation
AW	all weather
AWG	American Wire Gage
AWS	American War Standard
AWS	American Welding Society
AZ	azimuth

B

Abbreviation	Term
B TO B	back to back
B&S	bell and spigot
B&S	Brown and Sharp Wire Gage
B&W	black and white
B/L	bill of lading
B/M	bill of material
BA	backlash allowance
BAB	babbit
BACT	bacteriological
BAF	baffle
BAG	baggage
BAL	balance
BAL	ballistic
BARO	barometer
BAT	battery
BAY	bayonet
BB	bulletin board
BBB	body bound bolts
BBL	barrel
BBRG	ball bearing
BC	back-connected
BC	between centers
BC	binary counter
BC	bolt circle
BC	bottom chord
BCD	binary coded decimal
BCI	binary coded information
BCL	base contour line
BD	binary decoder
BD	board
BDG	binding
BDGH	binding head
BDL	bundle
BDN	bend down
BDR	binder
BDY	boundary
BE	Baume
BETW	between
BEV	bevel
BF	beat frequency
BF	boiler feed
BF	both faces
BFG	buffing
BFO	beat-frequency oscillator
BFP	boiler feed pump
BFR	before
BFR	buffer
BFW	boiler feed water
BG	back gear
BGP	barrier, grease proof
BH	Brinell hardness
BHD	bulkhead
BHN	Brinell hardness number
BHP	boiler horsepower
BHP	brake horsepower
BI	black iron
BIL	blue indicating light
BIN	binary
BIOL	biological
BIT	binary digit
BITUM	bituminous
BK	black (altn abbr)
BK	brake
BKLT	booklet
BL	base line
BL	bend line
BL	billet
BL	blade
BL	blue (altn abbr)
BL	building line
BLDG	building
BLK	black
BLK	blank
BLKG	blanking
BLO	blower
BLR	boiler
BLST	ballast
BLT	bolt
BLTIN	built-in
BLU	blue
BLVD	boulevard
BLW	below
BM	beam
BM	bench mark
BMEP	brake mean effective pressure
BMTLC	bimetallic
BND	bonded
BNDG	bonding
BNDZ	bonderize
BNG	bending
BNR	burner
BNSH	burnish
BO	black out
BO	blowoff
BOC	blowout coil
BOLS	bolster
BOT	bottom
BP	back pressure
BP	base pitch
BP	base plate
BP	between perpendiculars
BP	blueprint
BP	boiling point
BPD	barrels per day
BPH	barrels per hour
BPRF	bulletproof
BPS	bits per second
BR	bedroom
BR	bend radius
BR	branch

BR	brown (altn abbr)
BR STD	British Standard
BRC	brace
BRCH	broach
BRD	braided
BRG	bearing
BRK	break
BRK	brick
BRKR	breaker
BRKT	bracket
BRN	brown
BRS	brass
BRZ	braze
BRZ	bronze
BRZG	brazing
BS	both sides
BSC	basic
BSHG	bushing
BSKT	basket
BSMT	basement
BSTR	booster
BT	bent
BT	bus tie
BTFL	butterfly
BTL	bottle
BTN	button
BTRY	battery
BTU	British thermal unit
BTWLD	buttweld
BU	bureau
BU	bushel
BULL	bulletin
BUP	bend up
BUS	business
BUSH	bushing (altn abbr)
BUTN	butane
BUZ	buzzer
BW	bandwidth
BW	both ways
BWG	Birmingham Wire Gage
BWP	barrier, waterproof
BWS	beveled wood siding
BYNT	buoyant
BYP	bypass

C

C	Celsius
3/C	three conductor
C TO C	center to center
C/W	carrier wave
CA	cable
CA	clear aperture
CAB	cabinet
CAL	caliber
CAL	calibration
CAL	calorie
CALC	calculate
CAM	camber
CAM	checkout and automatic monitoring
CANC	cancel
CAND	candelabra
CANV	canvas
CAP	capacitor
CAP	capacity
CAP	capital
CARB	carburize
CARR	carrier
CAS	castle
CAT	catalog
CAT	category
CATH	cathode
CATV	community antenna television
CAV	cavity
CB	catch basin
CB	center of buoyancy
CB	circuit breaker
CB	continuous blowdown
CBAL	counterbalance
CBD	carbide
CBORE	counterbore
CBS	call box station
CC	close coupled
CC	closing coil
CC	color code
CC	common collector
CC	cubic centimeter
CCB	concrete block
CCG	cargo center of gravity
CCN	contract change notice
CCR	control contactor
CCT	constant current transformer
CCTV	closed-circuit television
CCW	counterclockwise
CD	cable duct
CD	candle
CD	card
CD	center distance
CD	cold-drawn
CD	common digitizer
CD	cord
CD	current density
CD PL	cadmium plate
CDR	critical design review
CDRL	contract data requirements list
CD RDR	card reader
CDEL	constant delivery
CDP	compressor discharge pressure
CDR	current directional relay
CDRILL	counterdrill
CDS	cold-drawn steel
CDW	chilled drinking water
CE	common emitter
CE	commutator end
CEA	circular error average
CEI	contract end item
CEL	celestial
CEM	cement
CEP	circular error probability
CER	ceramic
CERMET	ceramic metal element
CERT	certify
CET	cumulative elapsed time
CF	carrier frequency
CF	cathode follower
CF	cement floor
CF	center of flotation
CF	centrifugal force
CF	cold finished
CF	cooling fan
CFE	contractor-furnished equipment
CFL	calibrated focal length
CFLG	counterflashing
CFM	cubic feet per minute
CFP	contractor-furnished property
CFS	cold finished steel
CFS	cubic feet per second
CG	center of gravity
CG	chain grate
CGS	centimeter-gram-second
CGU	ceramic glazed units
CH	case harden
CH	chain
CH	chief
CH	choke
CHAM	chamfer
CHAN	channel
CHAP	chapter
CHAR	character
CHAS	chassis
CHC	choke coil
CHD	chord
CHEM	chemical
CHG	change (altn abbr)
CHG	charge

Abbreviation	Meaning
CHGOV	change-over
CHGR	charger
CHK	check
CHKR	checker
CHLD	chilled
CHNG	change
CHR	character register
CHR	chromium
CHRST	characteristic
CHW	chilled water
CI	cast iron
CI	circuit interrupter
CID	component identification
CIL	clear indicating light
CIN	carrier input
CIN	code identification number
CINE	cinematographic
CIP	cast-iron pipe
CIP	compressor inlet pressure
CIR	circle
CIRC	circular
CIT	compressor inlet temperature
CK	cork
CKBD	cork board
CKD	completely knocked down
CKPT	cockpit
CKT	check template
CKT	circuit
CKT BKR	circuit breaker (altn abbr)
CL	carload
CL	center line
CL	class
CL	clearance
CL	close
CL	closing
CL	clutch
CLASS	classification
CLB	center line bend
CLD	cooled
CLE	closed end
CLG	ceiling
CLG	cooling
CLK	clock
CLKG	caulking
CLN	clean
CLO	closet
CLP	clamp
CLPR	caliper
CLR	clear
CLR	collar
CLR	color
CLR	cooler
CLR	current-limiting resistor
CLRG	collector ring
CLTG	collecting
CLV	clevis
CLWG	clear wire glass
CMBD	combined
CMD	command
CMET	coated metal
CMF	coherent memory filter
CMFLR	cam follower
CMIL	circular mil
CML	current-mode logic
CMP	corrugated metal pipe
CMPD	compound
CMPLM	complement
CMPNT	component
CMPS	centimeters per second
CMPSN	composition
CMPST	composite
CMPT	compute
CMPTG	computing
CMPTR	computer
CMR	common-mode rejection
CMRLR	cam roller
CMS	current-mode switching
CN	change notice
CNCL	concealed
CNCR	concurrent
CNCTRC	concentric
CNCV	concave
CND	conduit
CNDCT	conductor
CNDS	condensate
CNTBRD	centerboard
CNTD	contained
CENTFGL	centrifugal
CNTNR	container
CNTOR	contactor
CNTR	counter
CNVC	convenience
CNVR	conveyor
CO	change order
CO	cleanout
CO	company
CO	cutoff
CO	cutout
COAM	coaming
COAX	coaxial
COE	cab over engine
COED	computer-operated electronic display
COEF	coefficient
COL	column
COLIM	collimator
COLL	collector
COM	common
COMB	combination
COMB	combustion
COML	commercial
COMM	communication
COMM	commutator
COMP	companion
COMP	compensator
COMPL	complete
COMPR	composition roof
COMPR	compressor (altn abbr)
COMPSG	compensating
COMPT	compartment
COMPTR	comparator
COMSAT	communications satellite
COMSN	commission
CONC	concentrate
CONC	concentric (altn abbr)
CONC	concrete
COND	condenser
COND	conductor (altn abbr)
CONF	confidential
CONN	connection
CONN	connector
CONST	constant
CONSTR	construction
CONT	contact
CONT	contents
CONT	continued
CONT	control
CONT	controller
CONTR	contract
CONTR	contractor
CONV	convention
COOL	coolant
COORD	coordinate
COP	copper
COR	corner
CORP	corporation
CORR	correct
CORR	corrugate
COT	cotter
COV	cover
COV	cutout valve
COWL	cowling
CP	candlepower
CP	center of pressure
CP	center punch
CP	cesspool
CP	chemically pure

Abbreviation	Term
CP	circular pitch
CP	circular polarization
CP	coefficient of performance
CP	cold-punched
CP	command pulse
CP	cone point
CP	constant pressure
CP	control point
CPD	contact potential difference
CPE	circular probable error
CPFF	cost plus fixed fee
CPI	cost plus incentive
CPL	cement plaster
CPL	couple
CPLD	coupled
CPLG	coupling
CPLR	coupler
CPM	critical-path method
CPNTR	carpenter
CPPD	capped
CPRSN	compression
CPRSR	compressor
CPS	circuit package schematic
CPSL	capsule
CPSN	capstan
CPT	control power transformer
CPUNCH	counterpunch
CQ	commercial quality
CR	cathode ray
CR	change request
CR	cold-rolled
CR	control relay
CR	control room
CR	credit
CR	crystal rectifier
CR	current relay
CR MOLY	chrome molybdenum
CR VAN	chrome-vanadium
CRAM	card random access memory
CRCLT	circulate
CRCMF	circumference
CRE	corrosion-resistant
CRG	carriage
CRIT	critical
CRK	crank
CRM	chrome
CRN	crown
CRO	cathode-ray oscilloscope
CRP	crimp
CRPL	chromium plate
CRS	coarse
CRS	cold-rolled steel
CRSN	corrosion
CRSV	corrosive
CRSVR	crossover
CRT	cathode-ray tube
CRTG	cartridge
CRV	curve
CRYST	crystalline
CS	carbon steel
CS	case
CS	cast steel
CS	Commercial Standard
CS	control switch
CSD	cold side
CSD	constant-speed drive
CSG	casing
CSHAFT	crankshaft
CSK	countersink
CSKH	countersunk head
CSL	console
CSTG	casting
CSTL	castellate
CSTR	canister
CSW	control power switch
CT	center tap
CT	ceramic tile
CT	current transformer
CTC	cam timing contact
CTD	coated
CTF	ceramic-tile floor
CTG	coating
CTG	crating
CTG	cutting
CTL	central
CTN	carton
CTR	center
CTR	contour
CTRG	centering
CTRS	contrast
CTSK	countersunk
CTT	card-to-tape tape
CTWT	counterweight
CU	crystal unit, piezoelectric
CU	cubic
CULV	culvert
CUM	cumulative
CUP	cupboard
CUR	current
CURT	curtain
CUSH	cushion
CUST	customer
CV	check valve
CVNTL	conventional
CVRSN	conversion
CVX	convex
CW	clockwise
CW	cold water
CW	continuous wave
CWP	circulating water pump
CWT	hundredweight
CY	cycle
CYL	cylinder, cylindrical
CYN	cyanide

D

Abbreviation	Term
1/2DP	half dog point
D/A	digital-to-analog
DAC	digital-to-analog converter
DACT	direct acting
DAD	double-acting door
DAT	datum
DB	decibel
DB	distribution box
DB	double braid (insul)
DB	dry bulb
DBAO	digital block AND-OR gate
DBB	detector back bias
DBC	diameter bolt circle
DBF	demodulator band filter
DBHP	drawbar horsepower
DBL	double
DBP	drawbar pull
DBUT	data base update time
DC	direct current
DC	double contact
DCD	decode
DCDR	decoder
DCKG	docking
DCL	door closer
DCN	design change notice
DCN	drawing change notice
DCR	drawing change request
DCTP	duct type
DDA	digital differential analyzer
DDR	direct drive
DE	drive end
DEC	December
DEC	decimal
DECAL	decalcomania
DECR	decrease
DED	dedendum
DEF	defective

DEF	definition
DEFT	definite time
DEG	degree
DEGUSG	degaussing
DEL	delineation
DENS	density
DEPT	department
DERIV	derivative
DES	designation
DET	detail
DET	detector
DETN	detention
DEV	development
DEVN	deviation
DF	double feeder
DF	Douglas fir
DF	drinking fountain
DF	drive fit
DF	drop forge
DFT	draft
DFTG	drafting
DFTR	deflector
DFTSMN	draftsman
DGTL	digital
DHMY	dehumidify
DHW	double-hung windows
DIA	diameter
DIAG	diagonal
DIAG	diagram
DIAPH	diaphragm
DIEL	dielectric
DIFF	difference
DIFF	differential
DIFFR	diffraction
DIM	dimension
DIR	direct
DIR	direction, directional
DISC	disconnect
DISCH	discharge
DISM	dismantle
DISP	dispatcher
DISP	dispenser
DISPL	displacement
DIST	distance
DIST	district
DISTR	distributor
DIV	division
DJ	drill jig
DK	dark
DK	deck
DL	dead load
DL	delay line
DLX	deluxe
DLY	delay
DM	demand meter
DN	down
DNA	does not apply
DNG	dining
DO	ditto
DOC	document
DOM	domestic
DOZ	dozen
DP	depth
DP	dew point
DP	diametral pitch
DP	double-pole
DP	dripproof
DP	dual purpose
DPDT	double-pole double-throw
DPG	damping
DPH	diamond-pyramid hardness
DPLR	doppler
DPLXR	duplexer
DPM	data-processing machine
DPNL	distribution panel
DPO	dripproof open
DPP	dripproof protected
DPR	diazoprint
DPR	double pulse ranging
DPS	dripproof semienclosed
DPST	double-pole single-throw
DPT	dripproof totally-enclosed
DPV	dry pipe valve
DR	door
DR	drafting request
DR	dram
DR	drill
DR	drive
DRM	drafting room manual
DRS	dressed (lumber)
DRTL	diode resistor transistor logic
DRVR	driver
DS	disconnect switch
DS	domestic service
DS&R	data storage and retrieval
DSCONT	discontinue
DSGN	design
DSL	diesel
DSPL	display
DSPL	disposal
DSTL	distill
DT	decay time
DT	double-throw
DT	dual tires
DT	dust tight
DTD	dated
DTL	diode transistor logic
DTR	definite-time relay
DTR	demand-totalizing relay
DUP	duplicate
DVM	digital voltmeter
DVTL	dovetail
DW	distilled water
DWG	drawing
DWL	dowel
DWN	drawn
DWT	deadweight
DWT	pennyweight
DX	duplex
DYHR	dehydrator
DYN	dynamic
DYNMT	dynamometer

E

E	east
E	emitter (electron device)
E&SP	equipment and spare parts
EA	each
ECC	eccentric
ECC	error correction code
ECD	estimated completion date
ECL	emitter-coupled logic
ECN	engineering change notice
ECO	electron-coupled oscillator
ECO	electronic checkout
ECO	engineering change order
ECP	engineering change proposal
ECR	engineering change request
ED	edition
EDGW	edgewise
EDP	electronic data processing
EDR	equivalent direct radiation
EDT	electrical discharge tube
EF	extra fine (threads)
EFF	effective

EFF	efficiency
EFL	effective focal length
EFL	equivalent focal length
EG	for example
EGT	exhaust gas temperature
EHF	extremely high frequency
EHP	effective horsepower
EL	elastic limit
EL	elevation
ELAS	elastic
ELB	elbow
ELEC	electric, electrical
ELEK	electronic
ELEM	elementary
ELEV	elevator
ELEX	electronics
ELIG	eligible
ELMCH	electromechanical
ELONG	elongation
ELP	elliptical
ELPC	electroluminescent-photoconductive
ELPNEU	electropneumatic
ELV	electrically operated valve
EM	electromagnetic
EM	engineering memorandum
EM	expanded metal
EMB	emboss
EMER	emergency
EMF	electromotive force
EMPL	employee
EMT	electrical metallic tubing
EMTR	emitter
EMU	electromagnetic units
EMUL	emulsion
EN	engineering notice
ENAM	enamel
ENCL	enclosure
ENG	engine
ENG	English
ENGR	engineer
ENGRG	engineering
ENTR	entrance
ENV	envelope
ENVIR	environment
EO	engineering order
EOLM	electro-optical light modulator
EP	electrical panel
EP	epitaxial planar
EP	explosion-proof
EP	extreme pressure
EPT	external pipe thread
EPWR	emergency power
EQ	equal (altn abbr)
EQ	equation
EQ SP	equally spaced (altn abbr)
EQL	equal
EQL SP	equally spaced
EQPT	equipment
EQUIL	equilibrium
EQUIV	equivalent
ERECT	erection
ERN	engineering release notice
ERR	error
ERRC	error correction
ES	electrostatic
ESHP	equivalent shaft horsepower
ESK	engineering sketch
ESP	especially
ESPEC	electrical specifications
ESR	equivalent series resistance
ESS	electronic switching system
EST	estimate
ESU	electrostatic units
ET	elapsed time
ETC	and so forth, et cetera
EVAC	evacuator
EVAP	evaporator
EVM	electronic voltmeter
EVOM	electronic volt-ohmmeter
EWO	engineering work order
EX	example
EX	extra
EXAM	examination
EXC	except
EXC	excitation
EXCH	exchange
EXCL	exclusive
EXCTO	exciter
EXEC	executive
EXER	exercise
EXH	exhaust
EXH	exhibit
EXP	expansion
EXP	expense
EXP	experiment
EXP	exposed
EXP	express
EXP JT	expansion joint
EXT	extension
EXT	exterior
EXT	external
EXT	extinguisher
EXTD	extrude
EXTM	extreme
EXTR	extrude (altn abbr)

F

F	Fahrenheit
F	force
FA	final assembly
FAB	fabricate
FABL	fire alarm bell
FABX	fire alarm box
FACIL	facility
FACT	fully automatic compiler-translator
FAO	finish all over
FAX	facsimile
FB	flat bar
FB	fuse block
FBCK	firebrick
FBM	board foot (measure)
FBR	feedback resistance
FBR	fiber
FBRBD	fiberboard
FBRS	fibrous
FC	file cabinet
FC	front-connected
FCB	free cutting brass
FCG	facing
FCSG	focusing
FCST	forecast
FCT	filament center tap
FCTN	function
FCTY	factory
FD	feed
FD	floor drain
FD	forced draft
FDB	forced-draft blower
FDBK	feedback
FDC	fire-department connection
FDI	field discharge
FDM	frequency division multiplex
FDN	foundation
FDP	full dog point
FDPL	fluid pressure line
FDR	feeder
FDR	fire door
FDRY	foundry

FDW	feed water
FEB	February
FED	Federal
FEM	female
FET	field-effect transistor
FEXT	fire extinguisher
FF	flip flop
FF	following
FFC	flip flop complementary
FFL	female flared
FFL	flip flop latch
FFL	front focal length
FFP	firm fixed price
FG	filament ground
FGD	forged
FGN	foreign
FH	fire hose
FH	flat head (altn abbr)
FH	full hard
FHC	fire-hose cabinet
FHP	fractional horsepower
FHP	friction horsepower
FHR	fire-hose rack
FHT	fully heat-treated
FHY	fire hydrant
FIC	frequency interference control
FIG	figure
FIIN	Federal item identification number
FIL	filament
FIL	fillet
FIL	fillister
FILH	fillister head
FIN	financial
FIN	finish (altn abbr)
FL	flashing
FL	flat
FL	floor
FL	flow
FL	fluid
FL	flush
FL	focal length
FL	footlambert
FL/W	flash welding
FLD	field
FLDG	folding
FLEX	flexible
FLG	flange
FLG	flooring
FLH	flat head
FLMB	flammable
FLMPRF	flameproof
FLMT	flushmount
FLN	flatten
FLOT	flotation

FLR	filler
FLRD	flared
FLRT	flow rate
FLSW	flow switch
FLT	flight
FLT	float
FLTD	fluted
FLTG	floating
FLTP	flush type
FLTR	filter
FLUOR	fluorescent
FLW	flat washer
FM	fathom
FM	field manual
FM	fire main
FM	frequency modulation
FMAN	foreman
FMS	free-machining steel
FNI	fan in
FNO	fan out
FNPT	fusion point
FNSH	finish
FOC	focal
FORG	forging
FORM	formula
FOUO	for official use only
FP	fireplace
FP	flat point
FP	freezing point
FP	fungus proof
FPI	fluorescent penetrant inspection
FPL	fire plug
FPM	feet per minute
FPRF	fireproof
FPS	feet per second
FPS	foot-pound-second (system)
FPT	female pipe thread
FR	failure rate
FR	fast release (relay)
FR	field reversing
FR	frame
FR	from
FR	front
FRAC	fractional
FREQ	frequency
FRES	fire resistant
FRICT	friction
FRMR	former
FRT	freight
FRWK	framework
FRZR	freezer
FS	far side
FS	Federal Specification

FS	finish specification (number)
FS	full size
FSBL	fusible
FSC	Federal supply class
FSC	full scale
FSN	Federal Stock Number
FST	forged steel
FSTNR	fastener
FT	feet
FT LB	foot-pound
FTD	fitted
FTF	flared tube fitting
FTG	fitting
FTG	footing
FTHRD	female thread
FTP	fuel transfer pump
FU	fuse
FUBX	fuse box
FUND	fundamental
FUR	furnace
FURN	furnish
FUS	fuselage (altn abbr)
FUSLG	fuselage
FV	flush valve
FV	front view
FV	full voltage
FW	fire wall
FW	fresh water
FWD	forward
FWP	feed water pump
FWP	fresh-water pump
FXD	fixed
FXP	fixed point
FXTR	fixture
FYIG	for your information and guidance
FZ	fuze

G

G	gram
G	gravity
G	green (altn abbr)
G	grid
GA	gage
GAL	gallons
GALV	galvanize
GALVI	galvanized iron
GALVS	galvanized steel
GAR	garage
GAS	gasoline
GAS/W	gas weld
GBG	garbage
GD	guard
GEN	generator

GENL	general
GEOL	geological
GFE	government-furnished equipment
GFLD	generator field
GI	gray iron
GIL	green indicating lamp
GL	grade line
GL	green light
GLB	glass block
GLD	gold
GLOSS	glossary
GLPG	glowplug
GLT	guide light
GLV	globe valve
GLZ	glaze
GM	geometric mean
GMBL	gimbal
GMET	gun metal
GMLDG	garnish molding
GMT	Greenwich mean time
GMV	guaranteed minimum value
GND	ground
GNLTD	granulated
GOV	governor
GOVT	government
GOX	gaseous oxygen
GP	general purpose
GP	group
GPC	gypsum-plaster ceiling
GPH	gallons per hour
GPH	graphite
GPM	gallons per minute
GPW	gypsum-plaster wall
GR	gear
GR	grade
GR	grain
GR	gross
GRA	gray
GRAD	gradient
GRAN	granite
GRBX	gearbox
GRC	gearcase
GRD	grind
GRDTN	graduation
GRG	gearing
GRL	grille
GRN	green
GROM	grommet
GRPH	graphic
GRS	grease
GRTG	grating
GRV	groove
GRVD	grooved
GRWT	gross weight
GSIL	German silver
GSKT	gasket
GSU	glazed structural units
GSV	globe stop valve
GSWR	galvanized steel wire rope
GT	gastight
GT	grease trap
GT	gross ton
GTRB	gas turbine
GTV	gate valve
GUT	gutter
GLV	gravel
GVW	gross vehicle weight
GWB	gypsum wallboard
GWT	glazed wall tile
GY	gray (altn abbr)
GYM	gymnasium
GYP	gypsum
GYRO	gyroscope
GZ	ground zero

H

1/2H	half-hard
HAZ	hazardous
HB	hose bib
HC	holding coil
HCL	horizontal center line
HCS	high-carbon steel
HD	hard
HD	head
HD	heavy duty
HD CR	hard chromium
HD DRN	hard-drawn
HDBK	handbook
HDG	heading
HDL	handle
HDLG	handling
HDLS	headless
HDN	harden
HDNS	hardness
HDR	header
HDST	headset
HDW	hardware
HDWC	hardware cloth
HDWD	hardwood
HEX	hexagon
HEX HD	hexagonal head
HEX SOCH	hexagonal socket head
HF	holding fixture
HGBN	herringbone
HGR	hanger
HGT	height
HI FI	high fidelity
HIPRI	high priority
HKP	hookup
HLCL	helical
HLDG	holding
HLDR	holder
HLNG	headlining
HLP	helper
HLXA	helix angle
HNDWL	hand wheel
HNG	hanging
HNG	hinge
HNTG	hunting
HNYCMB	honeycomb
HOL	hollow
HON	honorary
HORIZ	horizon (altn abbr)
HORIZ	horizontal
HOSP	hospital
HP	high pressure
HP	horsepower
HP HR	horsepower hour
HQ	headquarters
HR	hot rolled
HR	hour
HRPO	hot rolled, pickled and oiled
HRS	hot-rolled steel
HRZN	horizon
HS	high speed
HSE	house
HSG	housing
HSHLD	household
HSR	high-speed reader
HSS	high-speed steel
HST	hoist
HSTH	hose thread
HT	heat
HT	high tension
HT	hollow tile
HT RES	heat resisting
HT TR	heat treat
HTCI	high-tensile cast iron
HTD	heated
HTG	heating
HTM	high temperature
HTR	heater
HV	high velocity
HV	high voltage
HVR	high-voltage regulator
HVY	heavy
HW	half-wave
HW	hot water
HWC	hot-water circulating
HWH	hot-water heater
HWY	highway
HYB	hybrid
HYD	hydraulic (altn abbr)

Abbreviation	Term
HYDM	hydrometer
HYDR	hydraulic

I

Abbreviation	Term
I/O	input/output
IAS	indicated air speed
IB	instruction book
IBID	the same as
IC	instruction card
IC	integrated circuit
IC	internal combustion
ICF	inspection check fixture
ICS	Intercommunication System
ICT	inspection check template
ICT	international critical tables
ICW	interrupted continuous wave
ID	induced draft
ID	inside diameter
ID LT	identification light
ID NO	identification number
IDENT	identification
IDL	idler
IDP	integrated data processing
IDX	index
IE	that is
IF	inside frosted
IF	instruction folder
IF	intermediate frequency
IFL	inflatable
IG	inertial guidance
IGN	ignition
IGN	ignitron
IGNTR	ignitor
IGS	inertial guidance system
IGV	inlet guide vane
IHP	indicated horsepower
IL	index list
ILLUM	illuminate
ILLUS	illustrate
IMAG	imaginary
IMEP	indicated mean effective pressure
IMIT	imitation
IMPD	impedance
IMPF	imperfect
IMPRG	impregnate
IMPL	implement
IMPLR	impeller
IMPLS	impulse
IMPROV	improvement
IMPRSN	impression
IN	inch
IN LB	inch-pound
INBD	inboard
INC	incorporated
INCAND	incandescent
INCIN	incinerator
INCL	inclusive
INCLR	intercooler
INCLS	inclosure
INCM	incoming
INCND	incendiary
INCOH	incoherent
INCR	increase
IND	indicating
IND	induction
IND	industry
IND LP	indicating lamp
INDEP	independent
INDL	industrial
INF	infinite
INFO	information
INIT	initial
INJ	injection
INOP	inoperative
INORG	inorganic
INP	input
INQ	inquiry
INRT	inert
INRTG	inert gas
INRTL	inertial
INS	inside
INSOL	insoluble
INSP	inspection
INST	instantaneous
INST	institute
INST	instrument (altn abbr)
INSTL	installation
INSTM	instrumentation
INSTPN	instrument panel
INSTR	instruction
INSTR	instrument
INSUL	insulation
INT	interrupter (relay)
INTEG	integrating
INTEN	intensity
INTK	intake
INTL	internal
INTL	international
INTRO	introduction
INV	inverter
IOM	interoffice memorandum
IP	initial point
IP	instruction plate
IP	instruction pulse
IP	intermediate pressure
IP	iron pipe
IPB	illustrated parts breakdown
IPM	interruptions per minute
IPS	inches per second
IPS	Internal Pipe Standard
IPS	interruptions per second
IPS	iron pipe size
IPT	internal pipe thread
IPT	iron pipe thread
IR	infared
IR	inside radius
IRREG	irregular
IS	island
ISGN	insignia
ISO	International Organization for Standardization
ISO	isometric
ITR	inverse time relay
IV	initial velocity

J

Abbreviation	Term
JAN	January
JAN	Joint Army-Navy
JB	junction box
JC	janitor closet
JC	joint compound
JCT	junction
JFET	junction field-effect transistor
JKT	jacket
JMPR	jumper
JNL	journal
JO	job order
JOG	joggle
JR	junior
JT	joint
JUL	July
JUN	June

K

Abbreviation	Term
K	Kelvin
K	kip (1000 lb)
KD	kiln-dried
KD	knock down
KN	knot
KN SW	knife switch

KO	knockout
KPL	kick plate
KST	keyseat
KVAH	kilovoltampere hour
KWY	keyway
KYBD	keyboard

L

L	left
L	long
LA	lead angle
LA	lightning arrester
LAB	laboratory
LAD	ladder
LANG	language
LAQ	lacquer
LAT	latitude
LAU	laundry
LAV	lavatory
LB	pound
LB FT	pound-force foot
LB IN	pound-force inch
LBL	label
LBR	lumber
LBRY	library
LBYR	labyrinth
LC	inductance-capacitance
LC	laundry chute
LC	lead-covered
LC	learning curve
LC	load center
LC	lower case
LCL	local
LCTN	location
LD	leading
LD MK	land mark
LDG	landing
LDG	loading
LDR	leader
LE	leading edge
LF	low frequency
LG	long (altn abbr)
LGC	logic
LGE	large
LH	left hand
LIC	license
LIC	linear integrated circuit
LIM	limit
LIN	lineal
LIN	linear
LIQ	liquid
LK WASH	lock washer
LKD	locked
LKGE	linkage
LKNT	locknut
LKR	locker
LKROT	locked rotor
LKT	lookout
LKWR	lockwire
LL	live load
LL	low level
LLYP	longleaf yellow pine
LMTR	limiter
LNG	lining
LNTL	lintel
LOA	length overall
LONG	longitude
LOS	line of sight
LOX	liquid oxygen
LP	low pressure
LPW	lumen per watt
LR	load ration
LR	load resistor (relay)
LR	long range
LS	left side
LS	limestone
LS	loudspeaker
LT	lead time
LTD	limited
LTG	lighting
LTHR	leather
LTR	letter
LTSW	light switch
LUB	lubricate
LUBT	lubricant
LUM	luminous
LV	leave
LV	low voltage
LVD	louvered door
LVL	level
LVO	louver opening
LVP	low-voltage protection
LVR	lever
LVR	louver
LWB	long wheelbase
LWC	lightweight concrete
LWIC	lightweight insulating concrete
LWL	low-water line
LWR	lower
LYR	layer
LYT	layout

M

MACH	machine
MAG	magazine
MAG	magnetic
MAGN	magnetron
MAGTD	magnitude
MAH	mahogany
MAINT	maintenance
MAJ	major
MAL	malleable
MALF	malfunction
MAM	milliammeter
MAN	manual (altn abbr)
MANF	manifold
MAR	March
MAR	marine
MARG	margin
MASU	machined surface
MAT	matrix
MAX	maximum
MBL	mobile
MBR	mechanical buffer register
MBR	member
MBT	main ballast tank
MC	manhole cover
MC	medicine cabinet
MC	momentary contact
MC	multiple-contact
MCD	metal-covered door
MCH	mail chutes
MCHRY	machinery
MCHST	machinist
MCM	magnetic core memory
MCSW	motor circuit switch
MCW	metal casement window
MDC	motor direct connected
MDGT	midget
MDL	middle
MDL	module
MDM	medium
MDN	median
MDRL	mandrel
MDS	minimum discernible signal
MDSE	merchandise
MEAS	measure
MECH	mechanical
MED	medical
MED	medium (altn abbr)
MED	microelectronic device
MEM	memory
MEMO	memorandum
MEP	mean effective pressure
MER	meridian
MET	metal
MET	metallurgical
METF	metal flashing
METG	metal grill
METP	metal partition
METRL	meteorological
MEZZ	mezzanine

MF	main feed
MF	microfilm
MF	motor field
MFD	manufactured
MFG	manufacturing
MFLRD	male flared
MFR	manufacture
MG	motor generator
MG	multigage
MGC	manual gain control
MGN	magneto
MGR	manager
MGS	metal gravel stop
MGT	management
MH	manhole
MHD	magnetohydrodynamics
MHE	materials handling equipment
MHT	mean high tide
MI	malleable iron
MI	mile
MIC	micrometer
MIC	microphone
MIC	monolithic integrated circuit
MICAM	microammeter
MICR	magnetic ink character recognition
MID	middle (altn abbr)
MIL	military
MILSPEC	military specification
MIL-STD	military standard (book)
MIN	minimum
MIN	minor
MIN	minute
MISC	miscellaneous
MIX	mixture
MJ	mastic joint
MK	mark
MKD	marked
MKGS	markings
MKR	marker
MKS	meter-kilogram-second
ML	material list
ML	mold line
ML	monolithic
MLD	molded
MLDG	molding
MLT	mean low tide
MMC	maximum material condition
MMF	magnetomotive force
MMOD	micromodule
MN	main
MNG	managing
MNL	manual
MNPWR	manpower
MNRL	mineral
MO	month
MO	motor operated
MOD	model
MOD	modification
MOD	modulator
MOL	molecule
MOL WT	molecular weight
MOM	momentary
MON	monitor
MON	monument
MORT	morse taper
MOST	metal-oxide-semiconductor transistor
MOT	motor
MP	melting point
MPG	miles per gallon
MPH	miles per hour
MPI	magnetic particle inspection
MPT	male pipe thread
MR	marble
MRD	metal rolling door
MS	manuscript
MS	military standard (sheet)
MSCR	machine screw
MSG	message
MSL	mean sea level
MSNRY	masonry
MST	machine steel
MSW	master switch
MT	mean tide
MT	mean time
MT	metal threshold
MT	mount
MTBF	mean time between failure
MTD	mounted
MTG	meeting
MTG	mounting
MTLC	metallic
MTN	motion
MTR	meter
MTTF	mean time to failure
MTZ	motorized
MU	multiple unit
MUF	muffler
MULT	multiple
MULTH	multilith
MULTR	multiplier
MUW	music wire
MV	muzzle velocity
MVBL	movable
MVG	moving
MVT	movement
MWG	Music Wire Gage
MWP	membrane waterproofing
MWV	maximum working voltage
MXFL	mixed flow
MXG	mixing
MXT	mixture

N

N	north
NA	not applicable
NA	not available
NAR	narrow
NAS	National Aircraft Standards
NAT	natural
NATL	national
NAUT	nautical
NAV	naval
NB	narrow band
NB	no bias (relay)
NBFU	National Board of Fire Underwriters
NBS	National Bureau of Standards
NC	National coarse (thread)
NC	no change
NC	no connection
NC	normally closed
NC	numerical control
NCH	notched
NCOMBL	noncombustible
NDL	needle
NDRO	nondestructive read-out
NDT	nondestructive testing
NEC	National Electrical Code
NEC	necessary
NECS	National Electrical Code Standards
NEF	National extra fine (thread)
NEG	negative
NEIL	neon indicating light
NEMA	National Electrical
NESC	National Electric Safety Code
NEUT	neutral
NF	National fine (thread)
NF	National formulary
NF	near face

NF	noise frequency
NFL	navy standard flange
NFSD	nonfused
NG	narrow gage
NG	no good
NGO	National gas outlet (thread)
NI SIL	nickel-silver
NIP	nipple
NIS	not in stock
NK	neck
NKL	nickel
NLNR	nonlinear
NM	noise meter
NM	nonmetallic
NMAG	nonmagnetic
NMI	nautical miles
NO	normally open
NO	number
NOI	not otherwise identified
NOM	nominal
NOMEN	nomenclature
NONFLMB	nonflammable
NONSTD	nonstandard
NONSYN	nonsynchronous
NOP	number of passes
NOR	notice of revision
NORM	normal
NORM	normalize
NOS	nor otherwise specified
NOV	November
NOZ	nozzle
NP	nickle plated
NPL	nameplate
NPN	negative-positive-negative (transistor)
NPRN	neoprene
NPS	Navy Primary Standards
NPSC	National straight pipe thread in pipe couplings
NPSF	National straight pipe thread for dry seal pressure tight joints
NPSH	National straight pipe thread for hose couplings and nipples
NPSL	National straight pipe thread for locknuts and locknut pipe threads
NPSM	National straight pipe thread for mech joints
NPT	National taper pipe thread
NPTF	National taper pipe thread for dry seal pressure tight joints
NPTR	National taper pipe thread for railing fixtures
NR	nonreactive (relay)
NR	nuclear reactor
NS	National special (thread)
NS	near side
NSS	Navy Secondary Standards
NST	nonslip tread
NTFY	notify
NTN	neutron
NTP	normal temperature and pressure
NTS	not to scale
NTWK	network
NTWT	net weight
NUC	nuclear
NUM	numeral, numerical
NWG	National Wire Gage
NWT	nonwatertight
NYL	nylon
O	
O	orange (altn abbr)
O&R	overhaul and repair
O/P	ozalid print
OA	overall
OBJ	object
OBJV	objective
OBL	oblique
OBS	obsolete
OBSV	observation
OC	on center
OC	overcurrent
OCB	oil circuit breaker
OCC	occupation
OCLD	oil cooled
OCLNR	oil cleaner
OCLR	oil cooler
OCR	optical character recognition
OCR	overcurrent relay
OCT	octagon
OCT	octane
OCT	October
OD	olive drab
OD	outside
OD	ouside diameter
ODP	open dripproof
ODPP	open dripproof protected
OE	open end
OEM	original equipment manufacturer
OF	outside face
OFCE	office
OFCL	official
OFR	overfrequency relay
OFS	offset
OFT	outfit
OGL	obscure glass
OGR	outgoing repeater
OGT	outgoing trunk
OGT	outlet gas temperature
OH	open hearth
OHM	ohmmeter
OI	oil immersed
OI	oil insulated
OIFC	oil-insulated fan-cooled
OIL	orange indicating light
OISC	oil-insulated self-cooled
OIWC	oil-immersed water-cooled
OLVL	oil level
OML	outside mold line
OMNI	omnidirection
OP	oil pump
OP	oilproof
OPA	opaque
OPLG	oil plug
OPN	operation
OPNG	opening
OPP	octal print punch
OPP	opposite
OPT	optical
OPT	optimum
OPT	optional (altn abbr)
OPTL	optional
OR	output register
OR	outside radius
ORD	order
ORD	ordnance
ORF	orifice
ORG	organic
ORG	organization
ORIENT	orientation
ORIG	origin
ORLY	overload relay
ORN	orange
ORN	ornament
OS	oil switch
OS	otherwise specified
OS&D	over, short and damage
OSC	oscillator
OSCG	oscillating

OSP	operating steam pressure
OSR	output shift register
OT	oiltight
OT&E	operational test and evaluation
OTR	outer
OUT	outgoing
OUT	outlet
OUT	output
OUT	outside
OUTBD	outboard
OV	open ventilated
OV	over
OVBD	overboard
OVFL	overflow
OVH	oval head
OVHD	overhead
OVHG	overhanging
OVHL	overhaul
OVLD	overload
OVP	oval point
OVSP	overspeed
OVTR	overtravel
OVV	overvoltage
OWF	optimum working frequency
OWGL	obscure wire glass
OXD	oxide
OXD	oxidized
OXY	oxygen
OZ	ounce
OZ IN	ounce-inch

P

3PH	three phase
P	page
P	pitch
P	plate (electron device)
P	pole
P&L	power and lighting
P&O	pickled and oiled
P-D	punch-die
P-P	peak to peak
P/I	production illustration
P/O	part of
PA	pending availability
PA	power amplifier
PA	pressure angle
PA	profile angle
PA	public address
PA	pulse amplifier
PABX	private automatic branch exchange
PAC	primary address code
PACE	performance and cost evaluation
PAM	pulse-amplitude modulation
PAN	panoramic
PAN	pantry
PANT	pantograph
PAR	parallel (altn abbr)
PARA	paragraph
PARK	parkerized
PARV	paravane
PASS	passage
PASS	passenger
PAT	patent
PATPEND	patent pending
PATT	pattern
PAX	private automatic exchange
PB	pull box
PB	push button
PB STA	push button station
PBD	paperboard
PBD	pressboard
PBX	private branch exchange
PC	parts catalog
PC	piece
PC	pitch circle
PC	printed circuit
PC&H	packing, crating and handling
PC MK	piece mark
PCB	power circuit breaker
PCB	printed circuit board
PCF	pounds per cubic foot
PCH	punch
PCHT	parchment
PCM	pulse-code modulation
PCP	program change proposal
PCT	percent
PCTM	pulse-count modulation
PD	pitch diameter
PD	potential difference
PD	pulse doppler
PDIO	photodiode
PDISCH	pump discharge
PDISPL	pump displacement
PDM	pulse-duration modulation
PDR	power directional relay
PDS	purchasing department specification
PE	probable error
PE	professional engineer
PEC	photoelectric cell
PED	pedestal
PEJ	premolded expansion joint
PELEC	photoelectric
PEN	penetrate
PEND	pending
PENT	pentode
PEP	planar epitaxial passivated (transistor)
PER	person
PERC	percussion
PERF	perforate
PERF	performed
PERI	perimeter
PERM	permanent
PERMB	permeability
PERP	perpendicular
PERS	personnel
PERSP	perspective
PERT	performance evaluation and review technique
PES	photoelectric scanner
PETRO	petroleum
PEU	polyeurethane
PF	power factor
PF	profile
PF	pulse frequency
PFD	preferred
PFM	power factor meter
PFM	pulse-frequency modulation
PG	pressure gage
PGMT	pigment
pH	hydrogen ion concentration
PH	phase
PH BRZ	phosphor bronze
PHEN	phenolic
PHM	phase meter
PHOC	photocopy
PHONO	phonograph
PHOS	phosphate
PHP	propeller horsepower
PHR	phrase
PHR	preheater
PHSP	phase-splitter
PHT	preheat
PHYS	physical
PI	point of impact
PI	point of intersection
PI	programmed instruction
PIL	purple indicating light
PIM	pulse interval modulation
PIN	pinion
PIN	position indicator

PIN	positive-intrinsic-negative (transistor)
PIX	picture
PJTN	projection
PJTR	projector
PK	pack
PK	peak
PKG	package
PKG	packing
PKT	pocket
PKWY	parkway
PL	parting line (castings)
PL	parts list
PL	pitch line
PL	place
PL	plain
PL	plate
PL	plug
PL	property line
PL	proportional limit
PLAS	plaster
PLAT	platinum
PLATF	platform
PLBLK	pillow block
PLCY	policy
PLD	pay load
PLD	plated
PLG	piling
PLGL	plate glass
PLGR	plunger
PLK	plank
PLL	pallet
PLMB	plumbing
PLR	pillar
PLRT	polarity
PLS	pulse
PLSN	pulsation
PLSTC	plastic
PLT	pilot
PLT	procurement lead time
PLTG	plating
PLTR	plotter
PLVRZD	pulverized
PLYPH	polyphase
PLYWD	plywood
PLZ	polarize
PLZD	polarized
PLZN	polarization
PM	permanent magnet (speaker)
PM	phase modulation
PM	pilot motor
PM	preventive maintenance
PMD	program monitoring and diagnosis
PME	precision measuring equipment
PMFLT	pamphlet
PMP	pump
PMT	payment
PMVR	prime mover
PN	part number
PNEU	pneumatic
PNH	pan head
PNL	panel
PNP	positive-negative-positive (transistor)
PNT	paint
PO	post office
PO	power oscillator
PO	production order
PO	purchase order
POI	program of instruction
POL	petroleum, oils and lubricants
POL	polar
POL	polish
POLTHN	polyethylene (insul)
POLEST	polyester
PORC	porcelain
PORM	plus or minus
PORT	portable
POS	position (altn abbr)
POS	positive
POSN	position
POT	potential
POT	potentiometer
PP	piping
PP	power plant
PP	pressureproof
PP	push-pull
PPG	pipe plug
PPI	plan position indicator
PPLN	pipeline
PPM	parts per million
PPM	pulse-position modulation
PPR	paper
PPS	pulses per second
PPT	poppet
PPT	precipitate
PPT	product positioning time
PR	pair
PR	pipe rail
PR	pressure ratio
PR	purchase request
PRAC	practice
PRBNT	prebent
PRC	pierce
PRC	point of reverse curve
PRCH	precharge
PRCMT	procurement
PRCN	precision
PRCS	process
PRCST	precast
PREAMP	preamplifier
PRED	prediction
PREF	prefocused
PREFAB	prefabricated
PREFMD	preformed
PRELIM	preliminary
PREP	preparation
PRESS	pressure
PREV	previous
PRF	proof
PRF	pulse-repetition frequency
PRFM	performance
PRFRD	proofread
PRFT	press fit
PRGM	program
PRGMR	programmer
PRI	primary
PRI	priority
PRIN	principal
PRL	parallel
PRM	prime
PRMLD	premolded
PRMTR	parameter
PRNTG	printing
PROB	probability
PROB	problem
PROC	procedure
PROC	proceedings
PROC	procurement (altn abbr)
PROCU	processing unit
PROD	product
PROD	production
PROF	professional
PROG	progress
PROJ	project
PROP	propeller
PROP	property
PROT	protective
PROTR	protractor
PROV	provision
PRP	purple
PRPHL	peripheral
PRPLN	propulsion
PRPNE	propane
PRPSD	proposed
PRPSL	proposal
PRR	pulse-repetition rate
PRS	press
PRSD	pressed

PRSRZ	pressurize
PRSTC	prosthetic
PRT	pulse-repetition time
PRV	pressure reducing valve
PS	polystyrene
PS	process specification (number)
PSF	pound-force per square foot
PSI	pound-force per square inch
PSIA	pound-force per square inch absolute
PSIG	pound-force per square inch gage
PSIV	passive
PSL	pipe sleeve
PSL	pressure seal
PSM	prism
PSNR	positioner
PST	point of spiral tangent
PSTL	postal
PSU	power supply unit
PSV	preserve
PSVT	passivate
PSW	potential switch
PSWBD	power switchboard
PT	part
PT	pint
PT	pipe tap
PT	pneumatic tube
PT	point
PT	point of tangency
PT	potential transformer
PT	pulse time
PTD	painted
PTD	printed
PTFE	polytetrafluorethylene
PTL	pintle
PTM	pulse-time modulation
PTN	partition
PTO	power takeoff
PTR	printer
PTY	party
PU	pickup
PU	power unit
PUB	publication (altn abbr)
PUBN	publication
PUL	pulley
PULV	pulverizer
PUR	purifier
PURCH	purchase
PV	plan view
PVC	polyvinyl chloride
PVT	pivot
PVT	pressure-volume-temperature
PW	pulse width
PWB	printed wiring board
PWC	process water cooler
PWD	powder
PWM	pulse width modulation
PWR	power
PWRH	powerhouse
PYR	pyramid

Q

Q&A	question and answer
QA	quality assurance
QA	quick-acting
QB	quick break
QC	quality control
QDISC	quick disconnect
QDRNT	quadrant
QIK	quick
QOD	quick-opening device
QR	quire
QRY	quarry
QT	quart
QTR	quarter
QTY	quantity
QTZ	quartz
QUAD	quadrangle
QUAD	quadrant (altn abbr)
QUAL	qualitative
QUANT	quantitative
QUEST	question
QUOT	quotation

R

1/4RD	quarter round
R	radius
R	ratio
R	red
R	right
R&D	research and development
R-W	read-write (head)
R/W	right of way
RA	rayon (insul)
RA	reduction of area
RAACT	radioactive
RACT	reverse acting
RAD	radius
RAD HAZ	radiation hazard
RAD INSP	radiographic inspection
RAD OPR	radio operator
RADL	radiological
RADN	radiation
RALW	radioactive liquid waste
RB	relay block
RB	rubber base
RBE	relative biological effectiveness
RBN	radio beacon
RBN	ribbon
RBR	rubber
RC	reinforced concrete
RC	remote control
RC	resistance-capacitance
RCC	reader common contact
RCC	rubber-covered cable
RCD	record
RCD	reverse current device
RCDG	recording
RCDR	recorder
RCHT	ratchet
RCL	recall
RCN	recreation
RCNDT	recondition
RCO	reactor core
RCP	reinforced concrete pipe
RCP	right-circular polarization
RCPT	receipt
RCPT	receptacle
RCPTN	reception
RCTL	resistor-capacitor-transistor logic
RCTN	reaction
RCVD	received
RCVG	receiving
RCVR	receiver
RCVY	recovery
RD	read
RD	register drive
RD	road
RD	roof drain
RD	root diameter
RD	round (altn abbr)
RDC	radiac
RDCN	reduction
RDCR	reducer
RDF	radio direction finding
RDG	ridge
RDH	round head
RDL	radial
RDM	recording demand meter
RDNG	reading
RDOUT	readout
RDR	reader
RDTR	radiator

Abbreviation	Meaning
RDY	ready
REAC	reactance
REAC	reactor
REC	recess, recessed
RECIP	reciprocating
RECIRC	recirculate
RECM	recommend
RECP	receptacle (altn abbr)
RECT	rectangular
RECT	rectifier
REDSG	redesignate
REDWM	redrawm
REF	refer
REF	reference
REF	refinery
REFL	reference line
REFL	reflector
REFR	refrigerator
REG	register (altn abbr)
REG	regular (altn abbr)
REG	regulator (altn abbr)
REGEN	regenerative
REIMB	reimburse
REINF	reinforce
REJ	reject
REL	relative
REL	release (altn abbr)
RELOC	relocated
REM	removable
REORG	reorganize
REPL	replace
REPRO	reproduction
REQ	request
REQD	required
REQN	requisition
REQT	requirement
RES	research
RES	resistance
RES	resistor
RES FLD	residual field
RESC	rescind
RESID	residual
RESIL	resilient
RESN	resonant
RESP	respectively
RESP	responsible
RESTR	restorer
RETP	retape
RETR	retractable
RETRO	retroactive
RETRO	retrograde
RETROG	retrogressive
REV	review
REV	revise
REV	revision
REV	revolution
REWK	rework
RF	radio frequency
RF	raised face
RFC	radio-frequency choke
RFG	roofing
RFGT	refrigerant
RFI	radio-frequency interference
RFLX	reflex
RFND	refined
RFP	request for proposal
RFRC	refractory
RFS	regardless of feature size
RGD	rigid
RGH	rough
RGLR	regular
RGLT	regulating
RGLTR	regulator
RGN	region
RGNG	rigging
RGTR	register
RH	relative humidity
RH	right hand
RH	Rockwell hardness
RH DR	right-hand drive
RHEO	rheostat
RI	reflective insulation
RIL	red indicating light
RIOT	real-time input-output transducer
RIV	river
RK	rack
RKR	rocker
RL	resistance-inductance
RLC	resistance-inductance-capacitance
RLD	rolled
RLF	relief
RLG	railing
RLR	roller
RLR BRG	roller bearing
RLSE	release
RLV	relieve
RLY	relay
RM	raw material
RM	room
RMR	reamer
RMS	root mean square
RMT	remote
RND	round
RNDM	random
RNG	range
RNWBL	renewable
RO	rough opening
RO	runout
ROT	rotate
ROUT	routine
RPDR	reproducer
RPLSN	repulsion
RPM	revolutions per minute
RPR	repair
RPRT	report
RPS	revolutions per second
RPT	repeat
RPTR	repeater
RPVNTV	rust preventative
RR	railroad
RR	relief radii
RR	roll roofing
RS	right side
RSD	raised
RSD	rolling steel door
RSPTR	respirator
RSPV	respective
RSQ	resque
RST	reinforcing steel
RSTG	roasting
RSTPF	rustproof
RSVR	reservoir
RT	radio telephone
RT	raintight
RT	rate
RT	real time
RT	receiver-transmitter
RTC	reader tape contact
RTCL	reticle
RTD	resistance temperature detector
RTE	route
RTG	rating
RTL	resistor-transistor logic
RTM	recording tachometer
RTN	return
RTNG	retaining
RTNR	retainer
RTR	rotor
RTRSW	rotary switch
RTRY	rotary
RU	reproducing unit
RUB	rubber (altn abbr)
RUD	rudder
RV	rear view
RV	relief valve
RVLG	revolving
RVM	reactive voltmeter
RVS	reverse
RVSBL	reversible
RVT	rivet
RVTD	riveted
RWG	Roebling Wire Gage
RWND	rewind

RWY	runway
RY	railway
RYN	rayon

S

S	south
S2S	surfaced or dressed two sides
S/R	send and receive
SA	spectrum analyzer
SA	stress anneal
SAC	sprayed acoustical ceiling
SAE	Society of Automotive Engineers
SAF	safety
SAL	salinity
SALV	salvage
SAN	sanitary
SAR	storage address register
SAT	saturate
SATC	suspended acoustical-tile ceiling
SB	service bulletin
SB	sideband
SB	silver braze
SB	sleeve bearing
SB	stove bolt
SB	stuffing box
SBK	single-beam klystron
SBM	submersible
SBSTR	substrate
SBU	silver brazing union
SC	scale
SC	single contact
SC	smooth contour
SCC	single conductor cable
SCD	screen door
SCDP	steel cadmium plated
SCE	source
SCH	socket head
SCHED	schedule
SCHEM	schematic
SCI	science
SCINT	scintillator
SCLER	scleroscope
SCNG	scanning
SCNR	scanner
SCR	screw
SCR	semiconductor controlled rectifier
SCR	silicon controlled rectifier
SCR GT	screen gate
SCRDB	screwed bonnet
SCRN	screen
SCT	structural clay tile
SCTR	sector
SCTY	security
SD	shower drain
SD BL	sand blast
SDG	siding
SDL	saddle
SDR	sender
SE	single end
SE	special equipment
SEC	second
SEC	secondary
SECT	section
SEG	sealing
SEG	segment
SEJ	sliding expansion joint
SEL	select
SEL	selector
SELS	selsyn
SEMICOND	semiconductor
SEN	semienclosed
SENS	sensitive
SEP	separator
SEP	September
SEQ	sequence
SER	serial
SER	series
SERNO	serial number
SERR	serrated
SERV	service (altn abbr)
SERVO	servomechanism
SETLG	settling
SEW	sewage
SF	semifinished
SF	slip fit
SF	spot face
SFB	solid fiberboard
SFC	specific fuel consumption
SFDR	single feeder
SFGD	safeguard
SFO	spot face other side
SFT	shaft
SG	structural glass
SGD	signed
SGL	single
SGW	security guard window
SH	scleroscope hardness
SH	shackle
SH	sheet
SH	sheeting
SH	shield (electron device)
SH	shower
SH	shunt
SH ABS	shock absorber
SH BL	shot blast
SH TR	shunt trip
SH & T	shower and toilet
SHCR	shipping container
SHF	shift
SHF	super-high frequency
SHK	shank
SHK	shock
SHL	shell
SHL	shellac
SHLD	shield, shielding
SHLDR	shoulder
SHP	shaft horsepower
SHPBD	shipboard
SHPNG	shipping
SHPRF	shakeproof
SHPT	shipment
SHT	short
SHT DN	shut down
SHTG	shortage
SHTHG	sheathing
SHV	sheave
SI	salinity indicator
SI	standing instruction
SI	International System of Units (Systeme International d'Unites)
SIF	single face
SIF	selective identification feature
SIL	silence
SIL	silver
SILS	silver solder
SIM	similar
SIM	simulator
SIMP	specific impulse
SIT	situation
SJ	slip joint
SK	sink
SK	sketch
SKD	skilled
SKT	socket
SL	sea level
SL	sliding
SLBL	soluble
SLD	sealed
SLD	sliding door
SLDR	solder
SLFLKG	self-locking
SLFSE	self-sealing
SLFTPG	self-tapping
SLGR	slinger
SLSP	slow speed
SLT	self-loading tape
SLT	skylight

SLT slate
SLTD slotted
SLV sleeve
SLVT solvent
SLYP shortleaf yellow pine
SM small
SM smooth
SMLS seamless
SMY summary
SN siren
SND sound
SNDG sending
SNDPRF soundproof
SNL standard nomenclature list
SNLG signaling
SNO stock number
SNR signal-to-noise ratio
SNSR sensor
SNTR sinter
SO shop order
SO stock order
SO stop order
SOL solenoid
SOL solid
SOLN solution
SOLV solenoid valve
SOV shutoff valve
SP single-pole
SP soil pipe
SP space
SP spare
SP special purpose
SP specific
SP speed
SP splashproof
SP spool
SP standpipe
SP GR specific gravity
SP HT specific heat
SP PH split phase
SP SW single-pole switch
SP VOL specific volume
SPCL special
SPCR spacer
SPDL spindle
SPDOM speedometer
SPDT single-pole double throw
SPEC specification
SPEC specimen
SPFT desk, single-pedestal flat-top
SPH space heater
SPHER spherical
SPK spark
SPKR speaker
SPL special (altn abbr)
SPL spiral
SPLC splice
SPLN spline
SPLTR splitter
SPLX simplex
SPLY supply
SPM strokes per minute
SPNR spanner
SPNSN suspension
SPR spring
SPR sprinkler
SPRKT sprocket
SPRT support
SPST single-pole single throw
SPTW desk, single-pedestal typewriter
SQ square
SQCG squirrel cage
SQH square head
SQW square wave
SR selenium rectifier
SR senior
SR shift register
SR slip ring
SR slow release (relay)
SR stateroom
SRCH search
SRLY series relay
SRVL survival
SS semisteel
SS service sink
SS straight shank
SSB single sideband
SSBO single swing blocking oscillator
SSCR setscrew
SSCRN silkscreen
SSD subsoil drain
SSF saybolt seconds furol
SSK soil stack
SSR slate-shingle roof
SSS solid-state switching
SST stainless steel
SSTU seamless steel tubing
SSU saybolt seconds universal
ST sawtooth
ST single-throw
ST steam
ST store
ST street
ST BRZ statuary bronze
ST PR static pressure
STA station
STA stationary
STAB stabilizer
STAN stanchion
START starter
STBD starboard
STBLN stabilization
STBSCP stroboscope
STBY standby
STC sensitivity time control
STD standard
STDF standoff
STDY steady
STDZN standardization
STER sterilizer
STFG stuffing
STG stage
STG starting
STGEN steam generator
STGR stringer
STIF stiffener
STIR stirrup
STK stack
STK stock
STL steel
STL studio-transmitter link
STLR semitrailer
STLT stellite
STN stone
STNG sustaining
STNLS stainless
STOR storage
STP stamp
STPD stripped
STPR stepper
STPR stripper
STR straight
STR strainer
STR stream
STR strength
STRBK strongback
STRD stranded
STRG steering
STRG strong
STRK stroke
STRL structural
STRLN streamline
STRM storeroom
STRUCT structure
STTR stator
STU step-up
STW storm water
STWP steam working pressure
STWY stairway
SUBASSY subassembly
SUBJ subject
SUBMG submerged

SUBMIN subminiature
SUBQ subsequent
SUBST substitute
SUBSTA substation
SUBSTR substructure
SUBTR subtraction
SUCT suction
SUDT silicon unilateral diffused transistor
SUF sufficient
SUFF suffix
SUPERHET superheterodyne
SUPERSTR superstructure
SUPHTR superheater
SUPPL supplement
SUPPR suppressor
SUPSD superseded
SUPT supersensitive
SUPV supervise
SURF surface
SURG surgical
SURV survey
SUSP suspend
SV safety valve
SV self-ventilated
SV side view
SVCE service
SVMTR servomotor
SVO servo
SW salt water
SW short wave
SW spot weld
SW switch
SWB short wheel base
SWBD switchboard
SWG British Standard Wire Gage
SWG Stubs-Wire Gage
SWG swage
SWG swing
SWGD swinging door
SWGR switchgear
SWOV switchover
SWP safe working pressure
SWP sweep
SWR standing wave ratio (voltage)
SWR switch rails
SWS switch stand
SWSG security window screen and guard
SWT sweat
SWT switch ties
SWVL swivel
SYM symbol
SYMM symmetrical
SYMP symposium
SYN synchronous
SYNC synchronizer
SYNTH synthetic
SYS system

T

T tee
T teeth
T toilet
T&G tongue and groove
T/B title block
TAB tabulate
TAC tactical
TACH tachometer
TAF trim after forming
TARP tarpaulin
TAS true air speed
TB technical bulletin
TB terminal board
TBD to be determined
TBE threaded both ends
TBG tubing
TBL table
TBLR tumbler
TBO time between overhaul
TC tape core
TC temperature compensating
TC terra cotta
TC thermocouple
TC time constant
TC timed closing
TC top chord
TC trip coil
TCC turbine close coupled
TCD thyratron core driver
TCG tune-controlled gain
TCH temporary construction hole
TCU tape control unit
TD temperature differential
TD time delay
TD transmitter distributor
TDC time-delay closing
TDCM transistor driver core memory
TDG twist drill gage
TDL tunnel-diode logic
TDM tandem
TDO time-delay opening
TDX thermal demand transmitter
TE thermal element
TE totally enclosed
TE trailing edge
TECH technical
TECH technician
TEFC totally enclosed fan cooled
TEJ transverse expansion joint
TEL telephone
TELB telephone booth
TEM temper
TEMP temperature
TEMP temporary
TEMPL template
TER terrazzo
TER tertiary
TERM terminal
TERR territory
TET tetrode
TF thin film
TF thread forming
TF tile floor
TFL Teflon (insul)
TFT thin-film transistor
TGC transmit gain control
TGL toggle
TGT tail gate
TGT target
TH thermoid
THC thermal converter
THD thread
THD total harmonic distortion
THEOR theoretical
THERM thermometer
THERMO thermostat
THK thick
THKNS thickness
THMS thermistor
THP thrust horsepower
THPFB treated hard-pressed fiberboard
THR thrust
THRM thermal
THRMSTC thermostatic
THROT throttle
THRT throat
THRU through
THTR theater
THYR thyristor
TIC temperature indicating controller
TIM time meter
TIP turbine inlet pressure
TIR total indicator reading
TIT turbine inlet temperature

TJP	turbo-jet propulsion
TK	tank
TKR	tanker
TKT	ticket
TL	thrust line
TLG	telegraph
TLG	tilting
TLK	test link
TLLD	total load
TLM	telemeter
TLMY	telemetry
TLU	table lookup
TLY	tally
TM	technical manual
TM	temperature meter
TM	tone modulation
TM	training manual
TM	transverse magnetic
TMB	tumble
TMBR	timber
TMD	timed
TMG	timing
TMPD	tempered
TMR	timer
TMS	time and motion study
TMS	type model series
TMX	telemeter transmitter
TN	technical note
TN	thermonuclear
TN	tone
TN	train
TN	true north
TND	tinned
TNG	tongue
TNG	training
TNL	tunnel
TNLDIO	tunnel diode
TNR	thinner
TNSL	tensile
TNSN	tension
TNTV	tentative
TO	takeoff
TO	technical order
TO	time opening
TOB BRZ	tobin bronze
TOE	thread one end
TOL	tolerance
TOPG	topping
TOPO	topography
TOR	totalizing relay
TORNL	torsional
TOT	total
TOT	turbine outlet temperature
TP	tie plate
TP	timing point
TP	true position
TP	true profile
TP DR	tape drive
TPG	tapping
TPI	tape phase inverter
TPI	teeth per inch
TPI	threads per inch
TPL	triple
TPM	telemetry processor module
TPR	taper
TPR	teleprinter
TPS	task parameter synthesizer
TPSN	transposition
TPT	tappet
TPTG	tuned plate tuned grid
TPTOL	true position tolerance
TR	technical report
TR	trace
TR	transmitter receiver
TR	transom
TR	truss
TRAJ	trajectory
TRANS	transfer (altn abbr)
TRANS	transparent
TRANSP	transportation
TRANSV	transverse
TRB	treble
TRCR	tracer
TRD	tread
TRD	turbine reduction drive
TRF	tuned radio frequency
TRFC	traffic
TRG	trailing
TRH	truss head
TRI	triode
TRIB	tributary
TRIG	trigger
TRK	track
TRK	truck
TRK	trunk
TRKG	tracking
TRLR	trailer
TRLY	trolley
TRMR	trimmer
TRNBKL	turnbuckle
TRND	turned
TRNGL	triangle
TRNPS	transpose
TRNR	trainer
TRNSN	transition
TRNTBL	turntable
TROPR	troposphere
TRP	tripped
TRPL	terneplate
TRQ	torque
TRSBR	transcriber
TRSN	torsion
TRTD	treated
TRTMT	treatment
TRUN	trunnion
TRVLG	traveling
TRX	triplex
TS	taper shank
TS	Telegraph System
TS	tensile strength
TS	time switch
TS	tool steel
TS	tube sheet
TSL	typesetting lead
TSTG	testing
TSTR	tester
TSTRZ	transistorized
TSW	temperature switch
TT	teletype
TT	total time
TT	triple thermoplastic (insul)
TTL	transistor-transistor logic
TTNG	tightening
TTY	teletypewriter
TUN	tuning
TUNG	tungsten
TUR	turret
TURB	turbine
TVG	time variation of gain
TVM	tachometer voltmeter
TVM	transistor voltmeter
TW	temperature well
TW	traveling wave
TW	twisted
TW	typewriter
TWMBK	traveling-wave multiple-beam klystron
TWR	tower
TWSB	twin sideband
TWT	traveling-wave tube
TWX	teletypewriter exchange
TWY	taxiway
TXTL	textile
TYP	typical

U

U&L	upper and lower
U/O	used on
U/W	used with
UAR	use as required
UBL	unbleached
UC	unit cooler

UCL	upper control limit
UF	ultrasonic frequency
UFN	until further notice
UFR	underfrequency relay
UGND	underground
UH	unit heater
UHF	ultrahigh frequency
UJT	unijunction transistor
UK	unit check
ULI	Underwriters Laboratories, Inc.
ULT	ultimate
UMB	umbilical
UMUS	unbleached muslin
UN	unified
UN	union
UN	unit
UNAUTHD	unauthorized
UNB	union bonnet
UNC	unified coarse (thread)
UNCOND	unconditionally
UNCTD	uncoated
UND	under
UNDC	undercurrent
UNDEF	undefined
UNDETM	undetermined
UNDF	underfrequency
UNDV	undervoltage
UNEF	unified extra fine (thread)
UNF	unified fine (thread)
UNFIN	unfinished
UNIF	uniform
UNIV	universal
UNK	unknown
UNL	unloading
UNLIM	unlimited
UNS	unified special (thread)
UOS	unless otherwise specified
UPDFT	updraft
UPR	upper
UPS	uninterrupted power supply
UPWD	upward
UR	unsatisfactory report
UR	urinal
UR	utility room
URG	urgent
URT	upright
US	undersize
US	Uniform System (lens making)
USAS	USA Standard
USASI	United States of America Standards Institute (now ANSI)
USB	upper sideband
USC&GS	United States Coast and Geodetic Survey
USG	United States Gage
USGS	United States Geological Survey
USP	United States Pharmacopoeia
USS	United States Standard
UTIL	utility
UTN	utensil
UTRTD	untreated
UTS	universal time standards
UV	ultraviolet
UVD	undervoltage device
UWT	unit weight
UWTR	underwater

V

V	valve
V	violet (altn abbr)
V	volt
VAC	vacant
VAC	vacuum
VAC	volt, alternating current
VAL	valley
VAL	value
VAM	voltammeter
VAP PRF	vaporproof
VAR	variable
VAR	variance
VARN	varnish
VARR	visual aural radio range
VB	valve box
VC	vitrified clay
VC	voice coil
VCB	vertical location of center of buoyancy
VCG	vertical location of center of gravity
VCL	vertical center line
VCO	voltage controlled oscillator
VCT	vitrified clay tile
VCT	voltage control transfer
VD	vandyke
VD	vault door
VD	void
VD	voltage drop
VDC	volt, direct current
VDET	voltage detector
VDI	vertical display indicator
VE	ventilating equipment
VEG	vegetable
VEH	vehicle
VEL	vellum
VEL	velocity
VENT	ventilate
VENT	ventilator
VERB	verbatim
VERIF	verification
VERN	vernier
VERT	vertical
VEST	vestibule
VF	variable frequency
VF	voice frequency
VFL PT	variable floating point
VFO	variable frequency oscillator
VH	vent hole
VH	Vickers hardness
VHF	very high frequency
VI	viscosity index
VIB	vibrate, vibration
VID	video
VIDAMP	video amplifier
VIDF	video frequency
VIL	village
VIMP	vertical impulse
VINT	video integration
VIO	violet
VIP	visual identification point
VIS	visual
VISC	viscosity
VISMR	viscometer
VIT	vitreous
VIZ	namely
VLF	very low frequency
VLMTRC	volumetric
VLR	very long range
VLT	volute
VM	velocity modulation
VM	voltmeter
VNXL	vaneaxial
VO	voice
VOL	volume
VOM	volt ohm milliammeter
VP	vent pipe
VPR	vaporize
VR	voltage regulator
VRFY	verify
VRIS	varistor
VRL	vertical reference line
VRLY	voltage relay
VRYG	varying
VS	vapor seal
VS	vent stack
VS	versus

VS	voltmeter switch
VSB	vestigial sideband
VSBL	visible
VST	valve seat
VSTM	valve stem
VSWR	voltage standing wave ratio
VT	vacuum tube
VT	vaportight
VT	voice tube
VTM	voltage-tunable magnetron
VTR	video tape recorder
VTVM	vacuum-tube voltmeter
VTX	vertex

W

1/W	one way
W	watt
W	west
W	white (altn abbr)
W	wide
W&M GA	Washburn and Moen Gage
W/	with (comb form)
W/D	withdrawn
W/O	without
W/W	winding to winding
WAF	width across flats
WAG	wagon
WARR	warranty
WB	wet bulb
WB	wheel base
WB	wide band
WB	work bench
WBF	wood block floor
WBG	webbing
WBR	word buffer register
WC	water closet
WCHR	water chiller
WCLD	water-cooled
WCR	watercooler
WD	watt demand meter
WD	wind
WD	wood
WD	wood door
WD	word
WD	work day
WDF	wood door and frame
WDF	woodruff
WDG	winding
WDO	window
WEA	weather
WEAT	weathertight
WF	wash fountain
WFR	wafer
WFS	wood furring strips
WG	waveguide
WG	wedge
WG	window guard
WG	wing
WH	wall hydrant
WH	water heater
WH	watt-hour
WHL	wheel
WHM	watt-hour meter
WHSE	warehouse
WHT	white
WI	wrought iron
WIL	white indicating lamp
WJ	water jacket
WK	week
WK	work
WKG	working
WKR	wrecker
WKS	workshop
WL	water line
WL	wavelength
WL	white light
WL	wind load
WL	wire list
WLD	welded
WLDMT	weldment
WLDR	welder
WM	water meter
WM	wattmeter
WM	wire mesh
WMGR	worm gear
WN	winch
WND	wound
WP	waste pipe
WP	water pump
WP	weatherproof (insul)
WP	wet process
WPFC	waterproof fan cooled
WPG	waterproofing
WPG	wiping
WPM	words per minute
WPR	working pressure
WR	wall receptacle
WR	washroom
WR	weather resistant
WR	wrench
WR	writer
WRG	wiring
WRK	wrecking
WRN	warning
WRT	wrought
WS	water surface
WS	weather stripping
WS	wrought steel
WSHG	washing
WSHLD	windshield
WSHR	washer
WSL	weather seal
WSP	working steam pressure
WSR	wood-shingle roof
WT	weight
WTG	waiting
WTHPRF	weatherproof
WTRTT	watertight
WTRZ	winterize
WU	window unit
WU	work unit
WV	wall vent
WV	working voltage
WW	wire way
WW	wire wound

X

XARM	cross arm
XBAR	crossbar
XCVR	transceiver
XDCR	transducer
XFMR	transformer
XFR	transfer
XHAIR	crosshair
XHVY	extra heavy
XLTR	translator
XLWB	extra-long wheel base
XMSN	transmission
XMT	transmit
XMTD	transmitted
XMTG	transmitting
XMTR	transmitter
XPL	explosive
XSECT	cross section
XSTR	extra strong
XSTR	transistor
XTAL	crystal
XTLO	crystal oscillator
XX STR	double extra strong

Y

Y	yellow (altn abbr)
YD	yard
YEL	yellow
YIL	yellow indicating lamp
YLT	yellow light
YP	yield point
YR	year
YS	yield strength

Z

Z	zone
ZA	zero adjuster

Basic Weld Symbols and Their Location Significance

LOCATION SIGNIFICANCE	FILLET	PLUG OR SLOT	SPOT OR PROJECTION	SEAM	{FLASH OR UPSET} SQUARE	GROOVE V	GROOVE BEVEL	GROOVE U	GROOVE J
ARROW SIDE									
OTHER SIDE									
BOTH SIDES		NOT USED	NOT USED	NOT USED					
NO ARROW SIDE OR OTHER SIDE SIGNIFICANCE	NOT USED	NOT USED			NOT USED EXCEPT FOR FLASH OR UPSET WELDS	NOT USED	NOT USED	NOT USED	NOT USED

Supplementary Symbols

WELD ALL AROUND	FIELD WELD	MELT-THRU	CONTOUR FLUSH	CONTOUR CONVEX	CONTOUR CONCAVE

Location of Elements of a Welding Symbol

Typical Welding Symbols

BACK OR BACKING WELD SYMBOL

SURFACING WELD SYMBOL INDICATING BUILT-UP SURFACE

DOUBLE FILLET WELDING SYMBOL

CHAIN INTERMITTENT FILLET WELDING SYMBOL

STAGGERED INTERMITTENT FILLET WELDING SYMBOL

SINGLE-V-GROOVE WELDING SYMBOL

Supplementary Symbols Used with Welding Symbols

WELD-ALL-AROUND SYMBOL

FIELD WELD SYMBOL

MELT-THRU SYMBOL

Basic Joints—Identification of Arrow Side and Other Side of Joint and

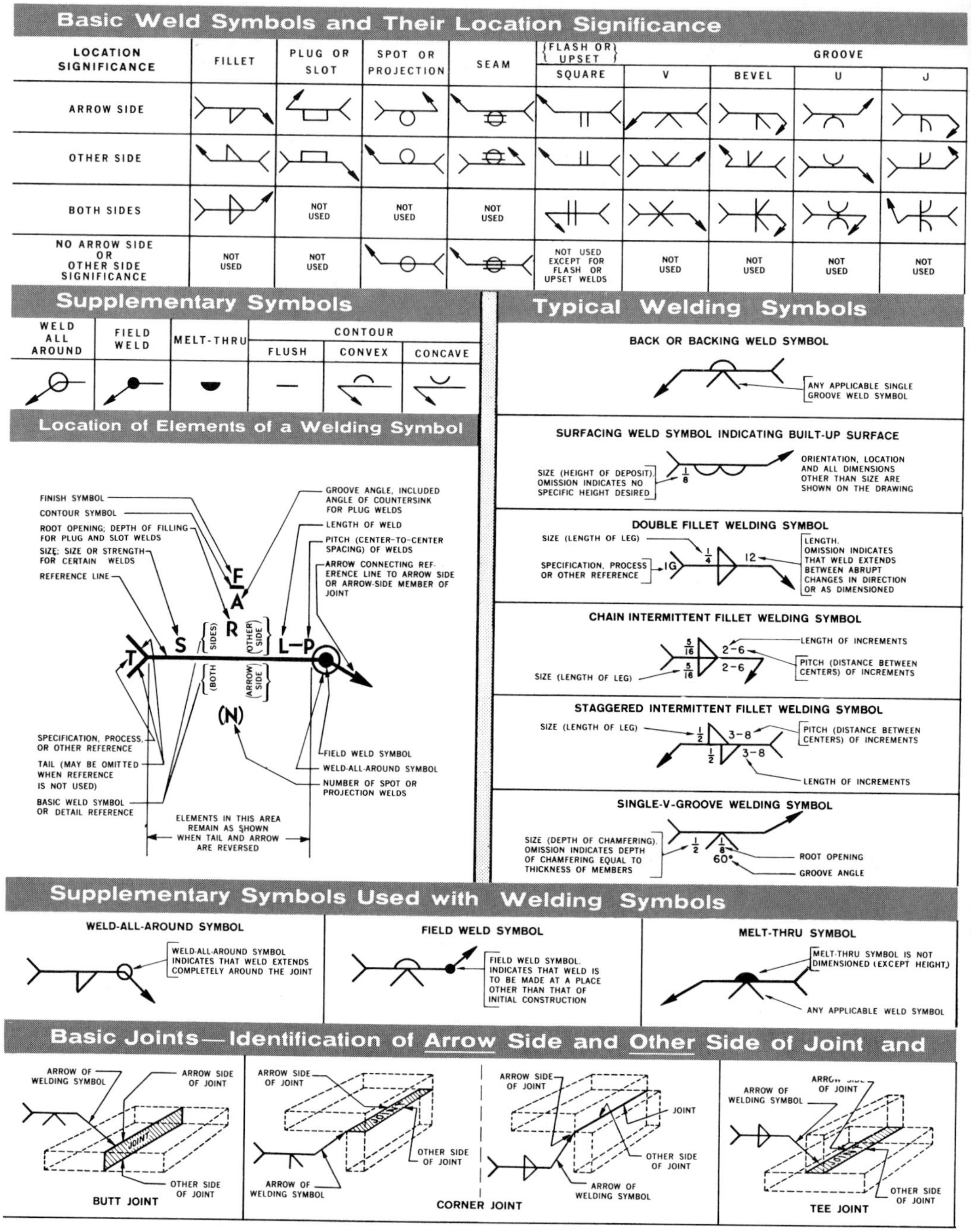

DESIGNATION OF WELDING PROCESSES BY LETTERS

CAW ... Carbon-Arc Welding
CW ... Cold Welding
DB ... Dip Brazing
DFW ... Diffusion Welding
EBW ... Electron Beam Welding
EW ... Electroslag Welding
EXW ... Explosion Welding
FB ... Furnace Brazing
FCAW ... Flux Cored Arc Welding
FOW ... Forge Welding
FRW ... Friction Welding

FW ... Flash Welding
GMAW ... Gas Metal-Arc Welding
GTAW ... Gas Tungsten-Arc Welding
IB ... Induction Brazing
IRB ... Infrared Brazing
IW ... Induction Welding
LBW ... Laser Beam Welding
OAW ... Oxyacetylene Welding
OHW ... Oxyhydrogen Welding
PAW ... Plasma-Arc Welding
PEW ... Percussion Welding
PGW ... Pressure Gas Welding

RB ... Resistance Brazing
RPW ... Projection Welding
RSEW ... Resistance-Seam Welding
RSW ... Resistance-Spot Welding
SAW ... Submerged Arc Welding
SMAW ... Shielded Metal-Arc Welding
SW ... Stud Welding
TB ... Torch Brazing
TW ... Thermit Welding
USW ... Ultrasonic Welding
UW ... Upset Welding

STANDARD WELDING SYMBOLS

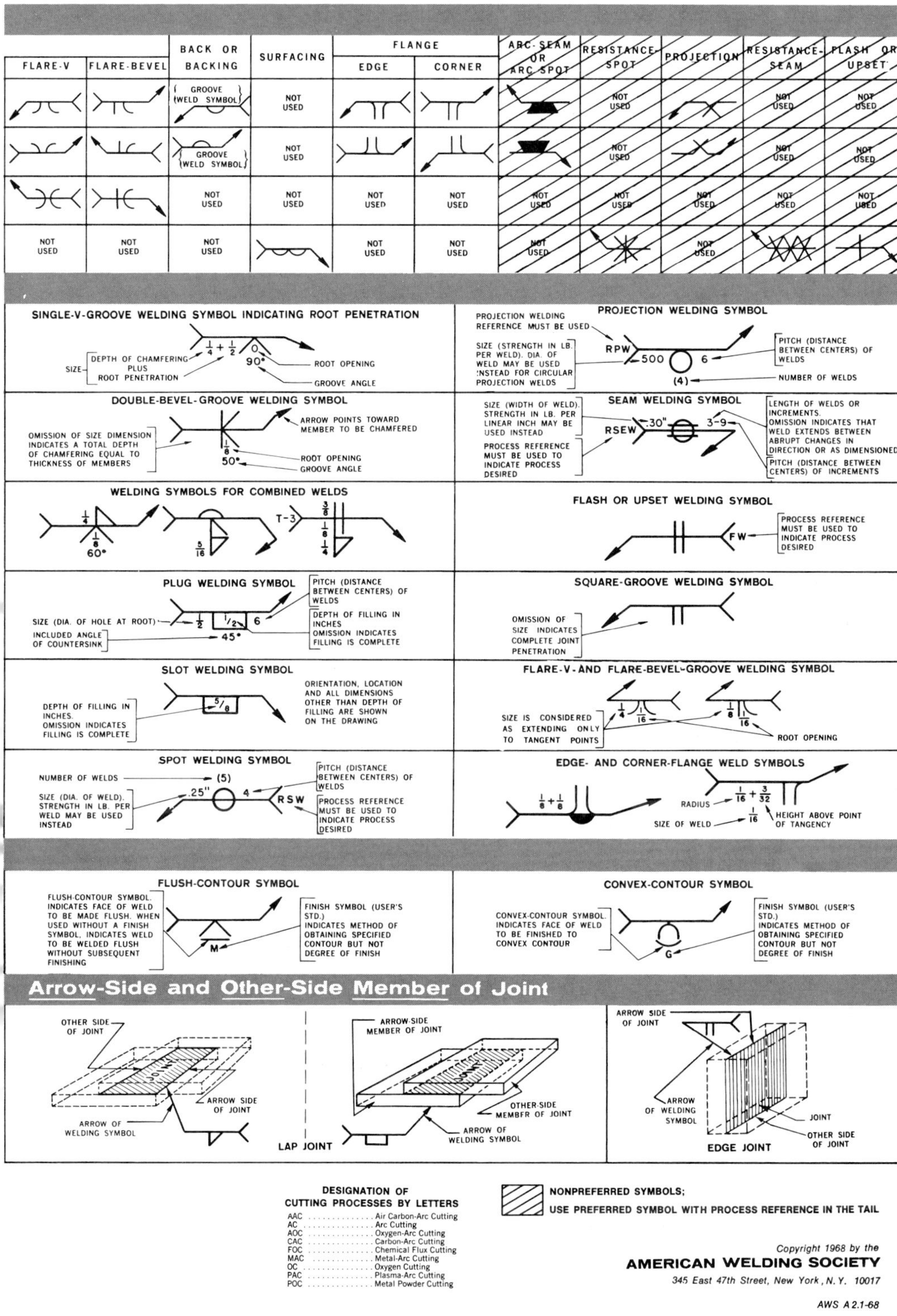

GRAPHICAL SYMBOLS FOR PIPE FITTINGS & VALVES

	FLANGED	SCREWED	BELL & SPIGOT	WELDED	SOLDERED
JOINT					
CONNECTING PIPE					
EXPANSION					
UNION					
REDUCER					
ELBOW					
90-DEGREE					
TURNED DOWN					
TURNED UP					
TEE					
(OUTLET UP)					
(OUTLET DOWN)					
CROSS					
GLOBE VALVE					
GATE VALVE					
COCK					
CHECK VALVE					
SAFETY VALVE					

EXTRACTED FROM ANSI Z32.2.3 – 1949 (R 1953)

TOPOGRAPHIC MAP SYMBOLS

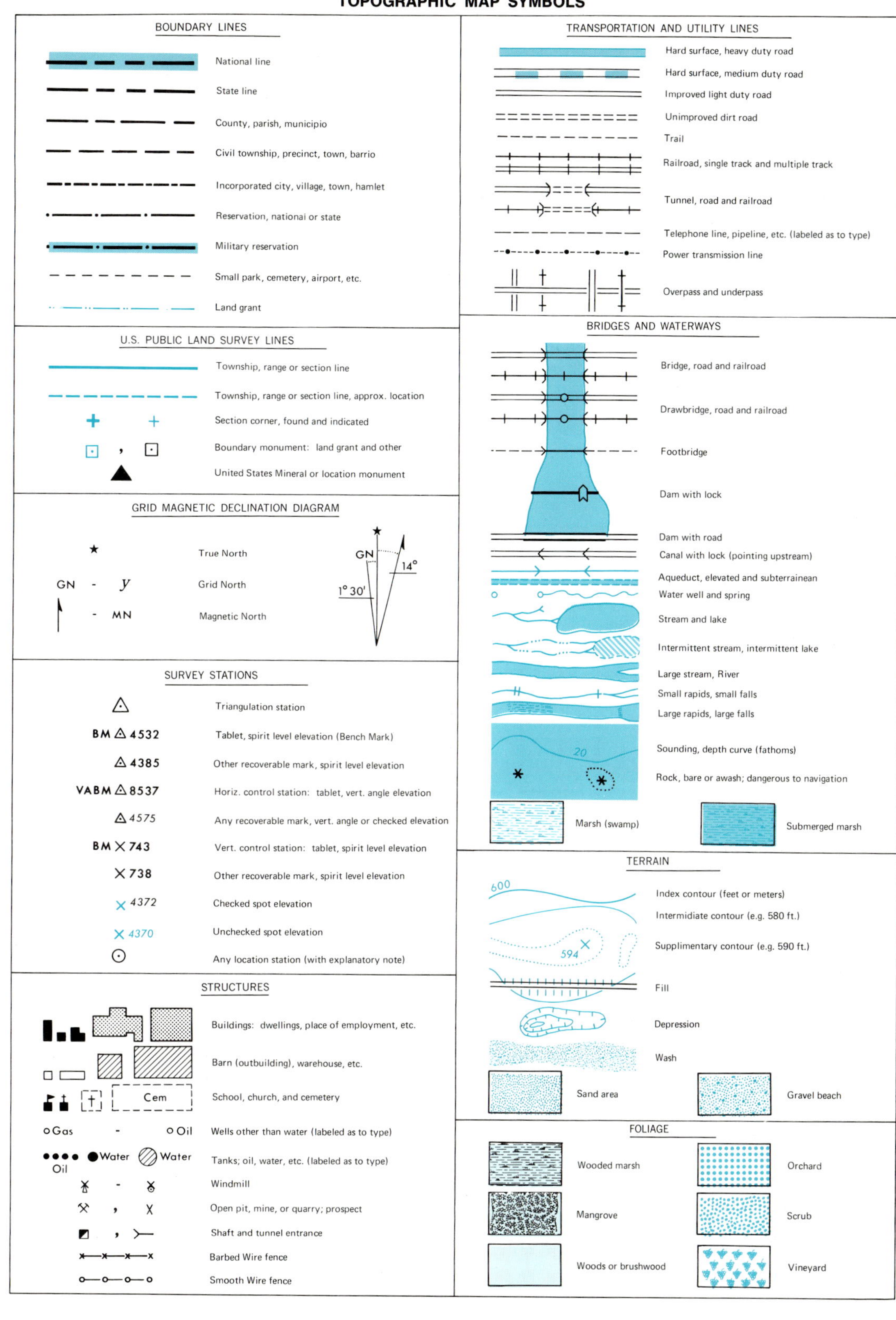

GRAPHICAL SYMBOLS FOR PIPING

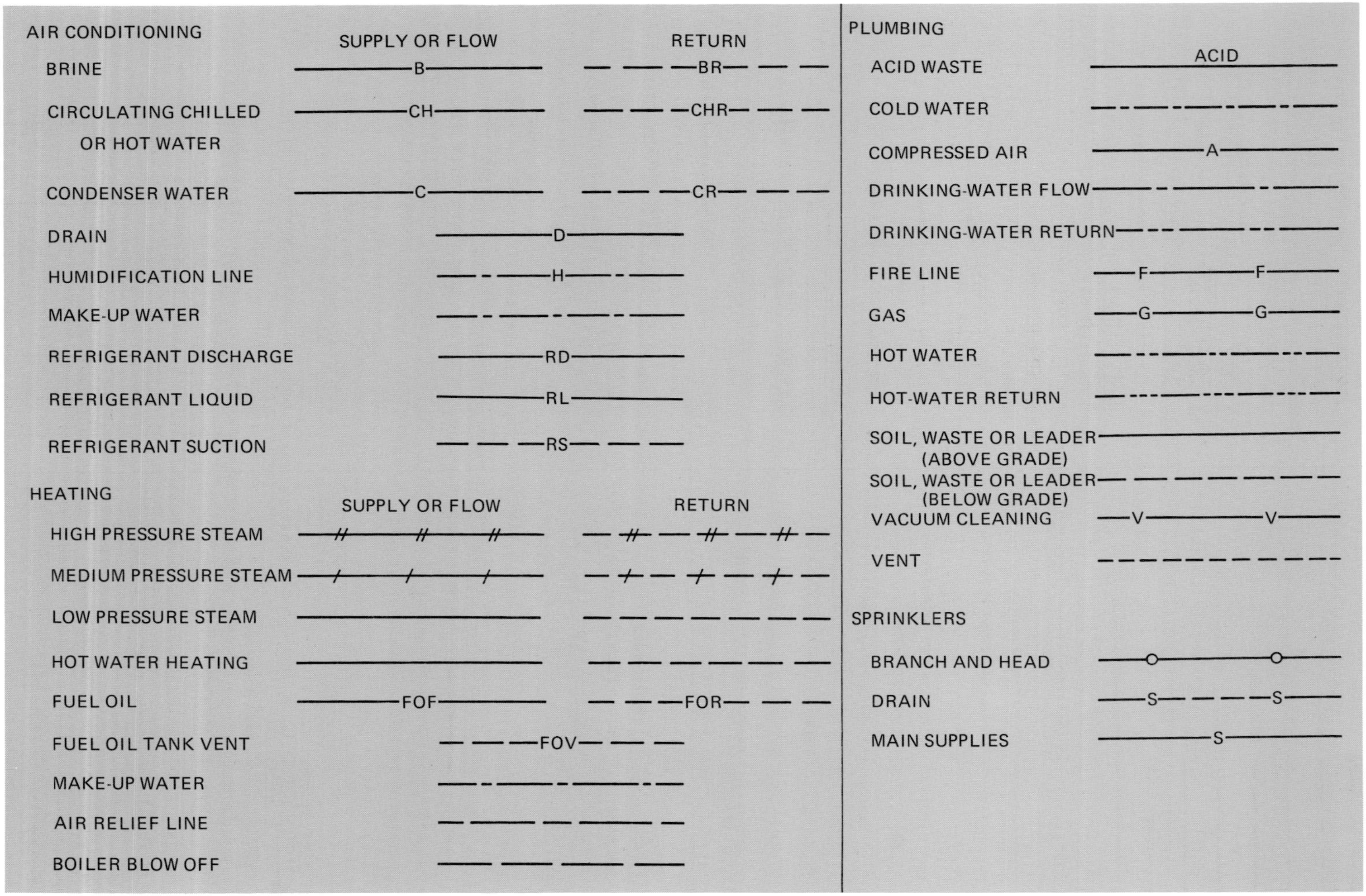

AIR CONDITIONING

	SUPPLY OR FLOW	RETURN
BRINE	B	BR
CIRCULATING CHILLED OR HOT WATER	CH	CHR
CONDENSER WATER	C	CR
DRAIN	D	
HUMIDIFICATION LINE	H	
MAKE-UP WATER		
REFRIGERANT DISCHARGE	RD	
REFRIGERANT LIQUID	RL	
REFRIGERANT SUCTION	RS	

HEATING

	SUPPLY OR FLOW	RETURN
HIGH PRESSURE STEAM		
MEDIUM PRESSURE STEAM		
LOW PRESSURE STEAM		
HOT WATER HEATING		
FUEL OIL	FOF	FOR
FUEL OIL TANK VENT	FOV	
MAKE-UP WATER		
AIR RELIEF LINE		
BOILER BLOW OFF		

PLUMBING

ACID WASTE	ACID
COLD WATER	
COMPRESSED AIR	A
DRINKING-WATER FLOW	
DRINKING-WATER RETURN	
FIRE LINE	F F
GAS	G G
HOT WATER	
HOT-WATER RETURN	
SOIL, WASTE OR LEADER (ABOVE GRADE)	
SOIL, WASTE OR LEADER (BELOW GRADE)	
VACUUM CLEANING	V V
VENT	

SPRINKLERS

BRANCH AND HEAD	
DRAIN	S S
MAIN SUPPLIES	S

EXTRACTED FROM ANSI 232.2.3 - 1949 (R 1953)

GRAPHICAL SYMBOLS FOR HEATING, VENTILATION, & AIR CONDITIONING

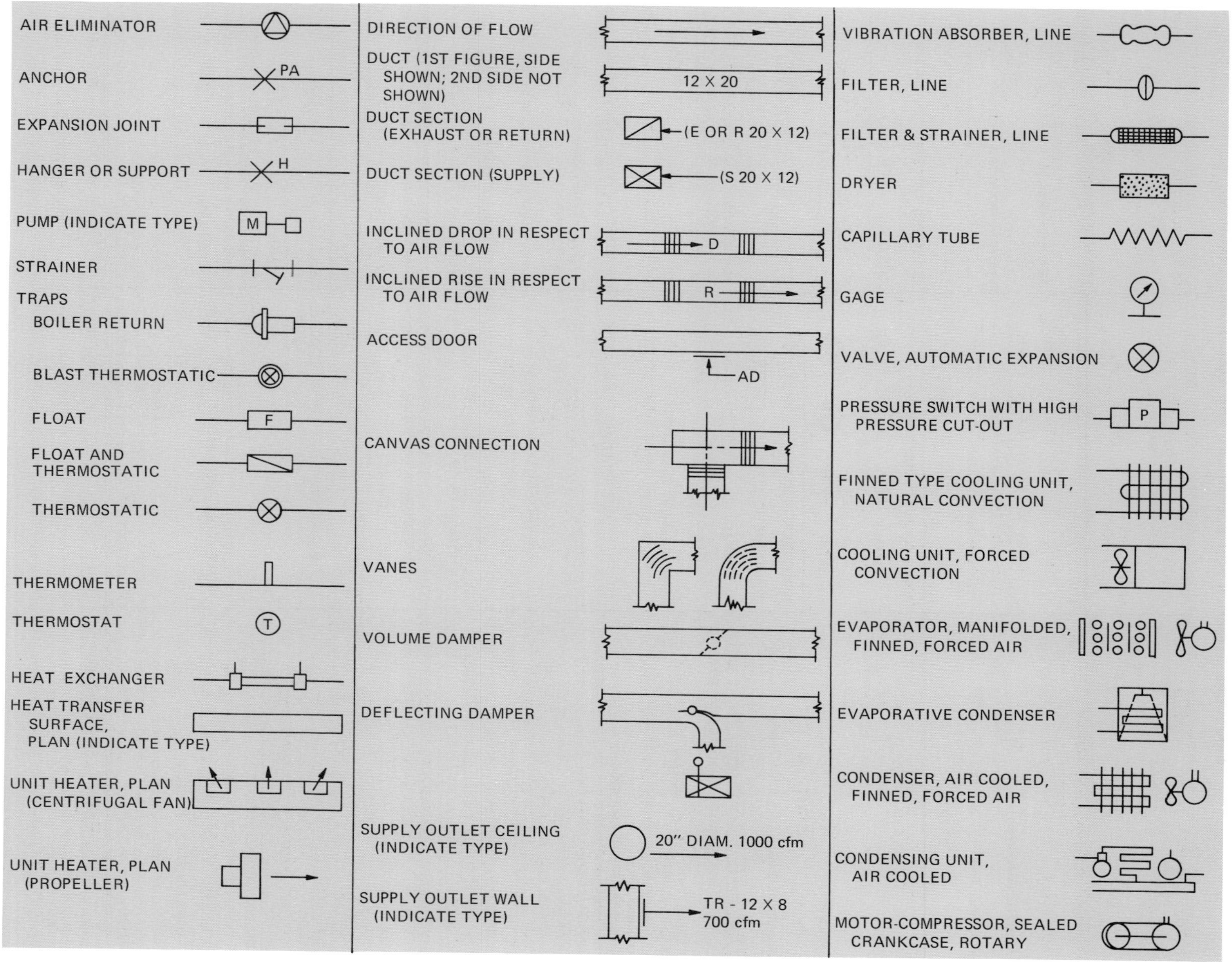

AIR ELIMINATOR
ANCHOR (PA)
EXPANSION JOINT
HANGER OR SUPPORT (H)
PUMP (INDICATE TYPE) (M)
STRAINER
TRAPS
BOILER RETURN
BLAST THERMOSTATIC
FLOAT (F)
FLOAT AND THERMOSTATIC
THERMOSTATIC
THERMOMETER
THERMOSTAT (T)
HEAT EXCHANGER
HEAT TRANSFER SURFACE, PLAN (INDICATE TYPE)
UNIT HEATER, PLAN (CENTRIFUGAL FAN)
UNIT HEATER, PLAN (PROPELLER)

DIRECTION OF FLOW
DUCT (1ST FIGURE, SIDE SHOWN; 2ND SIDE NOT SHOWN) (12 X 20)
DUCT SECTION (EXHAUST OR RETURN) (E OR R 20 X 12)
DUCT SECTION (SUPPLY) (S 20 X 12)
INCLINED DROP IN RESPECT TO AIR FLOW (D)
INCLINED RISE IN RESPECT TO AIR FLOW (R)
ACCESS DOOR (AD)
CANVAS CONNECTION
VANES
VOLUME DAMPER
DEFLECTING DAMPER
SUPPLY OUTLET CEILING (INDICATE TYPE) (20" DIAM. 1000 cfm)
SUPPLY OUTLET WALL (INDICATE TYPE) (TR - 12 X 8 700 cfm)

VIBRATION ABSORBER, LINE
FILTER, LINE
FILTER & STRAINER, LINE
DRYER
CAPILLARY TUBE
GAGE
VALVE, AUTOMATIC EXPANSION
PRESSURE SWITCH WITH HIGH PRESSURE CUT-OUT (P)
FINNED TYPE COOLING UNIT, NATURAL CONVECTION
COOLING UNIT, FORCED CONVECTION
EVAPORATOR, MANIFOLDED, FINNED, FORCED AIR
EVAPORATIVE CONDENSER
CONDENSER, AIR COOLED, FINNED, FORCED AIR
CONDENSING UNIT, AIR COOLED
MOTOR-COMPRESSOR, SEALED CRANKCASE, ROTARY

EXTRACTED FROM ANSI Z32.2.4 - 1949 (R 1953)

GRAPHICAL SYMBOLS FOR ELECTRICAL AND ELECTRONIC DIAGRAMS

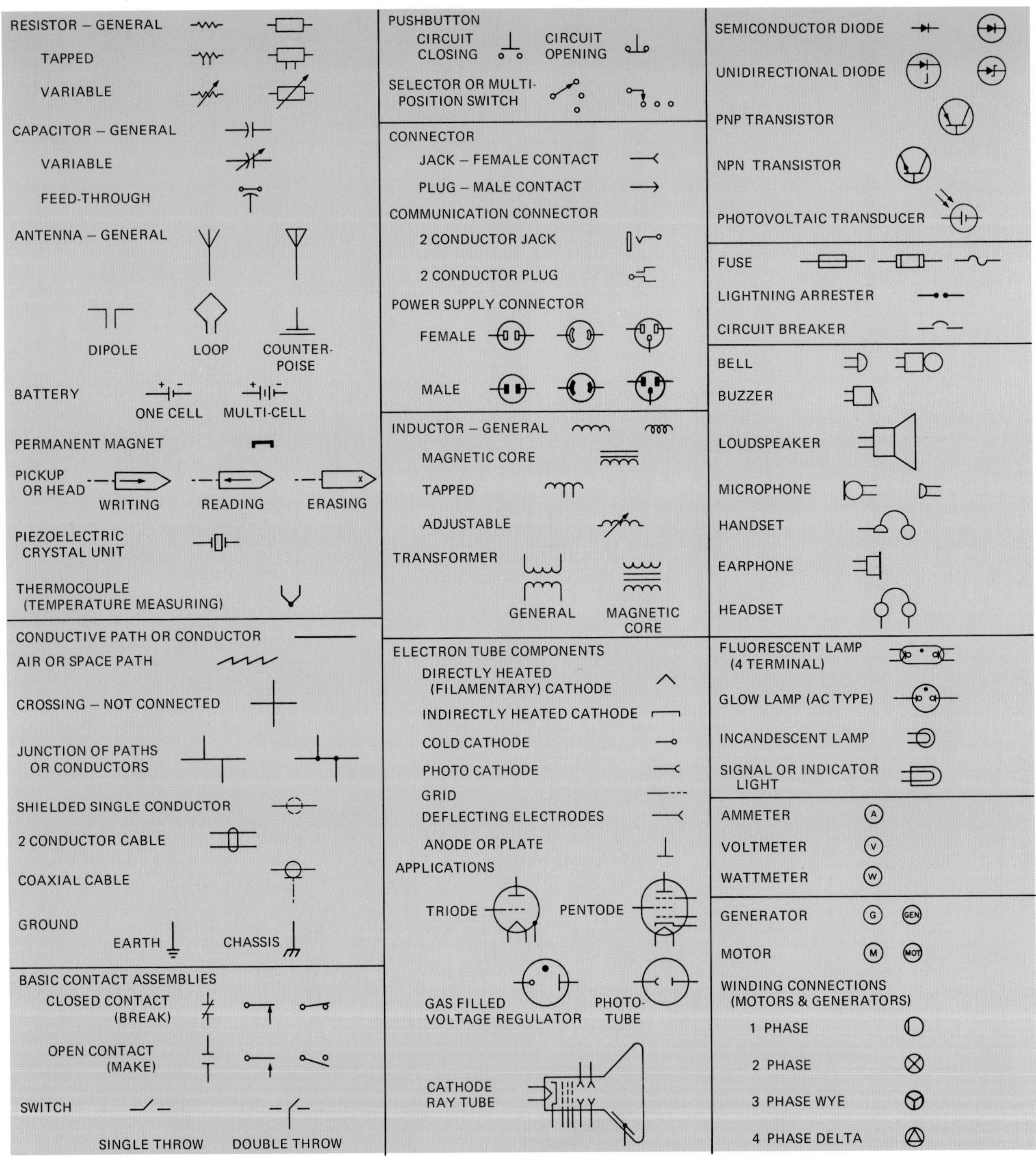

EXTRACTED FROM ANSI Y32.2 – 1970

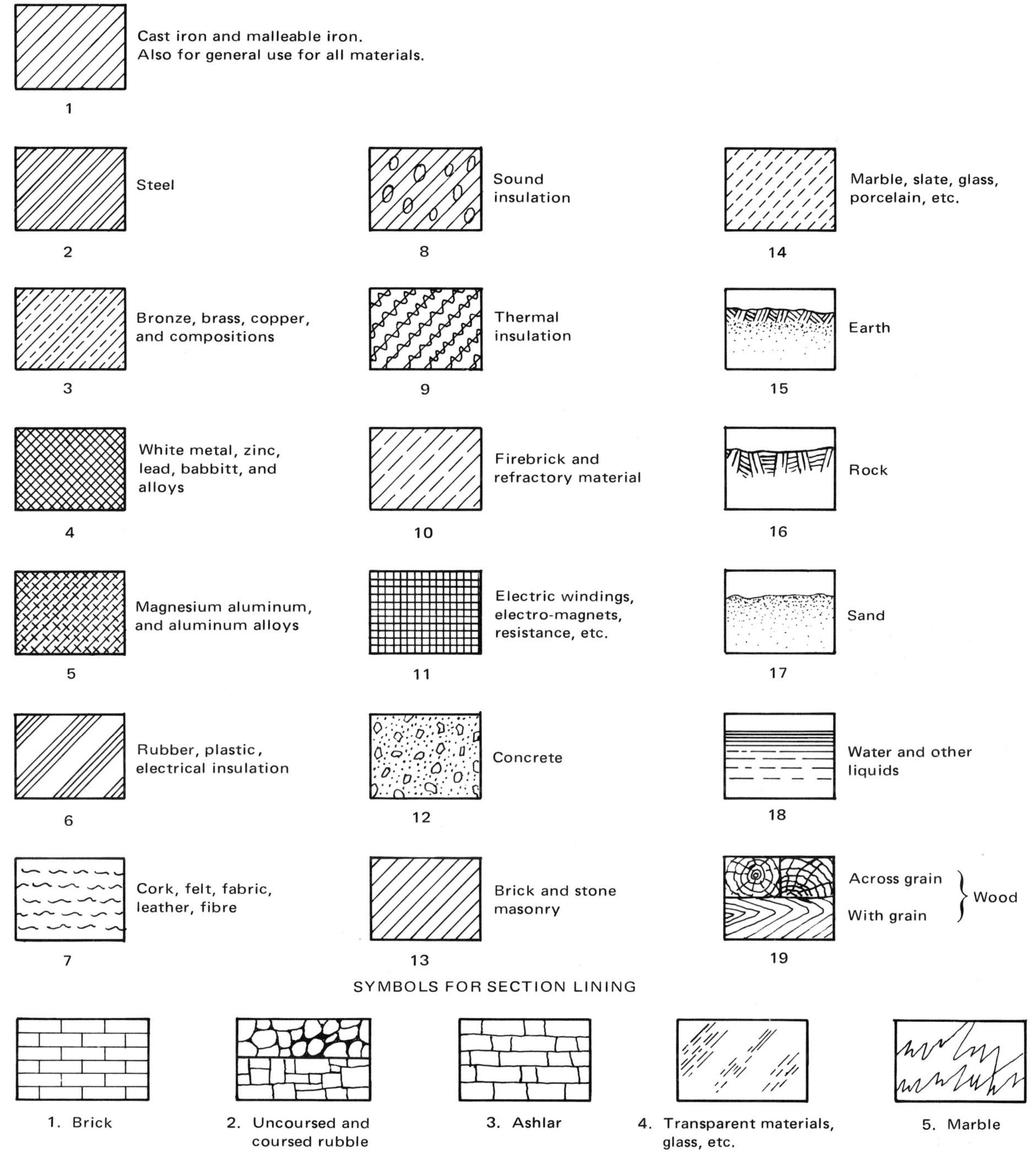

SYMBOLS FOR SECTION LINING

1. Brick
2. Uncoursed and coursed rubble
3. Ashlar
4. Transparent materials, glass, etc.
5. Marble

SYMBOLS FOR OUTSIDE VIEWS

Extracted from ANSI Y14.2—1957.

measurements and units

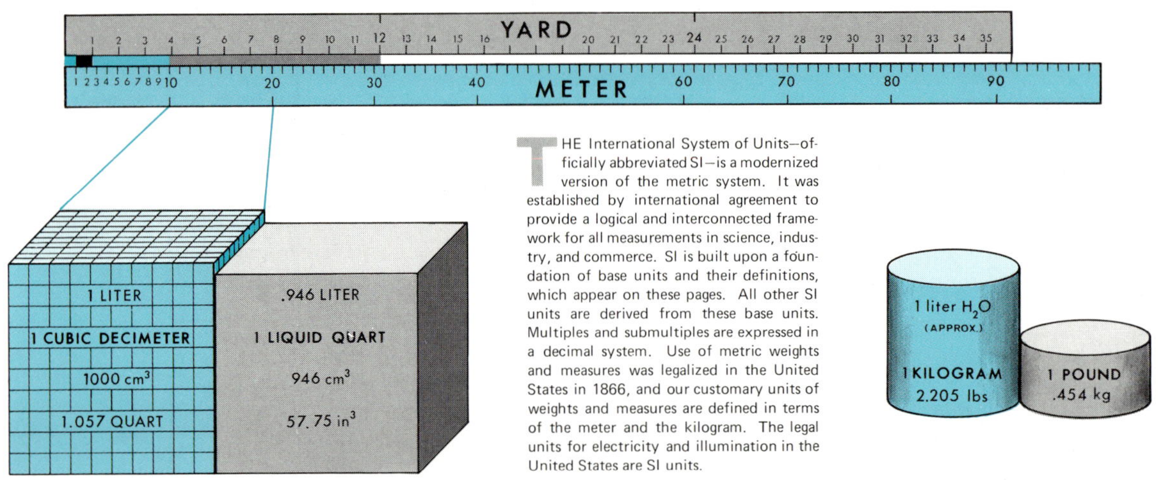

THE International System of Units—officially abbreviated SI—is a modernized version of the metric system. It was established by international agreement to provide a logical and interconnected framework for all measurements in science, industry, and commerce. SI is built upon a foundation of base units and their definitions, which appear on these pages. All other SI units are derived from these base units. Multiples and submultiples are expressed in a decimal system. Use of metric weights and measures was legalized in the United States in 1866, and our customary units of weights and measures are defined in terms of the meter and the kilogram. The legal units for electricity and illumination in the United States are SI units.

THE SIX BASIC UNITS OF MEASUREMENT

LENGTH, TIME, MASS, TEMPERATURE, ELECTRIC CURRENT AND LUMINOUS INTENSITY

Length—*METER*—m The meter is defined as 1,650,763.73 wavelengths in vacuum of the orange-red line of the spectrum of Krypton-86.

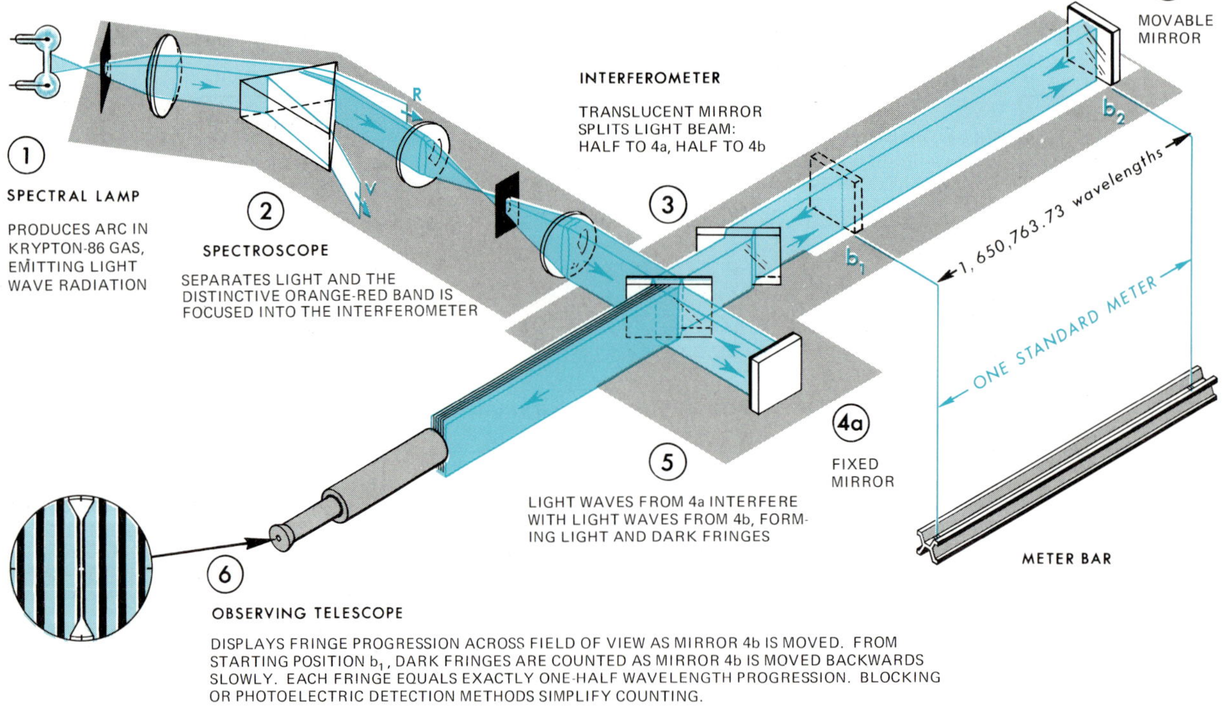

(1) SPECTRAL LAMP

PRODUCES ARC IN KRYPTON-86 GAS, EMITTING LIGHT WAVE RADIATION

(2) SPECTROSCOPE

SEPARATES LIGHT AND THE DISTINCTIVE ORANGE-RED BAND IS FOCUSED INTO THE INTERFEROMETER

R

V

INTERFEROMETER

(3) TRANSLUCENT MIRROR SPLITS LIGHT BEAM: HALF TO 4a, HALF TO 4b

(4b) MOVABLE MIRROR

(4a) FIXED MIRROR

b_1

b_2

1,650,763.73 wavelengths

ONE STANDARD METER

METER BAR

(5) LIGHT WAVES FROM 4a INTERFERE WITH LIGHT WAVES FROM 4b, FORMING LIGHT AND DARK FRINGES

(6) OBSERVING TELESCOPE

DISPLAYS FRINGE PROGRESSION ACROSS FIELD OF VIEW AS MIRROR 4b IS MOVED. FROM STARTING POSITION b_1, DARK FRINGES ARE COUNTED AS MIRROR 4b IS MOVED BACKWARDS SLOWLY. EACH FRINGE EQUALS EXACTLY ONE-HALF WAVELENGTH PROGRESSION. BLOCKING OR PHOTOELECTRIC DETECTION METHODS SIMPLIFY COUNTING.

Time—*SECOND*—s

The *second* is defined as the duration of 9,192,631,770 cycles of the radiation associated with a specified transition of the cesium atom. It is realized by tuning an oscillator to the resonance frequency of the cesium atoms as they pass through a system of magnets and a resonant cavity into a detector.

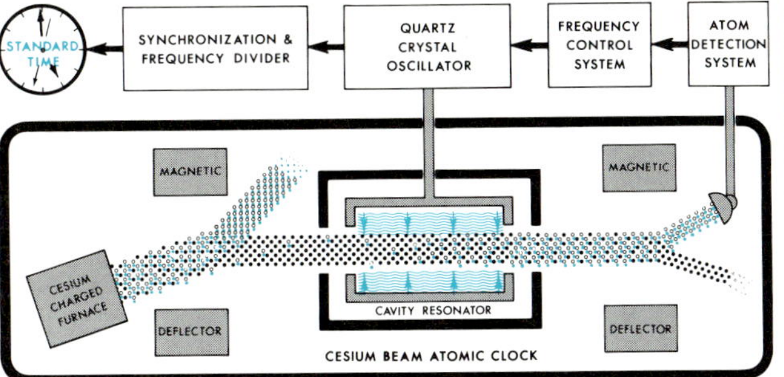

In the *atomic clock*, an electric furnace excites cesium atoms into a jet with thermal velocities, beamed into a magnetic field. Cesium, an alkali metal, has a one valence electron about a positive charge nucleus. It therefore consists of two spinning magnets, each in the field of the other, so that the atom experiences a magnetic torque when influenced by an external electromagnetic field. Entering the field, the atoms undergo a deflection according to their energy state. High-energy atoms (black dots) are selected to pass into the cavity resonator while most of the low energy atoms (color dots) are diverted away. A quartz crystal oscillates the cavity and when in resonance with the periodic frequency of the cesium atom, it causes some of the high energy atoms to emit energy and fall to the lower energy level. Flowing into another magnetic field, these low-energy atoms are deflected onto a detector and produce a positive ion current. A control system monitors this current closely and constantly tunes the crystal to the exact transition frequency. An accuracy of 1 part in 10^{11}, or 1 sec in 3000 yrs is realized.

Temperature—*KELVIN*—K

The thermodynamic or *Kelvin* scale has its origin or zero point at absolute zero and has a fixed point at the triple point of water defined as 273.16 kelvins, 0.01° Celsius, which is approximately 32.02° on the Fahrenheit scale.

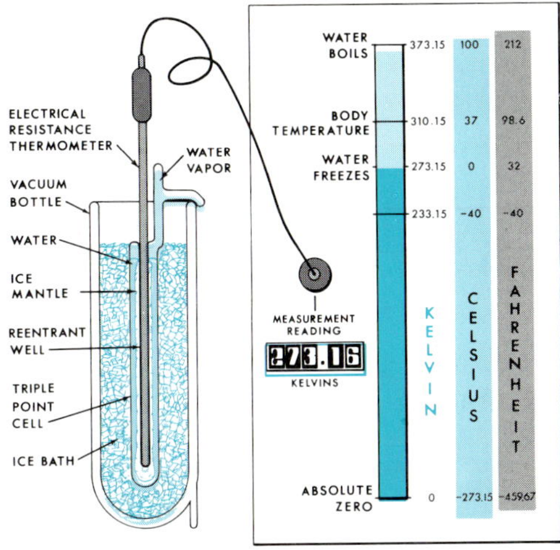

The *triple point cell*, an evacuated glass cylinder filled with pure water, is used to define a known fixed temperature. When the cell is cooled until a mantle of ice forms around the reentrant well, the temperature at the interface of solid, liquid, and vapor is 0.01°C. Thermometers to be calibrated are placed in the reentrant well.

Mass—*KILOGRAM*—kg

INTERNATIONAL PROTOTYPE KILOGRAM

The standard for the unit of mass, the *kilogram*, is a cylinder of platinum-iridium alloy kept by the International Bureau of Weights and Measures at Paris. A duplicate in the custody of the National Bureau of Standards serves as the mass standard for the United States. This is the only base unit still defined by an artifact.

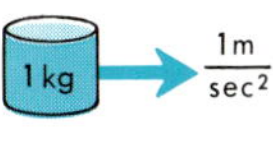

1 NEWTON

Closely allied to the concept of mass is that of force. The SI unit of force is the *newton* (N). A force of 1 newton, when applied for 1 second, will give to a 1 kilogram mass a speed of 1 meter per second (an acceleration of 1 meter per second per second).

Electric Current—*AMPERE*—A

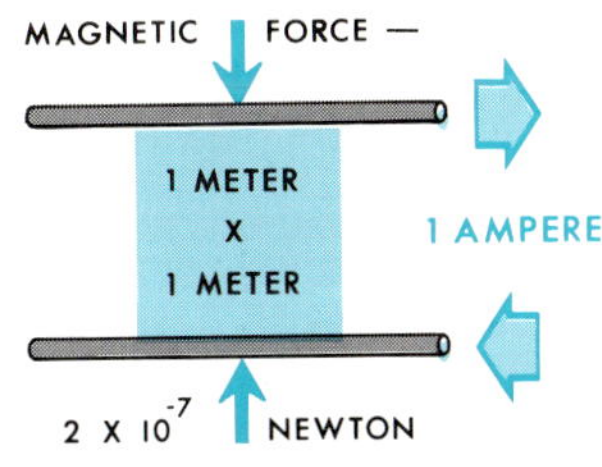

The *ampere* is defined as the magnitude of the current that, when flowing through each of two long parallel wires separated by one meter in free space, results in a force between the two wires (due to their magnetic fields) of 2×10^{-7} newton for each meter of length.

Luminous Intensity—*CANDELA*—cd

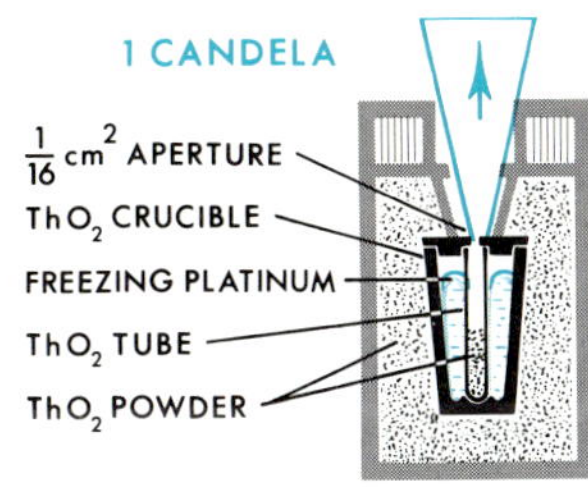

The *candela* is defined as the luminous intensity of 1/600,000 of a square meter of a radiating cavity at the temperature of freezing platinum (2042 K). A Thorium oxide tube (ThO_2) is used as a black body that absorbs all radiation falling upon it, becoming incandescent with the constant radiation during the long solidification process.

Decimal Inch
Decimal Equivalents of Common Fractions

1. Values or sizes that reflect common fractional increments shall be expressed as decimal equivalents of the fractional increments. These values should be expressed to 2, 3, or 4 decimal places as illustrated below. The number of decimal places will be determined by the tolerance required.

2. To avoid needless perpetuation of odd decimal numbers, this chart should not be used for new work. Instead, a value or size should be chosen having a final digit that is zero or an even number.

Fraction to Decimal Conversion Chart

4ths	8ths	16ths	32nds	64ths	to 2 places	to 3 places	to 4 places	4ths	8ths	16ths	32nds	64ths	to 2 places	to 3 places	to 4 places
				1/64	**0.02**	0.016	0.0156					33/64	**0.52**	0.516	0.5156
			1/32		**0.03**	0.031	0.0312				17/32		**0.53**	0.531	0.5312
				3/64	**0.05**	0.047	0.0469					35/64	**0.55**	0.547	0.5469
		1/16			**0.06**	0.062	0.0625			9/16			**0.56**	0.562	0.5625
				5/64	**0.08**	0.078	0.0781					37/64	**0.58**	0.578	0.5781
			3/32		**0.09**	0.094	0.0938				19/32		**0.59**	0.594	0.5938
				7/64	**0.11**	0.109	0.1094					39/64	**0.61**	0.609	0.6094
	1/8				**0.12**	0.125	0.1250		5/8				**0.62**	0.625	0.6250
				9/64	**0.14**	0.141	0.1406					41/64	**0.64**	0.641	0.6406
			5/32		**0.16**	0.156	0.1562				21/32		**0.66**	0.656	0.6562
				11/64	**0.17**	0.172	0.1719					43/64	**0.67**	0.672	0.6719
		3/16			**0.19**	0.188	0.1875			11/16			**0.69**	0.688	0.6875
				13/64	**0.20**	0.203	0.2031					45/64	**0.70**	0.703	0.7031
			7/32		**0.22**	0.219	0.2188				23/32		**0.72**	0.719	0.7188
				15/64	**0.23**	0.234	0.2344					47/64	**0.73**	0.734	0.7344
1/4					**0.25**	0.250	0.2500	3/4					**0.75**	0.750	0.7500
				17/64	**0.27**	0.266	0.2656					49/64	**0.77**	0.766	0.7656
			9/32		**0.28**	0.281	0.2812				25/32		**0.78**	0.781	0.7812
				19/64	**0.30**	0.297	0.2969					51/64	**0.80**	0.797	0.7969
		5/16			**0.31**	0.312	0.3125			13/16			**0.81**	0.812	0.8125
				21/64	**0.33**	0.328	0.3281					53/64	**0.83**	0.828	0.8281
			11/32		**0.34**	0.344	0.3438				27/32		**0.84**	0.844	0.8438
				23/64	**0.36**	0.359	0.3594					55/64	**0.86**	0.859	0.8594
	3/8				**0.38**	0.375	0.3750		7/8				**0.88**	0.875	0.8750
				25/64	**0.39**	0.391	0.3906					57/64	**0.89**	0.891	0.8906
			13/32		**0.41**	0.406	0.4062				29/32		**0.91**	0.906	0.9062
				27/64	**0.42**	0.422	0.4219					59/64	**0.92**	0.922	0.9219
		7/16			**0.44**	0.438	0.4375			15/16			**0.94**	0.938	0.9375
				29/64	**0.45**	0.453	0.4531					61/64	**0.95**	0.953	0.9531
			15/32		**0.47**	0.469	0.4688				31/32		**0.97**	0.969	0.9688
				31/64	**0.48**	0.484	0.4844					63/64	**0.98**	0.984	0.9844
1/2					**0.50**	0.500	0.5000	1					**1.00**	1.000	1.0000

(a) Omit zero to left of decimal point where used on drawings.

Extracted from ANSI B87.1—1965

American National Standard
Letter Symbols for Units Used in Science and Technology

Symbols for Units

Unit	Symbol	Notes
ampere	A	SI unit of electric current
ampere per meter	A/m	SI unit of magnetic field strength
angstrom	Å	1 Å = 10^{-10} m
atmosphere, standard	atm	1 atm = 101,325 N/m^2
atto	a	SI prefix for 10^{-18}
barrel	bbl	1 bbl = 9,702 in^3 = 0.15899 m^3
British thermal unit	Btu	
calorie	cal	1 cal = 4.1868 J
candela	cd	SI unit of luminous intensity
centi	c	SI prefix for 10^{-2}
centimeter	cm	
coulomb	C	SI unit of electric charge
cubic centimeter	cm^3	
cubic foot	ft^3	
cubic inch	in^3	
cubic meter	m^3	
curie	Ci	1 Ci = 3.7×10^{10} disintegrations per second. Unit of activity in the field of radiation dosimetry.
cycle per second	Hz, c/s	See hertz. The name hertz is internationally accepted for this unit; the symbol Hz is preferred to c/s.
day	d	
decibel	dB	
degree (plane angle)	. . . °	
degree (temperature):		
degree Celsius	°C	Note that there is no space between the symbol ° and the letter. The use of the word *centigrade* for the Celsius temperature scale was abandoned by the Conférence Générale des Poids et Mesures in 1948.
degree Fahrenheit	°F	
degree Kelvin		See Kelvin
degree Rankine	°R	
dyne	dyn	
electronvolt	eV	
erg	erg	
farad	F	SI unit of capacitance
foot	ft	
foot per second	ft/s	
foot pound-force	ft·lb_f	

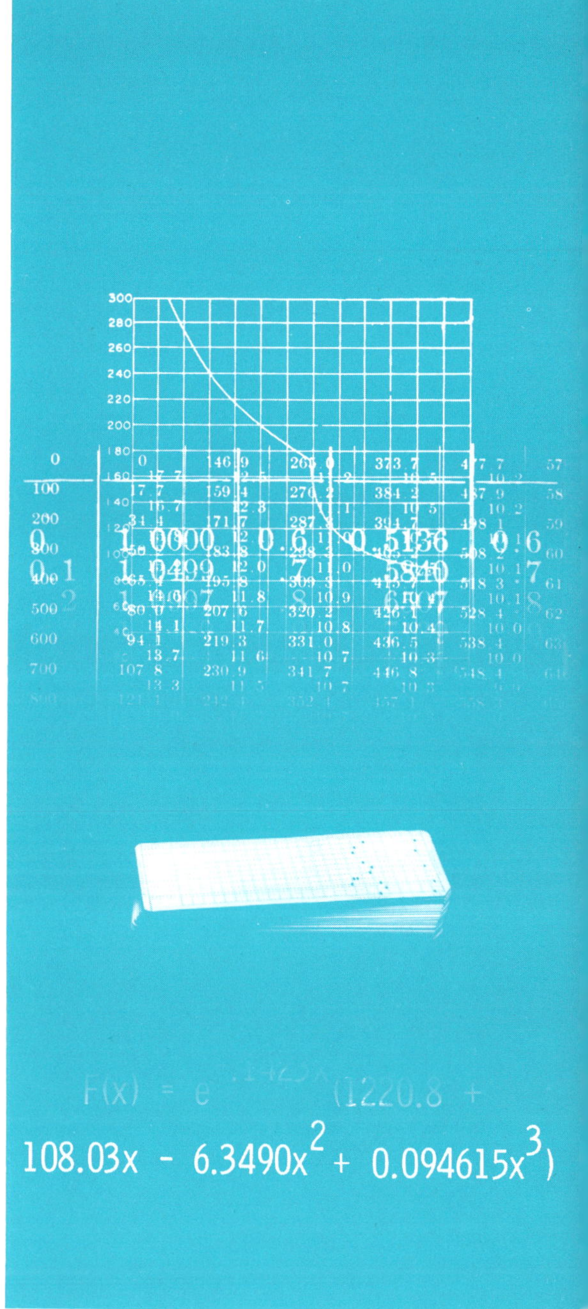

American National Standard
Letter Symbols for Units Used in Science and Technology

Symbols for Units (Cont'd)

Unit	Symbol	Notes
gallon	gal	The gallon, quart, and pint differ in the US and the UK, and their use in science and technology is deprecated.
gauss	G	The gauss is the electromagnetic CGS unit of magnetic flux density. Use of SI unit, the tesla, is preferred.
gram	g	
henry	H	SI unit of inductance
hertz	Hz	SI unit of frequency
horsepower	hp	The horsepower is an anachronism in science and technology. Use of the SI unit of power (the watt) is preferred.
hour	h	
joule	J	SI unit of energy
kelvin	K	In 1967 the CGPM gave the name *kelvin* to the SI unit of temperature which had formerly been called *degree Kelvin* and assigned it the symbol K (without the symbol °)
kilo	k	SI prefix for 10^3
kilogram	kg	SI unit of mass
kilogram-force	kg_f	In some countries the name *kilopond* (kp) has been adopted for this unit.
lambert	L	$1\ L = (1/\pi)\ cd/cm^2$ A CGS unit of luminance. One lumen per square centimeter leaves a surface whose luminance is one lambert in all directions within a hemisphere. Use of the SI unit of luminance, the candela per square meter, is preferred.
liter	l	$1\ l = 10^{-3}\ m^3$
lumen	lm	SI unit of luminous flux
mega	M	SI prefix for 10^6
megahertz	MHz	
meter	m	SI unit of length
mho	mho	CIPM has accepted the name *siemens* (S) for this unit and will submit it to the 14th CGPM for approval.

American National Standard
Letter Symbols for Units Used in Science and Technology

Symbols for Units (Cont'd)

Unit	Symbol	Notes
micro	μ	SI prefix for 10^{-6}
micron	μm	The name *micron* was abrogated by the Conférence Générale des Poids et Mesures, 1967.
mile (statute)	mi	1 mi = 5,280 ft
mile per hour	mi/h	Although use of mph as an abbreviation is common, it should not be used as a symbol.
milli	m	SI prefix for 10^{-3}
minute (time)	min	Time may also be designated by means of superscripts as in the following example: $9^h46^m30^s$.
mole	mol	SI unit of amount of substance
nano	n	SI prefix for 10^{-9}
newton	N	SI unit of force
newton per square meter	N/m^2	SI unit of pressure or stress; see pascal.
oersted	Oe	The oersted is the electromagnetic CGS unit of magnetic field strength. Use of the SI unit, the ampere per meter, is preferred.
ohm	Ω	SI unit of resistance
pascal	Pa	Pa = N/m^2 Si unit of pressure or stress. This name accepted by the CIPM in 1969 for submission to the 14th CGPM.
pico	p	SI prefix for 10^{-12}
poise	P	SI unit of absolute viscosity
pound	lb	
pound-force	lb_f	The symbol lb, without a subscript, may be used for pound-force where no confusion is foreseen.
pound-force per square inch	lb_f/in^2	Although use of the abbreviation psi is common, it should not be used as a symbol. Refer to note on pound-force regarding subscript to the symbol.
radian	rad	SI unit of plane angle
revolution per minute	r/min	Although use of rpm as an abbreviation is common, it should not be used as a symbol.

American National Standard
Letter Symbols for Units Used in Science and Technology

Symbols for Units (Cont'd)

Unit	Symbol	Notes
revolution per second	r/s	
roentgen	R	Unit of exposure in the field of radiation dosimetry
second (time)	s	SI unit of time
siemens	S	$S = \Omega^{-1}$
slug	slug	1 slug = 14.5939 kg
steradian	sr	SI unit of solid angle
stokes	St	SI unit of dynamic viscosity
tesla	T	$T = N/(A \cdot m) = Wb/m^2$ Si unit of magnetic flux density (magnetic induction)
ton	ton	1 ton = 2,000 lb
volt	V	SI unit of voltage
voltampere	VA	IEC name and symbol for the SI unit of apparent power
watt	W	SI unit of power
watt-hour	Wh	
weber	Wb	$Wb = V \cdot s$ SI unit of magnetic flux
yard	yd	

conversion tables

The number in parentheses following a value in the table indicates the power of 10 by which this value should be multiplied. Thus, 6.214(−6) means 6.214×10^{-6}.

1. Length Equivalents

cm	in	ft	m	mi*	
1	3.937(−1)	3.281(−2)	1.0(−2)	6.214(−6)	cm
2.540	1	8.333(−2)	2.54(−2)	1.578(−5)	in
3.048(+1)	1.2(+1)	1	3.048(−1)	1.894(−4)	ft
1.0(+2)	3.937(+1)	3.281	1	6.214(−4)	m
1.609(+5)	6.336(+4)	5.280(+3)	1.609(+3)	1	mi

Additional Measures	
Metric:	1 km = 10^3 m
	1 mm = 10^{-3} m
	1 μm = 10^{-6} m (micron)
	1 Å = 10^{-10} m (angstrom)
English:	1 mil = 10^{-3} in.
	1 yd = 3.0 ft
	1 rod = 5.5 yd = 16.5 ft
	1 furlong = 40 rod = 660 ft

*mile.

2. Area Equivalents

m^2	in^2	ft^2	acres	mi^2	
1	1.55(+3)	1.076(+1)	2.471(−4)	3.861(−7)	m^2
6.452(−4)	1	6.944(−3)	1.594(−7)	2.491(−10)	in^2
9.290(−2)	1.44(+2)	1	2.296(−5)	3.587(−8)	ft^2
4.047(+3)	6.273(+6)	4.356(+4)	1	1.562(−3)	acres
2.590(+6)	4.018(+9)	2.788(+7)	6.40(+2)	1	mi^2

Additional Measures

1 hectare = 10^4 m^2

3. Volume Equivalents

cm^3	in^3	ft^3	gal (U.S.)	
1	6.103(−2)	3.532(−5)	2.642(−4)	cm^3
1.639(+1)	1	5.787(−4)	4.329(−3)	in^3
2.832(+4)	1.728(+3)	1	7.481	ft^3
3.785(+3)	2.31(+2)	1.337(−1)	1	gal (U.S.)

Additional Measures

Metric: 1 liter = 10^3 cm^3
1 m^3 = 10^6 cm^3

English: 1 quart = 0.250 gal (U.S.)
1 bushel = 9.309 gal (U.S.)
1 barrel = 42 gal (U.S.)
(petroleum measure only)
1 imperial gal = 1.20 gal (U.S.) approx.
1 board-foot (wood) = 144 in^3
1 chord (wood) = 128 ft^3

4. Mass Equivalents

kg	slug	lb_m*	g	
1	6.85(−2)	2.205	1.0(+3)	kg
1.46(+1)	1	3.22(+1)	1.46(+4)	slug
4.54(−1)	3.11(−2)	1	4.54(+2)	lb_m
1.0(−3)	6.85(−5)	2.205(−3)	1	g

5. Force Equivalents

N**	lb_f†	dyn††	kg_f*	g_f*	poundal*	
1	2.248(−1)	1.0(+5)	1.019(−1)	1.019(+2)	7.234	N
4.448	1	4.448(+5)	4.54(−1)	4.54(+2)	3.217(+1)	lb_f
1.0(−5)	2.248(−6)	1	1.02(−6)	1.02(−3)	7.233(−5)	dyn
9.807	2.205	9.807(+5)	1	1.0(+3)	7.093(+1)	kg_f
9.807(−3)	2.205(−3)	9.807(+2)	1.0(−3)	1	7.093(−2)	g_f
1.382(−1)	3.108(−2)	1.383(+4)	1.410(−2)	1.410(+1)	1	poundal

Additional Measures

1 metric ton = 10^3 kg_f = 2.205 × 10^3 lb_f
1 pound troy = 0.8229 lb_f
1 oz† = 6.25 × 10^{-2} lb_f
1 oz troy = 6.857 × 10^{-2} lb_f

*Not recommended.
**Newton.
†Avoirdupois.
††Dyne.

6. Velocity and Acceleration Equivalents

Velocity					*Acceleration*				
cm/s	ft/s	mi/h (mph)	km/h		cm/s^2	ft/s^2	$\bar{g}$*		Additional Measures
1	3.281(−2)	2.237(−2)	3.60(−2)	cm/s	1	3.281(−2)	1.019(−3)	cm/s^2	1 knot = 1.152 miles/hr
3.048(+1)	1	6.818(−1)	1.097	ft/s	3.048(+1)	1	3.109(−2)	ft/s^2	
4.470(+1)	1.467	1	1.609	mi/h	9.807(+2)	3.217(+1)	1	$\bar{g}$	
2.778(+1)	9.113(−1)	6.214(−1)	1	km/h					

*Standard acceleration of gravity.

7. Pressure Equivalents

					Head†			
dyn/cm^2	N/m^2	lb_f/in^2 (psi)	lb/ft^2 (psf)	atm*	in (Hg)	ft (H_2O)		Additional Measures
1	1.0(−1)	1.45(−5)	2.089(−3)	9.869(−7)	2.953(−5)	3.349(−5)	dyn/cm^2	1 bar = 1 $dyne/cm^2$
1.0(+1)	1	1.45(−4)	2.089(−2)	9.869(−6)	2.953(−4)	3.349(−4)	N/m^2	1 pascal = 1 N/m^2
6.895(+4)	6.895(+3)	1	1.44(+2)	6.805(−2)	2.036	2.309	lb_f/in^2	
4.788(+2)	4.788(+1)	6.944(−3)	1	4.725(−4)	1.414(−2)	1.603(−2)	lb/ft^2	
1.013(+6)	1.013(+5)	1.47(+1)	2.116(+3)	1	2.992(+1)	3.393(+1)	atm	
3.386(+4)	3.386(+3)	4.912(−1)	7.073(+1)	3.342(−2)	1	1.134	in (Hg)	
2.986(+4)	2.986(+3)	4.331(−1)	6.237(+1)	2.947(−2)	8.819(−1)	1	ft (H_2O)	

*Standard atmospheric pressure.
†At std. gravity and 0°C for Hg, 15°C for H_2O.

8. Work and Energy Equivalents

J*	$ft\text{-}lb_f$	W-h	Btu**	Kcal†	kg-m		Additional Measures
1	7.376(−1)	2.778(−4)	9.478(−4)	2.388(−4)	1.020(−1)	J	1 Newton-meter = 1 J
1.356	1	3.766(−4)	1.285(−3)	3.238(−4)	1.383(−1)	$ft\text{-}lb_f$	1 erg = 1 dyne-cm = 10^{-7} J
3.60(+3)	2.655(+3)	1	3.412	8.599(−1)	3.671(+2)	W-h	1 cal = 10^{-3} kcal
1.055(+3)	7.782(+2)	2.931(−1)	1	2.520(−1)	1.076(+2)	Btu	1 therm = 10^{-5} Btu
4.187(+3)	3.088(+3)	1.163	3.968	1	4.269(+2)	Kcal	
9.807	7.233	2.724(−3)	9.295(−3)	2.342(−3)	1	kg-m	

*Joule.
**British thermal unit.
† = kilocalorie.

9. Power Equivalents

J/s	$ft\text{-}lb_f/s$	hp††	kW	Btu/h		Additional Measures
1	7.376(−1)	1.341(−3)	1.0(−3)	3.412	J/s	1 W = 10^{-3} kW
1.356	1	1.818(−3)	1.356(−3)	4.626	$ft\text{-}lb_f/s$	1 cal/s = 14.29 Btu/h
7.457(+2)	5.50(+2)	1	7.457(−1)	2.545(+3)	hp	1 poncelet = 100 kg-m/sec = 0.9807 kW
1.0(+3)	7.376(+2)	1.341	1	3.412(+3)	kW	1 ton of refrigeration = 1.2 × 10^4 Btu/h
2.931(−1)	2.162(−1)	3.930(−4)	2.931(−4)	1	Btu/h	

††Horsepower.

geometric figures

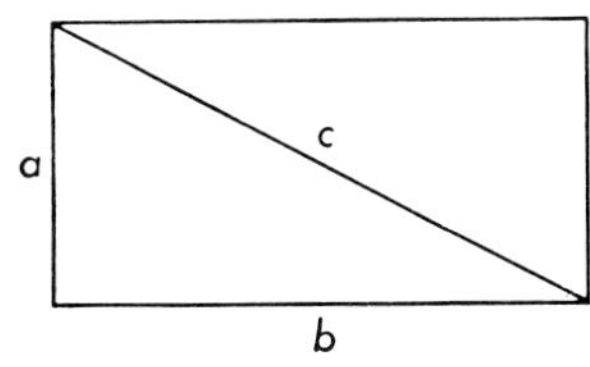

Rectangle

Area = (base)(altitude) = ab

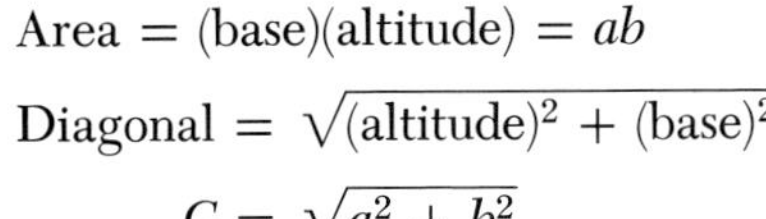

Diagonal = $\sqrt{(\text{altitude})^2 + (\text{base})^2}$

$C = \sqrt{a^2 + b^2}$

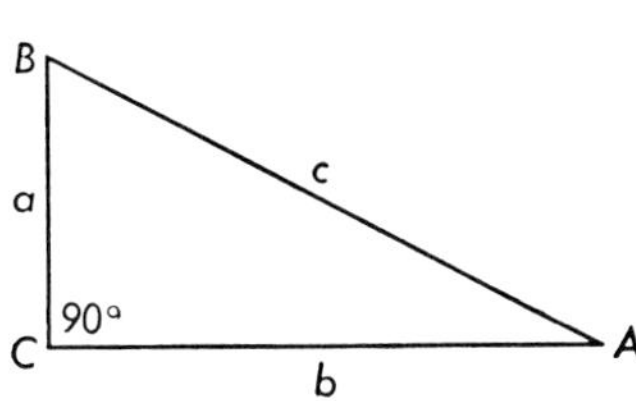

Right Triangle

Angle A + angle B = angle C = 90°

Area = $\frac{1}{2}$(base)(altitude)

Hypotenuse = $\sqrt{(\text{altitude})^2 + (\text{base})^2}$

$C = \sqrt{a^2 + b^2}$

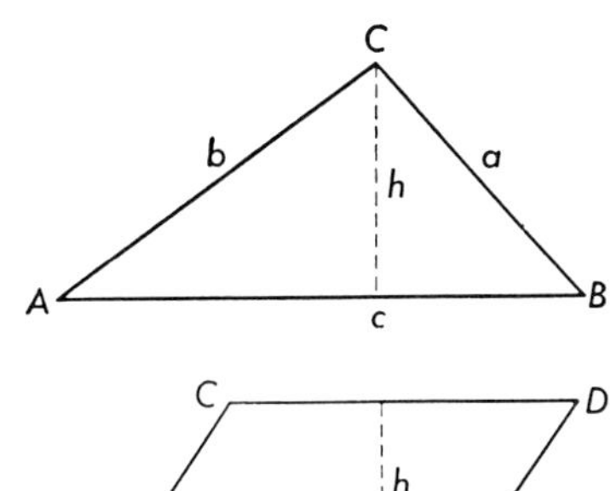

Any Triangle

Angles $A + B + C = 180°$
(Altitude h is perpendicular to base c)
Area = $\frac{1}{2}$(base)(altitude)

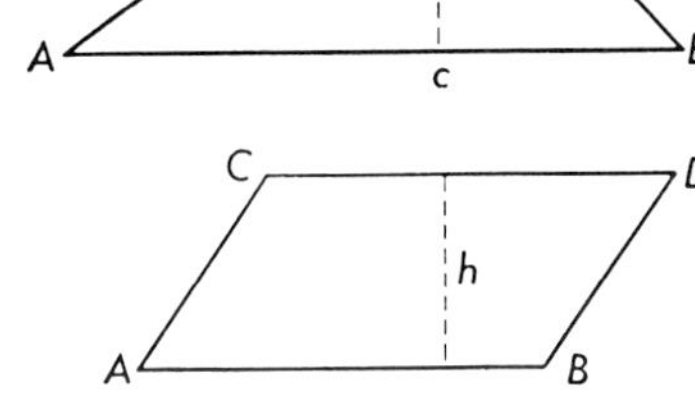

Parallelogram

Area = (base)(altitude)
Altitude h is perpendicular to base AB
Angles $A + B + C + D = 360°$

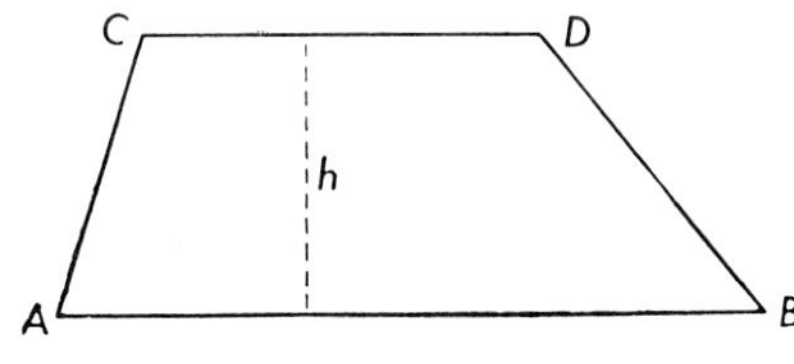

Trapezoid

Area = $\frac{1}{2}$(altitude)(sum of bases)
(Altitude h is perpendicular to sides AB and CD. Side AB is parallel to side CD.)

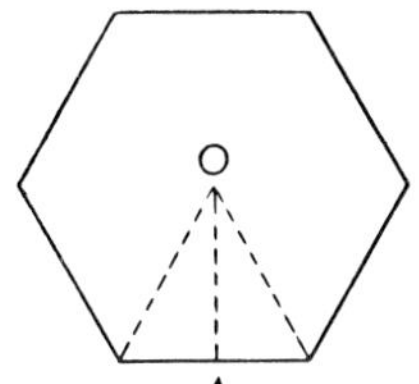

Regular Polygon

$$\text{Area} = \frac{1}{2}\begin{bmatrix}\text{Length of}\\ \text{one side}\end{bmatrix}\begin{bmatrix}\text{Number}\\ \text{of sides}\end{bmatrix}\begin{bmatrix}\text{Distance}\\ OA \text{ to}\\ \text{center}\end{bmatrix}$$

A regular polygon has equal angles and equal sides and can be inscribed in or circumscribed about a circle.

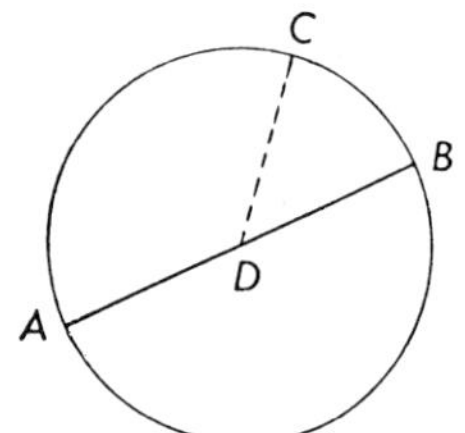

Circle

AB = diameter, CD = radius

$$\text{Area} = \pi(\text{radius})^2 = \frac{\pi(\text{diameter})^2}{4}$$

$$\text{Circumference} = \pi(\text{diameter})$$

$$C = 2\pi(\text{radius})$$

$$\frac{\text{arc } BC}{\text{circumference}} = \frac{\text{angle } BDC}{360^\circ}$$

$$1 \text{ radian} = \frac{180^\circ}{\pi} = 57.2958^\circ$$

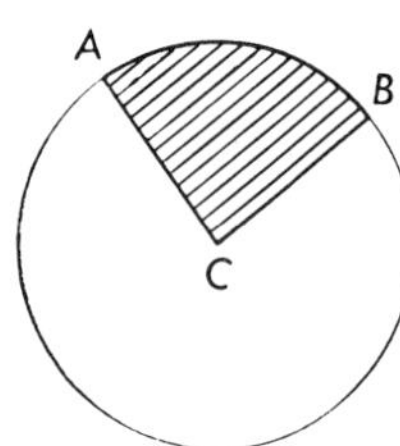

Sector of a Circle

$$\text{Area} = \frac{(\text{arc } AB)(\text{radius})}{2}$$

$$= \pi\frac{(\text{radius})^2(\text{angle } ACB)}{360^\circ}$$

$$= \frac{(\text{radius})^2(\text{angle } ACB \text{ in radians})}{2}$$

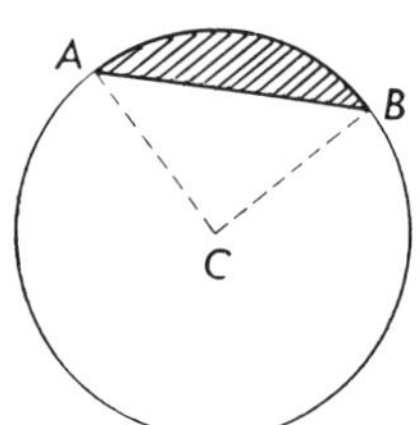

Segment of a Circle

$$\text{Area} = \frac{(\text{radius})^2}{2}\left[\frac{\pi(\measuredangle ACB^\circ)}{180} - \sin ACB^\circ\right]$$

$$\text{Area} = \frac{(\text{radius})^2}{2}\left[\measuredangle ACB \text{ in radians} - \sin ACB^\circ\right]$$

Area = area of sector ACB − area of triangle ABC

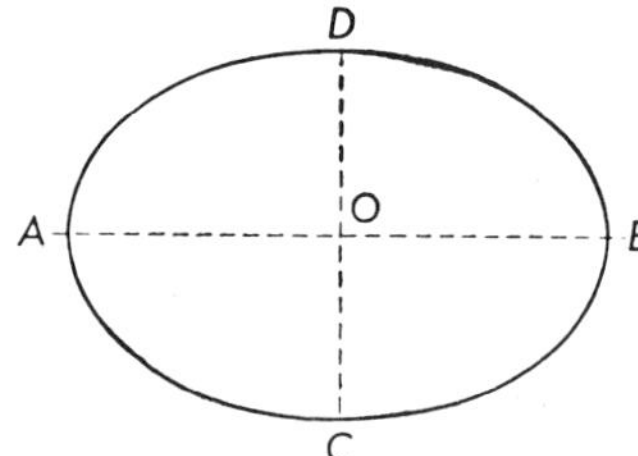

Ellipse

$$\text{Area} = \pi(\text{long radius } OA)(\text{short radius } OC)$$

$$\text{Area} = \frac{\pi}{4}(\text{long diameter } AB)(\text{short diameter } CD)$$

Volume and Center of Gravity Equations*

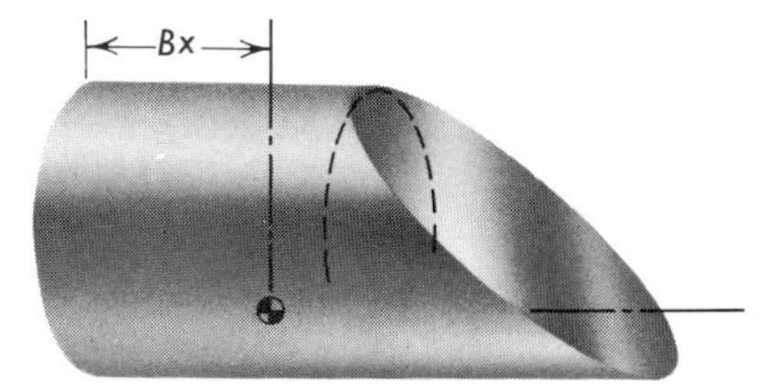

Volume equations are included for all cases. Where the equation for the CG (center of gravity) is not given, you can easily obtain it by looking up the volume and CG equations for portions of the shape and then combining values. For example, for the shape above, use the equations for a cylinder, Fig. 1, and a truncated cylinder, Fig. 10 (subscripts C and T, respectively, in the equations below). Hence taking moments

$$B_x = \frac{V_C B_C + V_T(B_T + L_C)}{V_C + V_T}$$

or

$$B_x = \frac{\left(\frac{\pi}{4}D^2L_C\right)\left(\frac{L_C}{2}\right) + \frac{\pi}{8}D^2L_T\left(\frac{5}{16}L_T + L_C\right)}{\frac{\pi}{4}D^2L_C + \frac{\pi}{8}D^2L_T}$$

$$B_x = \frac{L^2{}_C + L_T\left(\frac{5}{16}L_T + L_C\right)}{2L_C + L_T}$$

In the equations to follow, angle θ can be either in degrees or in radians.

Thus $\theta\ (\text{rad}) = \pi\theta/180\ (\text{deg}) = 0.01745\,\theta\ (\text{deg})$.

For example, if $\theta = 30$ deg in Fig. 3, then $\sin\theta = 0.5$ and

$$B = \frac{2R(0.5)}{3(30)(0.01745)} = 0.637R$$

Symbols used are:

B = distance from CG to reference plane,
V = volume,
D and d = diameter,
R and r = radius,
H = height,
L = length.

*Courtesy of Knoll Atomic Power Laboratory, Schenectady, New York, operated by the General Electric Company for the United States Atomic Energy Commission. Reprinted from *Product Engineering*—Copyright owned by McGraw-Hill.

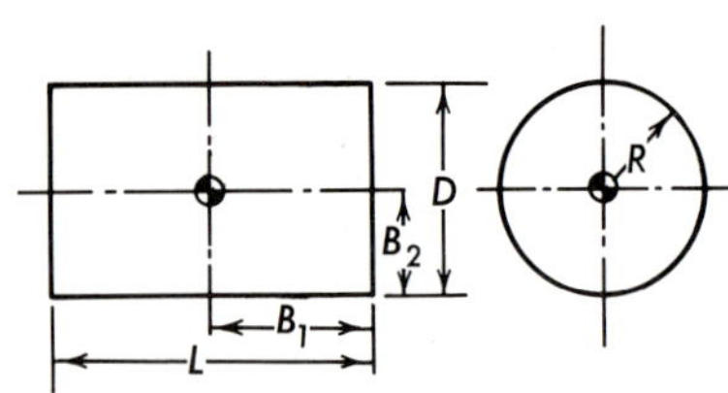

1. Cylinder

$$V = \frac{\pi}{4} D^2 L = 0.7854 D^2 L \qquad B_1 = L/2$$

$$B_2 = R$$

Area of cylindrical surface = (Perimeter of base)(perpendicular height)

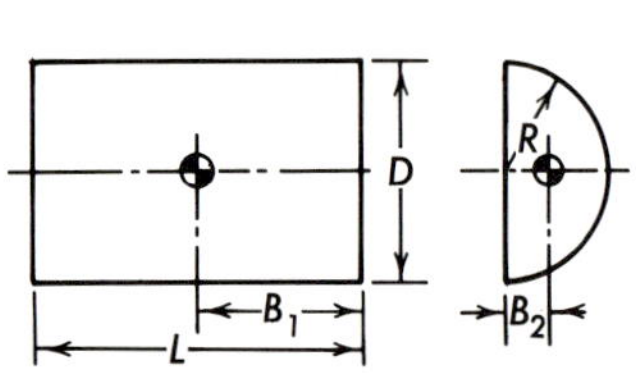

2. Half-cylinder

$$V = \frac{\pi}{8} D^2 L = 0.3927 D^2 L$$

$$B_1 = L/2 \qquad B_2 = \frac{4R}{3\pi} = 0.4244R$$

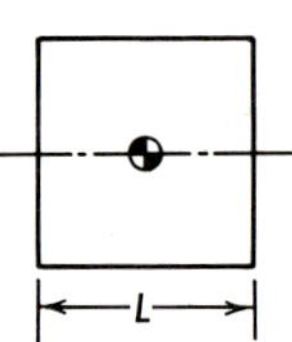

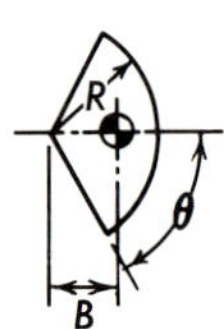

3. Sector of cylinder

$$V = \theta R^2 L \qquad B = \frac{2R \sin\theta}{3\theta}$$

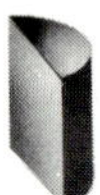

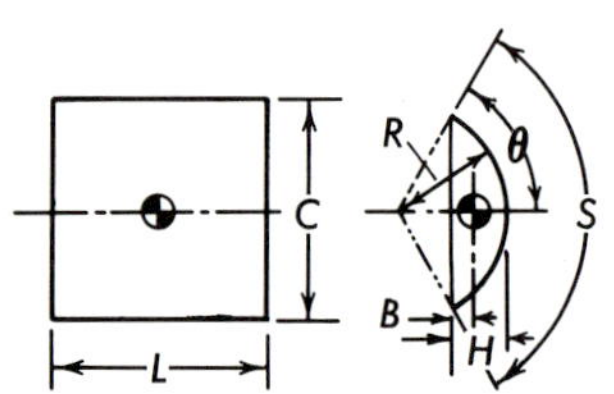

4. Segment of cylinder

$$V = LR^2(\theta - \tfrac{1}{2}\sin 2\theta)$$

$$V = 0.5L[RS - C(R - H)]$$

$$B = \frac{4R \sin^3\theta}{6\theta - 3\sin 2\theta}$$

$$S = 2R\theta$$

$$H = R(1 - \cos\theta)$$

$$C = 2R \sin\theta$$

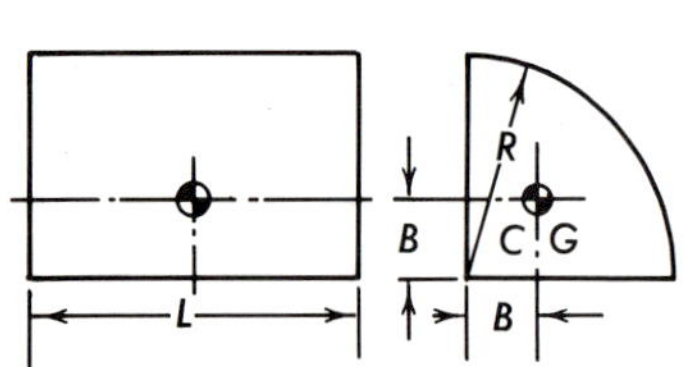

5. Quadrant of cylinder

$$V = \frac{\pi}{4} R^2 L = 0.7854 R^2 L$$

$$B = \frac{4R}{3\pi} = 0.4244R$$

6. Fillet or spandrel

$$V = \left(1 - \frac{\pi}{4}\right)R^2L = 0.2146R^2L$$

$$B = \frac{10 - 3\pi}{12 - 3\pi}R = 0.2234R$$

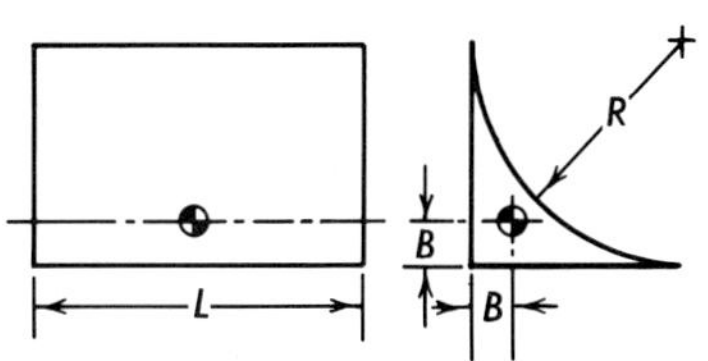

7. Hollow cylinder

$$V = \frac{\pi L}{4}(D^2 - d^2)$$

CG at center of part

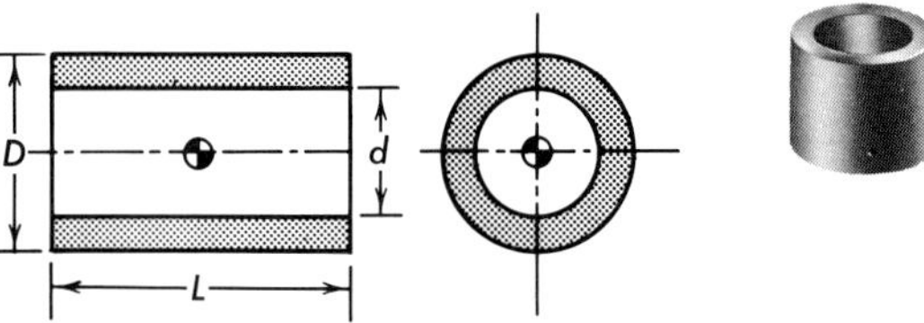

8. Half-hollow cylinder

$$V = \frac{\pi L}{8}(D^2 - d^2)$$

$$B = \frac{4}{3\pi}\left[\frac{R^3 - r^3}{R^2 - r^2}\right]$$

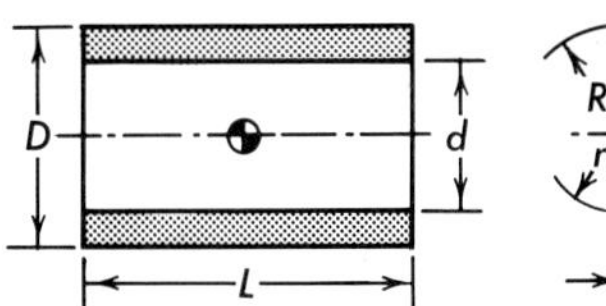

9. Sector of hollow cylinder

$$V = 0.01745(R^2 - r^2)\theta L$$

$$B = \frac{38.1972(R^3 - r^3)\sin\theta}{(R^2 - r^2)\theta}$$

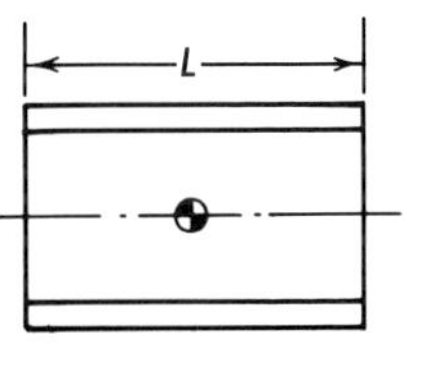

10. Truncated cylinder (with full circle base)

$$V = \frac{\pi}{8}D^2L = 0.3927D^2L$$

$$B_1 = 0.3125L$$

$$B_2 = 0.375D$$

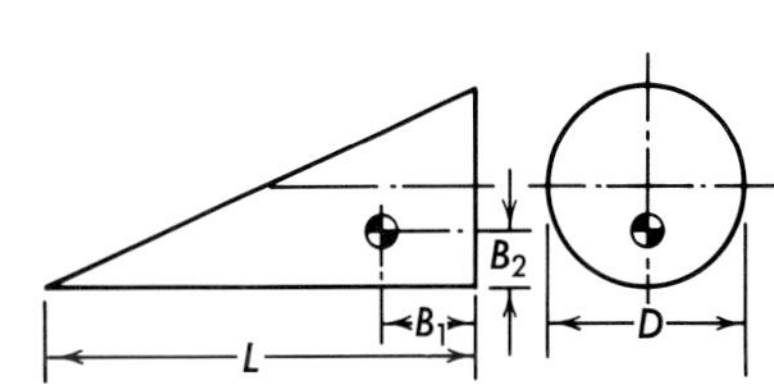

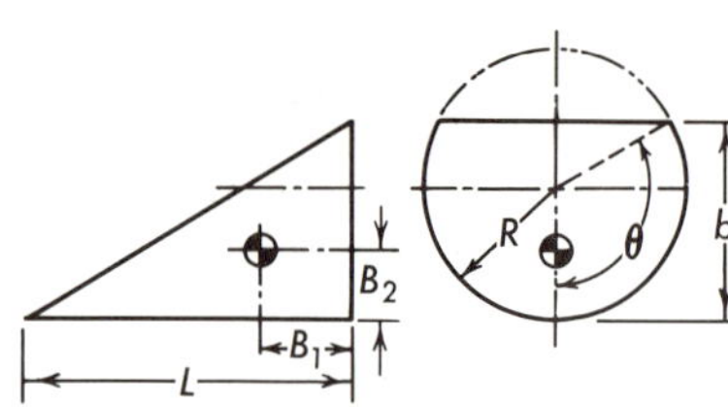

11. Truncated cylinder (with partial circle base)

$$b = R(1 - \cos\theta)$$

$$V = \frac{R^3L}{b}\left(\sin\theta - \frac{\sin^3\theta}{3} - \theta\cos\theta\right)$$

$$B_1 = \frac{L\left(\frac{\theta\cos^2\theta}{2} - \frac{5\sin\theta\cos\theta}{8} + \frac{\sin^3\theta\cos\theta}{12} + \frac{\theta}{8}\right)}{\left(1 - \cos\theta\right)\left(\sin\theta - \frac{\sin^3\theta}{3} - \theta\cos\theta\right)}$$

$$B_2 = \frac{2R\left(-\frac{\theta\cos\theta}{2} + \frac{\sin\theta}{2} - \frac{\theta}{8} + \frac{\sin\theta\cos\theta}{8} - N\right)}{\sin\theta - \frac{\sin^3\theta}{3} - \theta\cos\theta}$$

$$\text{where } N = \frac{\sin^3\theta}{6} - \frac{\sin^3\theta\cos\theta}{12}$$

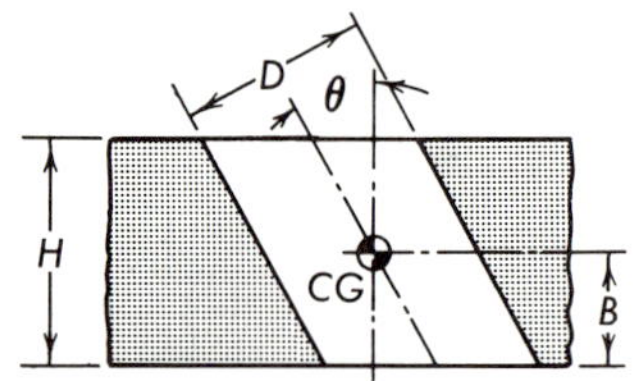

12. Oblique cylinder (or circular hole at oblique angle)

$$V = \frac{\pi}{4}D^2\frac{H}{\cos\theta} = 0.7854D^2H\sec\theta$$

$$B = \frac{H}{2} \qquad r = \frac{d}{2}$$

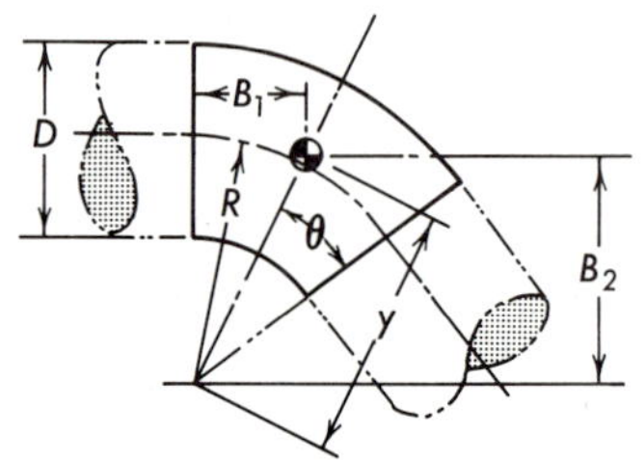

13. Bend in cylinder

$$V = \frac{\pi^2}{360}D^2R\theta = 0.0274D^2R\theta$$

$$y = R\left(1 + \frac{r^2}{4R^2}\right) \qquad \begin{aligned} B_1 &= y\tan\theta \\ B_2 &= y\cot\theta \end{aligned}$$

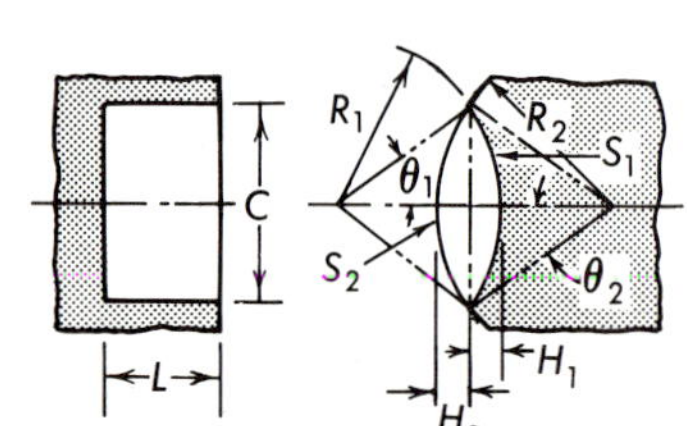

14. Curved groove in cylinder

$$\sin\theta_1 = \frac{C}{2R_1} \qquad \sin\theta_2 = \frac{C}{2R_2} \qquad S = 2R\theta$$

$$H_1 = R_1(1 - \cos\theta_1) \qquad H_2 = R_2(1 - \cos\theta_2)$$

$$V = L[R_1^2(\theta_1 - \tfrac{1}{2}\theta_1\sin 2\theta_1) + R_2^2(\theta_2 - \tfrac{1}{2}\theta_2\sin 2\theta_2)]$$

Compute CG of each part separately

15. Slot in cylinder

$$H = R(1 - \cos\theta) \qquad \sin\theta = \frac{C}{2R}$$

$$S = 2R\theta$$

$$V = L[CN + R^2(\theta - \tfrac{1}{2}\sin 2\theta)]$$

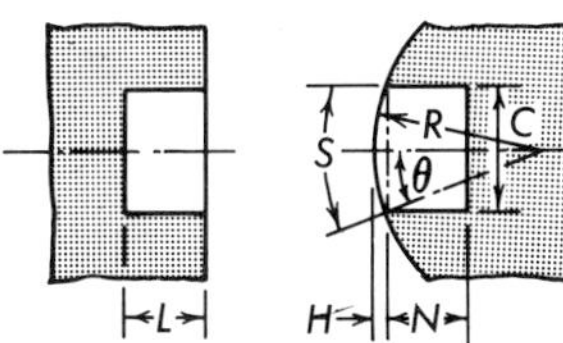

16. Slot in hollow cylinder

$$S = 2R\theta \qquad \sin\theta = \frac{C}{2R}$$

$$H = R(1 - \cos\theta)$$

$$V = L[CN - R^2(\theta - \tfrac{1}{2}\sin 2\theta)]$$

$$V = L\{CN - 0.5[RS - C(R - H)]\}$$

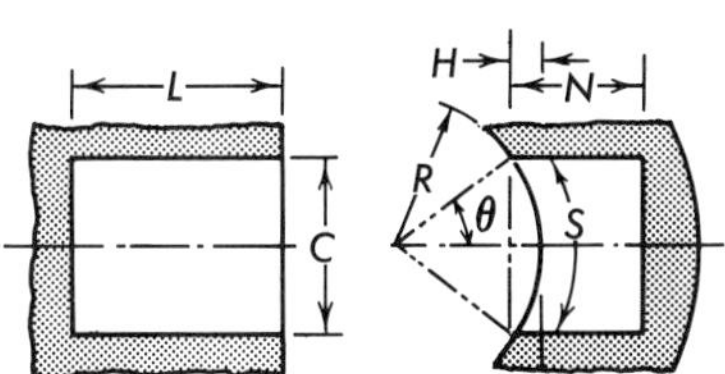

17. Curved groove in hollow cylinder

$$\sin\theta_1 = \frac{C}{2R_1} \qquad \sin\theta_2 = \frac{C}{2R_2} \qquad S = 2R\theta$$

$$H_1 = R_1(1 - \cos\theta_1)$$

$$H_2 = R_2(1 - \cos\theta_2)$$

$$V = L\{[R_2{}^2(\theta_2 - \tfrac{1}{2}\sin 2\theta_2)] - [R_1{}^2(\theta_1 - \tfrac{1}{2}\sin 2\theta_1)]\}$$

$$V = \frac{L}{2}\{[R_2S_2 - C(R_2 - H_2)] - [R_1S_1 - C(R_1 - H_1)]\}$$

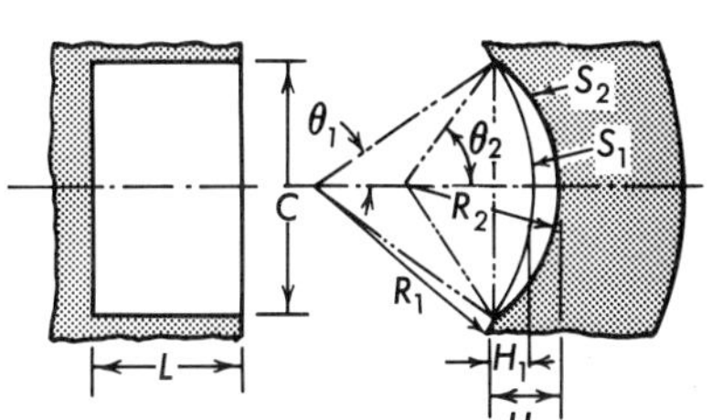

18. Slot through hollow cylinder

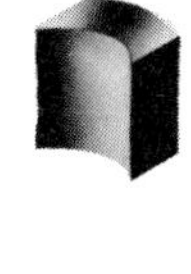

$$\sin\theta_1 = \frac{C}{R_1} \qquad \sin\theta_2 = \frac{C}{R_2}$$

$$S = 2R\theta$$

$$H_1 = R_1(1 - \cos\theta_1)$$

$$H_2 = R_2(1 - \cos\theta_2)$$

$$V = L\{CN + [R_1{}^2(\theta_1 - \tfrac{1}{2}\sin 2\theta_1)] - [R_2(\theta_2 - \tfrac{1}{2}\sin\theta_2)]\}$$

$$V = L\{CN + 0.5[R_1S_1 - C(R_1 - H_1)] - 0.5[R_2S_2 - C(R_2 - H_2)]\}$$

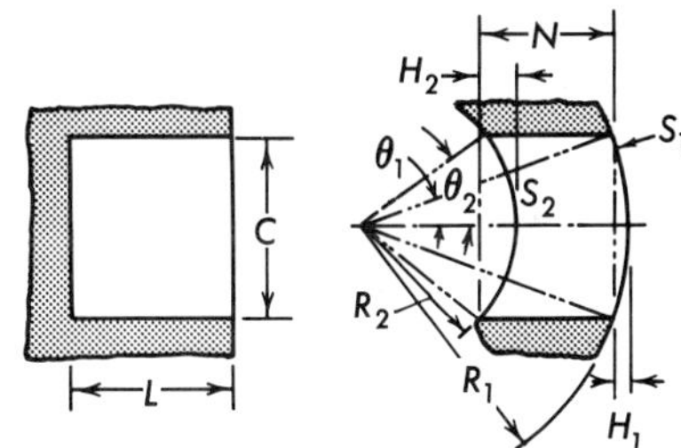

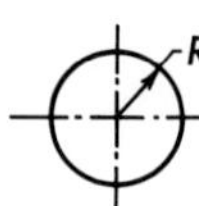

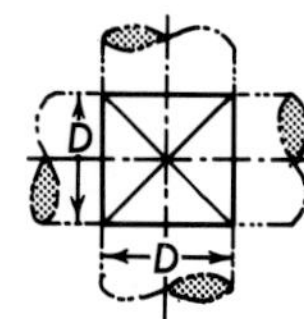

19. Intersecting cylinder (volume of junction box)

$$V = D^3\left(\frac{\pi}{2} - \frac{2}{3}\right) = 0.9041D^3$$

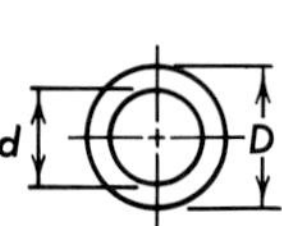

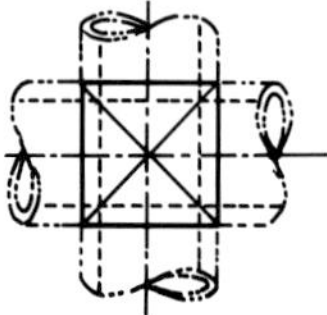

20. Intersecting hollow cylinders (volume of junction box)

$$V = \left(\frac{\pi}{2} - \frac{2}{3}\right)(D^3 - d^3) - \frac{\pi}{2}d^2(D - d)$$

$$V = 0.9041(D^3 - d^3) - 1.5708d^2(D - d)$$

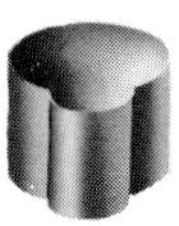
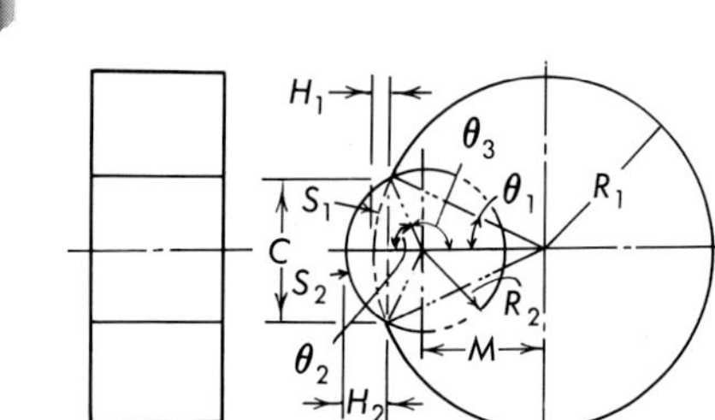

21. Intersecting parallel cylinders ($M < R_1$)

$$\theta_2 = 180° - \theta_3 \qquad \cos\theta_3 = \frac{R_2^2 + M^2 - R_1^2}{2MR_2}$$

$$\cos\theta_1 = \frac{R_1^2 + M^2 - R_2^2}{2MR_1}$$

$$H_1 = R_1(1 - \cos\theta_1)$$

$$S_1 = 2R_1\theta_1$$

$$V = L\{\pi R_1^2 + [R_2^2(\theta_2 - \tfrac{1}{2}\sin 2\theta_2)] - [R_1^2(\theta_1 - \tfrac{1}{2}\sin 2\theta_1)]\}$$

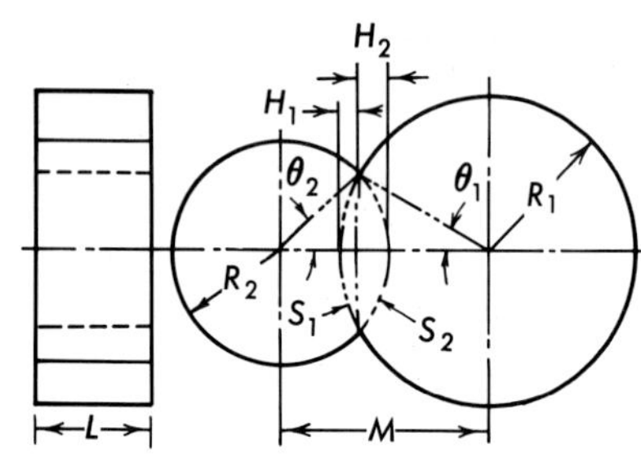

22. Intersecting parallel cylinders ($M > R_1$)

$$H_1 = R_1(1 - \cos\theta_1)$$

$$S_1 = 2R_1\theta_1$$

$$\cos\theta_1 = \frac{R_1^2 + M^2 - R_2^2}{2MR_1}$$

$$V = L\{[\pi(R_1^2 + R_2^2)] - [R_1^2(\theta_1 - \tfrac{1}{2}\sin 2\theta_1)] - [R_2^2(\theta_2 - \tfrac{1}{2}\sin 2\theta_2)]\}$$

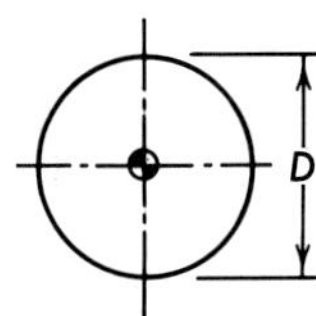

23. Sphere

$$V = \frac{\pi D^3}{6} = 0.5236D^3$$

$$\text{Area of surface} = 4\pi(\text{radius})^2 = \pi D^2$$

---

24. Hemisphere

$$V = \frac{\pi D^3}{12} = 0.2618D^3$$

$$B = 0.375R$$

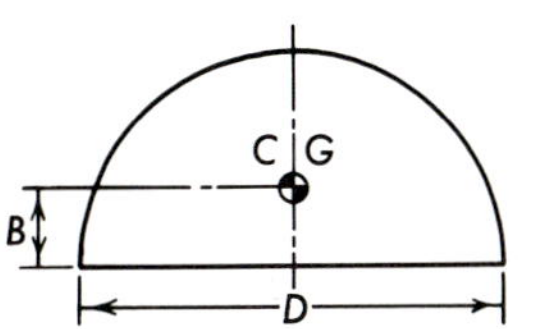

25. Spherical segment

$$V = \pi H^2\left(R - \frac{H}{3}\right)$$

$$B_1 = \frac{H(4R - H)}{4(3R - H)}$$

$$B_2 = \frac{3(2R - H)^2}{4(3R - H)}$$

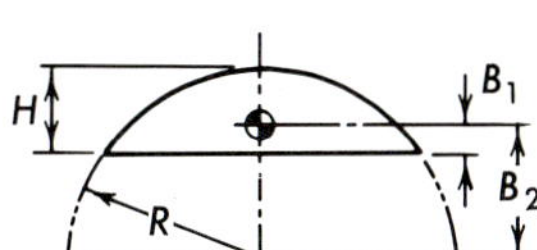

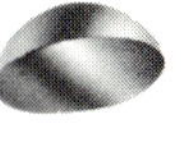

26. Spherical sector

$$V = \frac{2\pi}{3}R^2H = 2.0944R^2H$$

$$B = 0.375(1 + \cos\theta)$$

$$R = 0.375(2R - H)$$

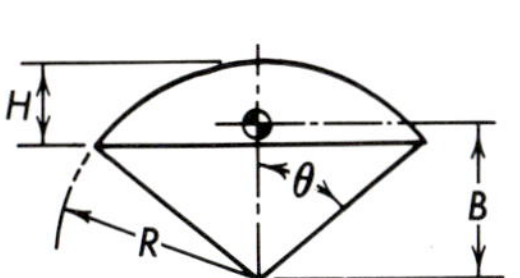

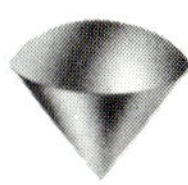

27. Shell of hollow hemisphere

$$V = \frac{2\pi}{3}(R^3 - r^3)$$

$$B = 0.375\left(\frac{R^4 - r^4}{R^3 - r^3}\right)$$

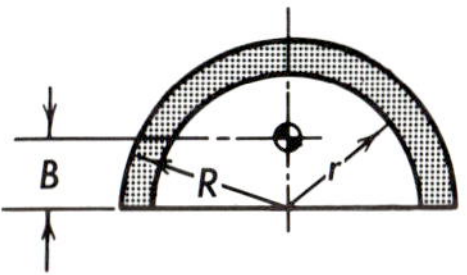

28. Hollow sphere

$$V = \frac{4\pi}{3}(R^3 - r^3)$$

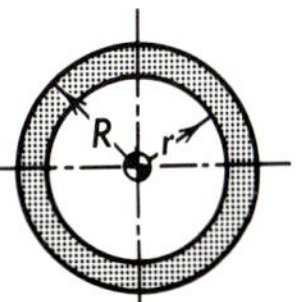

29. Shell of spherical sector

$$V = \frac{2\pi}{3}(R^2H - r^2h)$$

$$B = 0.375\left\{\frac{[R^2H(2R - H)] - [r^2h(2r - h)]}{R^2H - r^2h}\right\}$$

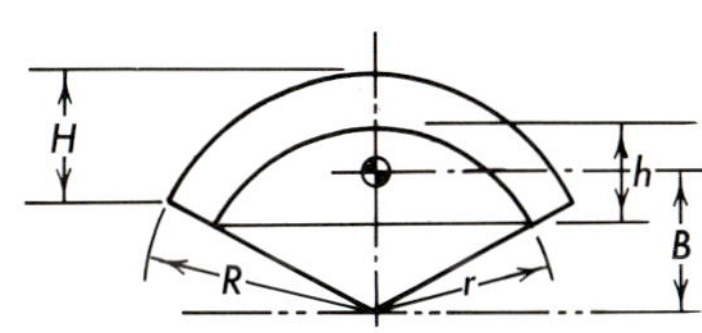

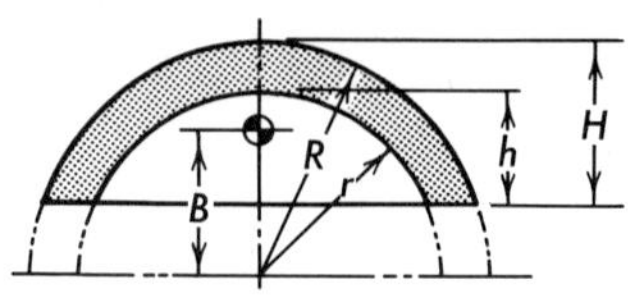

30. Shell of spherical segment

$$V = \pi\left[H^2\left(R - \frac{H}{3}\right) - h^2\left(r - \frac{h}{3}\right)\right]$$

$$B = \frac{3}{4}\left[\frac{\left(R - \frac{H}{3}\right)\frac{H^2(2R - H)^2}{3R - H} - \left(r - \frac{h}{3}\right)\frac{h^2(2r - h)^2}{3r - h}}{H^2\left(R - \frac{H}{3}\right) - h^2\left(r - \frac{h}{3}\right)}\right]$$

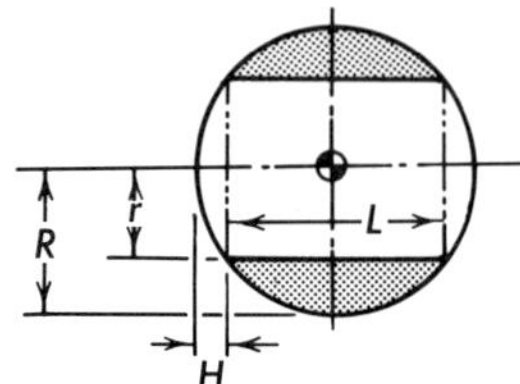

31. Circular hole through sphere

$$V = \pi\left[r^2L + 2H^2\left(R - \frac{H}{3}\right)\right] \qquad H = R - \sqrt{R^2 - r^2}$$

$$L = 2(R - H)$$

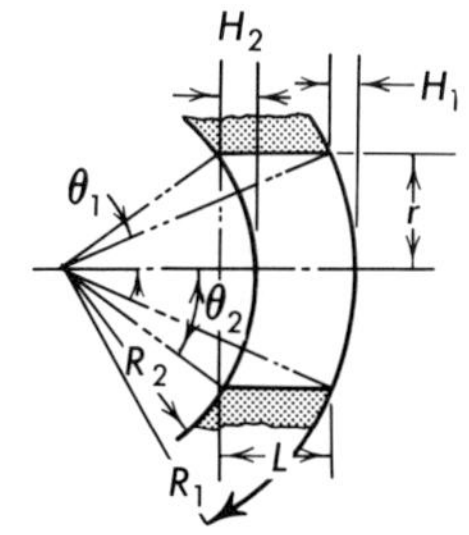

32. Circular hole through hollow sphere

$$V = \pi\left[r^2L + H_1{}^2\left(R_1 - \frac{H_1}{3}\right) - H_2{}^2\left(R_2 - \frac{H_2}{3}\right)\right]$$

$$\sin\theta_1 = \frac{r}{R_1} \qquad \sin\theta_2 = \frac{r}{R_2} \qquad H = R(1 - \cos\theta)$$

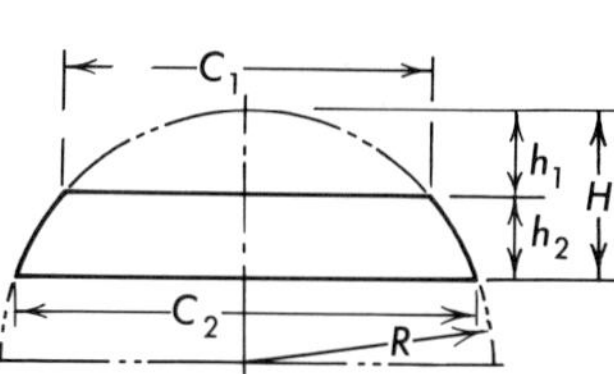

33. Spherical zone

$$V = \pi\left\{\left[H^2\left(R - \frac{H}{3}\right)\right] - \left[h_1{}^2\left(R - \frac{h_1}{3}\right)\right]\right\}$$

$$V = \frac{\pi h_2}{6}\left(\frac{3}{4}C_1{}^2 + \frac{3}{4}C_2{}^2 + h_2{}^2\right)$$

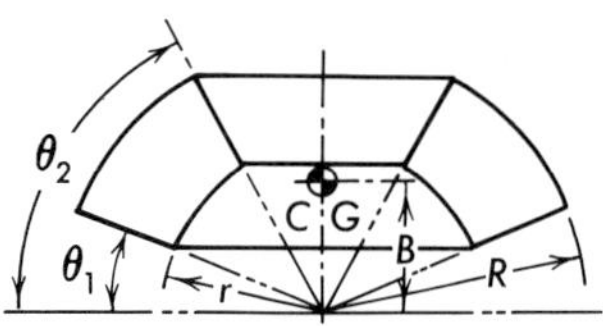

34. Conical hole through spherical shell

$$V = \frac{2\pi}{3}(R^3 - r^3)(\sin\theta_2 - \sin\theta_1)$$

$$B = \frac{0.375(R^4 - r^4)(\sin\theta_2 + \sin\theta_1)}{R^3 - r^3}$$

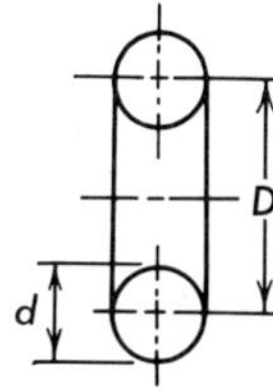

35. Torus

$$V = \tfrac{1}{4}\pi^2d^2D = 2.467d^2D$$

36. Hollow torus

$$V = \tfrac{1}{4}\pi^2 D(d_1^{\,2} - d_2^{\,2})$$

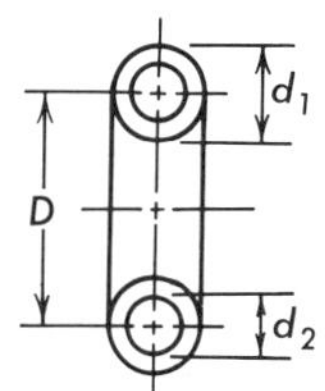

37. Bevel ring

$$V = \pi(R + \tfrac{1}{3}W)WH$$

$$B = H\left(\frac{\frac{R}{3} + \frac{W}{12}}{R + \frac{W}{3}}\right)$$

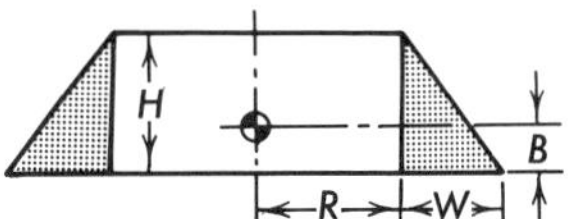

38. Bevel ring

$$B > \frac{H}{3}$$

$$V = \pi(R - \tfrac{1}{3}W)WH$$

$$B = H\left(\frac{\frac{R}{3} - \frac{W}{12}}{R - \frac{W}{3}}\right)$$

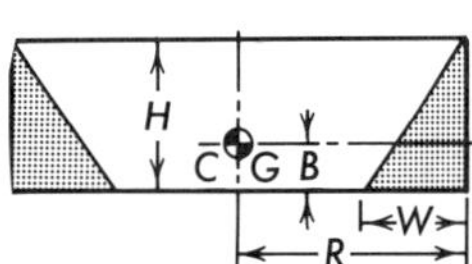

39. Quarter-torus

$$B < 0.4244R$$

$$V = \frac{\pi^2 R^2}{2}\left(r + \frac{4R}{3\pi}\right) = 4.9348R^2(r + 0.4244R)$$

$$B = \frac{4R}{3\pi}\left(\frac{r + \frac{3R}{8}}{r + \frac{4R}{3\pi}}\right) = \frac{0.4244Rr + 0.1592R^2}{r + 0.4244R}$$

40. Quarter-torus

$$V = \frac{\pi^2 R^2}{2}\left(r - \frac{4R}{3\pi}\right)$$

$$B = \frac{4R}{3\pi}\left(\frac{r - \frac{3R}{8}}{r - \frac{4R}{3\pi}}\right)$$

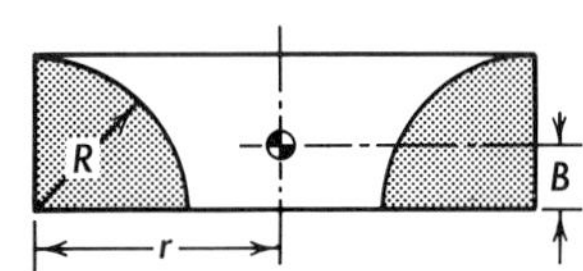

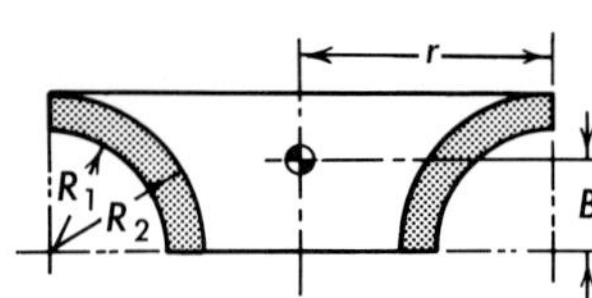

41. Curved shell ring

$$V = 2\pi\left[r - \frac{4}{3\pi}\left(\frac{R_2^3 - R_1^3}{R_2^2 - R_1^2}\right)\right]\frac{\pi}{4}(R_2^2 - R_1^2)$$

$$B = \frac{4}{3\pi}\left\{\frac{R_2^3\left(r - \frac{3}{8}R_2\right) - R_1^3\left(r - \frac{3}{8}R_1\right)}{(R_2^2 - R_1^2)\left[r - \frac{4}{3\pi}\left(\frac{R_2^3 - R_1^3}{R_2^2 - R_1^2}\right)\right]}\right\}$$

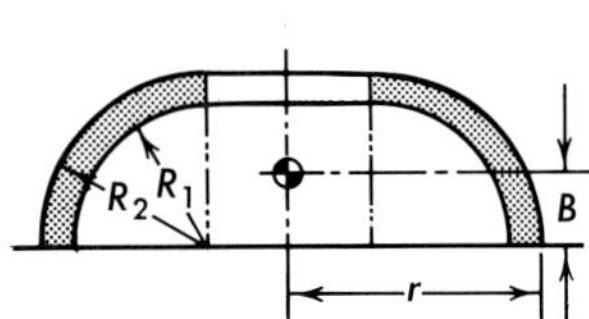

42. Curved shell ring

$$V = \frac{\pi^2}{2}\left[r(R_2^2 - R_1^2) + \frac{4}{3\pi}(R_2^3 - R_1^3)\right]$$

$$B = \frac{2}{\pi}\left[\frac{\frac{2r}{3}(R_2^3 - R_1^3) + \frac{1}{4}(R_2^4 - R_1^4)}{r(R_2^2 - R_1^2) + \frac{4}{3\pi}(R_2^3 - R_1^3)}\right]$$

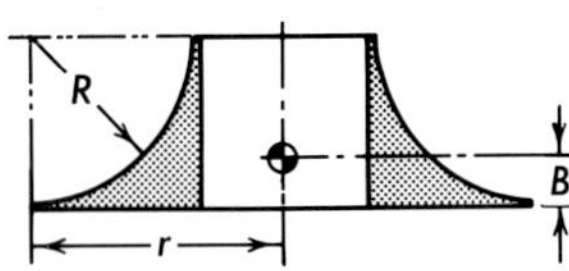

43. Fillet ring

$$V = 2\pi R^2\left[\left(1 - \frac{\pi}{4}\right)r - \frac{R}{6}\right]$$

$$B = R\left[\frac{\left(\frac{5}{6} - \frac{\pi}{4}\right)r - \frac{R}{24}}{\left(1 - \frac{\pi}{4}\right)r - \frac{R}{6}}\right]$$

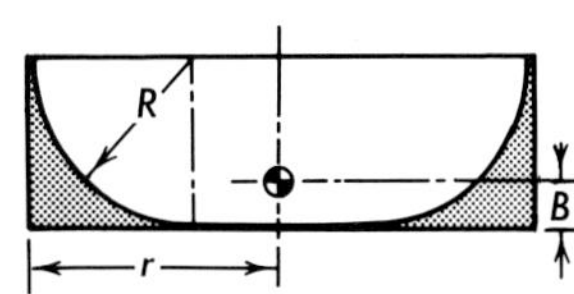

44. Fillet ring

$$V = 2\pi R^2\left[\left(1 - \frac{\pi}{4}\right)r - \left(\frac{5}{6} - \frac{\pi}{4}\right)R\right]$$

$$B = R\left[\frac{\left(\frac{5}{6} - \frac{\pi}{4}\right)r - \left(\frac{19}{24} - \frac{\pi}{4}\right)R}{\left(1 - \frac{\pi}{4}\right)r - \left(\frac{5}{6} - \frac{\pi}{4}\right)R}\right]$$

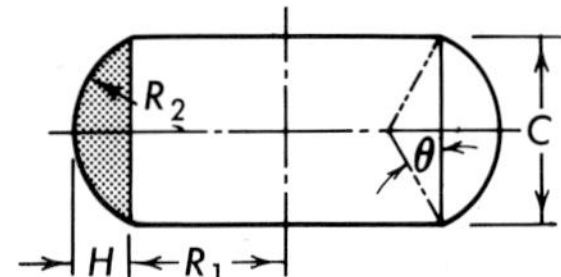

45. Curved-sector ring

$$V = 2\pi R_2^2\left[R_1 + \left(\frac{4\sin 3\theta}{6\theta - 3\sin 2\theta} - \cos\theta\right)R_2\right](\theta - 0.5\sin 2\theta)$$

46. Ellipsoidal cylinder

$$V = \frac{\pi}{4} AaL$$

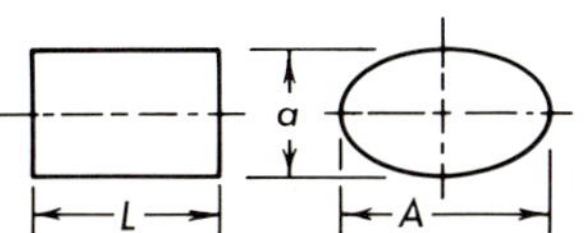

47. Ellipsoid

$$V = \tfrac{4}{3}\pi ACE$$

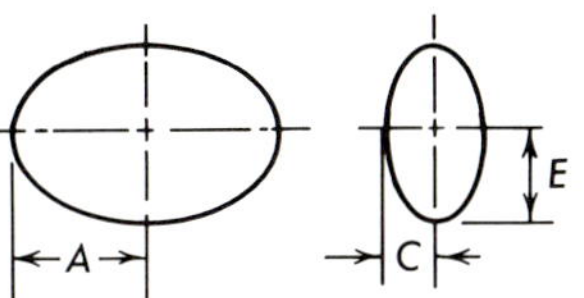

48. Paraboloid

$$V = \frac{\pi}{8} HD^2 \qquad B = \frac{1}{3} H$$

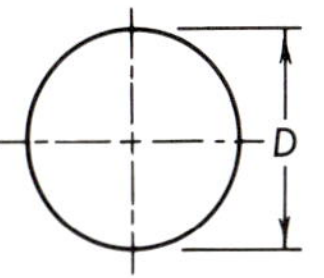

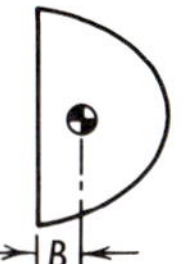

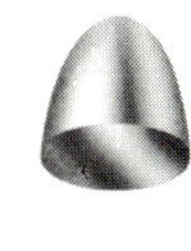

49. Pyramid (with base of any shape)

A = area of base $\qquad V = \tfrac{1}{3}AH \qquad B = \tfrac{1}{4}H$

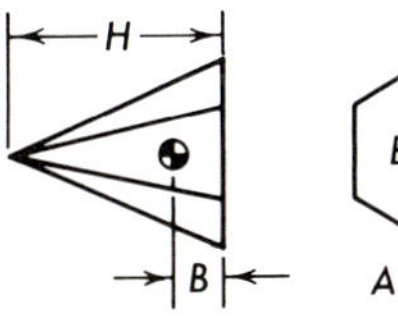

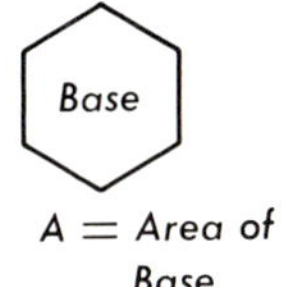

50. Frustum of pyramid (with base of any shape)

$$V = \tfrac{1}{3}H(A_1 + \sqrt{A_1A_2} + A_2)$$

$$B = \frac{H(A_1 + 2\sqrt{A_1A_2} + 3A_2)}{4(A_1 + \sqrt{A_1A_2} + A_2)}$$

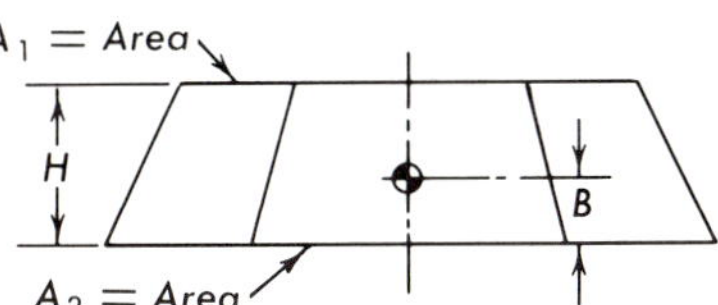

51. Cone

$$V = \frac{\pi}{12} D^2H \qquad B = \frac{1}{4} H$$

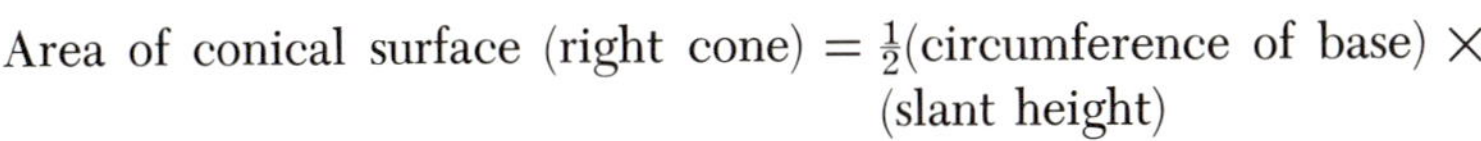
Area of conical surface (right cone) = $\tfrac{1}{2}$(circumference of base) × (slant height)

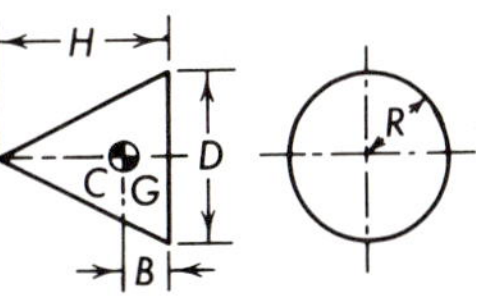

52. Frustum of cone

$$V = \frac{\pi}{12} H(D^2 + Dd + d^2)$$

$$B = \frac{H(D^2 + 2Dd + 3d^2)}{4(D^2 + Dd + d^2)}$$

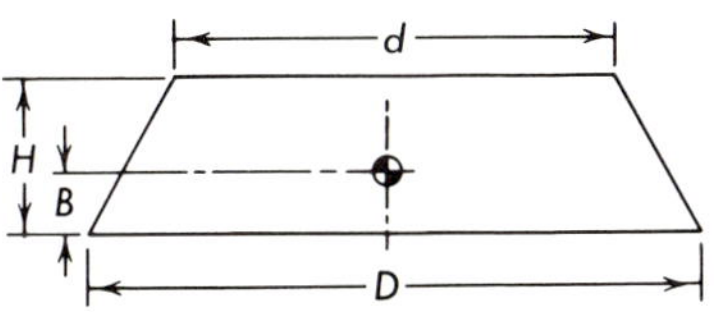

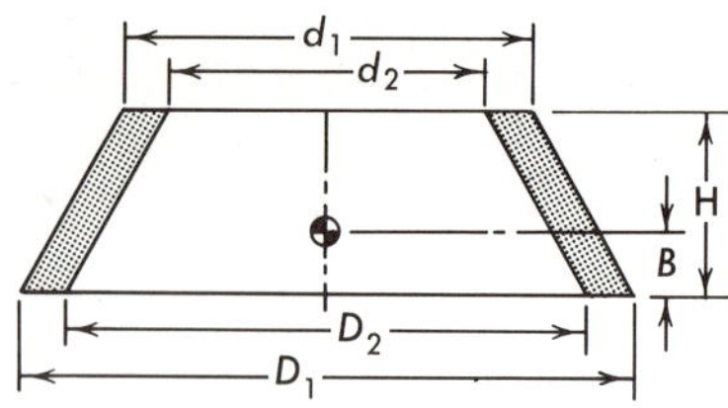

53. Frustum of hollow cone

$$V = 0.2618H[(D_1^2 + D_1d_1 + d_1^2) - (D_2^2 + D_2d_2 + d_2^2)]$$

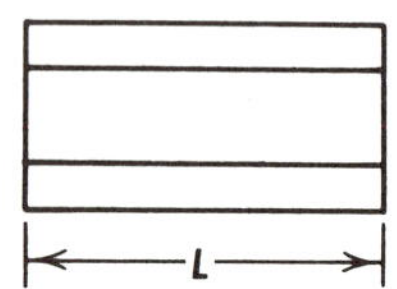

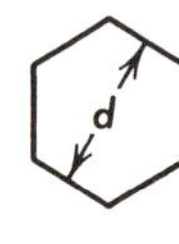

54. Hexagon

$$V = \frac{\sqrt{3}}{2}d^2L$$

$$V = 0.866d^2L$$

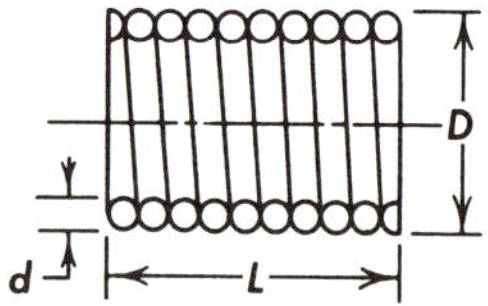

55. Closely packed helical springs

$$V = \frac{\pi^2 dL}{4}(D - d)$$

$$V = 2.4674(D - d)$$

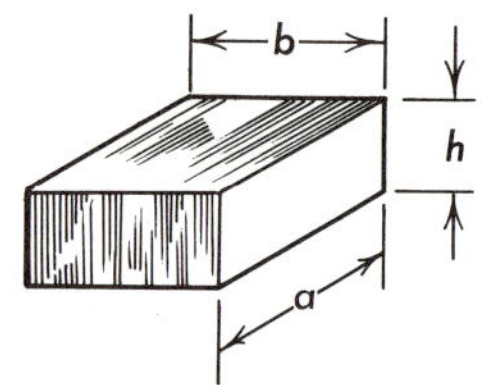

56. Rectangular prism

Volume = length × width × height

Volume = area of base × altitude

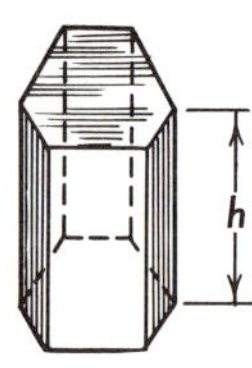

57. Any prism

(Axis either perpendicular or inclined to base)

Volume = (area of base)(perpendicular height)

Volume = (lateral length)(area of perpendicular cross section)

materials and basic manufacturing shapes III

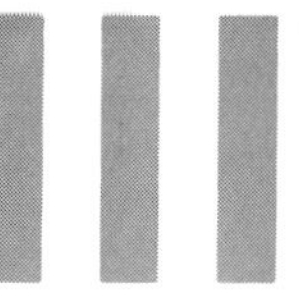

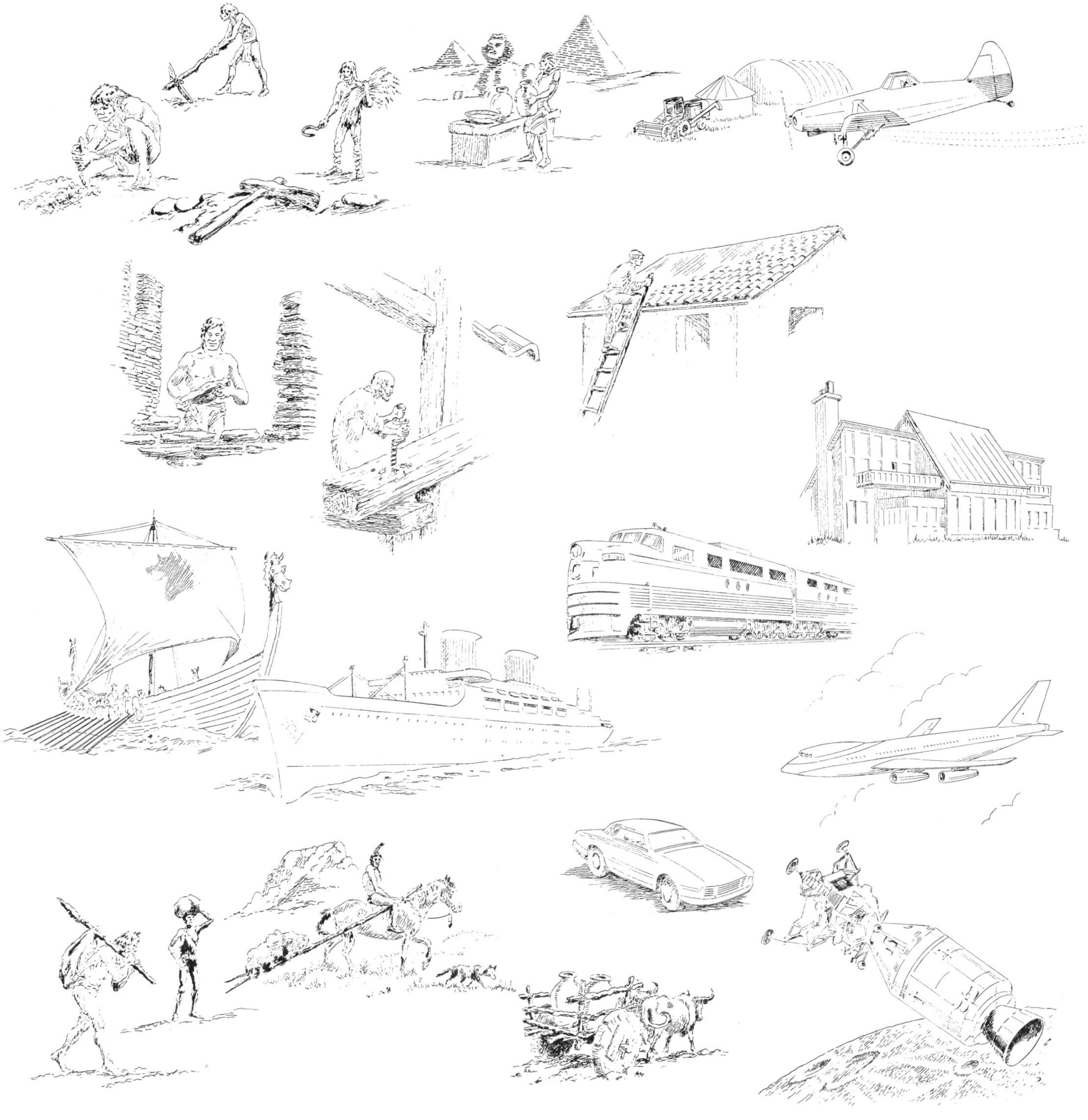

PROPERTIES, CHARACTERISTICS, AND USES OF TYPICAL ENGINEERING MATERIALS [1]

Material & Composition[2]	Specific Weight lb/ft^3	Modulus of Elasticity 10^6 psi	Tensile[3] Strength 10^3 psi	Elong.[3] at break % at 2"	Coeff. of Thermal Expansion $10^6 \times 1/°C$	Electr. Resist. 10^{-6} Ω-cm
CARBON STEELS	487	28.5-30			12.5	10
AISI 1020 (0.2 C, 0.45 Mn, 0.25 Si)			55-61	25-15		
AISI 1045 (0.45 C, 0.75 Mn)			82-94	16-12		
AISI 1090 (0.90 C, 0.40 Mn)			122	10		
ALLOY STEELS	487	28.5-30			~12	~9
AISI 1340 (1.6-1.9 Mn)			110-230	25-11		
AISI 4140 (.9 Mn, 1 Cr, .2 Mo)			100-225	25-10		
AISI 8640 (.9 Mn, .5 Cr, .2 Mo, .55 Ni)			120-240	23-10		
STAINLESS STEELS	495	28-29			17.3	72
304 (19 Cr, 9 Ni, 2 Mn)			85-185	60-8		
316 (17 Cr, 12 Ni, 2 Mn, 2-3 Mo)			90-150	50-8		
420 (16 Cr, 2 Ni, 1 Mn)			95-230	25-8		
CAST IRON	442				12.5	
A 159 (3.4-3.7 C, 2.3-2.8 Si)		13	25	0.5		67
A 571 (2.2-2.7 C, 2 Si, 4 Mn, 2.2 Ni)		25	65	30		30
CAST STEEL	480	28.5				
A 27 (0.3 C, .75 Mn, .6 Si, .5 Ni, .25 Cr)			65	24		
A 216 (.25 C, .7 Mn, .6 Si, .5 Ni, .25 Mo)			60	24		
COPPER ALLOYS						
Beryll. Copper (1.9 Be, .25 Co)	514	19	90-200	10-2	17	
Yellow Brass (65 Cu, 35 Zn)	534	15.5	46-128	65-3	19	7
Al-Bronze (91 Cu, 7Al, 2 Fe)	481	17	76-89	45-32	17	12
TITANIUM ALLOY (8 Al, 1 Mo, 1 V)	312	16.5	150-170	18-10	8.8	80
MAGNESIUM ALLOY (1 Zr, 3 Al)	110	6.5	37-42	21-14	26	9
ALUMINUM ALLOYS	170	10-10.5			23	4
1060 (99.6 Al)			10-20	43-6		
2024 (4.5 Cu, .6 Mn, 1.5 Mg)			65-70	18-6		
6061 (.6 Si, .25 Cu, 1 Mg, .2 Cr)			18-45	25-12		
PLASTICS						10^{12} Ω-cm
Acrylic	73.5	0.3-0.5	6-12	10-2	50-60	> 100
Epoxy	72	0.3-0.6	4-13	6-2	45-90	1-100
Glass Fiber Filament Reinforced	(136)	6.0-7.5	(130-200)			
Fluorocarbon	137	0.03-0.07	2-7	200-400	21-43	10^6
Nylon 6	71	0.4	9-12	200	74	
Polycarbonate	75	0.3-0.4	8-9.5	20-100	66	2×10^4
Poly-vinyl Chloride (PVC)	87	0.2-0.6	5-9	40-2	50-180	$1 - 10^4$
REFRACTORIES						
Concrete (Unreinforced)	140	2-5	0.3-0.8	~0		
Glass (Soda-Lime)	158	10	0.5-8	~0	8.5	1
Alumina (Al_2O_3)	247	50	25-35	~0	5.4	$10^2 - 10^3$
Graphite (C)	138	1	1-2		3.6	10^{-3} Ω-cm
WOOD						
Douglas Fir	32	1.1	1.6			
White Oak	46	1.5	2.1			

[1] From various sources, primarily Parker, E. R., *Materials Data Book*, McGraw-Hill 1967, and Weast, R. C., Ed. *Handbook for Applied Engineering Sciences*, The Chemical Rubber Co., 1970.

[2] Composition in percent.

[3] Depends on thermal treatment, usually hi-strength goes with lo-elongation and vice versa.

Comments	Major Characteristics	Major Uses
Mild Steel	Moderate strength, easy to machine & weld	Bolts, nails, tubing, structural steel
	Medium strength, moderate machineability	Forgings, basic machine parts
Hot Rolled	Hard, hi-strength, takes edge, hard to weld	Springs, cutting tools, drills, dies
Typically 0.4% C		
Manganese Steel	Good machineability, hi strength	Auto & farm implements, axles, shafts
Chrome-Moly Steel	Hardenable, good hi-temp strength	Gears, aircraft & machine tool parts
Nickel-Chrome-Moly Steel	Strength, corrosion resist, hi-temp	Heavy duty gears, cams, bolts
Austenitic	Non-magnetic, weldable	General purpose, food handling, ext. trim
Austenitic	Non-magnetic, hi-temp, hi corrosion resistance	Chemical & food process equipment
Martensitic	Magnetic, will take edge	Surgical instruments
Gray C.I.	Low cost, brittle, low strength	Casting where strength is not needed
Ductile C.I.	Easy to machine	Pumps, pressure parts, engine blocks
	Easy machine & weld, moderate strength	General purpose
	Hi-temp resistance	Furnace tubes & fittings
	Tough, flexible	Springs, diaphragms, bellows
	Cold workability, strength	Radiators, lamp fixtures
	Easy to form & forge, corrosion resistant	Machine parts
	High strength, low weight, costly	Aircraft & missile parts
Type AZ 31B	Medium strength, low weight, easy to machine	Structural, where weight is critical
Type 0 - H18	Soft, hi conductivity	Electrical conductors, chemical equipment
Type T3 - T86	Hi-strength, tough, easy machining	Screw machine products, aircraft parts
Type 0 - T6	Weldable & easy machining	Structures, general use
Plexiglas, Lucite	Transparent, easy machine & glue	Aircraft windows, lighting fixtures
	Strong, Hi-temp resistance	Potting electronic parts
	Hi strength, limited shapes, costly	Missile bodies, rocket motor cases
Teflon	Lo friction, hi temp resistance	Bearings, utensil liners, chemical reactors
	Lo friction, good strength & toughness	Bearings, gears, tubing
	Hi impact strength, moderate cost	Molded parts like crash helmets
	Low cost, moderate strength	Most molded parts--decorative, household, etc.
All refractories are 10-100 times stronger in compression than in tension	Strength depends on proportions & cure	Structural
	Ordinary glass	Windows, bottles, etc.
	Very stable, melts above 2000°C, hard	Rocket nozzles, abrasives, furnace parts
	Highest temp resist, melts above 2500°C	Heat shields for re-entry vehicles, furnaces
Strength in direction of fibers	Moderate strength, fast growth, little warpage	Buildings, concrete forms
	Dense, hard, uniform	Furniture, sills, floors

Specific Gravities and Specific Weights

Material	Average Specific Gravity	Average Specific Weight, lb_f/ft^3
Acid, sulfuric, 87%	1.80	112
Air, S.T.P.	0.001293	0.0806
Alcohol, ethyl	0.790	49
Aluminum, cast	2.65	165
Asbestos	2.5	153
Ash, white	0.67	42
Ashes, cinders	0.68	44
Asphalt	1.3	81
Babbitt metal, soft	10.25	625
Basalt, granite	1.50	96
Brass, cast-rolled	8.50	534
Brick, common	1.90	119
Bronze, 7.9 to 14% Sn	8.1	509
Cedar, white, red	0.35	22
Cement, Portland, bags	1.44	90
Chalk	2.25	140
Clay, dry	1.00	63
Clay, loose, wet	1.75	110
Coal, anthracite, solid	1.60	95
Coal, bituminous, solid	1.35	85
Concrete, gravel, sand	2.3	142
Copper, cast, rolled	8.90	556
Cork	0.24	15
Cotton, flax; hemp	1.48	93
Copper ore	4.2	262
Earth	1.75	105
Fir, Douglas	0.50	32
Flour, loose	0.45	28
Gasoline	0.70	44
Glass, crown	2.60	161
Glass, flint	3.30	205
Glycerine	1.25	78
Gold, cast-hammered	19.3	1205
Granite, solid	2.70	172
Gravel, loose, wet	1.68	105
Hickory	0.77	48
Ice	0.91	57
Kerosene	0.80	50
Lead	11.34	710
Leather	0.94	59
Limestone, solid	2.70	168
Limestone, crushed	1.50	95
Mahogany	0.70	44
Manganese	7.42	475
Marble	2.70	166
Mercury	13.56	845
Monel metal, rolled	8.97	555
Nickel	8.90	558
Oak, white	0.77	48
Oil, lubricating	0.91	57
Paper	0.92	58
Paraffin	0.90	56
Petroleum, crude	0.88	55
Pine, white	0.43	27
Redwood, California	0.42	26
Rubber	1.25	78
Sand, loose, wet	1.90	120
Sandstone, solid	2.30	144
Sea water	1.03	64
Silver	10.5	655
Steel, structural	7.90	490
Sulfur	2.00	125
Teak, African	0.99	62
Tin	7.30	456
Tungsten	19.22	1200
Turpentine	0.865	54
Water, 4°C	1.00	62.4
Water, snow, fresh fallen	0.125	8.0

Note: The value for the specific weight of water, which is usually used in problem solutions, is 62.4 lb_f/ft^3 or 8.34 lb_f/gal.

Properties of Pure Metals*

Metals	Specific Gravity	Modulus of Elasticity, millions of psi	Melting Point, °C	Coeff. of Expansion $10^{-6} \times 1/°C$**	Thermal Conductivity, kcal/hr-cm-°C†	Electrical Resistivity, $\mu\Omega$-cm
Aluminum	2.70	10	660	25	2.04	2.66
Beryllium	1.85	42	1286	12.0	1.87	4.0
Chromium	7.2	36	1860	5.9	0.77	13
Cobalt	8.9	30	1497	12.0	0.60	9
Copper	8.96	17	1082	16.5	3.42	1.673
Gold	19.32	11	1064	14.2	2.70	2.35
Iron	7.87	28.5	1538	12.0	0.69	9.7
Lead	11.35	2.0	327	29	0.30	20.6
Magnesium	1.74	6.4	649	25	1.37	4.45
Mercury	13.55	–	−39	–	0.072	98.4
Nickel	8.90	31	1452	13.1	0.77	6.85
Niobium (Columbium)	8.57	15	2470	7.0	0.45	13
Platinum	21.45	21	1772	9	0.63	10.5
Silicon	2.33	16	1410	5	0.72	1×10^5
Silver	10.50	10.5	961	20	3.67	1.59
Sodium	0.97	–	98	70	1.15	4.2
Tin	7.31	6	232	20	0.55	11.0
Titanium	4.54	16	1672	8.5	0.18	43
Tungsten	19.3	50	3400	4.5	1.53	5.65
Uranium	18.8	24	1130	13.3	0.21	30
Vanadium	6.1	19	1900	7.9	0.52	25
Zinc	7.0	12	420	34	0.99	5.92

*Compiled from various sources.
**Can also be written μin/in/°C.
†1 kcal/hr-cm-°C = 67.2 Btu/hr-ft-°F.

material specification systems

SAE Numbering System
for Wrought or Rolled Steel—SAE J402b
SAE Standard

Report of Iron and Steel Division approved January 1912 and last revised by Iron and Steel Technical Committee May 1969.

This SAE Standard is intended to supply a uniform means of designating wrought ferrous materials reported in SAE Standards and Recommended Practices.

Only compositions which conform to the SAE compositions given in the current SAE Handbook should bear the prefix "SAE."

A numeral index system is used to identify the compositions of the SAE steels, which system makes possible use of numerals on shop drawings and blueprints to describe partially the composition of the material.

The first digit indicates the type to which the steel belongs, that is, "1" indicates a carbon steel; "2" a nickel steel; and "3" a nickel–chromium steel. In the case of the simple alloy steels, the second digit generally indicates an alloy or alloy combination, and sometimes the approximate percentage of the predominant alloying element. Usually the last two or three digits indicate the approximate carbon content in "points" or hundredths of one per cent. Thus, "SAE 5135" indicates a chromium steel of approximately 1% chromium (0.80 to 1.05%) and 0.35% carbon (0.33 to 0.38%).

In some instances, in order to avoid confusion, it has been found necessary to depart from this system of identifying the approximate alloy composition of a steel by varying the second and third digits of the number. Instances of such departure are the steel numbers selected for several of the corrosion- and heat-resisting alloys and the triple alloy steels.

The basic numerals of the various types of SAE steel are given in the table.

Basic Numbering System for SAE Steels

Numerals and Digits	Type of Steel and Average Chemical Contents, %
	Carbon Steels
10XX	Plain Carbon (Mn 1.00% max)
11XX	Resulfurized
12XX	Resulfurized and Rephosphorized
15XX	Plain Carbon (max Mn range—over 1.00–1.65%)
	Manganese Steels
13XX	Mn 1.75
	Nickel Steels
23XX	Ni 3.50
25XX	Ni 5.00
	Nickel–Chromium Steels
31XX	Ni 1.25; Cr 0.65 and 0.80
32XX	Ni 1.75; Cr 1.07
33XX	Ni 3.50; Cr 1.50 and 1.57
34XX	Ni 3.00; Cr 0.77
	Molybdenum Steels
40XX	Mo 0.20 and 0.25
44XX	Mo 0.40 and 0.52
	Chromium–Molybdenum Steels
41XX	Cr 0.50, 0.80 and 0.95; Mo 0.12, 0.20, 0.25, and 0.30
	Nickel–Chromium–Molybdenum Steels
43XX	Ni 1.82; Cr 0.50 and 0.80; Mo 0.25
43BVXX	Ni 1.82; Cr 0.50; Mo 0.12 and 0.25; V 0.03 minimum
47XX	Ni 1.05; Cr 0.45; Mo 0.20 and 0.35
81XX	Ni 0.30; Cr 0.40; Mo 0.12
86XX	Ni 0.55; Cr 0.50; Mo 0.20
87XX	Ni 0.55; Cr 0.50; Mo 0.25
88XX	Ni 0.55; Cr 0.50; Mo 0.35
93XX	Ni 3.25; Cr 1.20; Mo 0.12
94XX	Ni 0.45; Cr 0.40; Mo 0.12
97XX	Ni 0.55; Cr 0.20; Mo 0.20
98XX	Ni 1.00; Cr 0.80; Mo 0.25

Numerals and Digits	Type of Steel and Average Chemical Contents, %
	Nickel–Molybdenum Steels
46XX	Ni 0.85 and 1.82; Mo 0.20 and 0.25
48XX	Ni 3.50; Mo 0.25
	Chromium Steels
50XX	Cr 0.27, 0.40, 0.50, and 0.65
51XX	Cr 0.80, 0.87, 0.92, 0.95, 1.00 and 1.05
501XX	Cr 0.50
511XX	Cr 1.02
521XX	Cr 1.45
	Chromium–Vanadium Steels
61XX	Cr 0.60, 0.80, and 0.95; V 0.10 and 0.15 minimum
	Tungsten–Chromium Steels
71XXX	W 13.50 and 16.50; Cr 3.50
72XX	W 1.75; Cr 0.75
	Silicon–Manganese Steels
92XX	Si 1.40 and 2.00; Mn 0.65, 0.82, and 0.85; Cr 0.00 and 0.65
	Low-Alloy High-Tensile Steels
9XX	Various
	Stainless Steels
	(Chromium–Manganese–Nickel)
302XX	Cr 17.00 and 18.00; Mn 6.50 and 8.75; Ni 4.50 and 5.00
	(Chromium–Nickel)
303XX	Cr 8.50, 15.50, 17.00, 18.00, 19.00, 20.00, 20.50, 23.00, 25.00
	Ni 7.00, 9.00, 10.00, 10.50, 11.00, 11.50, 12.00, 13.00, 13.50, 20.50, 21.00, 35.00
	(Chromium)
514XX	Cr 11.12, 12.25, 12.50, 13.00, 16.00, 17.00, 20.50, and 25.00
515XX	Cr 5.00
	Boron–Intensified Steels
XXBXX	B denotes Boron Steel
	Leaded–Steels
XXLXX	L denotes Leaded Steel

NOMENCLATURE FOR COMMON PRODUCT FEATURES

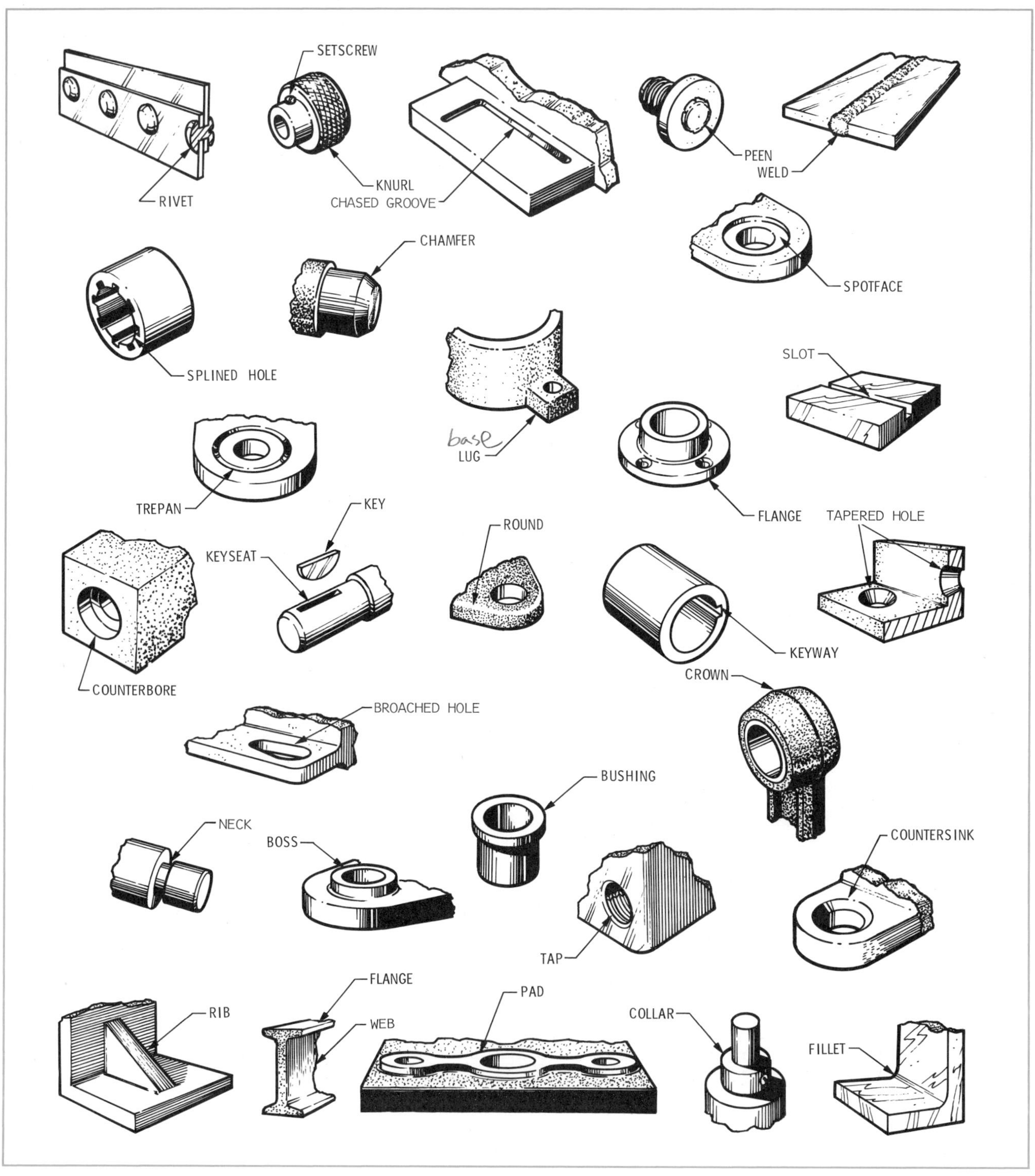

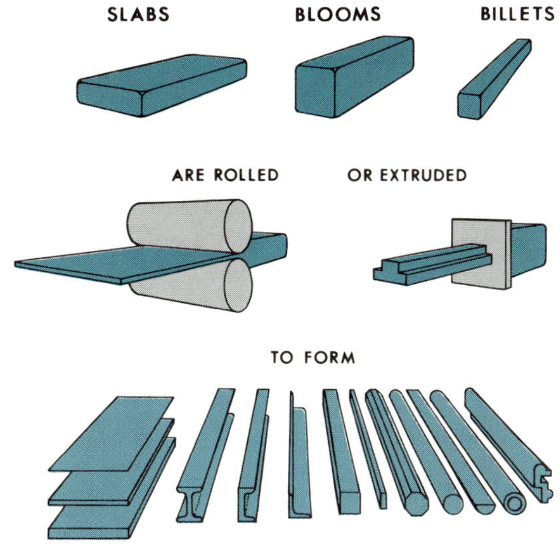

Structural Steel Nomenclature

The shapes listed in the tabulation are those that are most commonly available. Other shapes are also available from steel producers. The nomenclature for such unlisted shapes will be related to the shape profile in the same manner as the nomenclature for each listed shape is related to its profile. For the most generally used types of shapes, this relation is as follows:

1. W shapes are doubly symmetric wide flange shapes used as beams or columns whose inside flange surfaces are substantially parallel. A shape having essentially the same nominal weight and dimensions as a W shape listed in the tabulation but whose inside flange surfaces are not parallel may also be considered a W shape having the same nomenclature as the tabulated shape, provided its average flange thickness is essentially the same shape as the flange thickness of the W shape.

2. S shapes are doubly symmetric shapes produced in accordance with dimensional standards adopted in 1896 by the Association of American Steel Manufactures for American Standard beam shapes. The essential part of these standards is that the inside flange surfaces of American Standard beam shapes have approximately a 16–2/3% slope.

3. M shapes are doubly symmetric shapes that cannot be classified as W, S, or bearing pile shapes. (Although not included in the standard nomenclature tabulation, bearing piles are doubly symmetric wide flange shapes whose inside flange surfaces are essentially parallel and whose flange and web have essentially the same thickness).

4. C shapes are channels produced in accordance with dimensional standards adopted in 1896 by the Association of American Steel Manufactures for American Standard channels. The essential part of these standards is that the inside flange surfaces of American Standard channels have approximately a 16–2/3% slope.

5. L shapes are equal-leg and unequal-leg angles.

Shape	New Identifying Symbol
Wide Flange Shapes	W
Miscellaneous Shapes	M
American Standard Beams	S
American Standard Channels	C
Miscellaneous Channels	MC
Angles—Equal Legs	L
Angles—Unequal Legs	L
Bulb Angles	BL
Structural Tees Cut From Wide Flange Shapes	WT
Structural Tees Cut From Miscellaneous Shapes	MT
Structural Tees Cut From American Standard Beams	ST
Tees	T
Wall Tee	AT
Elevator Tees	ET
Zees	Z
Special Car Building Shapes	CZ

Courtesy American Institute of Steel Construction

STRUCTURAL STEEL SHAPES

HP Shapes

Designation	Depth d (in)	Flange Width b_f (in)	Flange Thickness t_f (in)	Web Thickness t_w (in)
HP 10 × 57	10	$10\frac{1}{4}$	$\frac{9}{16}$	$\frac{9}{16}$
× 42	$9\frac{3}{4}$	$10\frac{1}{8}$	$\frac{7}{16}$	$\frac{7}{16}$
HP 8 × 36	8	$8\frac{1}{8}$	$\frac{7}{16}$	$\frac{7}{16}$

M Shapes

Designation	Depth d (in)	Flange Width b_f (in)	Flange Thickness t_f (in)	Web Thickness t_w (in)
M 14 × 17.2	14	4	$\frac{1}{4}$	$\frac{3}{16}$
M 12 × 11.8	12	$3\frac{1}{8}$	$\frac{1}{4}$	$\frac{3}{16}$
M 10 × 29.1	$9\frac{7}{8}$	$5\frac{7}{8}$	$\frac{3}{8}$	$\frac{7}{16}$
× 22.9	$9\frac{7}{8}$	$5\frac{3}{4}$	$\frac{3}{8}$	$\frac{1}{4}$
M 10 × 9	10	$2\frac{3}{4}$	$\frac{3}{16}$	$\frac{3}{16}$
M 8 × 37.7	$8\frac{1}{8}$	8	$\frac{1}{2}$	$\frac{3}{8}$
× 34.3	8	8	$\frac{7}{16}$	$\frac{3}{8}$
× 32.6	8	8	$\frac{7}{16}$	$\frac{5}{16}$
M 8 × 22.5	8	$5\frac{3}{8}$	$\frac{3}{8}$	$\frac{3}{8}$
× 18.5	8	$5\frac{1}{4}$	$\frac{3}{8}$	$\frac{1}{4}$
M 8 × 6.5	8	$2\frac{1}{4}$	$\frac{3}{16}$	$\frac{1}{8}$
M 7 × 5.5	7	$2\frac{1}{8}$	$\frac{3}{16}$	$\frac{1}{8}$
M 6 × 33.75	$6\frac{1}{4}$	$6\frac{1}{8}$	$\frac{5}{8}$	$\frac{1}{2}$
× 22.5	6	6	$\frac{3}{8}$	$\frac{3}{8}$
× 20	6	6	$\frac{3}{8}$	$\frac{1}{4}$
M 6 × 4.4	6	$1\frac{7}{8}$	$\frac{3}{16}$	$\frac{1}{8}$
M 5 × 18.9	5	5	$\frac{7}{16}$	$\frac{5}{16}$
M 4 × 16.3	$4\frac{1}{4}$	4	$\frac{1}{2}$	$\frac{5}{16}$
× 13.8	4	4	$\frac{3}{8}$	$\frac{5}{16}$
× 13	4	4	$\frac{3}{8}$	$\frac{1}{4}$

W Shapes

Designation	Depth d (in)	Flange Width b_f (in)	Flange Thickness t_f (in)	Web Thickness t_w (in)
W 16 × 50	$16\frac{1}{4}$	$7\frac{1}{8}$	$\frac{5}{8}$	$\frac{3}{8}$
× 45	$16\frac{1}{8}$	7	$\frac{9}{16}$	$\frac{3}{8}$
× 40	16	7	$\frac{1}{2}$	$\frac{5}{16}$
× 36	$15\frac{7}{8}$	7	$\frac{7}{16}$	$\frac{5}{16}$
W 14 × 38	$14\frac{1}{8}$	$6\frac{3}{4}$	$\frac{1}{2}$	$\frac{5}{16}$
× 34	14	$6\frac{3}{4}$	$\frac{7}{16}$	$\frac{5}{16}$
× 30	$13\frac{7}{8}$	$6\frac{3}{4}$	$\frac{3}{8}$	$\frac{1}{4}$
W 12 × 36	$12\frac{1}{4}$	$6\frac{5}{8}$	$\frac{9}{16}$	$\frac{5}{16}$
× 31	$12\frac{1}{8}$	$6\frac{1}{2}$	$\frac{7}{16}$	$\frac{1}{4}$
× 27	12	$6\frac{1}{2}$	$\frac{3}{8}$	$\frac{1}{4}$
W 10 × 45	$10\frac{1}{8}$	8	$\frac{5}{8}$	$\frac{3}{8}$
× 39	10	8	$\frac{1}{2}$	$\frac{5}{16}$
× 33	$9\frac{3}{4}$	8	$\frac{7}{16}$	$\frac{5}{16}$
W 10 × 29	$10\frac{1}{4}$	$5\frac{3}{4}$	$\frac{1}{2}$	$\frac{5}{16}$
× 25	$10\frac{1}{8}$	$5\frac{3}{4}$	$\frac{7}{16}$	$\frac{1}{4}$
× 21	$9\frac{7}{8}$	$5\frac{3}{4}$	$\frac{5}{16}$	$\frac{1}{4}$
W 8 × 67	9	$8\frac{1}{4}$	$\frac{15}{16}$	$\frac{9}{16}$
× 58	$8\frac{3}{4}$	$8\frac{1}{4}$	$\frac{13}{16}$	$\frac{1}{2}$
× 48	$8\frac{1}{2}$	$8\frac{1}{8}$	$\frac{11}{16}$	$\frac{3}{8}$
× 40	$8\frac{1}{4}$	$8\frac{1}{8}$	$\frac{9}{16}$	$\frac{3}{8}$
× 35	$8\frac{1}{8}$	8	$\frac{1}{2}$	$\frac{5}{16}$
× 31	8	8	$\frac{7}{16}$	$\frac{5}{16}$
W 6 × 25	$6\frac{3}{8}$	$6\frac{1}{8}$	$\frac{7}{16}$	$\frac{5}{16}$
× 20	$6\frac{1}{4}$	6	$\frac{3}{8}$	$\frac{1}{4}$
× 15.5	6	6	$\frac{1}{4}$	$\frac{1}{4}$
W 5 × 18.5	$5\frac{1}{8}$	5	$\frac{7}{16}$	$\frac{1}{4}$
× 16	5	5	$\frac{3}{8}$	$\frac{1}{4}$
W 4 × 13	$4\frac{1}{8}$	4	$\frac{3}{8}$	$\frac{1}{4}$

Courtesy American Institute of Steel Construction, 1970.

STRUCTURAL STEEL SHAPES

S Shapes

Designation	Depth of Section d	Flange		Web Thickness t_w
		Width b_f	Thickness t_f	
	in	in	in	in
S 15 × 50	15	$5\frac{5}{8}$	$\frac{5}{8}$	$\frac{9}{16}$
× 42.9	15	$5\frac{1}{2}$	$\frac{5}{8}$	$\frac{7}{16}$
S 12 × 50	12	$5\frac{1}{2}$	$\frac{11}{16}$	$\frac{11}{16}$
× 40.8	12	$5\frac{1}{4}$	$\frac{11}{16}$	$\frac{7}{16}$
S 12 × 25	12	$5\frac{1}{8}$	$\frac{9}{16}$	$\frac{7}{16}$
× 31.8	12	5	$\frac{9}{16}$	$\frac{3}{8}$
S 10 × 35	10	5	$\frac{1}{2}$	$\frac{5}{8}$
× 25.4	10	$4\frac{5}{8}$	$\frac{1}{2}$	$\frac{5}{16}$
S 8 × 23	8	$4\frac{1}{8}$	$\frac{7}{16}$	$\frac{7}{16}$
× 18.4	8	4	$\frac{7}{16}$	$\frac{1}{4}$
S 7 × 20	7	$3\frac{7}{8}$	$\frac{3}{8}$	$\frac{7}{16}$
× 15.3	7	$3\frac{5}{8}$	$\frac{3}{8}$	$\frac{1}{4}$
S 6 × 17.25	6	$3\frac{5}{8}$	$\frac{3}{8}$	$\frac{7}{16}$
× 12.5	6	$3\frac{3}{8}$	$\frac{3}{8}$	$\frac{1}{4}$
S 5 × 14.75	5	$3\frac{1}{4}$	$\frac{5}{16}$	$\frac{1}{2}$
× 10	5	3	$\frac{5}{16}$	$\frac{3}{16}$
S 4 × 9.5	4	$2\frac{3}{4}$	$\frac{5}{16}$	$\frac{5}{16}$
× 7.7	4	$2\frac{5}{8}$	$\frac{5}{16}$	$\frac{3}{16}$
S 3 × 7.5	3	$2\frac{1}{2}$	$\frac{1}{4}$	$\frac{3}{8}$
× 5.7	3	$2\frac{3}{8}$	$\frac{1}{4}$	$\frac{3}{16}$

Misc. Channels

Designation	Depth of Section d	Flange		Web Thickness t_w
		Width b_f	Avg. Thickness t_f	
	in	in	in	in
MC 18 × 58	18	$4\frac{1}{4}$	$\frac{5}{8}$	$\frac{11}{16}$
× 51.9	18	$4\frac{1}{8}$	$\frac{5}{8}$	$\frac{5}{8}$
× 45.8	18	4	$\frac{5}{8}$	$\frac{1}{2}$
× 42.7	18	4	$\frac{5}{8}$	$\frac{7}{16}$
MC 12 × 50	12	$4\frac{1}{8}$	$\frac{11}{16}$	$\frac{13}{16}$
× 45	12	4	$\frac{11}{16}$	$\frac{11}{16}$
× 40	12	$3\frac{7}{8}$	$\frac{11}{16}$	$\frac{9}{16}$
× 35	12	$3\frac{3}{4}$	$\frac{11}{16}$	$\frac{7}{16}$
MC 10 × 41.1	10	$4\frac{3}{8}$	$\frac{9}{16}$	$\frac{13}{16}$
× 33.6	10	$4\frac{1}{8}$	$\frac{9}{16}$	$\frac{9}{16}$
× 28.5	10	4	$\frac{9}{16}$	$\frac{7}{16}$
MC 10 × 28.3	10	$3\frac{1}{2}$	$\frac{9}{16}$	$\frac{1}{2}$
× 25.3	10	$3\frac{1}{2}$	$\frac{1}{2}$	$\frac{7}{16}$
× 24.9	10	$3\frac{3}{8}$	$\frac{9}{16}$	$\frac{3}{8}$
× 21.9	10	$3\frac{1}{2}$	$\frac{1}{2}$	$\frac{5}{16}$
MC 7 × 22.7	7	$3\frac{5}{8}$	$\frac{1}{2}$	$\frac{1}{2}$
× 19.1	7	$3\frac{1}{2}$	$\frac{1}{2}$	$\frac{3}{8}$
MC 7 × 17.6	7	3	$\frac{1}{2}$	$\frac{3}{8}$
MC 6 × 18	6	$3\frac{1}{2}$	$\frac{1}{2}$	$\frac{3}{8}$
× 15.3	6	$3\frac{1}{2}$	$\frac{3}{8}$	$\frac{5}{16}$
MC 3 × 9	3	$2\frac{1}{8}$	$\frac{3}{8}$	$\frac{1}{2}$
× 7.1	3	2	$\frac{3}{8}$	$\frac{5}{16}$

Amer. Std. Channels

Designation	Depth of Section d	Flange		Web Thickness t_w
		Width b_f	Avg. Thickness t_f	
	in	in	in	in
C 15 × 50	15	$3\frac{3}{4}$	$\frac{5}{8}$	$\frac{11}{16}$
× 40	15	$3\frac{1}{2}$	$\frac{5}{8}$	$\frac{1}{2}$
× 33.9	15	$3\frac{3}{8}$	$\frac{5}{8}$	$\frac{3}{8}$
C 12 × 30	12	$3\frac{1}{8}$	$\frac{1}{2}$	$\frac{1}{2}$
× 25	12	3	$\frac{1}{2}$	$\frac{3}{8}$
× 20.7	12	3	$\frac{1}{2}$	$\frac{5}{16}$
C 10 × 30	10	3	$\frac{7}{16}$	$\frac{11}{16}$
× 25	10	$2\frac{7}{8}$	$\frac{7}{16}$	$\frac{1}{2}$
× 20	10	$2\frac{3}{4}$	$\frac{7}{16}$	$\frac{3}{8}$
× 15.3	10	$2\frac{5}{8}$	$\frac{7}{16}$	$\frac{1}{4}$
C 9 × 20	9	$2\frac{5}{8}$	$\frac{7}{16}$	$\frac{7}{16}$
× 15	9	$2\frac{1}{2}$	$\frac{7}{16}$	$\frac{5}{16}$
× 13.4	9	$2\frac{3}{8}$	$\frac{7}{16}$	$\frac{1}{4}$
C 8 × 18.75	8	$2\frac{1}{2}$	$\frac{3}{8}$	$\frac{1}{2}$
× 13.75	8	$2\frac{3}{8}$	$\frac{3}{8}$	$\frac{5}{16}$
× 11.5	8	$2\frac{1}{4}$	$\frac{3}{8}$	$\frac{1}{4}$
C 7 × 14.75	7	$2\frac{1}{4}$	$\frac{3}{8}$	$\frac{7}{16}$
× 12.25	7	$2\frac{1}{4}$	$\frac{3}{8}$	$\frac{5}{16}$
× 9.8	7	$2\frac{1}{8}$	$\frac{3}{8}$	$\frac{3}{16}$
C 6 × 13	6	$2\frac{1}{8}$	$\frac{5}{16}$	$\frac{7}{16}$
× 10.5	6	2	$\frac{5}{16}$	$\frac{5}{16}$
× 8.2	6	$1\frac{7}{8}$	$\frac{5}{16}$	$\frac{3}{16}$
C 5 × 9	5	$1\frac{7}{8}$	$\frac{5}{16}$	$\frac{5}{16}$
× 6.7	5	$1\frac{3}{4}$	$\frac{5}{16}$	$\frac{3}{16}$
C 4 × 7.25	4	$1\frac{3}{4}$	$\frac{5}{16}$	$\frac{5}{16}$
× 5.4	4	$1\frac{5}{8}$	$\frac{5}{16}$	$\frac{3}{16}$
C 3 × 6	3	$1\frac{5}{8}$	$\frac{1}{4}$	$\frac{3}{8}$
× 5	3	$1\frac{1}{2}$	$\frac{1}{4}$	$\frac{1}{4}$
× 4.1	3	$1\frac{3}{8}$	$\frac{1}{4}$	$\frac{3}{16}$

Courtesy American Institute of Steel Construction, 1970.

Wire and Sheet Metal Gages
In decimals of an inch

Name of Gage	*United States Standard Gage		The United States Steel Wire Gage	American or Brown & Sharpe Wire Gage	New Birmingham Standard Sheet & Hoop Gage	British Imperial or English Legal Standard Wire Gage	Birmingham or Stubs Iron Wire Gage	Name of Gage
Principal Use	Uncoated Steel Sheets and Light Plates		Steel Wire except Music Wire	Nonferrous Sheets and Wire	Iron and Steel Sheets and Hoops	Wire	Strips, Bands, Hoops, and Wire	Principal Use
Gage No.	Weight, oz/ft²	Approx. Thickness, In	Thickness, in					Gage No.
7/0's			0.4900		0.6666	0.500		7/0's
6/0's			0.4615	0.5800	0.625	0.464		6/0's
5/0's			0.4305	0.5165	0.5883	0.432	0.500	5/0's
4/0's			0.3938	0.4600	0.5416	0.400	0.454	4/0's
3/0's			0.3625	0.4096	0.500	0.372	0.425	3/0's
2/0's			0.3310	0.3648	0.4452	0.348	0.380	2/0's
1/0			0.3065	0.3249	0.3964	0.324	0.340	1/0
1			0.2830	0.2893	0.3532	0.300	0.300	1
2			0.2625	0.2576	0.3147	0.276	0.284	2
3	160	0.2391	0.2437	0.2294	0.2804	0.252	0.259	3
4	150	0.2242	0.2253	0.2043	0.250	0.232	0.238	4
5	140	0.2092	0.2070	0.1819	0.2225	0.212	0.220	5
6	130	0.1943	0.1920	0.1620	0.1981	0.192	0.203	6
7	120	0.1793	0.1770	0.1443	0.1764	0.176	0.180	7
8	110	0.1644	0.1620	0.1285	0.1570	0.160	0.165	8
9	100	0.1495	0.1483	0.1144	0.1398	0.144	0.148	9
10	90	0.1345	0.1350	0.1019	0.1250	0.128	0.134	10
11	80	0.1196	0.1205	0.0907	0.1113	0.116	0.120	11
12	70	0.1046	0.1055	0.0808	0.0991	0.104	0.109	12
13	60	0.0897	0.0915	0.0720	0.0882	0.092	0.095	13
14	50	0.0747	0.0800	0.0641	0.0785	0.080	0.083	14
15	45	0.0673	0.0720	0.0571	0.0699	0.072	0.072	15
16	40	0.0598	0.0625	0.0508	0.0625	0.064	0.065	16
17	36	0.0538	0.0540	0.0453	0.0556	0.056	0.058	17
18	32	0.0478	0.0475	0.0403	0.0495	0.048	0.049	18
19	28	0.0418	0.0410	0.0359	0.0440	0.040	0.042	19
20	24	0.0359	0.0348	0.0320	0.0392	0.036	0.035	20
21	22	0.0329	0.0317	0.0285	0.0349	0.032	0.032	21
22	20	0.0299	0.0286	0.0253	0.0313	0.028	0.028	22
23	18	0.0269	0.0258	0.0226	0.0278	0.024	0.025	23
24	16	0.0239	0.0230	0.0201	0.0248	0.022	0.022	24
25	14	0.0209	0.0204	0.0179	0.0220	0.020	0.020	25
26	12	0.0179	0.0181	0.0159	0.0196	0 018	0.018	26
27	11	0.0164	0.0173	0.0142	0.0175	0.0164	0.016	27
28	10	0.0149	0.0162	0.0126	0.0156	0.0148	0.014	28
29	9	0.0135	0.0150	0.0113	0.0139	0.0136	0.013	29
30	8	0.0120	0.0140	0.0100	0.0123	0.0124	0.012	30
31	7	0.0105	0.0132	0.0089	0.0110	0.0116	0.010	31
32	6.5	0.0097	0.0128	0.0080	0.0098	0.0108	0.009	32
33	6	0.0090	0.0118	0.0071	0.0087	0.0100	0.008	33
34	5.5	0.0082	0.0104	0.0063	0.0077	0.0092	0.007	34
35	5	0.0075	0.0095	0.0056	0.0069	0.0084	0.005	35
36	4.5	0.0067	0.0090	0.0050	0.0061	0.0076	0.004	36
37	4.25	0.0064	0.0085	0.0045	0.0054	0.0068		37
38	4	0.0060	0.0080	0.0040	0.0048	0.0060		38
39			0.0075	0.0035	0.0043	0.0052		39
40			0.0070	0.0031	0.0039	0.0048		40

*U. S. Standard Gage is officially a weight gage, in oz/ft² as tabulated. The approx. thickness shown is the "Manufacturers' Standard" of the American Iron and Steel Institute, based on steel as weighing 501.81 lb/ft³ (489.6 true weight plus 2.5 per cent for average overrun in area and thickness). The AISI standard nomenclature for flat-rolled carbon steel is as follows:

Classification of Rolled Stock

Thickness (Inches)	Width (Inches) To 3½ incl.	Over 3½ to 6	Over 6 to 8	Over 8 to 12	Over 12 to 48	Over 48
0.2300 & thicker	Bar	Bar	Bar	Plate	Plate	Plate
0.2299 to 0.2031	Bar	Bar	Strip	Strip	Sheet	Plate
0.2030 to 0.1800	Strip	Strip	Strip	Strip	Sheet	Plate
0.1799 to 0.0449	Strip	Strip	Strip	Strip	Sheet	Sheet
0.0448 to 0.0344	Strip	Strip				
0.0343 to 0.0255	Strip		Hot-rolled sheet and strip not generally produced in these widths and thicknesses			
0.0254 & thinner						

Courtesy American Institute of Steel Construction.

IV standard parts specifications

fastener nomenclature

A bolt is designed for assembly with a nut. A screw has features in its design that make it capable of being used in a tapped or other preformed hole in the work. Because of basic design, it is possible to use certain types of screws in combination with a nut. Any externally threaded fastener that has a majority of the design characteristics which assist its proper use in a tapped or other preformed hole is a screw, regardless of how it is used in its service application.

An externally threaded fastener that, because of head design or other feature, is prevented from being turned during assembly, and that can be tightened or released only by torquing a nut, is a bolt. (Example: round-head bolts, track bolts, plow bolts.)

An externally threaded fastener that has a thread form which prohibits assembly with a nut having a straight thread of multiple pitch length is a screw. (Example: wood screws, tapping screws.)

An externally threaded fastener that must be assembled with a nut to perform its intended service is a bolt. (Example: heavy-hex structural bolt.)

An externally threaded fastener that must be torqued by its head into a tapped or other preformed hole to perform its intended service is a screw. (Example: square-head set screw.)

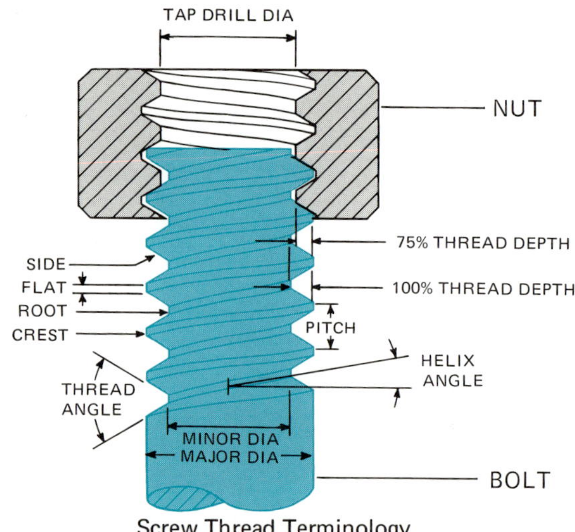

Screw Thread Terminology

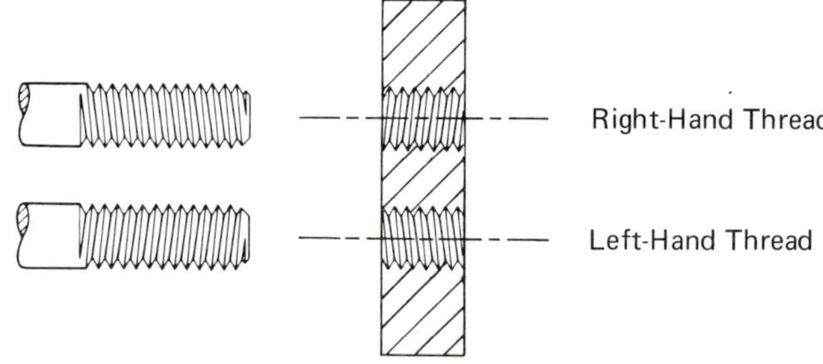

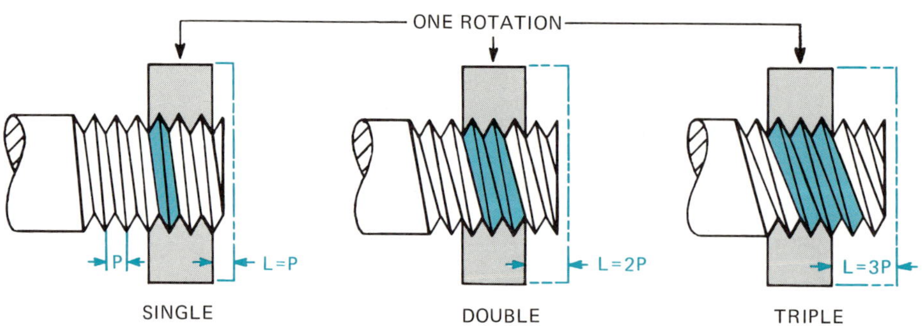

Single, Double, and Triple Threads

Extracted from ANSI B18.2.1-1965 and ANSI B18.12-1962.

Bolts

Bolt A bolt is an externally threaded fastener designed for insertion through holes in assembled parts, and is normally intended to be tightened or released by torquing a nut.

BENT BOLT A bent bolt is a cylindircal rod having one end threaded and the other end bent to some desired configuration; also, it may be a bent cylindrical rod having both ends threaded. Threads may be cut or rolled; points are usually plain sheared.

Eye Bolt, Closed Anchor Ring A closed anchor ring eye bolt is a bent bolt having a head in the form of a closed ring. An eye bolt may also be forged and have a flattened and pierced section, with or without a collar or shoulder under the head.

Hook Bolt, Right Angle Bend A right angle bend hook bolt is a bent bolt having the unthreaded end bent to form a right angle or hook.

U-Bolt, Round Bend A round bend U-bolt is a bent bolt having threads at both ends of the rod and the rod bent at the middle to a semicircle.

CARRIAGE BOLT A carriage bolt is a bolt having a thin circular head, an oval or flat top with one of various means under the head to prevent rotation of the fastener. Carriage bolts are supplied with nuts unless otherwise specified.

Countersunk Head Square Neck Bolt This bolt has a flat top, conical bearing surface, and a square shoulder under the head.

Round Head Fin Neck Bolt This bolt has two fins, 180° apart, under the head.

Round Head Ribbed Neck Bolt This bolt has a ribbed or serrated shoulder under the head.

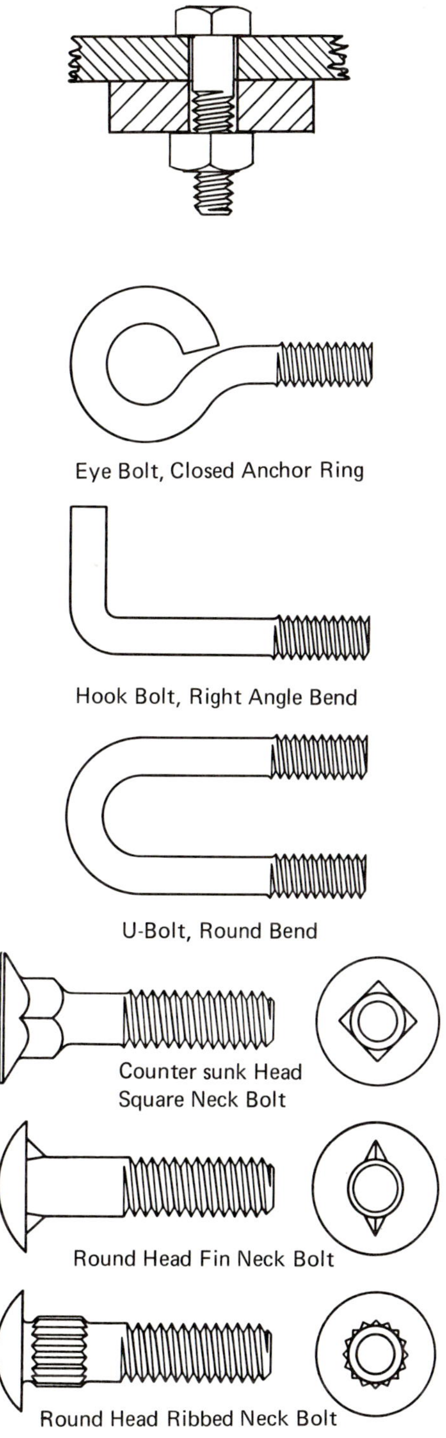

Round Head Short Square Neck Bolt This bolt has a short square shoulder under the head. It is designed for use in sheet metal where a full square shoulder would project through and present an obstruction.

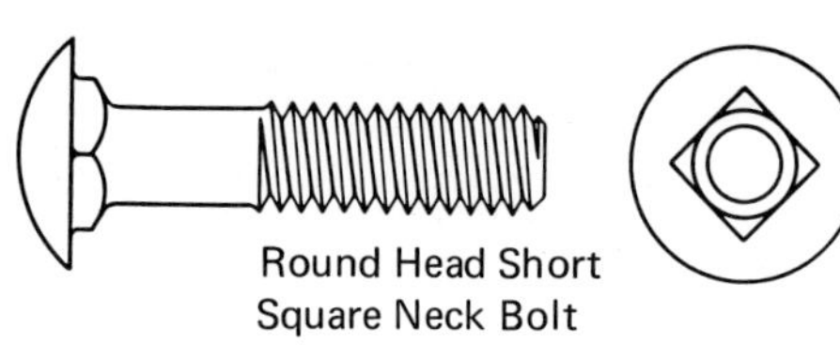

Round Head Short Square Neck Bolt

Round Head Square Neck Bolt This bolt has a square shoulder under the head. It is designed for use in wood.

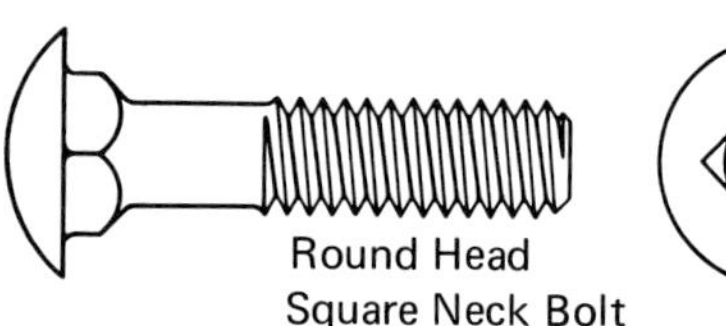
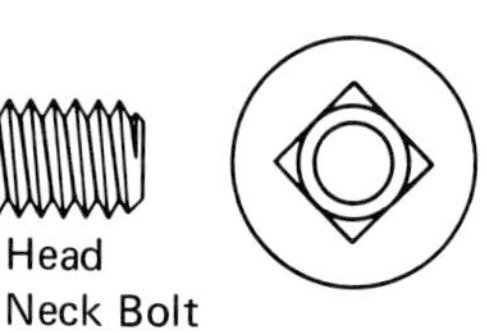

Round Head Square Neck Bolt

HANGER BOLT A hanger bolt is a threaded rod having a lag-bolt or wood screw thread and gimlet point at one end and a standard Unified thread at the other. The bolt may have a plain body or a ribbed shoulder body. It is designed for attaching hangers to woodwork. This bolt is sometimes known as a STAIR BOLT or WOOD SCREW STUD WITH THREADED END.

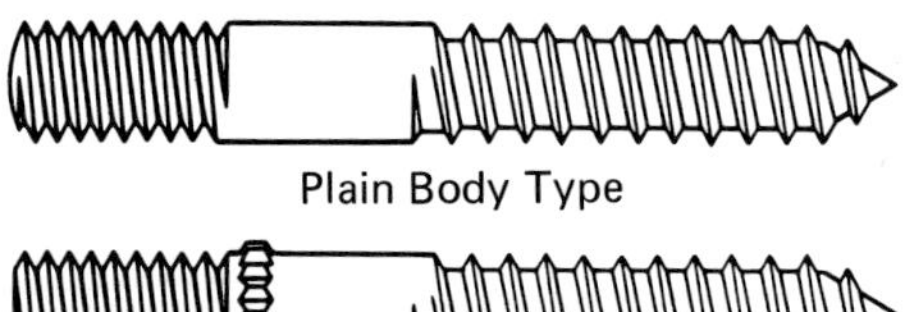

Plain Body Type

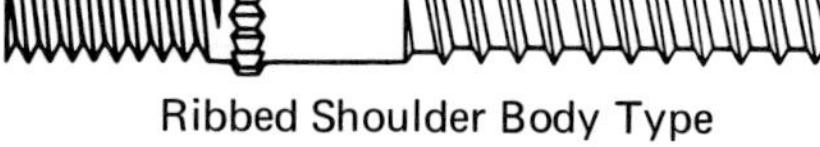

Ribbed Shoulder Body Type

HANGER BOLTS

HEXAGON HEAD BOLT A hexagon head bolt is a bolt having a hexagonal-shaped external wrenching head. It is available in several dimensional series, such as Finished Hexagon, Regular Hexagon, and Heavy Hexagon, and within these series in various grades with regard to materials, tolerances, and threads.

HEXAGON HEAD BOLT

HIGH-STRENGTH BOLT A high-strength bolt is a bolt developing specific high clamping loads at assembly. It may be of a type to clear the hole in the assembly or of a type that provides a bound-body fit in the assembly.

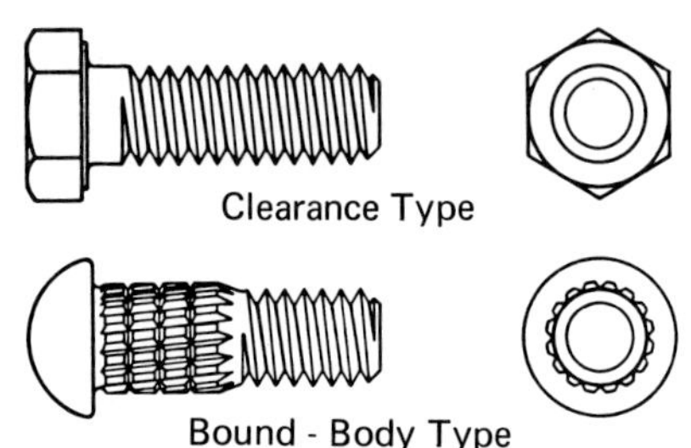

HIGH STRENGTH BOLTS

LAG BOLT*A lag bolt is a bolt having a square head, a gimlet or cone point, and a thin, sharp, coarse-pitch thread. It is designed for producing its own mating thread in wood or other resilient materials.

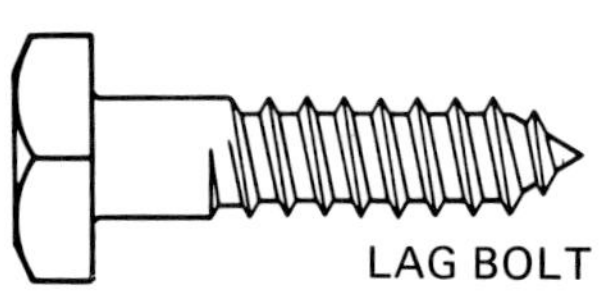

LAG BOLT

*More accurately called a lag screw.

MACHINE BOLT A machine bolt is a bolt having a conventional head, such as a square, hexagon, button, or countersunk type, and a cylindrical body below the head. It is designed for general use in machine and other types of construction. Machine bolts are supplied with nuts unless otherwise specified.

PLOW BOLT A plow bolt is a countersunk head bolt with head or shank designed to prevent rotation of the bolt and having coarse threads. The bolts are designed for fastening moldboards, plowshares, cultivator shoes, etc. Plow bolts are available in many types. Each type is available in a "regular head" for use on original equipment and a somewhat shallower "repair head" to fit a worn surface without grinding. Plow bolts are supplied with nuts unless otherwise specified.

Round Countersunk Head Square Neck Plow Bolt (No. 3 Head) This bolt has a round countersunk head with an 80° head angle and a short square neck to prevent rotation.

Round Countersunk Heavy Key Head Plow Bolt (No. 6 Head) This bolt has a round countersunk head with a 40° head angle and a triangular shaped key on one side to prevent rotation.

Round Countersunk Reverse Key Head Plow Bolt (No. 7 Head) This bolt has a round countersunk head with a 60° head angle and a rectangular key on one side to prevent rotation.

Square Countersunk Head Plow Bolt (No. 4 Head) This bolt has a square pyamidal shaped head with an 80° head angle in which the corners of the square prevent rotation.

SQUARE HEAD BOLT This bolt has a square-shaped external wrenching head of standard proportions.

STOVE BOLT A stove bolt is a former commercial standard having fractional sizes of 1/8-32, 5/32-28, 3/16-24, 7/32-22, and 1/4-18. It is now supplied as the equivalent machine screw with nut.

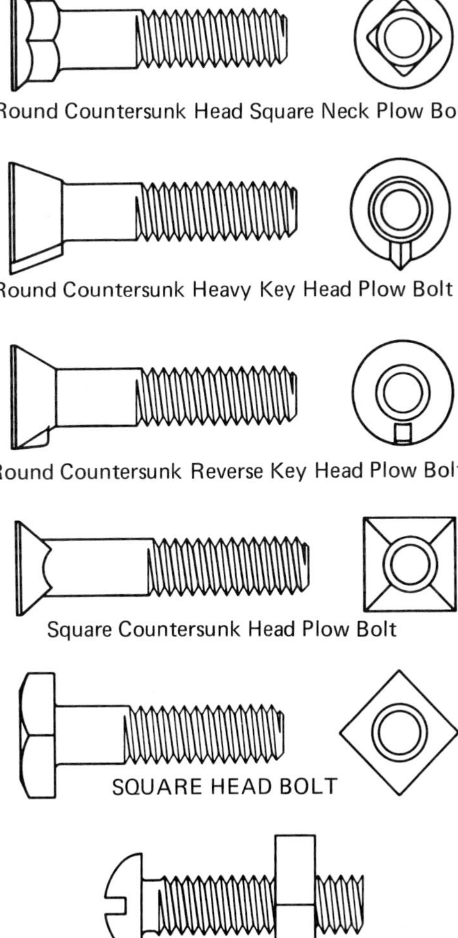

T-BOLT A T-bolt is a finished bolt with a square head. It is designed for holding fixtures and other accessories in the T-slots of machine tools.

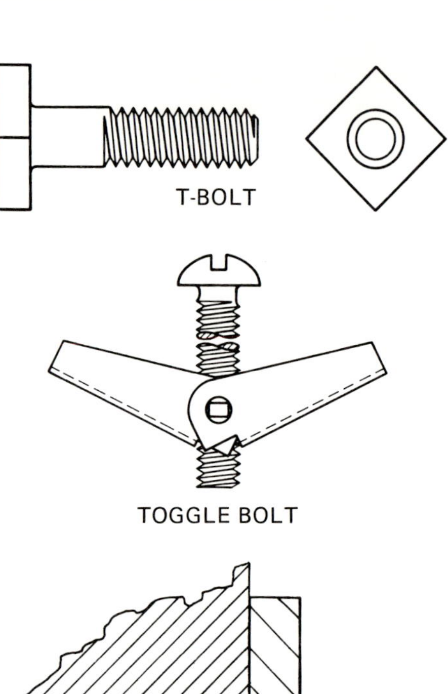

T-BOLT

TOGGLE BOLT

TOGGLE BOLT A toggle bolt is a bolt having a U-shaped wing rotatably attached to a nut so that it can be aligned with the shank and pushed through a hole. It is used as a fastener in a hole that is accessible only from one side. Toggle bolts are generally furnished with round, flat, or truss head slotted machine screws.

Studs

Stud A stud is a cylindrical rod of moderate length, threaded on either one or both ends or throughout its entire length

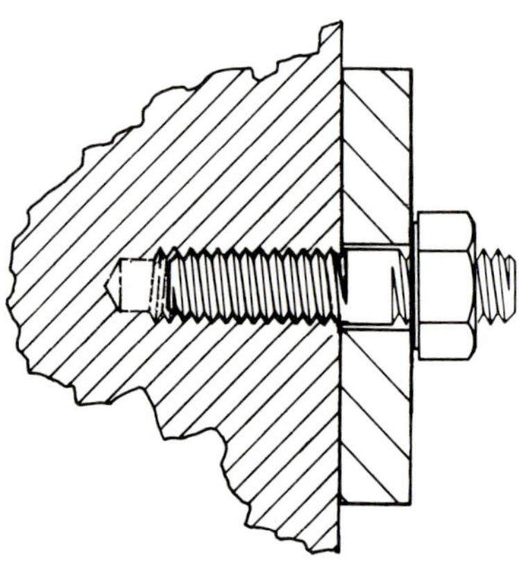

CONTINUOUS THREAD STUD This stud is threaded its entire length with conventional threads for the assembly of nuts on both ends. It is a variation of the Double End Stud (clamping type). It is known in some industries as BOLT STUD.

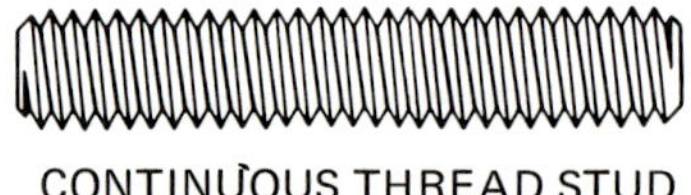

CONTINUOUS THREAD STUD

DOUBLE END STUD A double end stud is a stud threaded on both ends with a plain or unthreaded portion between the threaded ends. The double end stud is of two general types: the interference thread type and the clamping type.

Double End Stud (clamping type) This stud has conventional threads on both ends of the stud. It serves the function of clamping two bodies together with a nut on either end. It is known in some industries as STUD BOLT and BOLT STUD.

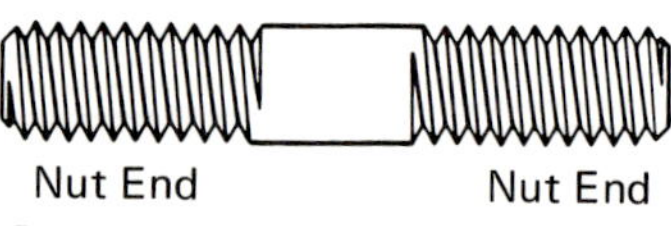

Double End Stud (clamping type)

Double End Stud (interference thread type) This stud has conventional threads on the "nut" end and threads on the "stud" end that will give an interference fit in the hole in which it is installed. Its application dictates that the tightness of the stud end thread in a tapped hole should not be disturbed by the removal of the nut from the nut end. The studs may have coarse, fine, or spaced series threads. The pitch diameter on the stud end is enlarged an amount depending on the individual application. It is also known as a TAP END STUD.

Stud End Nut End

Double End Stud (interference thread type)

SCREWS

Screw. A screw is an externally threaded fastener capable of being inserted into holes in assembled parts, of mating with a preformed internal thread or forming its own thread, and of being tightened or released by torquing the head.

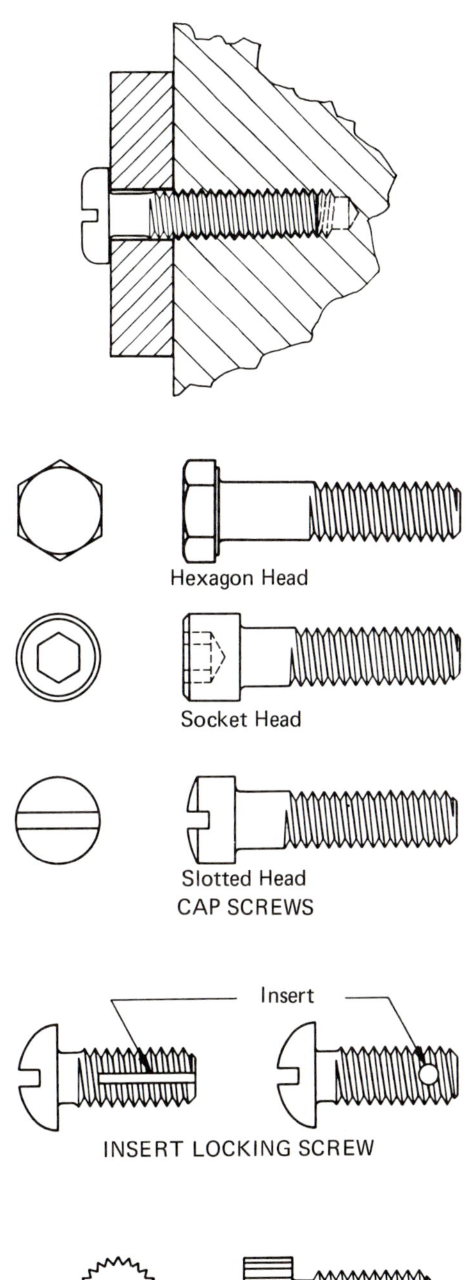

CAP SCREWS

INSERT LOCKING SCREW

CAP SCREW A cap screw is a screw having all surfaces machined or of an equivalent finish, closely controlled body diameter, and a flat chamfered point, with a wrench, slotted, recessed, or socket head of proportions and tolerances designed to assume full and proper loading when wrenched or driven into a tapped hole. Cap screws usually have hexagon, spline socket, hexagon socket, button, flat, fillister, or round head styles.

INSERT LOCKING SCREW An insert locking screw is a screw having a metallic or nonmetallic insert in the threaded portion.

INSERT SCREW An insert screw is a screw designed for permanent assembly of the head or shank within a cast or molded material, such as hard rubber, organic plastics, or die castings. The head or shank or both are provided with serrations, knurling, or other projections or indentations to prevent its rotation in the molded material.

INSERT SCREW

MACHINE SCREW A machine screw is a screw having a slotted, recessed, or wrenching head and threaded for assembly with a preformed internal thread. Machine screws are generally available in the following standard head styles: binding, fillister, 80° and 100° flat, flat trim, hexagon, hexagon washer, oval, oval trim, pan, round, and truss.

They are also made in numerous special head styles to suit particular requirements. They are generally furnished with plain points, but for special purposes may have chamfered, header, pilot, or other type points.

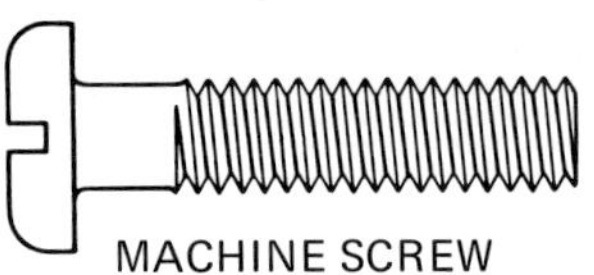

MACHINE SCREW

MACHINE SCREW WITH NUT has practically replaced the term STOVE BOLT.

METALLIC DRIVE SCREW (Type U) A metallic drive screw is a hardened screw having a blunt or sharp pilot point, single or multiples threads of steep lead angle, and generally furnished with a round or flat head.

It is used with a clearance hole in one of the parts to be fastened and designed for assembly by impact in sheet metal, castings, fiber, plastics, etc.

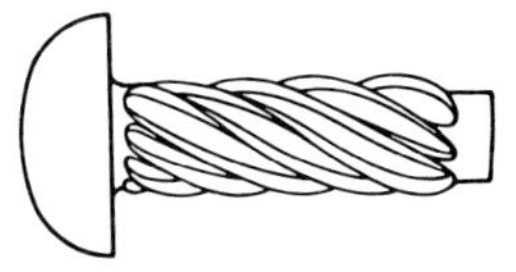

METALLIC DRIVE SCREW

ONE-WAY HEAD SCREW A one-way head screw is a round head screw that is slotted but has side clearances at diagonally opposite sides of the slot so that the screw can be driven only in the direction of assembly; designed to prevent tampering or theft.

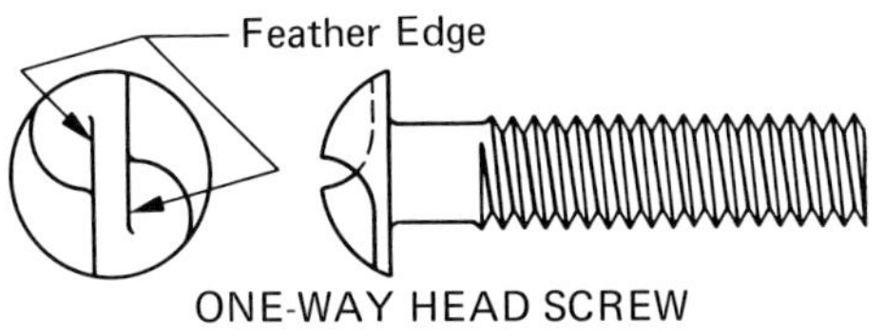

ONE-WAY HEAD SCREW

SET-SCREW A set screw is a hardened screw with or without a head, threaded the entire length and having a formed point designed to bear on a mating part. Set screws are regularly furnished in square head, headless slotted, hexagon socket and spline socket styles in combination with the POINT STYLES illustrated and described below.

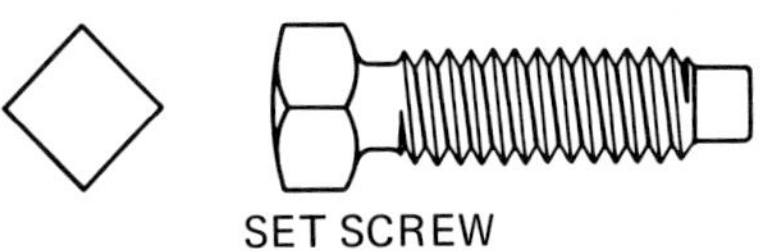

SET SCREW

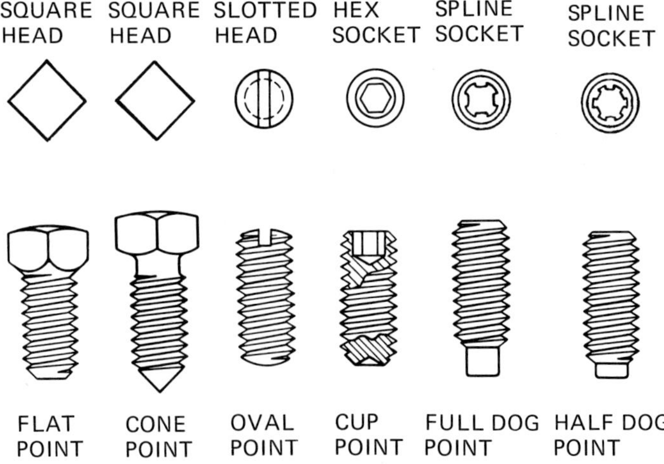

SHOULDER SCREW A shoulder screw is a slotted, flat fillister head screw having a cylindrical shoulder under the head to serve as a bearing or spacer.

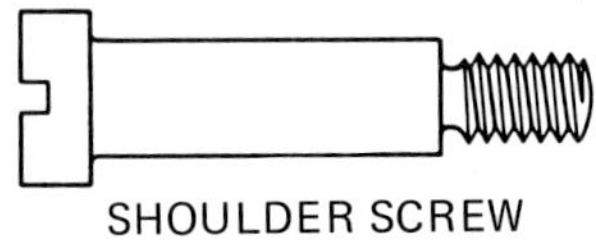

SHOULDER SCREW

SOCKET HEAD SHOULDER SCREW A socket head shoulder screw is a socket head screw having a cylindrical shoulder under the head to serve as a bearing or spacer, and a necked portion between the thread and the shoulder. (Formerly called STRIPPER BOLT.)

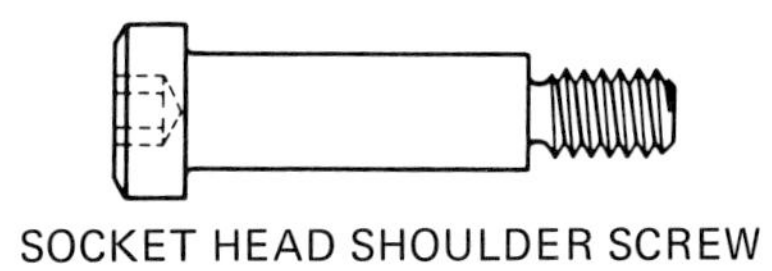

SOCKET HEAD SHOULDER SCREW

TAPPING SCREW A tapping screw is a screw having a slotted, recessed, or wrenching head designed to form or cut a mating thread in one or more of the parts to be assembled. Tapping screws are generally available in various combinations of the following head and screw styles: Fillister, flat, flat trim, hexagon, hexagon washer, oval, oval trim, pan, round, and truss head styles with thread-forming screws, Types AB, A, B, BP, and C, or thread-cutting screws, Types D, F, G, T, BF, and BT as illustrated and described below.

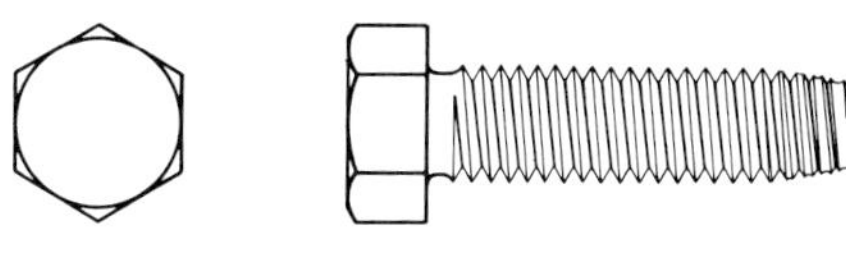

TAPPING SCREW

Type Designation of Tapping Screws and Metallic Drive Screws

Type	ANSI Standard	Manufacturer
	AB[1]	AB[1]
NOT RECOMMENDED – USE TYPE AB	A	A
	B	B
	BP	BP
	C	C
	D	1
	F	F
	G	G
	T	23
	BF	BF
	BT	25
DRIVE SCREW	U	U

[1] Formerly designated "Type BA"

THREAD LOCKING SCREW A thread locking screw is a screw having a thread designed to produce interference with its mating thread.

THREAD LOCKING SCREW

THUMB OR WING SCREW A thumb or wing screw is a screw having a flattened or wing shaped head, designed for manual turning without a driver or wrench.

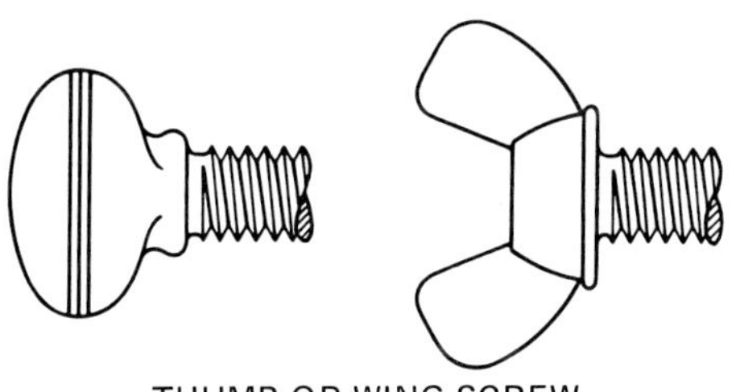

THUMB OR WING SCREW

WELDING SCREW A welding screw is a headed screw provided with lugs or weld projections on the top or underside of the head to facilitate attachment to a metal part by resistance welding.

WELDING SCREW

WOOD DRIVE SCREW A wood drive screw is a thread forming screw having a cone or pinch point, multiple threads of steep lead angle, a reduced diameter body, and generally available with flat, oval, or round head styles, designed for rapid assembly in wood.

WOOD DRIVE SCREW

WOOD SCREW A wood screw is a thread-forming screw having a slotted or recessed head, gimlet point, and a sharp crested, coarse pitch thread, and generally available with flat, oval and round head styles. It is designed to produce a mating thread when assembled into wood or other resilient materials.

WOOD SCREW

SCREW AND WASHER ASSEMBLY This is a preassembled screw and washer unit in which the washer is retained free to rotate under the screw head by the rolled thread. These units expedite assembly operations and assure the presence of a washer in each assembly. They are generally available in various combinations of head styles and washer types as indicated.

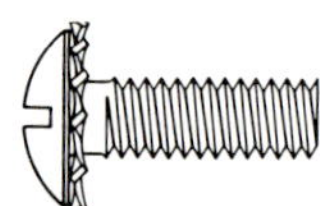
Truss Head Screw and External Tooth Lock Washer

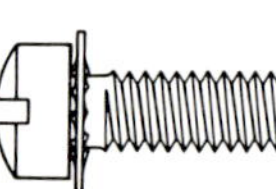
Fillister Head Screw and Internal Tooth Lock Washer

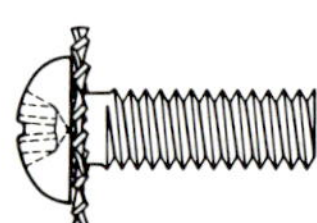
Round Head Screw and Internal-External Tooth Lock Washer

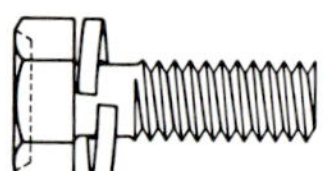
Hexagon Head Screw and Spring Lock Washer

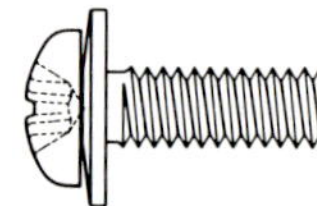
Pan Head Screw and Conical Spring Washer

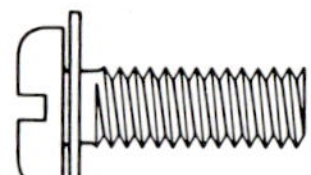
Pan Head Screw and Plain Flat Washer

SCREW AND WASHER ASSEMBLIES (SEMS)

NUTS

Nut. A nut is a block or sleeve having an internal thread designed to assemble with external thread on a bolt, screw, or other threaded part. It may serve as the fastening means, an adjusting means, a means for transmitting motion, or a means for transmitting power with a large mechanical advantage and nonreversible motion.

CAPTIVE NUT A captive nut consists of a threaded member, usually a square nut, held loosely in a shaped sheet metal box. The variations in mating assembly parts are usually over-come by this type nut since it can float laterally.

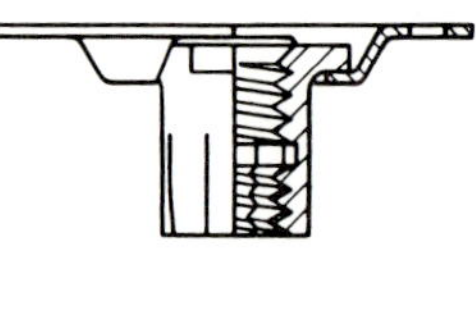
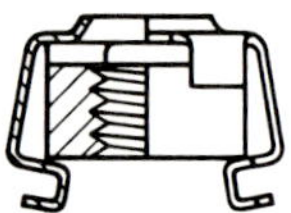
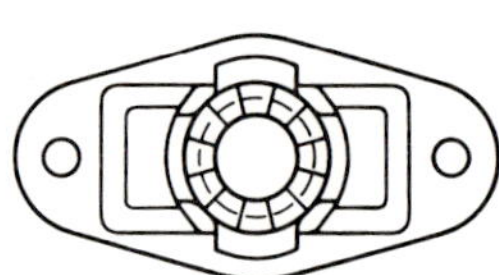
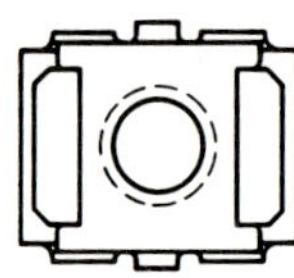

CAPTIVE NUTS

CASTLE NUT A castle nut is a slotted hexagon nut having a cylindrical portion at the slotted end equal in length to the slot depth and slightly smaller in diameter than the hexagon width.

This nut is designed for the insertion of a cotter pin to secure the nut in place when it is used with a drilled shank fastener. This nut was formerly known as a CASTELLATED NUT.

CLINCH NUT A clinch nut is a solid nut having a pilot or other feature to be inserted in a preformed hole. The pilot may be clinched, staked, or expanded to retain the nut and prevent rotation. It is available in a large variety of types, some of which are capable of piercing the holes for assembly. It is sometimes designated ANCHOR NUT.

CONDUIT NUT A conduit nut is a thin nut, usually stamped. It may be square with scalloped corners or hexagonal or octagonal in shape.

CROWN NUT A crown nut is a hexagon nut having an acorn-shaped top and a blind threaded hole.

Crown nuts are generally furnished in two types, high crown and low crown. It is sometimes designated as ACORN NUT or CAP NUT.

FLANGE NUT A flange nut is an intègral nut and washer designed for increased bearing area.

HEXAGON NUT A hexagon nut has a hexagonal base with or without a washer face. The six essentially rectangular sides serve as wrenching flats.

Hexagon nuts are available in various dimensional series, such as Finished Hexagon, Heavy Hexagon, Regular Hexagon, and in various thicknesses, such as standard, jam or thin, and thick, as shown in the illustrations. See MACHINE SCREW NUT.

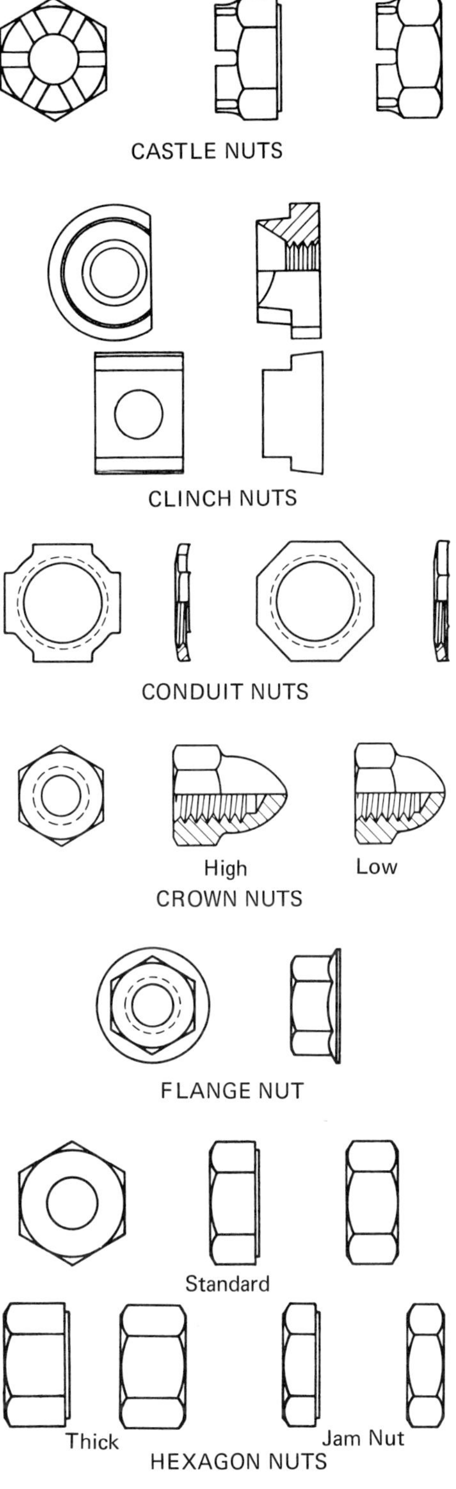

LOCK NUT There are two basically different types of lock nuts: (1) a prevailing torque type that resists relative bolt-nut movement with or without an axially applied load to the bolt-nut combination, and (2) a free-running type that exhibits a locking ability when there is an axial load applied to the base of the nut. The "locking" or stopping action of the nut is accomplished by thread deformation, or clamping, or by the addition of nonmetallic inserts. The free-running type usually has a design feature that adds to the elastic elongation of the bolt-nut combination.

MACHINE SCREW NUT A machine screw nut is a hexagon or square nut of proportions suitable for use with a machine screw.

PLATE NUT A plate nut is a nut consisting of an internally threaded unit and a plate, which is designed to hold the threaded unit in place relative to the work. The threaded unit may be integral with the plate or held by a retainer and may have conventional or locking threads. Two-piece plate nuts are generally of the floating type in which the threaded unit has a limited movement with respect to the plate and normal to the thread axis to facilitate alignment with the mating fastener. Plates may be of the following types: (1) the hole type, for riveting, nailing, or otherwise fastening the plate to work, (2) the boss type, having weld embossments for resistance welding, the plate to work (the embossments may be on the top of the plate–internal boss, or on the bottom of the plate–external boss), or (3) the prong type, having projections to grip soft materials such as wool.

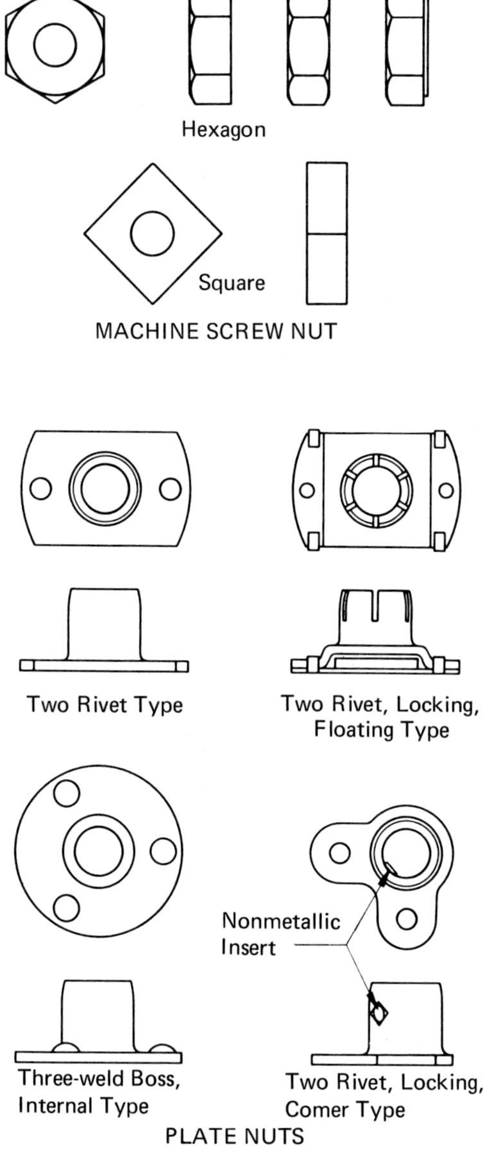

SLOTTED NUT A slotted nut is a hexagon nut having opposed slots through the centers of the flats. The slots are on the end opposite the bearing surface and are perpendicular to the axis of the nut. Slotted nuts are designed for the insertion of a cotter pin to secure the nut in place when it is used with a drilled shank fastener.

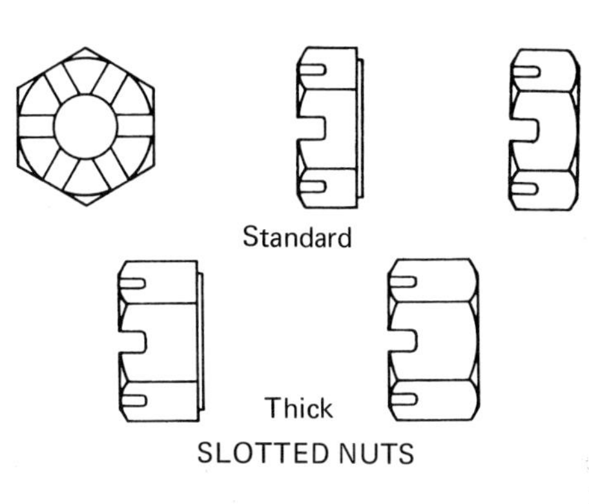

SLOTTED NUTS

SPRING NUT A spring nut is a nut fabricated from thin spring steel having an impression designed to accommodate the mating thread. It is used in place of a solid nut. Spring nuts are available in many shapes and styles.

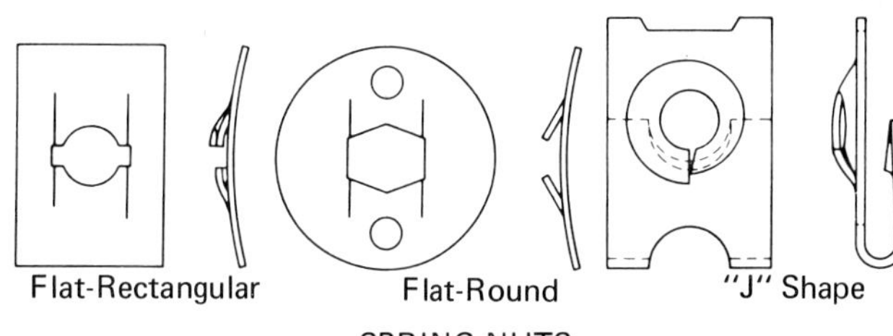

SPRING NUTS

SQUARE NUT A square nut is a solid nut with a square base, generally without a washer face. The four essentially rectangular sides serve as wrenching flats. These nuts are available in the regular and heavy series with varying proportions.

SQUARE NUT

STAMPED NUT A stamped nut is a hexagon nut, sometimes with an integral washer, stamped from thin spring steel, having prongs formed to engage the mating thread. It is used in place of a solid nut in low-stress applications or as a retaining nut against a solid nut.

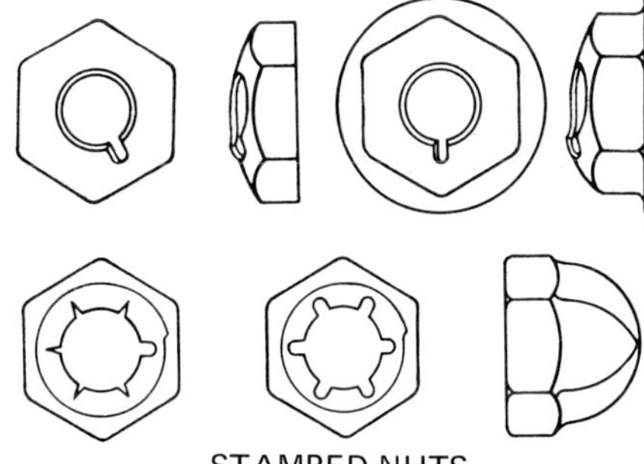
STAMPED NUTS

WELD NUT A weld nut is a solid nut provided with lugs, annular rings, or embossments to facilitate its attachment to a metal part by resistance welding.

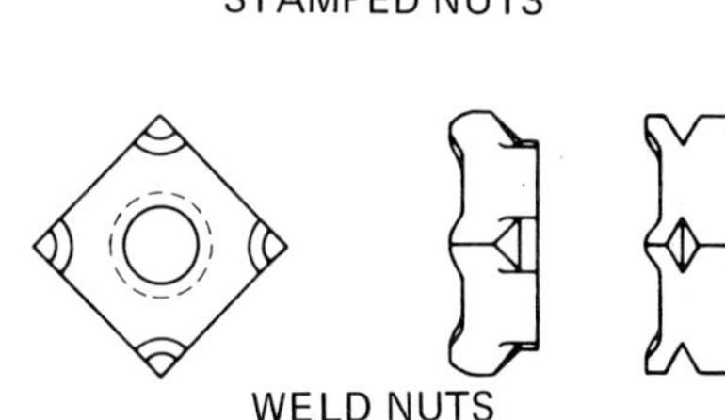
WELD NUTS

WING NUT A wing nut is a nut having "wings" designed for manual turning. It may be forged, machined, stamped, or cast. It is sometimes called THUMB NUT.

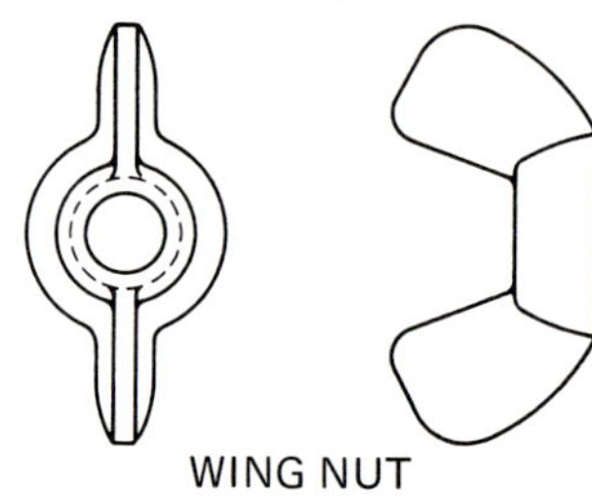

WING NUT

Washers

Washer A washer is a part usually thin, having a centrally located hole or partial slot. The washer performs various functions when assembled between the bearing surface of a fastener and the part being attached. Insulation, lubrication, spanning of large clearance holes, and improved stress distribution are a few design uses.

BEVEL WASHER A bevel washer is a flat, square, or circular washer with a definite taper between opposite bearing faces.

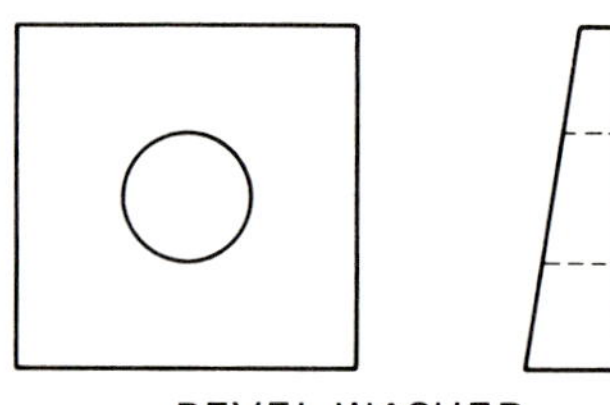

BEVEL WASHER

CONICAL SPRING WASHER A conical spring washer is a hardened circular steel washer formed with a slight dish and having edges sheared parallel to the center line. This type of washer is designed to store a large amount of energy and also provides a "scaling" effect as the sharp edges are tightened into the bearing surfaces. This washer is also known as BELLEVILLE WASHER and CONE LOCK WASHER.

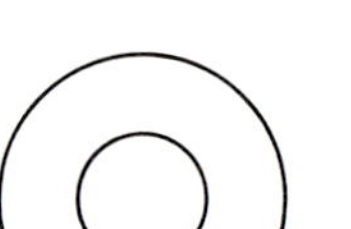
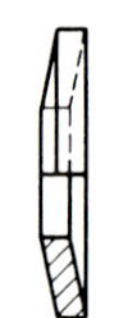

CONICAL SPRING WASHER

FINISH WASHER A finish washer is a formed circular washer designed to accommodate the head of a flat or oval head screw and provide additional bearing area on the material being fastened. Finish Washers are available in the Raised Type and Flush Type.

FLAT WASHER See PLAIN WASHER for definition.

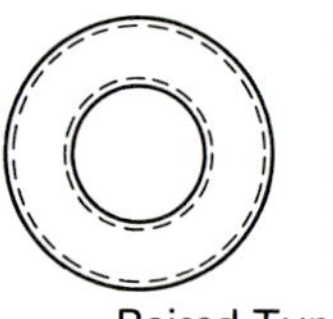
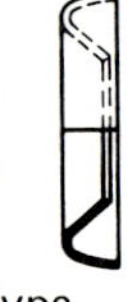
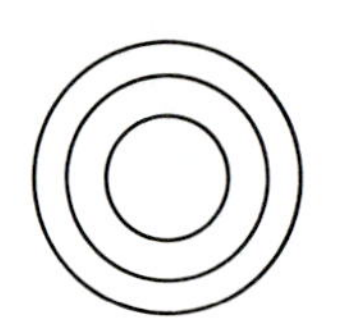

Raised Type Flush Type

FINISH WASHERS

LOCKPLATE A lockplate is a flat plate fastened to an assembled element with screws, or held by lanced ears. The lockplate provides projections that are bent into place against a flat of the screw head, effectively preventing rotation of the head.

Ears Bent

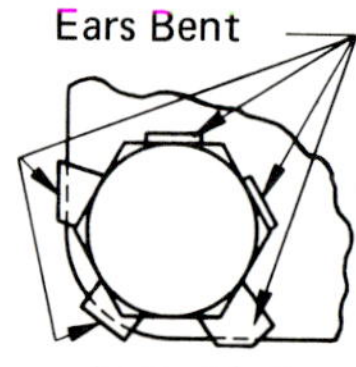

Assembled

LOCKPLATE

OPEN OR HORSESHOE WASHER An open or horseshoe washer is a flat circular washer having a slot of width equal to the hole diameter and extending from the hole to the periphery. It is designed for installation on or removal from the shank of the fastener without removing the fastener from the assembly. This washer is also known as a C-WASHER.

OPEN OR HORSESHOE WASHER

PLAIN WASHER A plain washer is a flat, circular, or square washer with a central hole designed to fit around a bolt or screw and under the head or nut.

PLAIN WASHER

RIVETING BURR A riveting burr is a small plain washer assembled with a small rivet before peening the end to provide a large area of contact on the part.

RIVETING BURR

SPRING LOCK WASHER A spring lock washer is a coiled, hardened, split circular washer having a slightly trapezoidal wire section. It is designed to serve as a spring take-up device to compensate for developed looseness and loss of tension between the parts of an assembly and to function as a hardened thrust bearing. The ends of the washer are designed to bite into the bearing surfaces.

SPRING LOCK WASHER

TOOTH LOCK WASHER A tooth lock washer is a hardened circular washer, having twisted or bent prongs or projections that are deformed when assembled. The prongs, on which the pressure is localized, resist loosening of the fastener. It is generally furnished in External Tooth, Internal Tooth, and Internal-External Tooth Types,

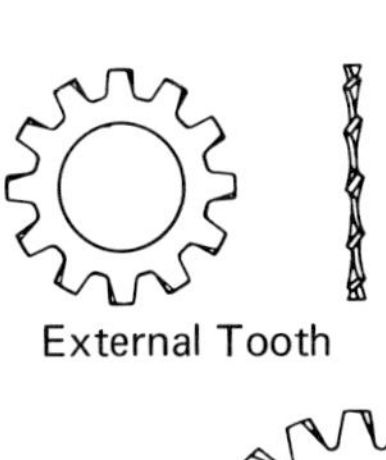

External Tooth

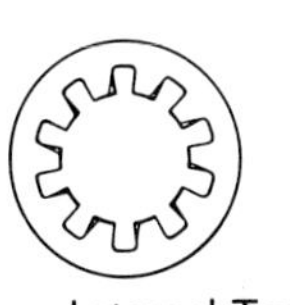

Internal Tooth

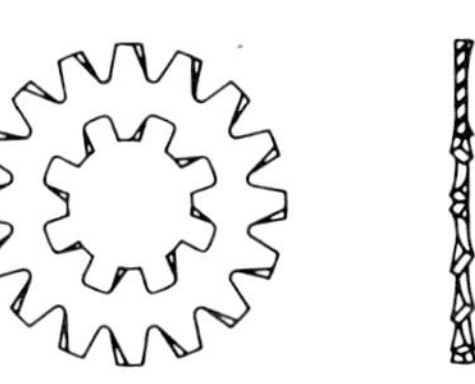

Internal - External Tooth

TOOTH LOCK WASHER

Rivets

Rivet A rivet is a headed metal fastener of malleable material used to join parts of structures and machines by inserting the shank through the aligned holes in each piece and forming a head on the headless end by upsetting.

LARGE RIVET A large rivet is a solid rivet having a body diameter of $\frac{1}{2}$ inch or more and a head of one of the following forms: button, high button, cone, countersunk, or pan. Large rivets are usually driven at forging heat.

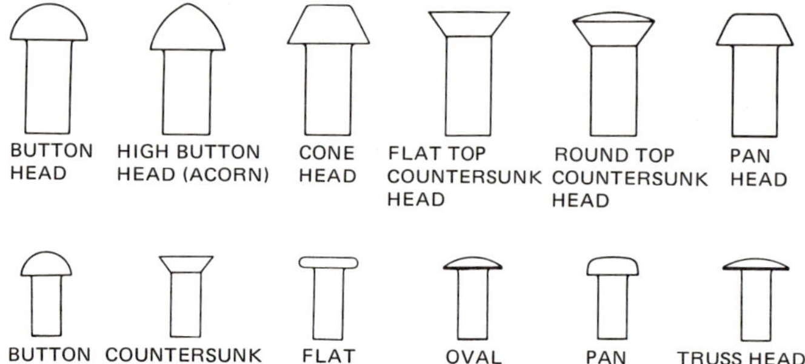

SMALL RIVET A small rivet is a rivet, usually solid, having a body diameter of less than $\frac{1}{2}$ inch and a head of one of the following forms: button, countersunk, flat, oval, pan, or truss. Small rivets are usually driven cold.

BELT RIVET A belt rivet is a small solid rivet having a flat top, countersunk head, and chamfer point. It is used with a riveting burr for joining leather.

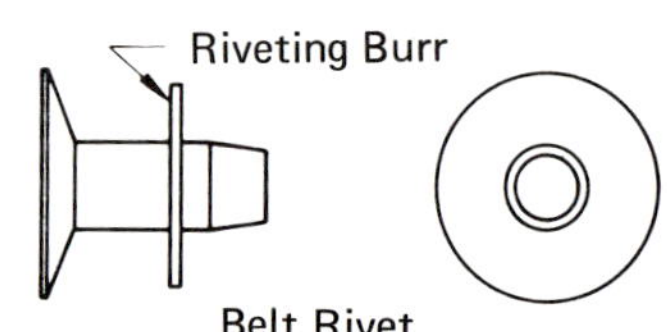

Belt Rivet

BLIND RIVET A blind rivet is a rivet designed for use where only one side of the work is accessible. The rivet will have a suitable means for expanding or forming the rivet end on the blind side, which is actuated after the rivet is inserted from the open side. Blind rivets are generally further categorized by the expanding or forming features such as: collar, drive pin, explosive, and pull stem. A few examples of the various types of blind rivets are shown in the accompanying illustrations.

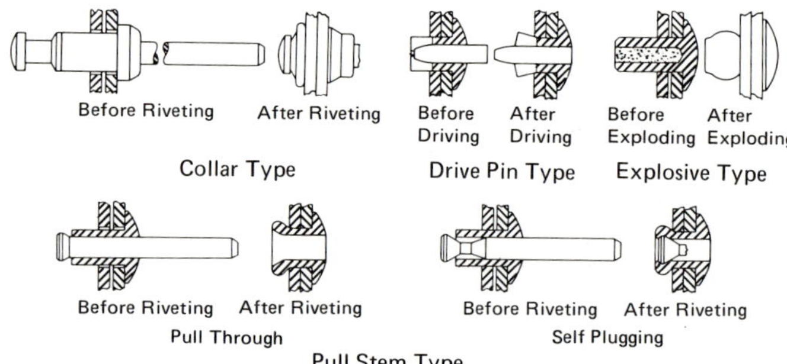

BLIND RIVET TYPES

BOILER RIVET A boiler rivet is a large rivet with a cone head.

COMPRESSION RIVET A compression rivet is a rivet consisting of two parts: a solid rivet and a deep-drilled tubular rivet. The diameters of the solid shank and the drilled hole are selected so as to produce a compression or pressed fit when the two parts are assembled. Most rivets of this type have flat heads. This rivet is sometimes referred to as CULTERY RIVET.

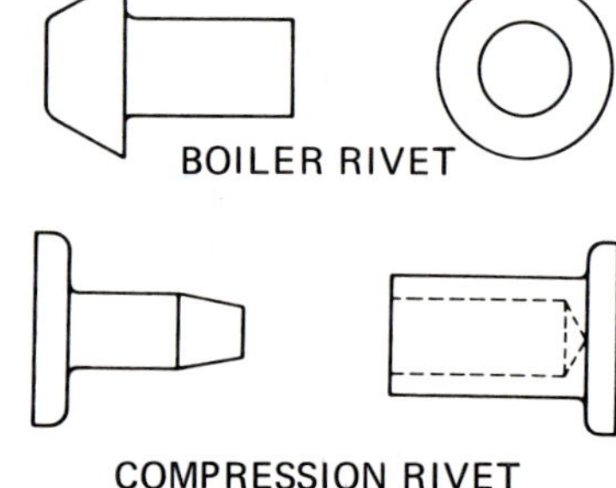

BOILER RIVET

COMPRESSION RIVET

COOPERS' RIVET A coopers' rivet is a small solid rivet having a flat top, countersunk head, and chamfer point. It is used for joining the ends of barrel hoops.

COOPERS' RIVET

SPLIT RIVET A split rivet is a small rivet having a split end for securing by spreading the ends. It is commonly furnished with an oval or countersunk head.

SPLIT RIVET

TINNERS' RIVET A tinners' rivet is a small solid rivet having a head of the same form as a flat head rivet but larger in diameter. It is designed for use in sheet-metal work.

TINNERS' RIVET

TUBULAR RIVET A tubular rivet is a small rivet having a coaxial cylindrical or tapered hole in the headless end. It is commonly furnished with a countersunk, flat, oval, or truss head. The top of the flat-top countersunk head may be slightly chamfered, as shown. Tubular rivets are designed to be secured by splaying or curling the end. They are further classified as semitubular, those having hole depths which do not exceed 112 per cent of the mean shank diameter, measured on the wall; and full tubular, those having hold depths which do exceed 112 per cent of the mean shank diameter.

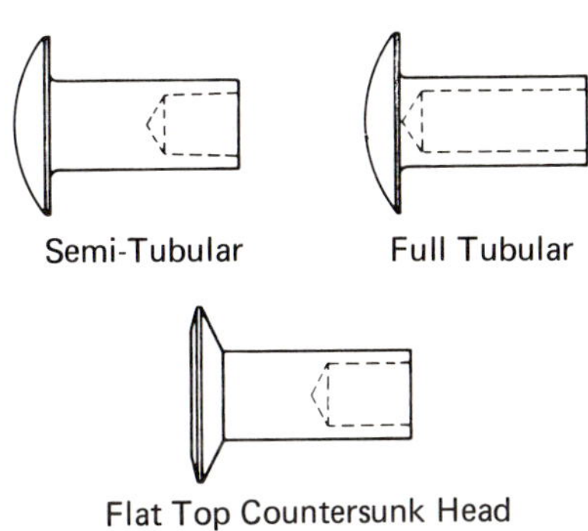

Flat Top Countersunk Head
With Chamfered Top
(Brake Lining Rivet)
TUBULAR RIVETS

PINS

Pin A pin is a straight cylindrical or tapered fastener, with or without a head, designed to perform a semipermanent attaching or locating function.

CLEVIS PIN A clevis pin is a solid pin having a cylindrical head on one end with a chamfer point and drilled hole for a cotter pin at the other; designed for use with clevises and rod ends.

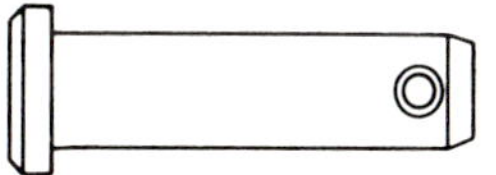
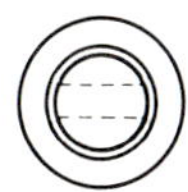
CLEVIS PIN

COTTER PIN A cotter pin is a double bodied pin formed from semicircular wire, a loop at one end of which provides a head. Available in various point styles as illustrated.

DOWEL PIN A dowel pin is a solid headless straight pin. Its diameter is closely controlled. Hardened and ground dowel pins have one end chamfered and the other end radiused to form a crown. Unhardened ground pins have both ends chamfered.

ESCUTCHEON PIN An escutcheon pin is a pin having a semi-spherical head with a flat bearing surface formed on one end and a long cone or pinch point on the other.

GROOVED PIN A grooved pin is a solid headless pin having controlled diameter and length, with multiple longitudinal grooves either rolled or pressed into the body, displacing the pin stock within predetermined limits. The grooves may extend over the full length or only portions of the pin and the ends are generally crowned. During installation, the pin material is pressed back within its elastic limits and the compressive forces actuated through the constraining action of the hole wall produce a force fit having very high resistance to loosening under shock and vibration.

SPRING PIN A spring pin is a hollow, headless pin having controlled length with rounded or chamfered ends, formed to a diameter somewhat greater than that of the hole into which it is to be assembled. Spring pins are available in two styles, slotted and coiled, as illustrated.

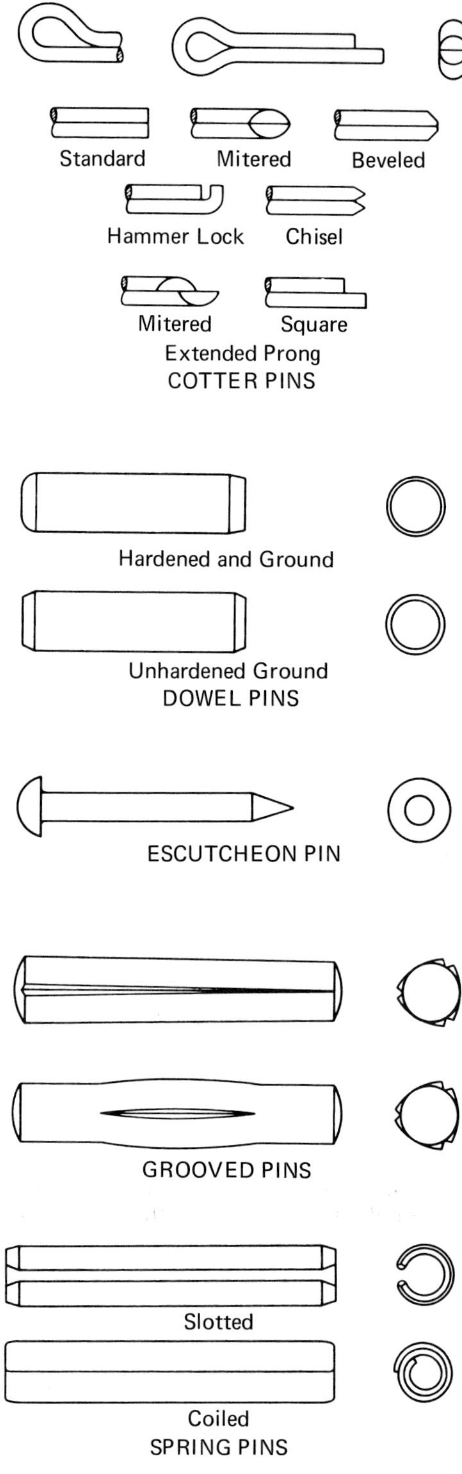

TAPER PIN A taper pin is a headless, solid pin having controlled diameter, length, and taper, with crowned ends.

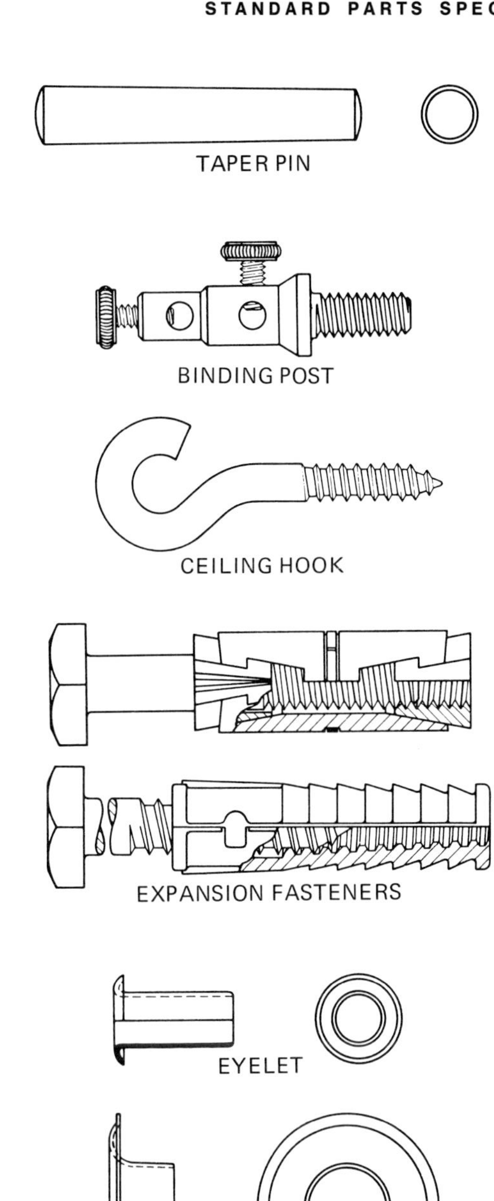

Miscellaneous Fasteners

BINDING POST A binding post is a special type of subassembly used for clamping or holding electrical conductors in a rigid position. It commonly consists of a screw having a collar head or body with one or more clamping screws.

CEILING HOOK A ceiling hook is a fastener similar to an open eye bolt except that it has lag threads.

EXPANSION FASTENER One type of expansion fastener consists of a machine bolt, an expansion shield, and an expander nut. The shield body expands in a wedgelike manner when the expander nut is tightened. This fastener is commonly used in fastening to masonry.

Another type of expansion fastener consists of a lag bolt and an internally threaded split sleeve. It is designed for fastening to stone or concrete by inserting the sleeve into a hole in the stone or concrete and expanding to a tight fit in the hole by turning the lag bolt.

EYELET An eyelet is a flanged tubular fastener designed for securing by curling or splaying the tubular end.

GROMMET A grommet is a large eyelet-type fastener designed for securing by curling the tubular end over a formed washer to provide strength in holes through resilient materials.

THREAD INSERT A thread insert is an internally threaded bushing designed to be molded in or inserted into soft or brittle materials to provide greater strength and minimize wear of threaded assembly.

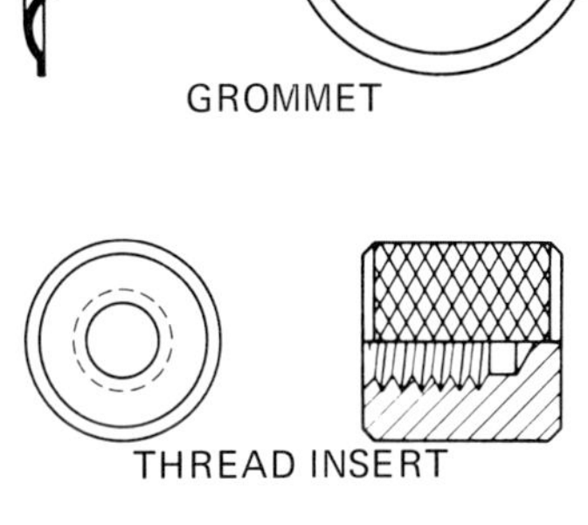

Helical Thread Insert A helical thread insert is a wire of diamond shaped cross section coiled in the form of a helix designed to form a thread insert when assembled into a tapped hole. The insert is designed to lock itself in the tapped hole.

Helical Thread Insert

Tapping Thread Insert A tapping thread insert is a thread insert having an external thread and cutting slots for cutting and forming a mating thread when assembled into an untapped drilled or cored hole.

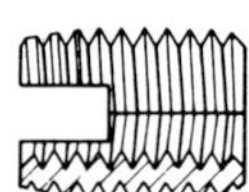

Tapping Thread Insert

TURNBUCKLE A turnbuckle is a loop or sleeve usually internally threaded with a left-hand thread at one end and a right-hand thread at the other end, intended for assembly with a threaded stud, eye, hook, or jaw at each end, used for applying tension to rods, wire rope, etc. Turnbuckles are sometimes made with a swivel feature at one end.

TURNBUCKLE

Unified Screw Threads

SIZES		Basic Major Diam.	Coarse UNC		Fine UNF		Extra-Fine UNEF	
Primary	Secondary		Thds. per in.	Tap Drill	Thds. per in.	Tap Drill	Thds. per in.	Tap Drill
0		0.0600			80	3/64		
	1	0.0730	64	53	72	53		
2		0.0860	56	50	64	50		
	3	0.0990	48	47	56	45		
4		0.1120	40	43	48	42		
5		0.1250	40	38	44	37		
6		0.1380	32	36	40	33		
8		0.1640	32	29	36	29		
10		0.1900	24	25	32	21		
	12	0.2160	24	16	28	14	32	13
1/4		0.2500	20	7	28	3	32	7/32
5/16		0.3125	18	F	24	I	32	9/32
3/8		0.3750	16	5/16	24	Q	32	11/32
7/16		0.4375	14	U	20	25/64	28	13/32
1/2		0.5000	13	27/64	20	29/64	28	15/32
9/16		0.5625	12	31/64	18	33/64	24	33/64
5/8		0.6250	11	17/32	18	37/64	24	37/64
	11/16	0.6875					24	41/64
3/4		0.7500	10	21/32	16	11/16	20	45/64
	13/16	0.8125					20	49/64
7/8		0.8750	9	49/64	14	13/16	20	53/64
	15/16	0.9375					20	57/64
1		1.0000	8	7/8	12	59/64	20	61/64
	1 1/16	1.0625					18	1
1 1/8		1.1250	7	63/64	12	1 3/64	18	1 5/64
	1 3/16	1.1875					18	1 9/64
1 1/4		1.2500	7	1 7/64	12	1 11/64	18	1 3/16
	1 5/16	1.3125					18	1 17/64
1 3/8		1.3750	6	1 7/32	12	1 19/64	18	1 5/16
	1 7/16	1.4375					18	1 3/8
1 1/2		1.5000	6	1 11/32	12	1 27/64	18	1 7/16
	1 9/16	1.5625					18	1 1/2
1 5/8		1.6250					18	1 9/16
	1 11/16	1.6875					18	1 5/8
1 3/4		1.7500	5	1 9/16				
2		2.0000	4 1/2	1 25/32				
2 1/4		2.2500	4 1/2	2 1/32				
2 1/2		2.5000	4	2 1/4				
2 3/4		2.7500	4	2 1/2				
3		3.0000	4	2 3/4				
3 1/4		3.2500	4	3				
3 1/2		3.5000	4	3 1/4				
3 3/4		3.7500	4	3 1/2				
4		4.0000	4	3 3/4				

All dimensions given in inches.
Tap drill sizes not American Standard.
Extracted from ANSI 81.1–1960.

Straight Shank Twist Drills

Dia. of Drill	Decimal Equiva-lent	Dia. of Drill	Decimal Equiva-lent	Dia. of Drill	Decimal Equiva-lent	Dia. of Drill	Decimal Equiva-lent	Dia. of Drill	Decimal Equiva-lent
97	0.0059	3/64	0.0469	19	0.166	21/64	0.3281	27/32	0.8438
96	0.0063	55	0.052	18	0.1695	*Q*	0.332	55/64	0.8594
95	0.0067	54	0.055	11/64	0.1719	*R*	0.339	7/8	0.875
94	0.0071	53	0.0595	17	0.173	11/32	0.3438	57/64	0.8906
93	0.0075	1/16	0.0625	16	0.177	*S*	0.348	29/32	0.9062
92	0.0079	52	0.0635	15	0.180	*T*	0.358	59/64	0.9219
91	0.0083	51	0.067	14	0.182	23/64	0.3594	15/16	0.9375
90	0.0087	50	0.070	13	0.185	*U*	0.368	61/64	0.9531
89	0.0091	49	0.073	3/16	0.1875	3/8	0.375	31/32	0.9688
88	0.0095	48	0.076	12	0.189	*V*	0.377	63/64	0.9844
87	0.010	5/64	0.0781	11	0.191	*W*	0.386	1	1.000
86	0.0105	47	0.0785	10	0.1935	25/64	0.3906	1 1/64	1.0156
85	0.011	46	0.081	9	0.196	*X*	0.397	1 1/32	1.0312
84	0.0115	45	0.082	8	0.199	*Y*	0.404	1 3/64	1.0469
83	0.012	44	0.086	7	0.201	13/32	0.4062	1 1/16	1.0625
82	0.0125	43	0.089	13/64	0.2031	*Z*	0.413	1 5/64	1.0781
81	0.013	42	0.0935	6	0.204	27/64	0.4219	1 3/32	1.0938
80	0.0135	3/32	0.0938	5	0.2055	7/16	0.4375	1 7/64	1.1094
79	0.0145	41	0.096	4	0.209	29/64	0.4531	1 1/8	1.125
1/64	0.0156	40	0.098	3	0.213	15/32	0.4688	1 9/64	1.1406
78	0.016	39	0.0995	7/32	0.2188	31/64	0.4844	1 5/32	1.1562
77	0.018	38	0.1015	2	0.221	1/2	0.500	1 11/64	1.1719
76	0.020	37	0.104	1	0.228	33/64	0.5156	1 3/16	1.1875
75	0.021	36	0.1065	*A*	0.234	17/32	0.5312	1 13/64	1.2031
74	0.0225	7/64	0.1094	15/64	0.2344	35/64	0.5469	1 7/32	1.2188
73	0.024	35	0.110	*B*	0.238	9/16	0.5625	1 15/64	1.2344
72	0.025	34	0.111	*C*	0.242	37/64	0.5781	1 1/4	1.250
71	0.026	33	0.113	*D*	0.246	19/32	0.5938	1 9/32	1.2812
70	0.028	32	0.116	*E* & 1/4	0.250	39/64	0.6094	1 5/16	1.3125
69	0.0292	31	0.120	*F*	0.257	5/8	0.625	1 11/32	1.3438
68	0.031	1/8	0.125	*G*	0.261	41/64	0.6406	1 3/8	1.375
1/32	0.0312	30	0.1285	17/64	0.2656	21/32	0.6562	1 13/32	1.4062
67	0.032	29	0.136	*H*	0.266	43/64	0.6719	1 7/16	1.4375
66	0.033	28	0.1405	*I*	0.272	11/16	0.6875	1 15/32	1.4688
65	0.035	9/64	0.1406	*J*	0.277	45/64	0.7031	1 1/2	1.500
64	0.036	27	0.144	*K*	0.281	23/32	0.7188	1 9/16	1.5625
63	0.037	26	0.147	9/32	0.2812	47/64	0.7344	1 5/8	1.625
62	0.038	25	0.1495	*L*	0.290	3/4	0.750	1 11/16	1.6875
61	0.039	24	0.152	*M*	0.295	49/64	0.7656	1 3/4	1.750
60	0.040	23	0.154	19/64	0.2969	25/32	0.7812	1 13/16	1.8125
59	0.041	5/32	0.1562	*N*	0.302	51/64	0.7969	1 7/8	1.875
58	0.042	22	0.157	5/16	0.3125	13/16	0.8125	1 15/16	1.9375
57	0.043	21	0.159	*O*	0.316	53/64	0.8281	2	2.000
56	0.0465	20	0.161	*P*	0.323				

Extracted from ANSI B94.11–1967.

BOLTS AND HEX CAP SCREWS

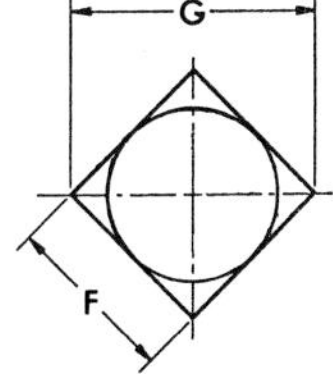

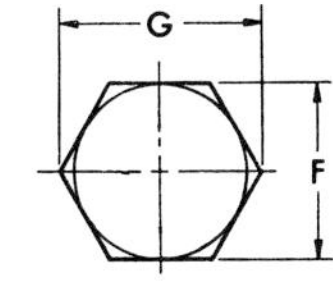

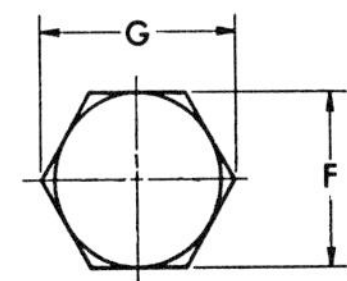

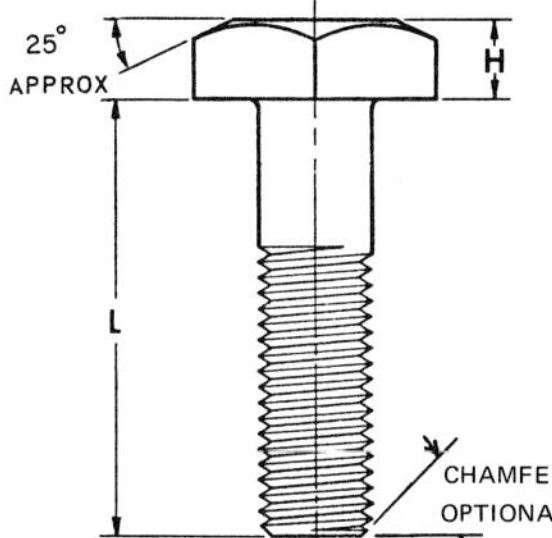

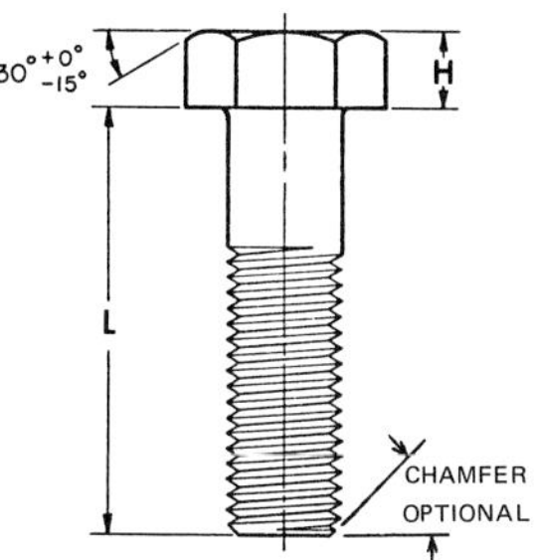

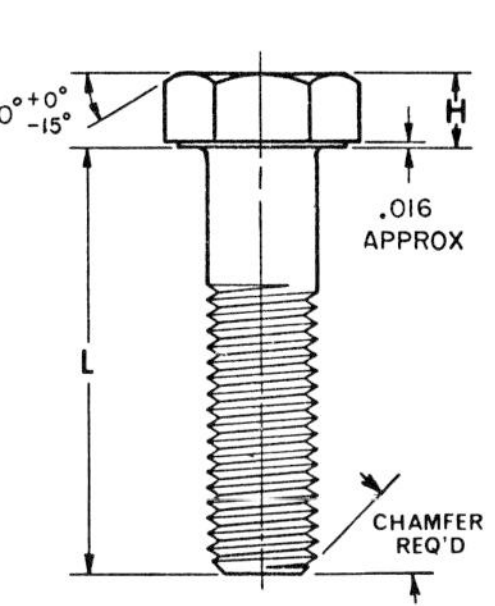

Nominal Size or Basic Product Dia.	Square Bolts			Hex Bolts		
	F	G	H	F	G	H
	Width Across Flats	Width Across Corners	Head Height	Width Across Flats	Width Across Corners	Head Height
	Basic	Max.	Basic	Basic	Max.	Basic
1/4 0.250	3/8	0.530	11/64	7/16	0.505	11/64
5/16 0.3125	1/2	0.707	13/64	1/2	0.577	7/32
3/8 0.375	9/16	0.795	1/4	9/16	0.650	1/4
7/16 0.4375	5/8	0.884	19/64	5/8	0.722	19/64
1/2 0.500	3/4	1.061	21/64	3/4	0.866	11/32
5/8 0.625	15/16	1.326	27/64	15/16	1.083	27/64
3/4 0.750	1 1/8	1.591	1/2	1 1/8	1.299	1/2
7/8 0.875	1 5/16	1.856	19/32	1 5/16	1.516	37/64
1 1.000	1 1/2	2.121	21/32	1 1/2	1.732	43/64
1 1/8 1.125	1 11/16	2.386	3/4	1 11/16	1.949	3/4
1 1/4 1.250	1 7/8	2.652	27/32	1 7/8	2.165	27/32
1 3/8 1.375	2 1/16	2.917	29/32	2 1/16	2.382	29/32
	2 1/4	3.182	1			
1 1/2 1.500				2 1/4	2.598	1
1 3/4 1.750				2 5/8	3.031	1 5/32
2 2.000				3	3.464	1 11/32

Nominal Size or Basic Product Dia.	Hex Cap Screws (Finished Hex Bolts)		
	F	G	H
	Width Across Flats	Width Across Corners	Head Height
	Basic	Max.	Basic
1/4 0.250	7/16	0.505	5/32
5/16 0.3125	1/2	0.577	13/64
3/8 0.375	9/16	0.650	15/64
7/16 0.4375	5/8	0.722	9/32
1/2 0.500	3/4	0.866	5/16
9/16 0.5625	13/16	0.938	23/64
5/8 0.625	15/16	1.083	25/64
3/4 0.750	1 1/8	1.299	15/32
7/8 0.875	1 5/16	1.516	35/64
1 1.000	1 1/2	1.732	39/64
1 1/8 1.125	1 11/16	1.949	11/16
1 1/4 1.250	1 7/8	2.165	25/32
1 3/8 1.375	2 1/16	2.382	27/32
1 1/2 1.500	2 1/4	2.598	15/16
1 3/4 1.750	2 5/8	3.031	1 3/32
2 2.000	3	3.464	1 7/32

All dimensions given in inches.
Bolt need not be finished on any surface except threads.
Top of head shall be flat and chamfered.
Minimum thread length shall be twice the basic bolt diameter plus 0.25 in for lengths up to and 6 in, and twice the basic diameter plus 0.50 in for lengths over 6 in.
Bolts too short for the formula thread length shall be threaded as close to the head as practical.
Threads shall be in the Unified coarse thread series (UNC series), Class 2A.
Extracted from ANSI B18.2.1 – 1965.

All dimensions given in inches.
Top of head shall be flat and chamfered.
Minimum thread length shall be twice the basic product diameter plus 0.25 in for lengths up to and including 6 in, and twice the basic diameter plus 0.50 in for lengths over 6 in. On products too short for minimum thread lengths, the distance from the bearing surface of the head to the head to the first complete thread shall not exceed the length of 2½ threads for sizes up to and including 1 in, and 3½ threads for sizes larger than 1 in.
Threads shall be in the Unified coarse, fine, or 8 thread series (UNC, UNF, or 8UN series), Class 2A.
Unification of fine thread products is limited to sizes 1 in and under.
Extracted from ANSI B18.2.1 – 1965.

NUTS

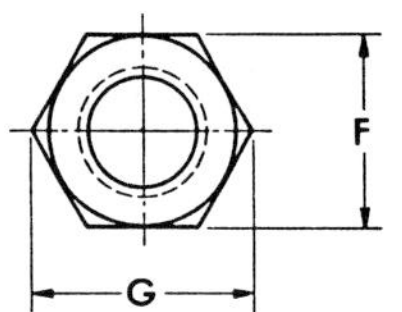

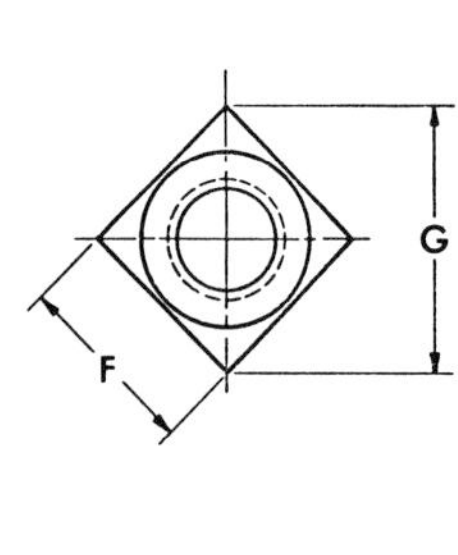

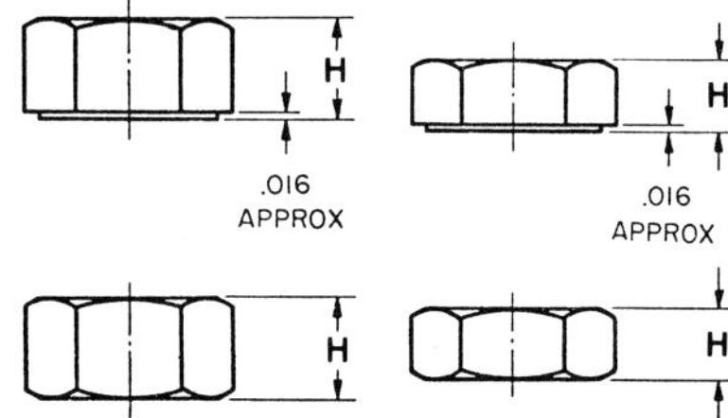

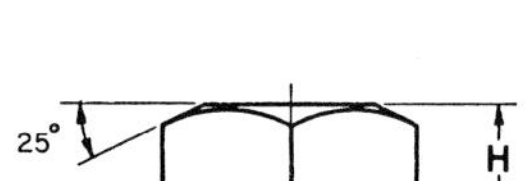

Square Nuts

Nominal Size or Basic Major Dia. of Thread		F Width Across Flats	G Width Across Corners	H Thickness
		Basic	Max.	Basic
1/4	0.250	7/16	0.619	7/32
5/16	0.3125	9/16	0.795	17/64
3/8	0.375	5/8	0.884	21/64
7/16	0.4375	3/4	1.061	3/8
1/2	0.500	13/16	1.149	7/16
5/8	0.625	1	1.414	35/64
3/4	0.750	1 1/8	1.591	21/32
7/8	0.875	1 5/16	1.856	49/64
1	1.000	1 1/2	2.121	7/8
1 1/8	1.125	1 11/16	2.386	1
1 1/4	1.250	1 7/8	2.652	1 3/32
1 3/8	1.375	2 1/16	2.917	1 13/64
1 1/2	1.500	2 1/4	3.182	1 5/16

All dimension given in inches
Tops of nuts shall be flat and chamfered or washer crowned.
Threads shall be in the Unified coarse thread series (UNC series), Class 2B.
Extracted from ANSI B18.2.2 – 1965.

Hex. Nuts and Hex. Jam Nuts

Nominal Size or Basic Major Dia. of Thread		F Width Across Flats	G Width Across Corners	H Thickness Hex. Nut	H Thickness Hex. Jam Nut
		Basic	Max.	Basic	Basic
1/4	0.250	7/16	0.505	7/32	5/32
5/16	0.3125	1/2	0.577	17/64	3/16
3/8	0.375	9/16	0.650	21/64	7/32
7/16	0.4375	11/16	0.794	3/8	1/4
1/2	0.500	3/4	0.866	7/16	5/16
9/16	0.5625	7/8	1.010	31/64	5/16
5/8	0.625	15/16	1.083	35/64	3/8
3/4	0.750	1 1/8	1.299	41/64	27/64
7/8	0.875	1 5/16	1.516	3/4	31/64
1	1.000	1 1/2	1.732	55/64	35/64
1 1/8	1.125	1 11/16	1.949	31/32	39/64
1 1/4	1.250	1 7/8	2.165	1 1/16	23/32
1 3/8	1.375	2 1/16	2.382	1 11/64	25/32
1 1/2	1.500	2 1/4	2.598	1 9/32	27/32

All dimensions given in inches.
Nuts in sizes up to and including $\frac{5}{8}$ in. shall be double chamfered or have washer faced bearing surface and chamfered top.
Threads shall be in the Unified coarse, fine, or 8 thread series (UNC, UNF or 8UN series), Class 2B.
Unification of fine-thread products is limited to sizes 1 in. and under.
Extracted from ANSI B18.2.2 – 1965.

CAP SCREWS

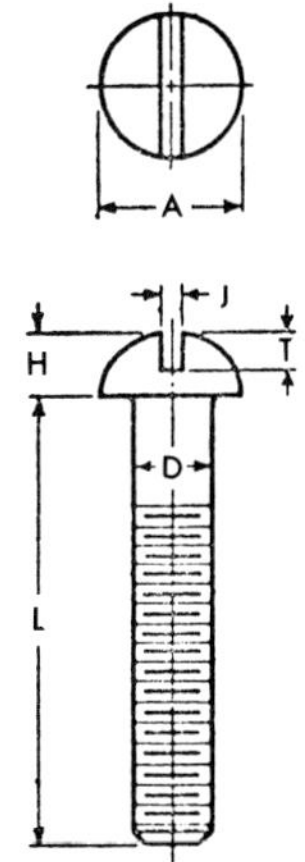

Round Head Cap Screws

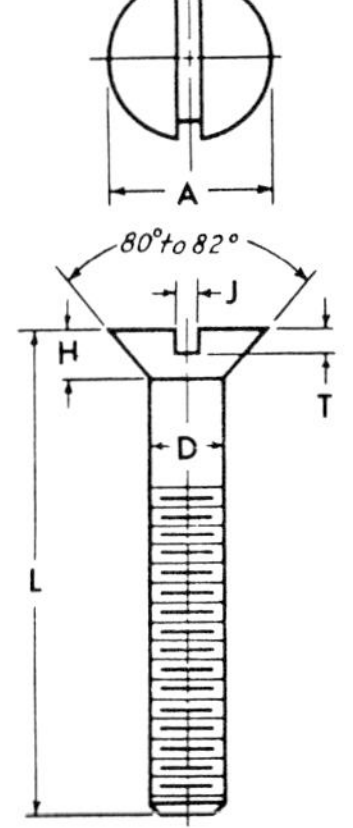

Flat Head Cap Screws

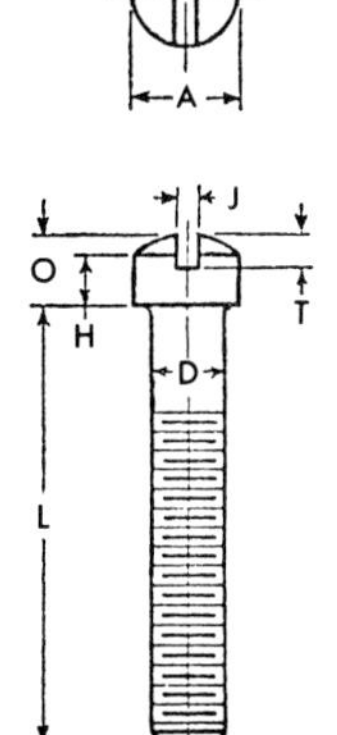

Fillister Head Cap Screws

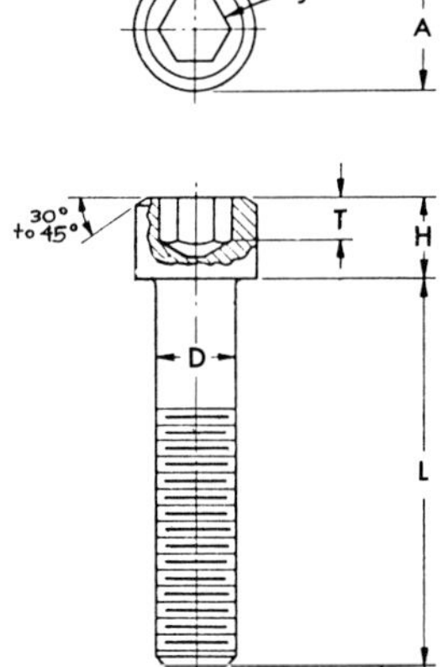

Hex. Socket Head Cap Screws

Round Head Cap Screws

Nominal Size	D Body Dia. Max.	A Head Dia. Max.	H Head Height Max.	J Slot Width Min.	T Slot Depth Min.
1/4	0.250	0.437	0.191	0.064	0.097
5/16	0.3125	0.562	0.245	0.072	0.126
3/8	0.375	0.625	0.273	0.081	0.138
7/16	0.4375	0.750	0.328	0.081	0.167
1/2	0.500	0.812	0.354	0.091	0.178
9/16	0.5625	0.937	0.409	0.102	0.207
5/8	0.625	1.000	0.437	0.116	0.220
3/4	0.750	1.250	0.546	0.131	0.278
7/8	0.875				
1	1.000				
1 1/8	1.125				
1 1/4	1.250				
1 3/8	1.375				
1 1/2	1.500				

Flat Head Cap Screws

Nominal Size	A Head Dia. Max	H Head Height Average	J Slot Width Min.	T Slot Depth Min.
1/4	0.500	0.140	0.064	0.045
5/16	0.625	0.177	0.072	0.057
3/8	0.750	0.210	0.081	0.068
7/16	0.8125	0.210	0.081	0.068
1/2	0.875	0.210	0.091	0.068
9/16	1.000	0.244	0.102	0.080
5/8	1.125	0.281	0.116	0.091
3/4	1.375	0.352	0.131	0.115
7/8	1.625	0.423	0.147	0.138
1	1.875	0.494	0.166	0.162
1 1/8	2.062	0.529	0.178	0.173
1 1/4	2.312	0.600	0.193	0.197
1 3/8	2.562	0.665	0.208	0.220
1 1/2	2.812	0.742	0.240	0.244

Fillister Head Cap Screws

Nominal Size	A Head Dia. Max.	H Head Height Max.	O Total Head Height Max.	J Slot Width Min.	T Slot Depth Min.
1/4	0.375	0.172	0.216	0.064	0.077
5/16	0.437	0.203	0.253	0.072	0.090
3/8	0.562	0.250	0.314	0.081	0.112
7/16	0.625	0.297	0.368	0.081	0.133
1/2	0.750	0.328	0.413	0.091	0.153
9/16	0.812	0.375	0.467	0.102	0.168
5/8	0.875	0.422	0.521	0.116	0.189
3/4	1.000	0.500	0.612	0.131	0.223
7/8	1.125	0.594	0.720	0.147	0.264
1	1.312	0.656	0.803	0.166	0.291
1 1/8					
1 1/4					
1 3/8					
1 1/2					

Hex. Socket Head Cap Screws

Nominal Size	A Head Dia. Max.	H Head Height Max.	J Hex. Socket Size Nom.	T Key Engagement Min.
1/4	0.375	0.250	3/16	0.120
5/16	0.469	0.312	1/4	0.151
3/8	0.562	0.375	5/16	0.182
7/16	0.656	0.438	3/8	0.213
1/2	0.750	0.500	3/8	0.245
9/16	0.938	0.625	1/2	0.307
5/8	1.125	0.750	5/8	0.370
3/4	1.312	0.875	3/4	0.432
7/8	1.500	1.000	3/4	0.495
1				
1 1/8	1.688	1.125	7/8	0.557
1 1/4	1.875	1.250	7/8	0.620
1 3/8	2.062	1.375	1	0.682
1 1/2	2.250	1.500	1	0.745

All dimensions given in inches.
Points shall be flat and chamfered.
For slotted-head cap screws the minimum length of thread shall be equal to 2D plus ¼ in. When too short, for the specified minimum thread length, the complete threads shall extend to within 2½ threads of the head.
The threads on slotted-head cap screws shall be coarse, fine, or 8-thread series, Class 2A.
The threads on hexagon socket head cap screws shall be Unified external threads with radius root: Class 3A UNRC and UNRF.
Series for screw sizes ¼ in through 1 in; Class 2A UNRC and UNRF Series for sizes over 1 in to 1½ in inclusive.
Hexagon socket head cap screws shall be designated by the following data in the sequence shown: Nominal size; threads per inch; length; product name; material; and protective coating, if required.
Extracted from ANSI B18.6.2 - 1956 (Slotted-Head Cap Screws); ANSI B18.3 – 1969 (Socket-Head Cap Screws).

MACHINE SCREWS

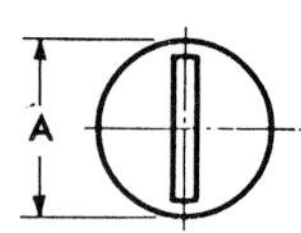

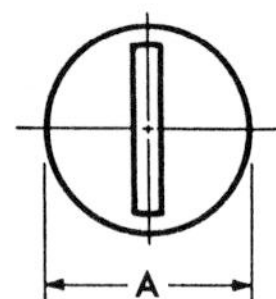

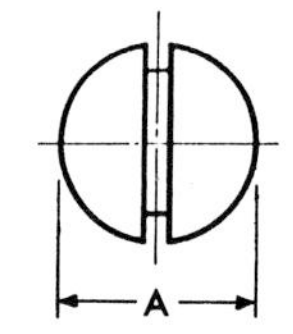

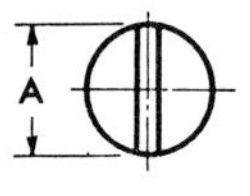

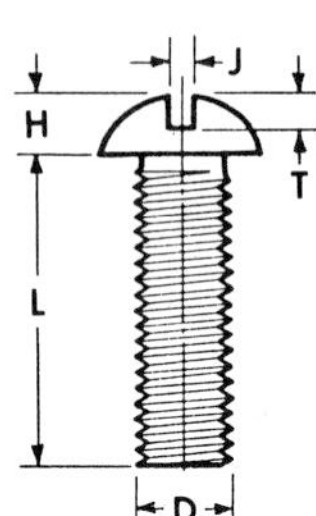

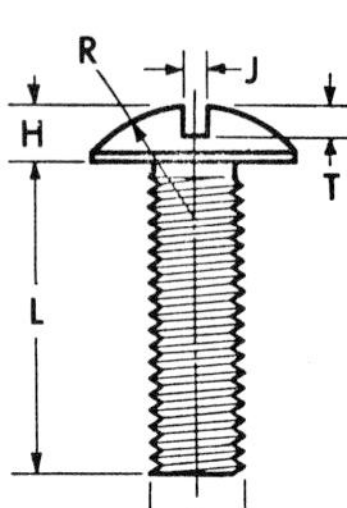

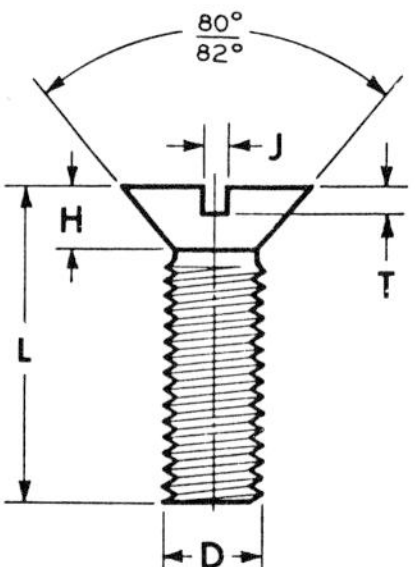

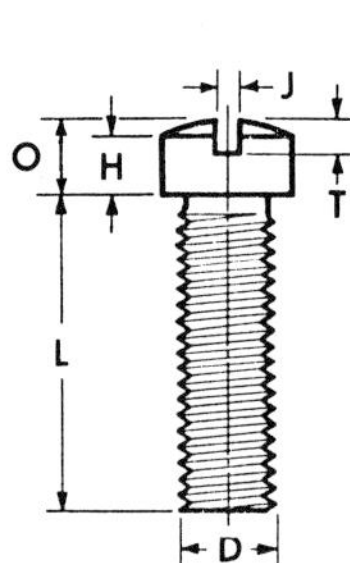

Nominal Size	Round Head Machine Screws					Truss Head Machine Screws					Flat Head Machine Screws				Fillister Head Machine Screws				
	D	A	H	J	T	A	H	J	T	R	A	H	J	T	A	H	O	J	T
	Diameter of Screw	Head Dia.	Head Height	Slot Width	Slot Depth	Head Dia.	Head Height	Slot Width	Slot Depth	Radius	Head Dia	Head Height Ref.	Slot Width	Slot Depth	Head Dia.	Head Side Height	Total Head Height	Slot Width	Slot Depth
	Basic	Max.	Max.	Min.	Min.	Max.	Max.	Min.	Min.	Max.	Max. Sharp		Min.	Min.	Max.	Max.	Max.	Min.	Min.
0	0.060	0.113	0.053	0.016	0.029	0.131	0.037	0.016	0.014	0.087	0.119	0.035	0.016	0.010	0.096	0.045	0.059	0.016	0.015
1	0.073	0.138	0.061	0.019	0.033	0.164	0.045	0.019	0.018	0.107	0.146	0.043	0.019	0.012	0.118	0.053	0.071	0.019	0.020
2	0.086	0.162	0.069	0.023	0.037	0.194	0.053	0.023	0.022	0.129	0.172	0.051	0.023	0.015	0.140	0.062	0.083	0.023	0.025
3	0.099	0.187	0.078	0.027	0.040	0.226	0.061	0.027	0.026	0.151	0.199	0.059	0.027	0.017	0.161	0.070	0.095	0.027	0.030
4	0.112	0.211	0.086	0.031	0.044	0.257	0.069	0.031	0.030	0.169	0.225	0.067	0.031	0.020	0.183	0.079	0.107	0.031	0.035
5	0.125	0.236	0.095	0.035	0.047	0.289	0.078	0.035	0.034	0.191	0.252	0.075	0.035	0.022	0.205	0.088	0.120	0.035	0.040
6	0.138	0.260	0.103	0.039	0.051	0.321	0.086	0.039	0.037	0.211	0.279	0.083	0.039	0.024	0.226	0.096	0.132	0.039	0.045
8	0.164	0.309	0.120	0.045	0.058	0.384	0.102	0.045	0.045	0.254	0.332	0.100	0.045	0.029	0.270	0.113	0.156	0.045	0.054
10	0.190	0.359	0.137	0.050	0.065	0.448	0.118	0.050	0.053	0.283	0.385	0.116	0.050	0.034	0.313	0.130	0.180	0.050	0.064
12	0.216	0.408	0.153	0.056	0.073	0.511	0.134	0.056	0.061	0.336	0.438	0.132	0.056	0.039	0.357	0.148	0.205	0.056	0.074
1/4	0.250	0.472	0.175	0.064	0.082	0.573	0.150	0.064	0.070	0.375	0.507	0.153	0.064	0.046	0.414	0.170	0.237	0.064	0.087
5/16	0.3125	0.590	0.216	0.072	0.099	0.698	0.183	0.072	0.085	0.457	0.635	0.191	0.072	0.058	0.518	0.211	0.295	0.072	0.110
3/8	0.375	0.708	0.256	0.081	0.117	0.823	0.215	0.081	0.100	0.538	0.762	0.230	0.081	0.070	0.622	0.253	0.355	0.081	0.133
7/16	0.4375	0.750	0.328	0.081	0.148	0.948	0.248	0.081	0.116	0.619	0.812	0.223	0.081	0.066	0.625	0.265	0.368	0.081	0.135
1/2	0.500	0.813	0.355	0.091	0.159	1.073	0.280	0.091	0.131	0.701	0.875	0.223	0.091	0.065	0.750	0.297	0.412	0.091	0.151
9/16	0.5625	0.938	0.410	0.102	0.183	1.198	0.312	0.102	0.146	0.783	1.000	0.260	0.102	0.077	0.812	0.336	0.466	0.102	0.172
5/8	0.625	1.000	0.438	0.116	0.195	1.323	0.345	0.116	0.162	0.863	1.125	0.298	0.116	0.088	0.875	0.375	0.521	0.116	0.193
3/4	0.750	1.250	0.547	0.131	0.242	1.573	0.410	0.131	0.182	1.024	1.375	0.372	0.131	0.111	1.000	0.441	0.612	0.131	0.226

All dimensions given in inches.
Unless otherwise specified, machine screws shall have plain sheared ends.
Threads on machine screws shall be UNC or UNF, Class 2A.
Screws up to and including 2 in in length shall have full form threads extending to within two threads of the bearing surface of the head or closer if practicable. Screws over 2 in in length shall have a minimum thread length of 1¾ in.
Extracted from ANSI B18.6.3 – 1962.

MACHINE SCREWS AND NUTS

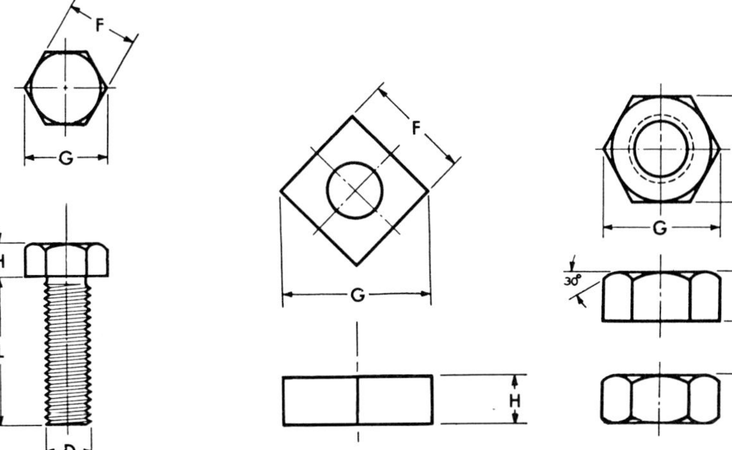

Hex. Head Machine Screws

Nominal Size	D Diameter of Screw Basic	F Width Across Flats Max.	G Width Across Corners Min.	H Head Height Max.
2	0.086	0.125	0.134	0.050
3	0.099	0.187	0.202	0.055
4	0.112	0.187	0.202	0.060
5	0.125	0.187	0.202	0.070
6	0.138	0.250	0.272	0.093
8	0.164	0.250	0.272	0.110
10	0.190	0.312	0.340	0.120
12	0.216	0.312	0.340	0.155
1/4	0.250	0.375	0.409	0.190
5/16	0.3125	0.500	0.545	0.230
3/8	0.375	0.562	0.614	0.295

All dimensions given in inches.
Unless otherwise specified, machine screws shall have plain sheared ends.
Threads on machine screws shall be UNC or UNF, Class 2A.
Screws up to and including 2 in in length shall have full form threads extending to within two threads of the bearing surface of the head or closer if practicable.
Screws over 2 in in length shall have a minimum thread length of 1 ¾ in.
Extracted from ANSI B18.6.3 – 1962.

Square and Hexagon Machine Screw Nuts

Nominal Size	Major Diameter of Thread Basic	F Width Across Flats Basic	G Width Across Corners Square Max.	G Width Across Corners Hex. Max.	H Thickness Nom.
0	0.060	5/32	0.221	0.180	3/64
1	0.073	5/32	0.221	0.180	3/64
2	0.086	3/16	0.265	0.217	1/16
3	0.099	3/16	0.265	0.217	1/16
4	0.112	1/4	0.354	0.289	3/32
5	0.125	5/16	0.442	0.361	7/64
6	0.138	5/16	0.442	0.361	7/64
8	0.164	11/32	0.486	0.397	1/8
10	0.190	3/8	0.530	0.433	1/8
12	0.216	7/16	0.619	0.505	5/32
1/4	0.250	7/16	0.619	0.505	3/16
5/16	0.3125	9/16	0.795	0.650	7/32
3/8	0.375	5/8	0.884	0.722	1/4

All dimensions given in inches.
Hexagon machine screw nuts shall have tops flat and chamfered. Bottoms are flat but for special purposes may be chamfered if so specified.
Square machine screw nuts shall have tops and bottoms flat without chamfer.
Threads in hexagon machine screw nuts shall be UNC or UNF, Class 2B; and in square machine screw nuts shall be UNC, Class 2B.
Extracted from ANSI B18.6.3 – 1962.

SET SCREWS

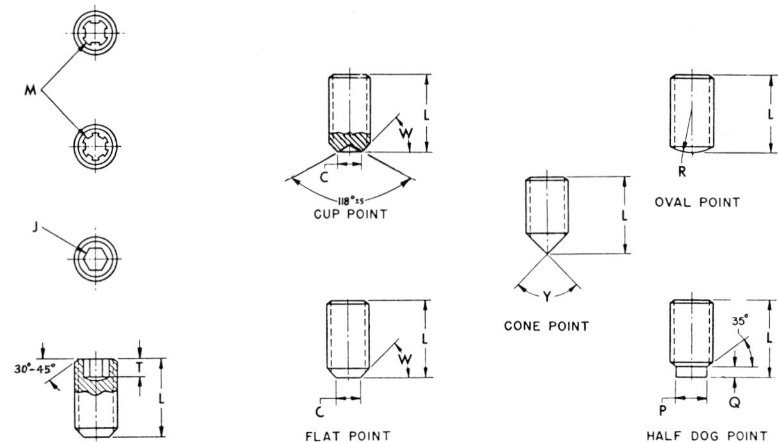

Hexagon and Spline Socket Set Screws

Nominal Size or Basic Screw Diameter		J	M	T		C	R	Y	P	Q
		Hexagon Socket Size	Spline Socket Size	Min. Key Engagement to Develop Functional Capability of Key		Cup and Flat Pt. Dia.	Oval Point Radius	Cone Point Angle 90° ±2° For these Nominal Lengths or Longer; 118° ±2° For Shorter Nominal Lengths	Half Dog Point	
				Hex. Socket	Spline Socket				Dia.	Length
		Nom.	Nom.	T_H Min.	T_S Min.	Max.	Basic		Max.	Max.
0	0.060	0.028	0.033	0.050	0.026	0.033	0.045	5/64	0.040	0.017
1	0.073	0.035	0.033	0.060	0.035	0.040	0.055	3/32	0.049	0.021
2	0.086	0.035	0.048	0.060	0.040	0.047	0.064	7/64	0.057	0.024
3	0.099	0.050	0.048	0.070	0.040	0.054	0.074	1/8	0.066	0.027
4	0.112	0.050	0.060	0.070	0.045	0.061	0.084	5/32	0.075	0.030
5	0.125	1/16 0.062	0.072	0.080	0.055	0.067	0.094	3/16	0.083	0.033
6	0.138	1/16 0.062	0.072	0.080	0.055	0.074	0.104	3/16	0.092	0.038
8	0.164	5/64 0.078	0.096	0.090	0.080	0.087	0.123	1/4	0.109	0.043
10	0.190	3/32 0.094	0.111	0.100	0.080	0.102	0.142	1/4	0.127	0.049
1/4	0.250	1/8 0.125	0.145	0.125	0.125	0.132	0.188	5/16	0.156	0.067
5/16	0.3125	5/32 0.156	0.183	0.156	0.156	0.172	0.234	3/8	0.203	0.082
3/8	0.375	3/16 0.188	0.216	0.188	0.188	0.212	0.281	7/16	0.250	0.099
7/16	0.4375	7/32 0.219	0.251	0.219	0.219	0.252	0.328	1/2	0.297	0.114
1/2	0.500	1/4 0.250	0.291	0.250	0.250	0.291	0.375	9/16	0.344	0.130
5/8	0.625	5/16 0.312	0.372	0.312	0.312	0.371	0.469	3/4	0.469	0.164
3/4	0.750	3/8 0.375	0.454	0.375	0.375	0.450	0.562	7/8	0.562	0.196
7/8	0.875	1/2 0.500	0.595	0.500	0.500	0.530	0.656	1	0.656	0.227
1	1.000	9/16 0.562	–	0.562	–	0.609	0.750	1 1/8	0.750	0.260
1 1/8	1.125	9/16 0.562	–	0.562	–	0.689	0.844	1 1/4	0.844	0.291
1 1/4	1.250	5/8 0.625	–	0.625	–	0.767	0.938	1 1/2	0.938	0.323
1 3/8	1.375	5/8 0.625	–	0.625	–	0.848	1.031	1 5/8	1.031	0.354
1 1/2	1.500	3/4 0.750	–	0.750	–	0.926	1.125	1 3/4	1.125	0.385
1 3/4	1.750	1 1.000	–	1.000	–	1.086	1.312	2	1.312	0.448
2	2.000	1 1.000	–	1.000	–	1.244	1.500	2 1/4	1.500	0.510

All dimensions given in inches.
Threads shall be Unified standard, Class 3A, UNC and UNF Series.
Hexagon and spline socket set screws shall be designated by the following data in the sequence shown: nominal size; threads per inch; length; product name; point style; material; and protective coating, if required.
W is normally 45° (30° minimum for very short screws).
Extracted from ANSI B18.3 – 1969.

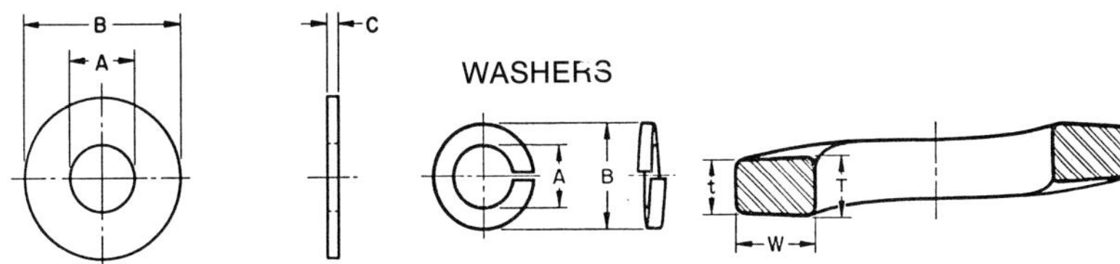

Preferred Sizes of Type A Plain Washers

Nominal Washer Size			Inside Dia. A	Outside Dia. B	Thickness C
–	–		0.078	0.188	0.020
–	–		0.094	0.250	0.020
–	–		0.125	0.312	0.032
No. 6	0.138		0.156	0.375	0.049
No. 8	0.164		0.188	0.438	0.049
No. 10	0.190		0.219	0.500	0.049
3/16	0.188		0.250	0.562	0.049
No. 12	0.216		0.250	0.562	0.065
1/4	0.250	N	0.281	0.625	0.065
1/4	0.250	W	0.312	0.734	0.065
5/16	0.312	N	0.344	0.688	0.065
5/16	0.312	W	0.375	0.875	0.083
3/8	0.375	N	0.406	0.812	0.065
3/8	0.375	W	0.438	1.000	0.083
7/16	0.438	N	0.469	0.922	0.065
7/16	0.438	W	0.500	1.250	0.083
1/2	0.500	N	0.531	1.062	0.095
1/2	0.500	W	0.562	1.375	0.109
9/16	0.562	N	0.594	1.156	0.095
9/16	0.562	W	0.625	1.469	0.109
5/8	0.625	N	0.656	1.312	0.095
5/8	0.625	W	0.688	1.750	0.134
3/4	0.750	N	0.812	1.469	0.134
3/4	0.750	W	0.812	2.000	0.148
7/8	0.875	N	0.938	1.750	0.134
7/8	0.875	W	0.938	2.250	0.165
1	1.000	N	1.062	2.000	0.134
1	1.000	W	1.062	2.500	0.165
1 1/8	1.125	N	1.250	2.250	0.134
1 1/8	1.125	W	1.250	2.750	0.165
1 1/4	1.250	N	1.375	2.500	0.165
1 1/4	1.250	W	1.375	3.000	0.165
1 3/8	1.375	N	1.500	2.750	0.165
1 3/8	1.375	W	1.500	3.250	0.180
1 1/2	1.500	N	1.625	3.000	0.165
1 1/2	1.500	W	1.625	3.500	0.180
1 5/8	1.625		1.750	3.750	0.180
1 3/4	1.750		1.875	4.000	0.180
1 7/8	1.875		2.000	4.250	0.180
2	2.000		2.125	4.500	0.180
2 1/4	2.250		2.375	4.750	0.220
2 1/2	2.500		2.625	5.000	0.238
2 3/4	2.750		2.875	5.250	0.259
3	3.000		3.125	5.500	0.284

All dimensions given in inches.
Preferred sizes are for the most part from series previously designated "Standard Plate" and "SAE." Where common sizes existed in the two series, the SAE size is designated "N" (narrow) and the Standard Plate "W" (wide). These sizes as well as all other sizes of Type A Plain Washers are to be ordered by ID, OD, and thickness dimensions.
Nominal Washer sizes are intended for use with comparable nominal screw or bolt sizes.
Extracted from ANSI B27.2 – 1965.

Regular Helical Spring Lock Washers

Nominal Washer Size		Inside Diameter A	Outside Diameter B	Thickness $\frac{T+t}{2}$
		Min.	Max.	Min.
No. 22	0.086	0.088	0.172	0.020
No. 3	0.099	0.101	1.195	0.025
No. 44	0.112	0.115	0.209	0.025
No. 5	0.125	0.128	0.236	0.031
No. 6	0.138	0.141	0.250	0.031
No. 8	0.164	0.168	0.293	0.040
No. 10	0.190	0.194	0.334	0.047
No. 12	0.216	0.221	0.377	0.056
1/4	0.250	0.255	0.489	0.062
5/16	0.312	0.318	0.586	0.078
3/8	0.375	0.382	0.683	0.094
7/16	0.438	0.446	0.779	0.109
1/2	0.500	0.509	0.873	0.125
9/16	0.562	0.572	0.971	0.141
5/8	0.625	0.636	1.079	0.156
11/16	0.688	0.700	1.176	0.172
3/4	0.750	0.763	1.271	0.188
13/16	0.812	0.826	1.367	0.203
7/8	0.875	0.890	1.464	0.219
15/16	0.938	0.954	1.560	0.234
1	1.000	1.017	1.661	0.250
1 1/16	1.062	1.080	1.756	0.266
1 1/8	1.125	1.144	1.853	0.281
1 3/16	1.188	1.208	1.950	0.297
1 1/4	1.250	1.271	2.045	0.312
1 5/16	1.312	1.334	2.141	0.328
1 3/8	1.375	1.398	2.239	0.344
1 7/16	1.438	1.462	2.334	0.359
1 1/2	1.500	1.525	2.430	0.375

All dimensions given in inches.
Formerly designated Medium Helical Spring Lock Washers.
Extracted from ANSI B27.1 – 1965.

WOOD SCREWS

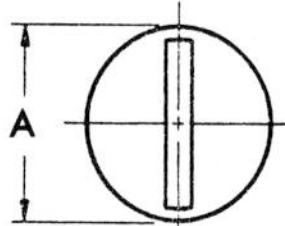

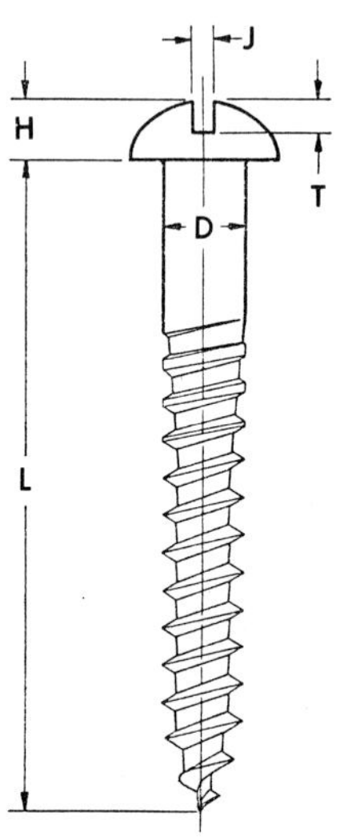

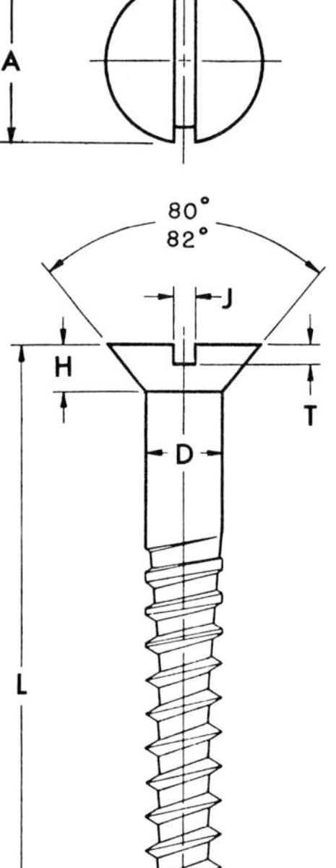

		D	A	H	J	T	A	H	J	T
			Slotted Round Head				Slotted Flat Head			
Nominal Size	Threads per Inch	Diameter of Screw	Head Dia.	Head Height	Slot Width	Slot Depth	Head Dia.	Head Height	Slot Width	Slot Depth
		Basic	Max.	Max.	Min.	Min.	Max. Sharp	Max.	Min.	Min.
0	32	0.060	0.113	0.053	0.016	0.029	0.119	0.035	0.016	0.010
1	28	0.073	0.138	0.061	0.019	0.033	0.146	0.043	0.019	0.012
2	26	0.086	0.162	0.069	0.023	0.037	0.172	0.051	0.023	0.015
3	24	0.099	0.187	0.078	0.027	0.040	0.199	0.059	0.027	0.017
4	22	0.112	0.211	0.086	0.031	0.044	0.225	0.067	0.031	0.020
5	20	0.125	0.236	0.095	0.035	0.047	0.252	0.075	0.035	0.022
6	18	0.138	0.260	0.103	0.039	0.051	0.279	0.083	0.039	0.024
7	16	0.151	0.285	0.111	0.039	0.055	0.305	0.091	0.039	0.027
8	15	0.164	0.309	0.120	0.045	0.058	0.332	0.100	0.045	0.029
9	14	0.177	0.334	0.128	0.045	0.062	0.358	0.108	0.045	0.032
10	13	0.190	0.359	0.137	0.500	0.065	0.385	0.116	0.050	0.034
12	11	0.216	0.408	0.153	0.056	0.073	0.438	0.132	0.056	0.039
14	10	0.242	0.457	0.170	0.064	0.080	0.491	0.148	0.064	0.044
16	9	0.268	0.506	0.187	0.064	0.087	0.544	0.164	0.064	0.049
18	8	0.294	0.555	0.204	0.072	0.094	0.597	0.180	0.072	0.054
20	8	0.320	0.604	0.220	0.072	0.101	0.650	0.196	0.072	0.059
24	7	0.372	0.702	0.254	0.081	0.116	0.756	0.228	0.081	0.069

All dimensions given in inches.
The length of thread on wood screws shall be equal to approximately $\frac{2}{3}$ of the screw length.
Extracted from ANSI B18.6.1 – 1961.

WOODRUFF KEYS AND KEYSEATS

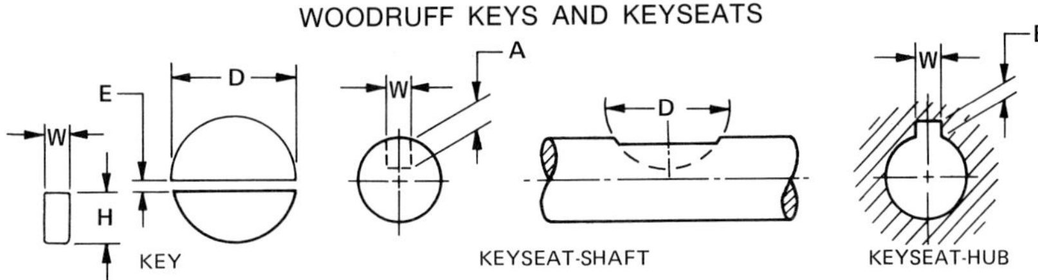

Key No.	Nominal Key Size W X D	Height of Key H (Max.)	Distance Below Center E	Keyseat in Shaft Depth A	Keyseat in Hub Depth B
204	1/16 X 1/2	0.203	3/64	0.1668	0.0372
304	3/32 X 1/2	0.203	3/64	0.1511	0.0529
404	1/8 X 1/2	0.203	3/64	0.1355	0.0685
305	3/32 X 5/8	0.250	1/16	0.1981	0.0529
405	1/8 X 5/8	0.250	1/16	0.1825	0.0685
505	5/32 X 5/8	0.250	1/16	0.1669	0.0841
406	1/8 X 3/4	0.313	1/16	0.2455	0.0685
506	5/32 X 3/4	0.313	1/16	0.2299	0.0841
606	3/16 X 3/4	0.313	1/16	0.2143	0.0997
507	5/32 X 7/8	0.375	1/16	0.2919	0.0841
607	3/16 X 7/8	0.375	1/16	0.2763	0.0997
807	1/4 X 7/8	0.375	1/16	0.2450	0.1310
608	3/16 X 1	0.438	1/16	0.3393	0.0997
808	1/4 X 1	0.438	1/16	0.3080	0.1310
1008	5/16 X 1	0.438	1/16	0.2768	0.1622
609	3/16 X 1 1/8	0.484	5/64	0.3853	0.0997
809	1/4 X 1 1/8	0.484	5/64	0.3540	0.1310
1009	5/16 X 1 1/8	0.484	5/64	0.3228	0.1622
810	1/4 X 1 1/4	0.547	5/64	0.4170	0.1310
1010	5/16 X 1 1/4	0.547	5/64	0.3858	0.1622
1210	3/8 X 1 1/4	0.547	5/64	0.3545	0.1935
811	1/4 X 1 3/8	0.594	3/32	0.4640	0.1310
1011	5/16 X 1 3/8	0.594	3/32	0.4328	0.1622
1211	3/8 X 1 3/8	0.594	3/32	0.4015	0.1935
812	1/4 X 1 1/2	0.641	7/64	0.5110	0.1310
1012	5/16 X 1 1/2	0.641	7/64	0.4798	0.1622
1212	3/8 X 1 1/2	0.641	7/64	0.4485	0.1935

All dimensions are in inches.
The key numbers indicate nominal key dimensions. The last two digits give the nominal diameter D in eighths of an inch and the digits preceding the last two give the nominal width W in thirty-seconds of an inch.
Extracted from ANSI B17.2 – 1967.

PARALLEL KEYS

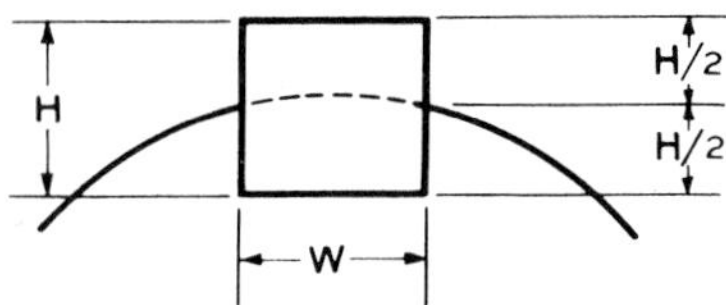

Key Size Versus Shaft Diameter

NOMINAL SHAFT DIAMETER		NOMINAL KEY SIZE			NOMINAL KEYSEAT DEPTH	
			Height, *H*		*H*/2	
Over	To (Incl)	Width, *W*	Square	Rectangular	Square	Rectangular
5/16	7/16	3/32	3/32		3/64	
7/16	9/16	1/8	1/8	3/32	1/16	3/64
9/16	7/8	3/16	3/16	1/8	3/32	1/16
7/8	1-1/4	1/4	1/4	3/16	1/8	3/32
1-1/4	1-3/8	5/16	5/16	1/4	5/32	1/8
1-3/8	1-3/4	3/8	3/8	1/4	3/16	1/8
1-3/4	2-1/4	1/2	1/2	3/8	1/4	3/16
2-1/4	2-3/4	5/8	5/8	7/16	5/16	7/32
2-3/4	3-1/4	3/4	3/4	1/2	3/8	1/4
3-1/4	3-3/4	7/8	7/8	5/8	7/16	5/16
3-3/4	4-1/2	1	1	3/4	1/2	3/8
4-1/2	5-1/2	1-1/4	1-1/4	7/8	5/8	7/16
5-1/2	6-1/2	1-1/2	1-1/2	1	3/4	1/2
6-1/2	7-1/2	1-3/4	1-3/4	1-1/2*	7/8	3/4
7-1/2	9	2	2	1-1/2	1	3/4
9	11	2-1/2	2-1/2	1-3/4	1-1/4	7/8
11	13	3	3	2	1-1/2	1
13	15	3-1/2	3-1/2	2-1/2	1-3/4	1-1/4
15	18	4		3		1-1/2
18	22	5		3-1/2		1-3/4
22	26	6		4		2
26	30	7		5		2-1/2

*Some key standards show 1-1/4 in. Preferred size is 1-1/2 in.

All dimensions given in inches.
For a stepped shaft, the size of a key is determined by the diameter of the shaft at the point of location of the key, regardless of the number of different diameters on the shaft.
Square keys are preferred through 6½ inch diameter shaft and rectangular keys for larger shafts. Sizes and dimensions in shaded area are preferred.
Extracted from ANSI B17.1 – 1967.

Welded and Seamless Steel Pipe and Pipe Threads

Size				Plain	Identification			American Standard Taper Pipe Thread NPT		
Nominal (in.)	O.D. (in.)	Wall Thickness (in.)	I.D. (in.)	End Weight (lb/ft)	API Standard	Standard (STD) X-strong(XS) XX-strong(XXS)	Schedule No.	Threads Per Inch	Tap Drill	Handtight Engagement
1/8	0.405	0.068	0.269	0.24	5L	STD	40	27	R	0.1615
		0.095	0.215	0.31	5L	XS	80			
1/4	0.540	0.088	0.364	0.42	5L	STD	40	18	7/16	0.2278
		0.119	0.302	0.54	5L	XS	80			
3/8	0.675	0.091	0.493	0.57	5L	STD	40	18	37/64	0.240
		0.126	0.423	0.74	5L	XS	80			
1/2	0.840	0.109	0.622	0.85	5L	STD	40			
		0.147	0.546	1.09	5L	XS	80	14	23/32	0.320
		0.188	0.464	1.31			160			
		0.294	0.252	1.71	5L	XXS				
3/4	1.050	0.113	0.824	1.13	5L	STD	40			
		0.154	0.742	1.47	5L	XS	80	14	59/64	0.339
		0.219	0.612	1.94			160			
		0.308	0.434	2.44	5L	XXS				
1	1.315	0.133	1.049	1.68	5L	STD	40			
		0.179	0.957	2.17	5L	XS	80	11.5	1 5/32	0.400
		0.250	0.815	2.84			160			
		0.358	0.599	3.66	5L	XXS				
1 1/4	1.660	0.140	1.380	2.27	5L	STD	40			
		0.191	1.278	3.00	5L	XS	80	11.5	1 1/2	0.420
		0.250	1.160	3.76			160			
		0.382	0.896	5.21	5L	XXS				
1 1/2	1.900	0.145	1.610	2.72	5L	STD	40			
		0.200	1.500	3.63	5L	XS	80	11.5	1 47/64	0.420
		0.281	1.338	4.86			160			
		0.400	1.100	6.41	5L	XXS				
		0.083	2.209	2.03	5L 5LX					
		0.109	2.157	2.64	5L 5LX					
		0.125	2.125	3.00	5L 5LX					
		0.141	2.093	3.36	5L 5LX					
		0.154	2.067	3.65	5L 5LX	STD	40			
		0.172	2.031	4.05	5LX					
2	2.375	0.188	1.999	4.39	5LX			11.5	2 7/32	0.436
		0.218	1.939	5.02	5L 5LX	XS	80			
		0.250	1.875	5.67	5LX					
		0.281	1.813	6.28	5LX					
		0.344	1.687	7.46			160			
		0.436	1.503	9.03	5L 5LX	XXS				

Extracted from ANSI B36.10—1970 and ANSI B2.1—1968
Tap drill sizes not American Standard.

MALLEABLE IRON SCREWED PIPE FITTINGS—150 Lb.

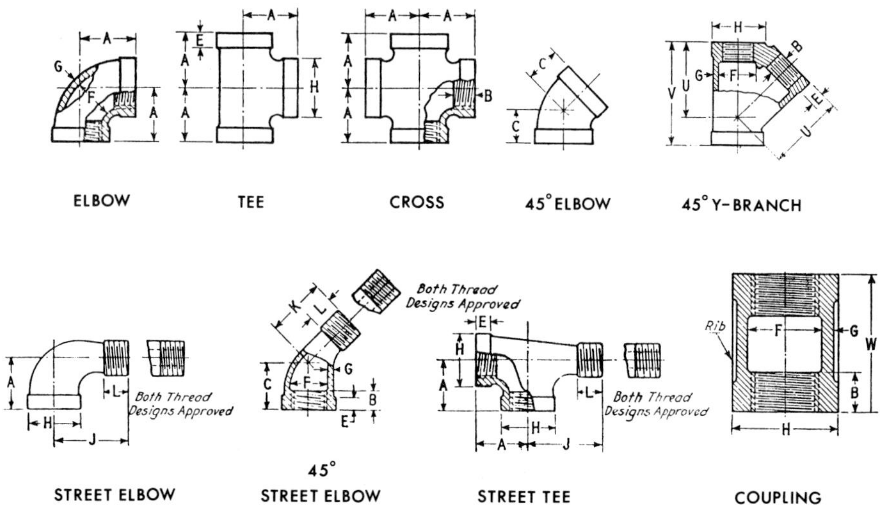

Nominal Pipe Size	Center to End, Elbows, Tees, and Crosses	Center to End, 45-deg Elbows	Length of Thread Min	Width of Band, Min	Inside Dia Min	Metal Thickness	Outside Diameter of Band Min	Center to Male End Elbows Tees	Center to Male End 45-deg Elbows	Center to End Outlet	End to End	Length of Straight Couplings
	A	*C*	*B*	*E*	*F*	*G*	*H*	*J*	*K*	*U*	*V*	*W*
1/8	0.69	....	0.25	0.200	0.405	0.090	0.693	*1.00	...			0.96
1/4	0.81	0.73	0.32	0.215	0.540	0.095	0.844	1.19	0.94			1.06
3/8	0.95	0.80	0.36	0.230	0.675	0.100	1.015	1.44	1.03	1.43	1.93	1.16
1/2	1.12	0.88	0.43	0.249	0.840	0.105	1.197	1.63	1.15	1.71	2.32	1.34
3/4	1.31	0.98	0.50	0.273	1.050	0.120	1.458	1.89	1.29	2.05	2.77	1.52
1	1.50	1.12	0.58	0.302	1.315	0.134	1.771	2.14	1.47	2.43	3.28	1.67
1 1/4	1.75	1.29	0.67	0.341	1.660	0.145	2.153	2.45	1.71	2.92	3.94	1.93
1 1/2	1.94	1.43	0.70	0.368	1.900	0.155	2.427	2.69	1.88	3.28	4.38	2.15
2	2.25	1.68	0.75	0.422	2.375	0.173	2.963	3.26	2.22	3.93	5.17	2.53
2 1/2	2.70	1.95	0.92	0.478	2.875	0.210	3.589	*3.86	2.57	4.73	6.25	2.88
3	3.08	2.17	0.98	0.548	3.500	0.231	4.285	*4.51	3.00	5.55	7.26	3.18
3 1/2	3.42	2.39	1.03	0.604	4.000	0.248	4.843					
4	3.79	2.61	1.08	0.661	4.500	0.265	5.401	5.69	3.70	6.97	8.98	3.69
5	4.50	3.05	1.18	0.780	5.563	0.300	6.583	*6.86	....			
6	5.13	3.46	1.28	0.900	6.625	0.336	7.767	*8.03	....			

All dimensions are given in inches.

*This dimension applies to street elbows only. Street tees are not made in these sizes.

Extracted from ANSI B16.3 – 1963.

WROUGHT STEEL BUTTWELDING PIPE FITTINGS

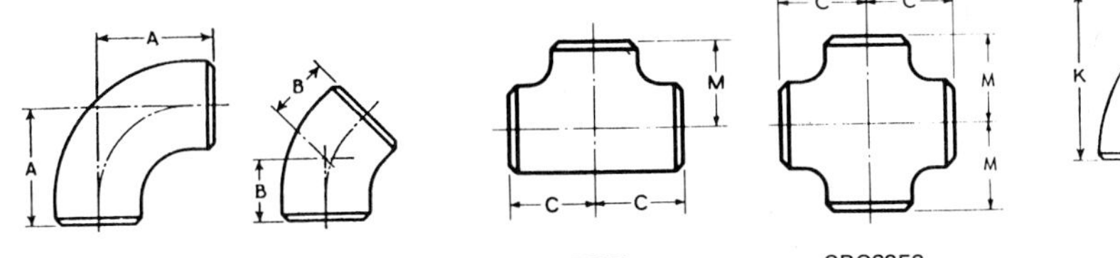

LONG RADIUS ELBOWS — TEES — CROSSES — LONG RADIUS RETURNS

Nominal Pipe Size	Outside Diameter at Bevel	Long Radius Elbows — Center to End: 90° Elbows A	Long Radius Elbows — Center to End: 45° Elbows B	Tees and Crosses — Center to End: Run C	Tees and Crosses — Center to End: Outlets* M	Long Radius Returns: Center to Center O	Long Radius Returns: Back to Face K
1/2	0.840	1 1/2	5/8	1	1	3	1 7/8
3/4	1.050	1 1/8	7/16	1 1/8	1 1/8	2 1/2	1 11/16
1	1.315	1 1/2	7/8	1 1/2	1 1/2	3	2 3/16
1 1/4	1.660	1 7/8	1	1 7/8	1 7/8	3 3/4	2 3/4
1 1/2	1.900	2 1/4	1 1/8	2 1/4	2 1/4	4 1/2	3 1/4
2	2.375	3	1 3/8	2 1/2	2 1/2	6	4 3/16
2 1/2	2.875	3 3/4	1 3/4	3	3	7 1/2	5 3/16
3	3.500	4 1/2	2	3 3/8	3 3/8	9	6 1/4
3 1/2	4.000	5 1/4	2 1/4	3 3/4	3 3/4	10 1/2	7 1/4
4	4.500	6	2 1/2	4 1/8	4 1/8	12	8 1/4
5	5.563	7 1/2	3 1/8	4 7/8	4 7/8	15	10 5/16
6	6.625	9	3 3/4	5 5/8	5 5/8	18	12 5/16
8	8.625	12	5	7	7	24	16 5/16
10	10.750	15	6 1/4	8 1/2	8 1/2	30	20 3/8
12	12.750	18	7 1/2	10	10	36	24 3/8
14	14.000	21	8 3/4	11	11	42	28
16	16.000	24	10	12	12	48	32
18	18.000	27	11 1/4	13 1/2	13 1/2	54	36
20	20.000	30	12 1/2	15	15	60	40
22	22.000	33	13 1/2	16 1/2	16 1/2	66	44
24	24.000	36	15	17	17	72	48

All dimensions given in inches.
*Outlet dimension "M" for run sizes 14 in. and larger is recommended but not mandatory.
Extracted from ANSI B16.9 – 1964.

V anthropometric tables

Table V-1 (All dimensions in inches)*

Measurement	Range	Mean	Standard Deviation	Percentiles				
				1st	5th	50th	95th	99th
Weight								
1. Weight (pounds)	104. - 265.	163.66	20.86	123.1	132.5	161.9	200.8	215.9
Body Lengths								
2. Stature	59.45 - 77.56	69.11	2.44	63.5	65.2	69.1	73.1	74.9
3. Nasal root height	56.30 - 73.23	64.95	2.39	59.4	61.0	65.0	68.9	70.7
4. Eye height	56.30 - 73.23	64.69	2.38	59.2	60.8	64.7	68.6	70.3
5. Tragion height	54.72 - 74.41	63.92	2.39	58.4	60.0	64.0	67.8	69.6
6. Cervicale height	50.39 - 66.93	59.08	2.31	53.7	55.3	59.2	62.9	64.6
7. Shoulder height	47.24 - 64.17	56.50	2.28	51.2	52.8	56.6	60.2	61.9
8. Suprasternale height	48.03 - 63.78	56.28	2.19	51.3	52.7	56.3	59.9	61.5
9. Nipple height	42.13 - 57.09	50.41	2.08	45.6	47.0	50.4	53.9	55.3
10. Substernale height	41.34 - 55.51	48.71	2.02	44.0	45.6	48.7	52.1	53.5
11. Elbow height	36.61 - 49.21	43.50	1.77	39.5	40.6	43.5	46.4	47.7
12. Waist height	34.65 - 48.82	42.02	1.81	37.7	39.1	42.1	45.0	46.4
13. Penale height	27.95 - 41.34	34.52	1.75	30.6	31.6	34.5	37.4	38.7
14. Wrist height	27.56 - 39.76	33.52	1.54	30.1	31.0	33.6	36.1	37.1
15. Crotch height (inseam)	26.77 - 38.19	32.83	1.73	29.3	30.4	32.8	35.7	37.0
16. Gluteal furrow height	25.20 - 37.01	31.57	1.62	27.9	29.0	31.6	34.3	35.5
17. Knuckle height	24.80 - 35.04	30.04	1.45	26.7	27.7	30.0	32.4	33.5
18. Kneecap height	15.75 - 23.23	20.22	1.03	17.9	18.4	20.2	21.9	22.7

*Adapted from H. T. E. Hertzberg, G. S. Daniels, and E. Churchill, *Anthropometry of Flying Personnel*—1950, WADC Technical Report 52-321, USAF, Wright Air Development Center, Wright-Patterson AFB, Ohio, September, 1954. It should be noted that these data represent measurements made on approximately 4,000 male USAF personnel and thus do not specifically represent the U.S. population at large.

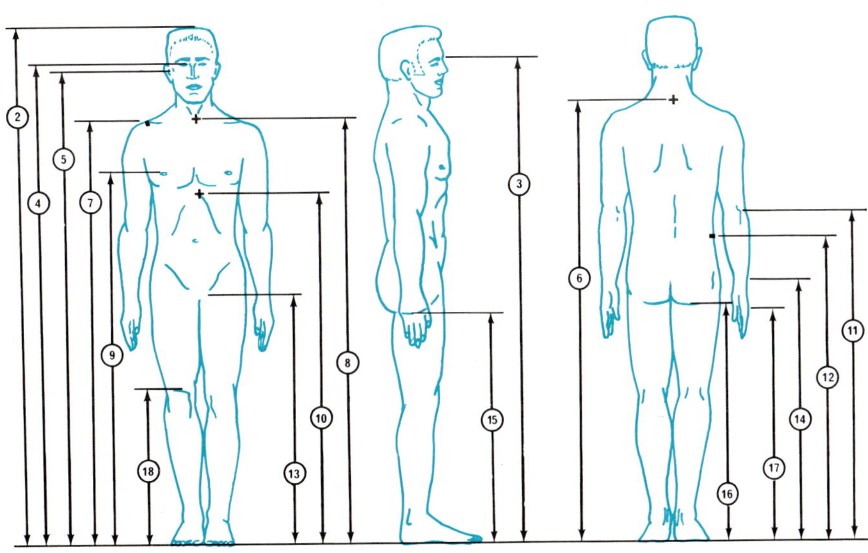

Table V-2 (All dimensions in inches)

Measurement	Range	Mean	Standard Deviation	Percentiles				
				1st	5th	50th	95th	99th
19. Sitting height	29.92 - 40.16	35.94	1.29	32.9	33.8	36.0	38.0	38.9
20. Eye	26.38 - 36.61	31.47	1.27	28.5	29.4	31.5	33.5	34.4
21. Shoulder	18.90 - 27.17	23.26	1.14	20.6	21.3	23.3	25.1	25.8
22. Waist height, sitting	6.30 - 12.99	9.24	0.76	7.4	7.9	9.3	10.4	10.9
23. Elbow rest height, sitting	4.33 - 12.99	9.12	1.04	6.6	7.4	9.1	10.8	11.5
24. Thigh clearance height	3.94 - 7.09	5.61	0.52	4.5	4.8	5.6	6.5	6.8
25. Knee height, sitting	17.32 - 24.80	21.67	0.99	19.5	20.1	21.7	23.3	24.0
26. Popliteal height, sitting	14.17 - 19.29	16.97	0.77	15.3	15.7	17.0	18.2	18.8
27. Buttock-knee length	18.50 - 27.56	23.62	1.06	21.2	21.9	23.6	25.4	26.2
28. Buttock-leg length	35.43 - 50.00	42.70	2.04	38.2	39.4	42.7	46.1	47.7
29. Shoulder-elbow length	11.42 - 18.11	14.32	0.69	12.8	13.2	14.3	15.4	15.9
30. Forearm-hand length	15.35 - 22.05	18.86	0.81	17.0	17.6	18.9	20.2	20.7
31. Span	58.27 - 82.28	70.80	2.94	63.9	65.9	70.8	75.6	77.6
32. Arm reach from wall	27.56 - 39.76	34.59	1.65	30.9	31.9	34.6	37.3	38.6
33. Maximum reach from wall	31.10 - 46.06	38.59	1.90	34.1	35.4	38.6	41.7	43.2
34. Functional reach	26.77 - 40.55	32.33	1.63	28.8	29.7	32.3	35.0	36.4

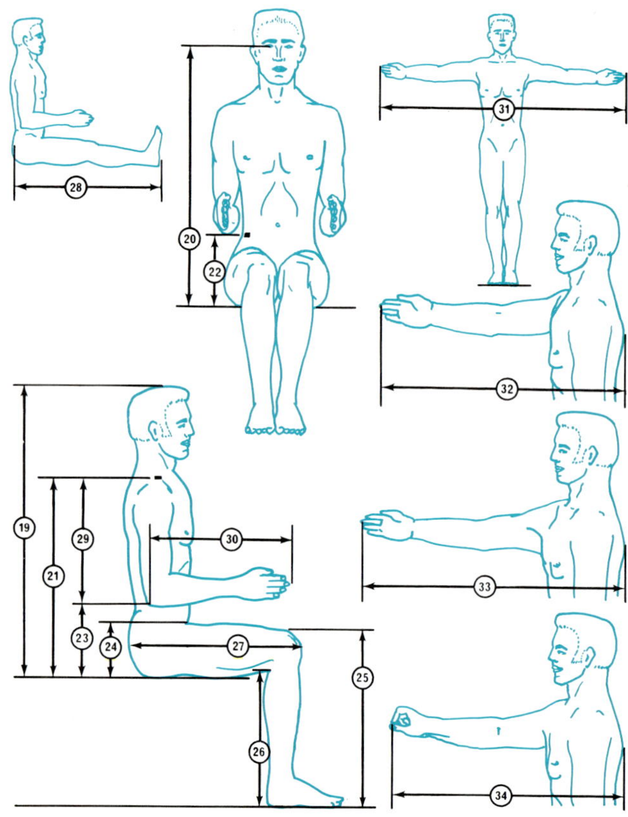

Table V-3 (All dimensions in inches)

Measurement	Range	Mean	Standard Deviation	Percentiles 1st	5th	50th	95th	99th
Body Breadths and Thicknesses								
35. Elbow-to-elbow breadth	11.42 - 23.62	17.28	1.42	14.5	15.2	17.2	19.8	20.9
36. Hip breadth, sitting	11.42 - 18.11	13.97	0.87	12.2	12.7	13.9	15.4	16.2
37. Knee-to-knee breadth	6.30 - 10.24	7.93	0.52	7.0	7.2	7.9	8.8	9.4
38. Biacromial diameter	12.60 - 18.50	15.75	0.74	14.0	14.6	15.8	16.9	17.4
39. Shoulder breadth	14.57 - 22.83	17.88	0.91	15.9	16.5	17.9	19.4	20.1
40. Chest breadth	9.45 - 15.35	12.03	0.80	10.4	10.8	12.0	13.4	14.1
41. Waist breadth	7.87 - 15.35	10.66	0.94	8.9	9.4	10.6	12.3	13.3
42. Hip breadth	8.27 - 15.75	13.17	0.73	11.3	12.1	13.2	14.4	15.2
43. Chest depth	6.69 - 12.99	9.06	0.75	7.6	8.0	9.0	10.4	11.1
44. Waist depth	5.51 - 11.81	7.94	0.88	6.3	6.7	7.9	9.5	10.3
45. Buttock depth	6.30 - 11.81	8.81	0.82	7.2	7.6	8.8	10.2	10.9
Circumferences and Body Surface Measurements								
46. Neck circumference	10.24 - 19.29	14.96	0.74	13.3	13.8	14.9	16.2	16.8
47. Shoulder circumference	35.43 - 56.69	45.25	2.43	40.2	41.6	45.1	49.4	51.5
48. Chest circumference	31.10 - 49.61	38.80	2.45	33.7	35.1	38.7	43.2	44.8
49. Waist circumference	24.41 - 47.24	32.04	3.02	26.5	27.8	31.7	37.5	40.1
50. Buttock circumference	29.92 - 46.85	37.78	2.29	33.0	34.3	37.7	41.8	43.5
51. Thigh circumference	14.57 - 28.74	22.39	1.74	18.3	19.6	22.4	25.3	26.4
52. Lower thigh circumference	11.81 - 23.23	17.33	1.41	14.2	15.1	17.3	19.6	20.9
53. Calf circumference	9.84 - 18.50	14.40	0.96	12.2	12.9	14.4	16.0	16.7
54. Ankle circumference	7.09 - 12.99	8.93	0.57	7.8	8.1	8.9	9.8	10.5

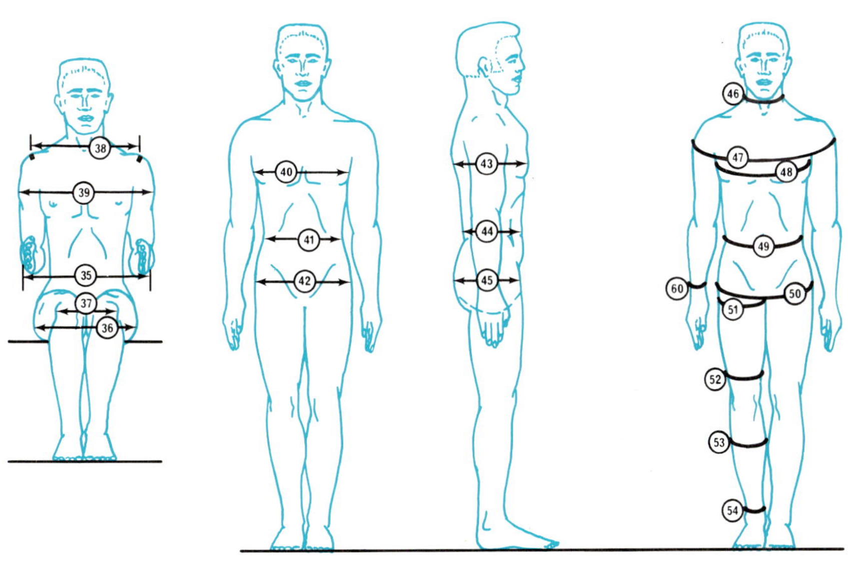

Table V-4 (All dimensions in inches)

Measurement	Range	Mean	Standard Deviation	Percentiles 1st	5th	50th	95th	99th
55. Scye circumference	11.02 - 22.83	18.09	1.38	15.1	16.1	18.0	20.5	21.8
56. Axillary arm circumference	7.87 - 16.54	12.54	1.10	10.2	10.9	12.4	14.4	15.2
57. Biceps circumference	8.27 - 16.93	12.79	1.07	10.5	11.2	12.8	14.6	15.4
58. Elbow circumference	8.27 - 15.35	12.26	0.80	10.7	11.1	12.2	13.6	14.3
59. Lower arm circumference	8.66 - 15.35	11.50	0.73	9.9	10.4	11.5	12.7	13.3
60. Wrist circumference	3.94 - 8.27	6.85	0.40	6.0	6.3	6.8	7.5	7.8
61. Sleeve inseam	15.35 - 24.80	19.83	1.14	17.1	18.0	19.8	21.7	22.6
62. Sleeve length	27.56 - 38.98	33.64	1.50	30.2	31.3	33.7	36.0	37.3
63. Anterior neck length	1.38 - 5.31	3.40	0.64	1.8	2.3	3.4	4.4	4.9
64. Posterior neck length	1.57 - 6.10	3.64	0.61	2.3	2.7	3.6	4.7	5.2
65. Shoulder length	4.33 - 8.66	6.77	0.56	5.5	5.9	6.8	7.7	8.1
66. Waist back	11.81 - 22.83	17.72	1.07	14.8	16.1	17.7	19.4	20.2
67. Waist front	10.63 - 21.26	15.24	1.12	12.3	13.5	15.2	17.0	18.1
68. Gluteal arc	7.87 - 17.32	11.71	0.92	9.7	10.4	11.7	13.1	14.8
69. Crotch length	20.08 - 38.19	28.20	2.00	23.7	25.1	28.2	31.6	33.5
70. Vertical trunk circumference	54.72 - 74.41	64.81	2.88	58.3	60.2	64.8	69.7	71.7
71. Interscye	12.20 - 24.41	19.62	1.40	16.3	17.3	19.6	22.0	22.9
72. Interscye maximum	17.72 - 27.17	22.85	1.33	19.8	20.7	22.9	25.1	26.0
73. Buttock circumference	33.46 - 52.36	41.74	2.82	36.1	37.4	41.5	46.7	49.3
74. Knee circumference	11.42 - 20.47	15.39	0.92	13.5	14.0	15.4	16.9	17.7

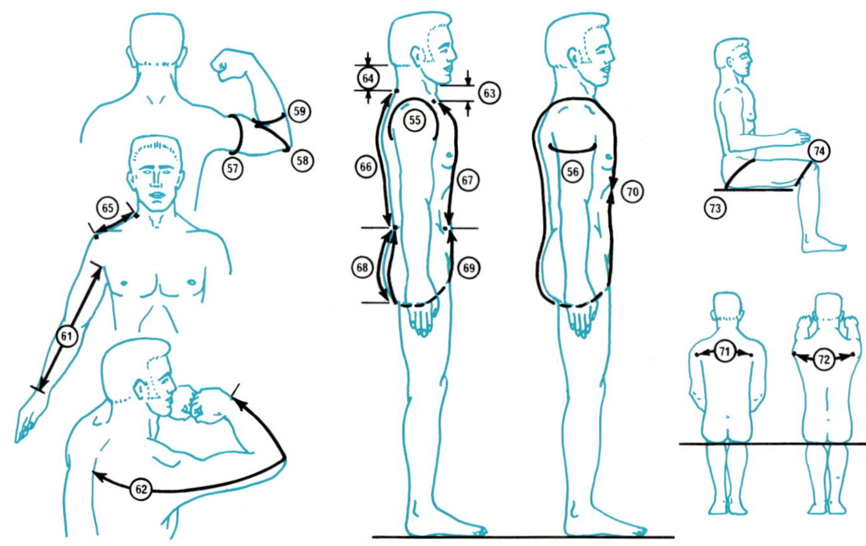

Table V-5 (All dimensions in inches)

Measurement	Range	Mean	Standard Deviation	Percentiles 1st	5th	50th	95th	99th
The Foot								
75. Foot length	8.86 - 12.24	10.50	0.45	9.5	9.8	10.5	11.3	11.6
76. Instep length	6.42 - 8.86	7.64	0.34	6.9	7.1	7.6	8.2	8.4
77. Foot breadth	3.19 - 4.65	3.80	0.19	3.40	3.50	3.78	4.10	4.36
78. Heel breadth	2.13 - 3.27	2.64	0.15	2.30	2.40	2.63	2.87	3.01
79. Bimalleolar breadth	2.44 - 3.58	2.95	0.15	2.61	2.70	2.95	3.19	3.32
80. Medial malleolus height	2.60 - 4.29	3.45	0.21	3.0	3.1	3.5	3.8	4.0
81. Lateral malleolus height	2.01 - 3.70	2.73	0.22	2.2	2.4	2.7	3.1	3.3
82. Ball of foot circumference	7.87 - 12.60	9.65	0.48	8.6	8.9	9.6	10.4	10.8
The Hand								
83. Hand length	5.87 - 8.74	7.49	0.34	6.7	6.9	7.5	8.0	8.3
84. Palm length	3.39 - 5.04	4.24	0.21	3.77	3.89	4.24	4.60	4.74
85. Hand breadth at thumb	3.23 - 4.76	4.07	0.21	3.59	3.73	4.08	4.42	4.57
86. Hand breadth at metacarpale	2.99 - 4.09	3.48	0.16	3.12	3.22	3.49	3.74	3.86
87. Thickness at metacarpale III	0.75 - 1.54	1.17	0.07	1.00	1.05	1.17	1.28	1.35
88. First phalanx III length	2.21 - 3.07	2.67	0.12	2.40	2.49	2.67	2.85	2.95
89. Finger diameter III	0.75 - 1.00	0.86	0.05	0.77	0.79	0.85	0.93	0.96
90. Grip diameter (inside)	1.37 - 2.63	1.90	0.14	1.52	1.62	1.83	2.05	2.16
91. Grip diameter (outside)	3.15 - 4.72	4.09	0.21	3.58	3.72	4.09	4.44	4.57
92. Fist circumference	7.09 - 13.39	11.56	0.57	10.2	10.7	11.6	12.4	12.8

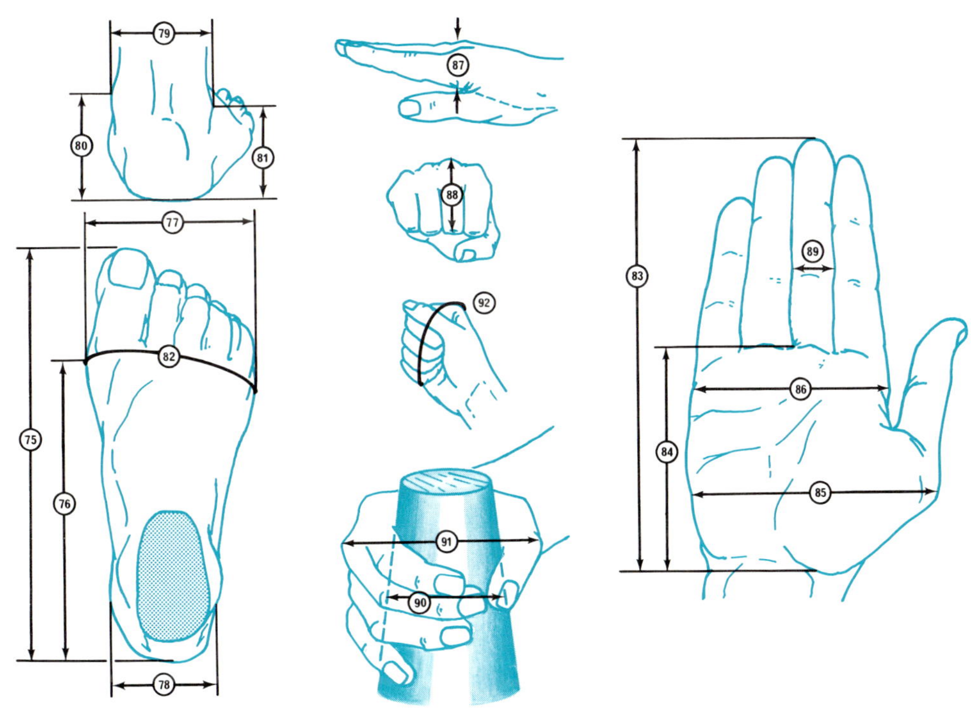

Table V-6
(All dimensions in inches)

Measurement	Range	Mean	Standard Deviation	Percentiles 1st	5th	50th	95th	99th
The Head and Face								
93. Head length	6.89 - 8.78	7.76	0.25	7.2	7.3	7.7	8.2	8.3
94. Head breadth	5.35 - 6.89	6.07	0.20	5.61	5.74	6.05	6.40	6.56
95. Minimum frontal diameter	3.54 - 5.00	4.35	0.19	3.88	4.04	4.35	4.68	4.80
96. Maximum frontal diameter	4.02 - 5.47	4.71	0.20	4.26	4.39	4.72	5.05	5.20
97. Bizygomatic diameter	4.72 - 6.22	5.55	0.20	5.07	5.21	5.54	5.88	6.02
98. Bigonial diameter	3.50 - 5.08	4.27	0.22	3.8	3.9	4.3	4.6	4.8
99. Bitragion diameter	4.76 - 6.30	5.60	0.21	5.1	5.3	5.6	5.9	6.1
100. Interocular diameter	0.87 - 1.65	1.25	0.10	1.03	1.09	1.25	1.42	1.50
101. Biocular diameter	3.19 - 4.45	3.78	0.17	3.38	3.48	3.78	4.06	4.19
102. Interpupillary distance	2.01 - 2.99	2.49	0.14	2.19	2.27	2.49	2.74	2.84
103. Nose length	1.46 - 2.56	2.01	0.14	1.69	1.79	2.00	2.23	2.33
104. Nose breadth	0.91 - 1.85	1.31	0.11	1.09	1.16	1.31	1.49	1.58
105. Nasal root breath	0.28 - 0.91	0.61	0.08	0.42	0.48	0.61	0.74	0.81
106. Nose protrusion	0.43 - 1.42	0.89	0.11	0.63	0.72	0.90	1.08	1.17
107. Philtrum length	0.35 - 1.46	0.77	0.14	0.48	0.54	0.76	0.98	1.09
108. Menton-Subnasale length	1.81 - 3.54	2.63	0.27	2.05	2.19	2.62	3.07	3.28
109. Menton-Crinion length	6.18 - 8.58	7.36	0.36	6.6	6.8	7.4	8.0	8.2
110. Lip-to-Lip distance	0.16 - 1.26	0.64	0.12	0.35	0.44	0.63	0.83	0.94
111. Lip length (Bichelion Dia.)	1.34 - 2.64	2.03	0.14	1.72	1.81	2.02	2.27	2.38
112. Ear length	1.69 - 3.15	2.47	0.16	2.08	2.21	2.47	2.73	2.85
113. Ear breadth	1.10 - 1.93	1.44	0.11	1.20	1.27	1.44	1.61	1.70
114. Ear length above tragion	0.79 - 1.61	1.17	0.11	0.92	0.99	1.17	1.35	1.42
115. Ear protrusion	0.31 - 1.54	0.84	0.14	0.55	0.63	0.83	1.10	1.23

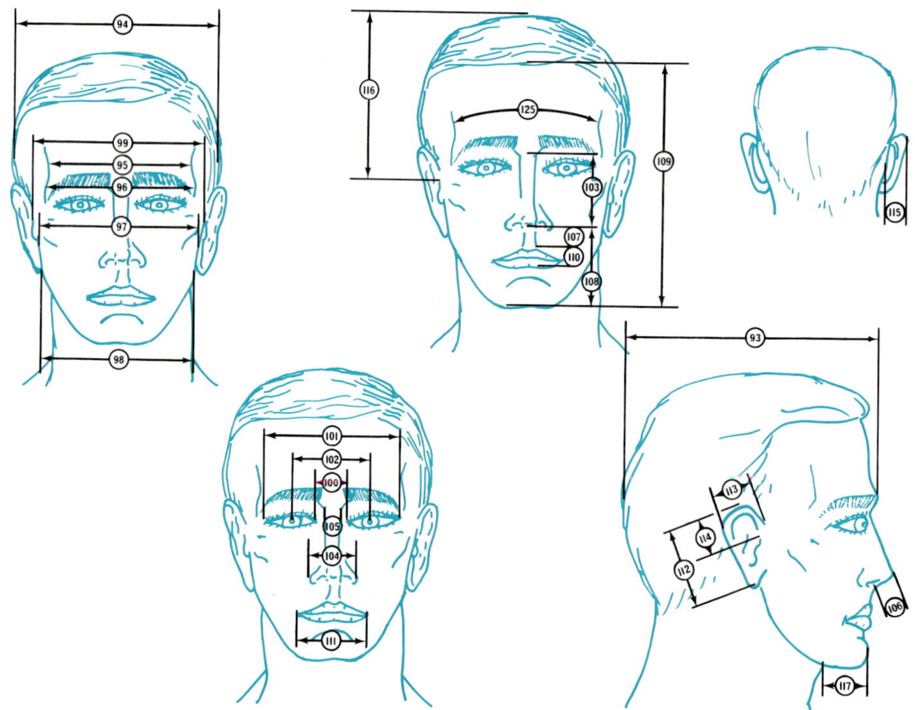

index